AR

ANNUAL REVIEW OF BIOPHYSICS AND BIOENGINEERING

ANNUAL REVIEW OF BIOPHYSICS AND BIOENGINEERING

VOLUME 13, 1984

DONALD M. ENGELMAN, *Editor*
Yale University

CHARLES R. CANTOR, *Associate Editor*
Columbia University

THOMAS D. POLLARD, *Associate Editor*
The Johns Hopkins University School of Medicine

ANNUAL REVIEWS INC. 4139 EL CAMINO WAY PALO ALTO, CALIFORNIA 94306 USA

ANNUAL REVIEWS INC.
Palo Alto, California, USA

International Standard Serial Number: 0084-6589
International Standard Book Number: 0-8243-1813-7
Library of Congress Catalog Card Number: 79-188446

TYPESET BY A.U.P. TYPESETTERS (GLASGOW) LTD., SCOTLAND
PRINTED AND BOUND IN THE UNITED STATES OF AMERICA

PREFACE

Biophysics has traditionally resisted attempts at precise definition. Some regard this circumstance as desirable and are content with the functional view that biophysics is whatever those who call themselves biophysicists do. It seems, however, that a delineation of boundaries may usefully serve as a guide to editorial policy. This volume is the transition from the *Annual Review of Biophysics and Bioengineering* to the *Annual Review of Biophysics*, and it is appropriate to consider once more whether some more sensible definition can be formulated.

Two branches of biophysics can be distinguished: direct application of physical concepts to the understanding of biological problems, and sophisticated use of physical measurements with a full understanding of their conceptual bases. The first category would include the use of mechanics, thermodynamics, quantum mechanics, mathematical methods, electrostatics, etc., to understand the structural and functional properties of macromolecules, macromolecular assemblies, cells, or organisms. In practice, this branch of biophysics is substantially smaller than the other. The second category is based on the distinction between the use of a physical method by biophysicists and by other biological scientists. An example might be electron microscopy, which is used productively by a wide range of biological scientists but which is practiced as a biophysical technique by relatively few. Thus, we consider biophysics to be a conceptual and methodological perspective, rather than a collection of specific topics in biology.

The *Annual Review of Biophysics* will focus its attention on the solution of biological problems and will present topics in molecular, cell, and organismic biology that have been effectively pursued in biophysical studies. Additionally, new methods and approaches will be presented when important biological applications are apparent. We hope that this focus will prove useful and that it will encourage the productive growth of the biophysical community.

THE EDITORS AND EDITORIAL COMMITTEE

Annual Review of Biophysics and Bioengineering
Volume 13, 1984

CONTENTS

(continued)

ANNUAL REVIEWS INC. is a nonprofit scientific publisher established to promote the advancement of the sciences. Beginning in 1932 with the *Annual Review of Biochemistry*, the Company has pursued as its principal function the publication of high quality, reasonably priced *Annual Review* volumes. The volumes are organized by Editors and Editorial Committees who invite qualified authors to contribute critical articles reviewing significant developments within each major discipline. The Editor-in-Chief invites those interested in serving as future Editorial Committee members to communicate directly with him. Annual Reviews Inc. is administered by a Board of Directors, whose members serve without compensation.

SOME RELATED ARTICLES APPEARING IN OTHER *ANNUAL REVIEWS*

From the *Annual Review of Biochemistry*, Volume 53 (1984)

Protein-DNA Recognition, C. O. Pabo and R. T. Sauer
Molybdenum in Nitrogenase, V. K. Shah, R. A. Ugalde, J. Imperial, and W. J. Brill
Principles That Determine the Structure of Proteins, C. Chothia
Structure of Ribosomal RNA, H. F. Noller
Protein–Nucleic Acid Interactions in Transcription: A Molecular Analysis, P. H. von Hippel, D. G. Bear, W. D. Morgan, and J. A. McSwiggen

From the *Annual Review of Microbiology*, Volume 37 (1983)

Crystalline Surface Layers on Bacteria, Uwe B. Sleytr and Paul Messner
Evolutionary Relationships in Vibrio and Photobacterium: A Basis for a Natural Classification, Paul Baumann, Linda Baumann, Marilyn J. Woolkalis, and Sookie S. Bang
Structure, Function, and Assembly of Cell Walls of Gram-Positive Bacteria, Gerald D. Shockman and John F. Barrett
Role of Proton Motive Force in Sensory Transduction in Bacteria, Barry L. Taylor

From the *Annual Review of Physical Chemistry*, Volume 34 (1983)

Nonequilibrium Molecular Dynamics, William G. Hoover
High Resolution Vibration-Rotation Spectroscopy, Alan G. Robiette and J. Lindsay Duncan
Pulsed-Nozzle, Fourier-Transform Microwave Spectroscopy of Weakly Bound Dimers, A. C. Legon
Pseudorotation: A Large Amplitude Molecular Motion, Herbert L. Strauss
Magnetic Field Effects on Reaction Yields in the Solid State: An Example From Photosynthetic Reaction Centers, Steven G. Boxer, Christopher E. D. Chidsey, and Mark G. Roelofs
Diffusion-Controlled Reactions, Daniel F. Calef and J. M. Deutch

From the *Annual Review of Physiology*, Volume 46 (1984)

Cochlear Mechanics, William S. Rhode
Relation of Receptor Potentials of Cochlear Hair Cells to Spike Discharges of Cochlear Neurons, T. F. Weiss
Patch Clamp Techniques for Studying Ionic Channels in Excitable Membranes, B. Sakmann and E. Neher
K^+ Channels Gated by Voltage and Ions, Ramon Latorre, Roberto Coronado, and Cecilia Vergara
Fluctuation Analysis of Sodium Channels in Epithelia, B. Lindemann
Gramicidin Channels, Olaf S. Andersen
Ion Channels in Liposomes, Christopher Miller
Magnetic Field Sensitivity in Animals, James L. Gould
Physiological Mechanisms for Spatial Filtering and Image Enhancement in the Sonar of Bats, J. A. Simmons and S. A. Kick

Ann. Rev. Biophys. Bioeng. 1984 13: 1–24

RAMAN SPECTROSCOPY OF THERMOTROPIC AND HIGH-PRESSURE PHASES OF AQUEOUS PHOSPHOLIPID DISPERSIONS[1]

Patrick T. T. Wong

Division of Chemistry, National Research Council of Canada, Ottawa, Ontario, Canada K1A OR6

Introduction

The lamellar bilayer structure of aqueous phospholipid dispersions is an integral part of biological membranes. Therefore, aqueous phospholipid bilayers have been extensively used as model membrane systems, and their physical and chemical properties have been the subject of extensive studies.

A wide variety of physical techniques has been used for these studies. Raman spectroscopy is one of the most powerful techniques and provides not only macroscopic but also microscopic information on the structure and dynamics of whole molecules as well as various functional groups. Many Raman spectral features of phospholipids are sensitive to the conformation of various functional groups, the interchain packing, the inter- and intrachain interactions, and the chain mobility.

A large number of Raman spectroscopic studies of aqueous phospholipid systems have been reported in the literature since the first one appeared in 1971 (53). The majority of this work has been carried out on phosphatidylcholine model membrane systems and concerns the thermotropic phase transition between the liquid crystal phase and the gel phase (main transition or acyl chain melting transition), particularly the changes in the conformation and critical temperature of this phase transition with or without perturbations from cholestrol, protein, polypeptid, metal ion, sonication, and others (33, 104).

[1] NRCC No. 22547.

0084–6589/84/0615–0001$02.00

Recently, some new phase transitions induced by temperature and pressure in aqueous phospholipid dispersions have been discovered (24, 106–109, 112, 113) and studied by the Raman spectroscopic method (57, 106–109, 112, 113). Additional Raman features have been used for monitoring structural and dynamical changes in various phase transitions (5, 7, 9, 14, 16, 49, 61, 65, 113). The correlation between the Raman features and the structural and the dynamical properties of phospholipid bilayers is better understood (60, 85, 87–89, 111, 113). These constitute the main objectives of the present review.

Raman Spectra of Fully Hydrated Phospholipids

The Raman spectrum of a phospholipid consists of bands and lines arising from transitions between the vibrational states of various types of vibrations of functional groups. Most of the bands and lines are due to the vibrations of the methylene chains in the lipid moiety. The Raman spectrum of a phospholipid represents a superposition of the spectrum of methylene chains and a few bands arising from the vibrational modes of the head group, methyl groups, and the C=C and =CH groups in the case of an unsaturated phospholipid. Therefore, the main features in the Raman spectra of phospholipids are analogous to those of long chain *n*-alkanes, fatty acids, and surfactants (54, 85, 100, 111). In the case of fully hydrated phospholipids, some additional bands which are due to the vibrational modes of water molecules appear in the Raman spectra (93). There is, however, only a little interference from these water bands in the Raman spectra of phospholipids.

Typical Raman spectra of fully hydrated phospholipids in the CH stretching modes region are shown in Figure 1. The origins of the features in this region are reasonably well established (15, 30, 31, 54, 59, 63, 85, 87, 90, 102). The bands near 2850 cm^{-1} and 2880 cm^{-1} are attributed to the symmetric, $\nu_s(CH)$, and the asymmetric, $\nu_{as}(CH_2)$, methylene stretching modes. The bands near 2930, 2960, and 3040 cm^{-1} are due to the CH stretching modes of the α-methylene group adjacent to the ester carbonyl group $\nu(\alpha\text{–}CH_2)$, the methyl end group $\nu_{as}(CH_3)$, and the choline methyl group $\nu_c(CH_3)$, respectively. The broad band underneath the $\nu_{as}(CH_2)$ band is the result of the Fermi resonance interaction between the methylene $\nu_s(CH_2)$ and the binary combinations of the methylene scissoring modes $\delta(CH_2)$.

The methylene deformation modes are located in the frequency region 1250–1500 cm^{-1} (77, 86). Typical Raman spectra in this region are shown in Figure 2. The methylene wagging mode $\omega(CH_2)$ is expected at $\sim 1370\ cm^{-1}$ (83, 84). However, it appears as an extremely weak band in the spectra. The spectral region around 1440 cm^{-1} consists primarily of the CH_2 scissoring

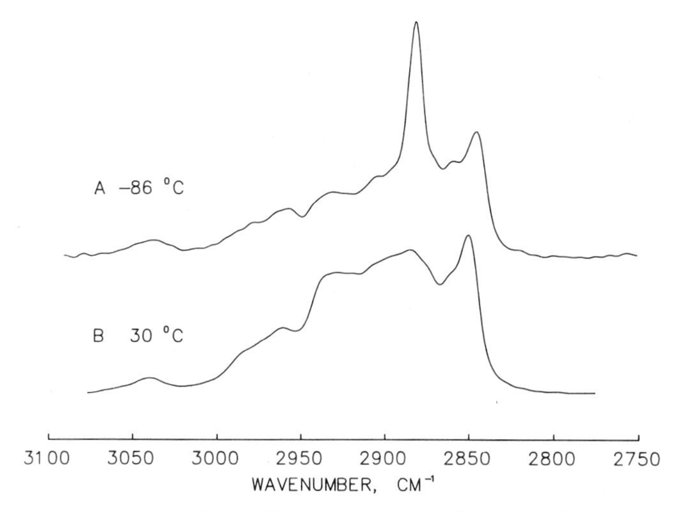

Figure 1 Raman spectra of the gel (*A*) and the liquid crystalline phase (*B*) of aqueous DMPC dispersions in the CH stretching region.

mode, $\delta(CH_2)$ (77, 86). The broad band on the high frequency side of the $\delta(CH_2)$ band has been interpreted as the result of the Fermi resonance interaction between $\delta(CH_2)$ and the binary combination of the CH_2 rocking modes, $\gamma(CH_2)$ (85). It has been shown that the deformation mode

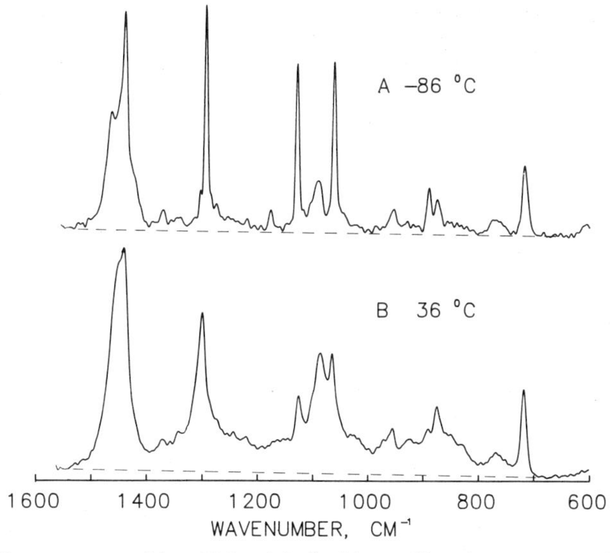

Figure 2 Raman spectra of the gel (*A*) and the liquid crystalline phase (*B*) of aqueous DMPC dispersions in the 600–1600 cm^{-1} region.

of the methyl end group is also in the frequency region of this broad band (102). The band at ~1300 cm^{-1} is due to the methylene twisting mode, $\tau(CH_2)$ (77, 86).

The skeletal C–C stretching modes of the methylene chains are in the frequency region 1050–1150 cm^{-1}. The spectra in this region for straight chain alkanes, fatty acids, and phospholipids have been studied in detail (9, 30, 54, 86, 94, 97, 103, 115) and have been used widely to monitor the conformation changes in the acyl chains of natural and model membranes (30, 31, 48, 53, 94, 103, 104, 115). Typical spectra in this region are also shown in Figure 2. The bands at 1062 cm^{-1} and ~1127 cm^{-1} are attributed to the asymmetric, ν_{as}(C–C), and the symmetric, ν_s(C–C), stretching modes of all-trans C–C bonds, respectively, whereas the band near 1090–1085 cm^{-1}, observed only at higher temperatures, is due to the skeletal C–C stretching mode of gauche segments, ν_g(C–C). The band near 1100 cm^{-1} arises from the symmetric stretching of the PO_2^- group, $\nu_s(PO_2^-)$.

The spectra below 1000 cm^{-1} are shown in Figure 3. The accordian-like longitudinal-acoustic mode of the methylene chains, LA, is expected below 400 cm^{-1}. Unfortunately, it overlaps with the strong water band and thus is seldom observed in fully hydrated phospholipids (15, 93). It appears as a weak band in the spectra of the gel phase of the fully hydrated phospholipids as seen in Figure 3. In this frequency region the strongest band is near 720 cm^{-1}, which is attributed to the symmetric CN stretching mode of the head group, ν_s(CN) (2, 13, 30). The weak bands near 760–860 cm^{-1} are

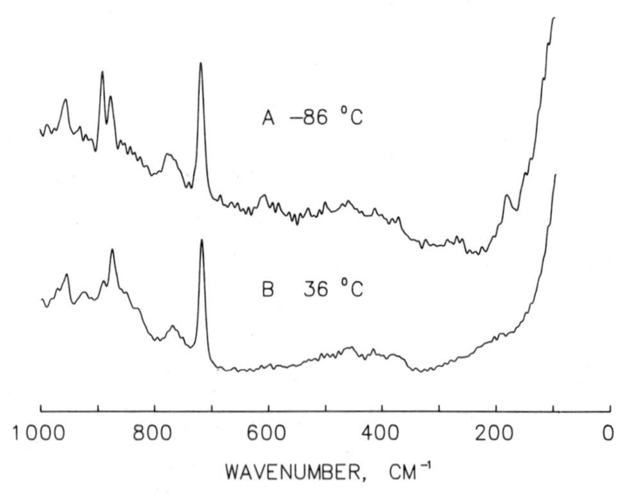

Figure 3 Raman spectra of the gel (*A*) and the liquid crystalline phase (*B*) of aqueous DMPC dispersions in the 0–1000 cm^{-1} region.

due to the O–P–O diester symmetric, $\nu_s(OPO)$, and asymmetric, $\nu_{as}(OPO)$, stretching modes (8, 9); those near 900 cm^{-1} are due to the acyl C_2–C_1 stretching modes (8).

Two very weak bands at ~1740 and ~1720 cm^{-1} are observed in the spectra of solid phospholipids. They arise from the carbonyl stretching modes of the ester linkage (7, 65) and become a broad band in fully hydrated phospholipids (16).

Until now, the qualitative method of group frequencies has been used to describe the Raman spectra of phospholipids. In this method, one associates a band in the Raman spectrum with a specific vibrational mode of an isolated chemical group (for instance, the CH_2 group) in phospholipid molecules. However, this approach cannot give information about the structural configuration of a long chain molecule, and it fails to interpret the broad and complex features in the spectra, although it is sometimes useful for characterizing chemical groups.

In a fully extended zig-zag methylene chain, the translational repeat unit along the chain is C_2H_4, and each repeat unit contains two CH_2 chemical groups. There are 3 atoms and thus 9 degrees of freedom in each chemical group. Consequently, 9 normal modes including 3 zero-frequency modes are expected for each CH_2 group. Each of these 9 modes of a CH_2 group in a methylene chain is not isolated and is coupled with the corresponding mode of the neighboring CH_2 groups to give various normal modes according to various possible phase differences (δ_c) between motions in neighboring CH_2 groups. The frequencies of these resulting normal modes depend on the phase difference. Consequently, there will be more frequencies for a specific CH_2 group vibration in a polymethylene chain than would be predicted by the group frequency approach. These resulting frequencies of each group vibration form a frequency branch and may be plotted as a function of phase difference $\delta_c(0–\pi)$ to give what is usually referred to as a dispersion curve. Because of the frequency dispersion of each normal mode of the CH_2 group in a methylene chain, a total of 9 dispersion curves are expected for a methylene chain, which have been calculated for various δ_c values between 0 and π (51, 73, 96, 97). At $\delta_c = 0$, all the CH_2 groups in the methylene chain vibrate in phase, whereas at $\delta_c = \pi$, the CH_2 groups vibrate out-of-phase between neighboring CH_2 groups. It is apparent from the dispersion curves that the longitudinal acoustic mode, the skeletal C–C stretching mode, the CH_2 rocking mode, and the CH_2 wagging mode are highly dispersed, i.e. the frequencies of these modes are considerably affected by the phase difference, δ_c.

For an infinite methylene chain only those normal modes at $\delta_c = 0$ and $\delta_c = \pi$ are optically active (Raman or infrared). For finite methylene chains with even or odd carbon atoms, however, some of the normal modes with δ_c

values between 0 and π are also potentially active in Raman scattering or infrared absorption. Whether even those will be active depends on the change of polarizability or dipole moment associated with them. According to Born-Karman's periodic condition (11), the allowed phase differences are

$$\delta_c = \frac{k\pi}{M},\ k = 0, 1, 2, \ldots, M-1,$$

where M is the number of CH_2 groups in a methylene chain. The relation between the optically allowed δ_c values, the symmetry species of the normal modes, and the selection rules for *n*-alkanes has been given by Snyder (82) and Tasumi et al (97).

The frequencies of the highly dispersed $\omega(CH_2)$, $\gamma(CH_2)$, ν(C–C), and LA modes are different at different allowed δ_c values. Therefore, there is more than one band for each of these modes in the Raman spectrum of a methylene chain, which gives rise to the so-called band progressions. However, the Raman band of $\omega(CH_2)$ is extremely weak, and $\gamma(CH_2)$ is Raman inactive or weakly active. Therefore, the progression of bands in the Raman spectra of polymethylene chains is only observed in the ν(C–C) and LA mode regions (68, 69, 78, 79, 103).

The relationship between mode frequencies and phase differences was first introduced to the interpretation of the Raman spectra of methylene chains in phospholipids by Lippert & Peticolas (54). In their treatment, however, they consider the phase differences between the successive translational repeat units of methylene chains, C_2H_4, instead of the successive chemical units, CH_2. There are 6 atoms in each repeat unit and thus the degrees of freedom in each C_2H_4 group are 18. Consequently, there are 18 N normal modes for a methylene chain with N repeat units, which are distributed in 18 frequency branches, and each of these frequency branches gives rise to a dispersion curve in the frequency-phase difference diagram as shown in Figure 1 of Reference 28. The allowed phase differences are

$$\phi = k\pi/N,\ k = 0, 1, 2, \ldots, N-1.$$

If a polymethylene chain is considered as one-dimensional crystal, the limit of the phase difference $\phi(0\text{–}\pi)$ may be called the first Brillouin zone. The relationship between this diagram and that resulting from the dispersion between neighboring CH_2 groups (51, 73, 96, 97) is that the dispersion curves on the basis of δ_c are those resulting from the unfolding of the coupled dispersion curves obtained in terms of ϕ. Nine of the 18 normal modes at $\phi = 0$ are equivalent to those at $\delta_c = \pi$. For the infinite methylene chain only those normal modes at $\phi = 0$ may be optically active, whereas the normal modes at both $\delta_c = 0$ and $\delta_c = \pi$ may be optically active. For finite methylene chains, some additional normal modes with allowed phase differences δ_c or ϕ between 0 and π are also optically active, which gives rise to the band progressions.

In principle, the allowed values of ϕ for a chain with one end fixed, such as methylene chains in phospholipids fixed by the presence of the head group), are different from those of a chain with free ends and have been shown to be $\phi = k\pi/(2N+1)$, $k = 1, 3, 5, \ldots, 2N-1$ (54, 103). However, the root-mean-square displacement of one end of a long extended chain, such as that in LA mode, is extremely small and the coupling of this small displacement to the head group is insignificant, so that the chain vibrates as if it had free ends (103).

In fully hydrated phospholipid systems, the Raman intensity of the LA mode is usually very weak and broad, and this band is often in the frequency region of a strong and broad water band (see Figure 3). Therefore, the band progression and even the strongest LA mode ($k = 1$) is seldom observed (103). The skeletal C–C stretching modes at $\delta_c = 0$ and $\delta_c = \pi$ are relatively strong and are well separated from other bands. However, those bands at the allowed values of δ_c between 0 and π are very weak and overlap with the relatively broad PO_2^- stretching band in phospholipids. Only a very weak band with the δ_c value indexed by $k = 3$ has been observed (103). Consequently, the band progression is not a major concern in the interpretation of Raman spectra of fully hydrated phospholipids. Nevertheless, the frequency dispersion concept is important in order to understand properly the origin of many spectral features in the Raman spectra of these systems. For instance, the Fermi resonance between the $\nu_s(CH_2)$ mode and a single binary combination of the $\delta(CH_2)$ mode at zone-center ($\phi = 0$) would give a single narrow band rather than the observed broad and complex Fermi resonance band. To interpret this broad feature properly, it is necessary to consider the resonance interaction between the zone-center $\nu_s(CH_2)$ mode and the continuum of binary combinations involving all the $\delta(CH_2)$ modes over the entire Brillouin zone (85, 87).

Furthermore, there is so-called perpendicular frequency dispersion (85) in aqueous phospholipid bilayers. The methylene chains in these systems are packed in a regular lattice. Consequently, the normal modes of the CH_2 groups could couple with the corresponding modes of the CH_2 groups in neighboring chains. The coupling of normal modes between neighboring chains with various phase differences gives rise to the perpendicular frequency dispersions. The Raman intensity shown in the frequency region between the $\nu_s(CH_2)$ and the $\nu_{as}(CH_2)$ bands in the Raman spectra of aqueous phospholipids is the result of the Fermi resonance arising from this type of frequency dispersion of the $\delta(CH_2)$ mode (85).

Correlation Between Raman Spectral Features and the Structural and Dynamical Properties in Aqueous Phospholipids

Raman spectral parameters, particularly the frequencies, widths, intensities, and shapes of Raman bands, are very sensitive to the structure and dynamics of phospholipid molecules in aqueous phospholipid dispersions.

The frequencies of normal modes of phospholipids are determined not only by the intramolecular force constants but also by the intermolecular interaction force constants. When the intermolecular interaction changes abruptly at first-order phase transitions, the frequencies of the normal modes shift discontinuously. Moreover, when the number of gauche C–C bonds increases in the methylene chains, the steric repulsion potential between neighboring chains becomes stronger as a result of the increase in the hard core radii of the chains (66), and thus the frequencies of the normal modes of the methylene chains shift to higher frequencies. Therefore, the normal mode frequencies are sensitive to the packing between neighboring chains as well as the intrachain conformation.

It is well accepted that a change in the width of the anisotropic component of Raman bands of a condensed phase is mainly due to the change in the reorientational fluctuations of molecules (6, 75). Therefore, the widths of the depolarized and weakly polarized Raman bands of phospholipids such as the $\nu_{as}(CH_2)$, $\delta(CH_2)$, $\tau(CH_2)$, ν_{as}(C–C), and ν_s(C–C) bands, which are dominated by the anisotropic component, can be used as a measure of the reorientational mobility of the phospholipid chains. On the other hand, the width of the isotropic component, which makes the dominant contribution to the intensity of a highly polarized Raman band such as the $\nu_s(CH_2)$ band, has been shown to increase with increasing interchain interaction (1).

For systems in which the orientation of the methylene groups is highly disordered along or perpendicular to the acyl chains as a result of the existence of a large number of gauche bonds or the reorientational fluctuations of the chains, the zone-center selection rule is relaxed. Consequently, all the vibrational modes in the Brillouin zone shown on the dispersion curves become optically active, although these disorder-allowed modes are weaker than the order-allowed modes (81). Therefore, for a highly disordered system, broad and weak bands will appear in the frequency regions of those highly dispersed modes. This is clearly seen in the Raman spectrum of the liquid crystalline phase of phospholipids in Figure 2. Similar disorder-allowed bands have been observed in a disordered micellar phase of sodium oleate (111). Moreover, in these disordered systems a variety of local structures coexist, and thus the static correlation field at different molecular sites is different (the site effect). Consequently, the distribution of frequencies leads to the broadening of the Raman bands (37, 38).

Some Raman bands of phospholipids are characteristic of the methylene chains with all-trans configurations, and some of them arise from the gauche configuration of the chains. Therefore, the relative Raman intensity change of these bands provides information on the conformational changes in the chains.

In addition to the above general spectral parameters, some Raman spectral features that are characteristic of specific frequency regions in the Raman spectra of fully hydrated phospholipids are commonly used for monitoring the structural and dynamical changes in these systems.

In the CH stretching region the features near the $\nu_s(CH_2)$ and $\nu_{as}(CH_2)$ bands are particularly important for the Raman spectroscopic study of phospholipid systems. With the aid of the Raman spectra of *n*-alkanes of known crystal structures (45, 85), it has been shown that the band shape of $\nu_s(CH_2)$ near 2845 cm^{-1} is extremely sensitive to the interchain packing of methylene chains in phospholipids. There is a single and narrow band near 2845 cm^{-1} for the packing of the chains in the hexagonal lattice. The $\nu_s(CH_2)$ band loses its height near 2845 cm^{-1}, gains some intensity near 2855 cm^{-1}, and becomes a broad band for the interchain structures with orthorhombic and monoclinic packing. There are two bands near 2845 cm^{-1} and 2855 cm^{-1} for the triclinic packing. In all these structures, the methylene chains are essentially extended and the $\nu_{as}(CH_2)$ mode exhibits a single narrow band near 2880 cm^{-1} on top of a broad Fermi resonance band. When there are a large number of C–C gauche conformers in the methylene chains, as in the cases of melted alkanes, liquid fatty acids, micellar solutions of surfactants, and the liquid crystalline phase of phospholipids, the narrow $\nu_{as}(CH_2)$ band near 2880 cm^{-1} disappears from the spectrum and only the broad Fermi resonance band remains (13, 63, 85, 102, 104, 111). Although the polarized spectra of these systems (85, 89, 111) demonstrate that the $\nu_{as}(CH_2)$ band does not disappear and only broadens and merges with the Fermi resonance band, nevertheless, the disappearance of the $\nu_{as}(CH_2)$ band in the unpolarized Raman spectra of aqueous phospholipids can be considered as a signal that the methylene chains in these systems change from extended structure to the liquid-like structure containing a large number of gauche conformers.

Another useful parameter obtained from the Raman spectrum in the CH stretching region is the peak height ratio between the $\nu_{as}(CH_2)$ and the $\nu_s(CH_2)$ bands, H(2880)/(2850). Isotopic dilution studies of *n*-alkanes at constant temperature (30, 61, 85) have shown that this peak height ratio may be qualitatively used as a measure of the amount of interchain interaction. Similar results have been observed in a series of studies on the pressure-enhanced interchain interaction at constant temperature in some aqueous phospholipid dispersions (108, 113) and in the micellar solution of sodium oleate (110).

It has been shown (85, 88), however, that the integrated intensity ratio between the $\nu_{as}(CH_2)$ and the $\nu_s(CH_2)$ bands is independent of environment; only the peak height ratio of these two bands is sensitive to the change in the interchain interaction; and the change in the peak heights is a result of the change in the widths of these bands, i.e. the peak heights are increased by the narrowing and decreased by the broadening of these bands. As indicated above, the width of the totally symmetric $\nu_s(CH_2)$ band increases with increasing interchain interaction, whereas the width of the depolarized $\nu_{as}(CH_2)$ band decreases with decreasing reorientational mobility of the chains and with the reduction of the number of gauche conformers in the chains. These inter- and intrachain ordering processes in turn increase the static interchain interaction and thus also broaden the $\nu_s(CH_2)$ band. Consequently, when the interchain interaction is increased by the change in environment, the intrachain ordering, or the damping of the interchain reorientational fluctuations, the peak height ratio between the $\nu_{as}(CH_2)$ and the $\nu_s(CH_2)$ bands, H(2880)/H(2850), increases as a result of the broadening of the $\nu_s(CH_2)$ band and the narrowing of the $\nu_{as}(CH_2)$ band. An attempt has been made to derive an order parameter from this peak height ratio as a quantitative parameter to determine the amount of interchain interaction and the closest packing probability in phospholipids (30). However, the general validity and the quantitative significance of this order parameter have been questioned (43, 85, 88).

The symmetric and asymmetric skeletal C–C stretching modes of the methylene chain with the all-trans conformation are at frequencies near 1130 cm^{-1} and 1060 cm^{-1}, respectively. The broad skeletal C–C stretching band around 1080 cm^{-1} is characteristic of the methylene chains in the disordered gauche state. The ratio between the trans and gauche band intensities is commonly used to estimate relative populations of the trans and gauche conformers in the aqueous phospholipids (30, 48, 53, 95, 103, 104, 115). Although the integrated intensities of these bands are more closely related to this relative population, frequently the peak heights of these bands are used, since a band profile analysis is required to obtain the integrated intensities of these bands and the peak heights vary more strongly with temperature than the integrated intensities, especially at a phase transition (94, 103).

In the case of phosphatidylcholines, the intensities of ν_s(C–C) and ν_{as}(C–C) have been used directly to estimate the change of conformational order; the C–N stretching band at ~720 cm^{-1} is used as a reference of intensity that is temperature independent with respect to frequency and intensity (31, 91, 92, 94, 103).

The intensity ratio between the trans and gauche C–C stretching modes has been used to derive a trans-parameter for phospholipids as a quantitative measure of the all-trans chain probability (30). It has been shown, however, that the trans-parameter exceeds the maximum theoretical value of unity in the presence of certain cations (43).

While the 1060 cm^{-1} trans band frequency is independent of the chain length and temperature, the 1130 cm^{-1} trans band has been found to be frequency sensitive to both chain length and temperature. The frequency of this band shifts toward lower values with decreasing chain length and increasing temperature (115).

Spectral changes in the $\delta(CH_2)$ band near 1450 cm^{-1} are sensitive to the changes in interchain structure (10, 72, 107, 113, 117). In polyethylene and *n*-alkanes with monoclinic and orthorhombic interchain structures in which the orientation of the methylene chains is highly ordered and is nearly perpendicular between the zig-zag planes of neighboring chains, a well-defined narrow band appears on the low-frequency side of the $\delta(CH_2)$ band near 1418 cm^{-1}. This band does not exist in *n*-alkanes with hexagonal and triclinic structures and has been assigned to the crystal field component of the A_{1g} scissoring mode (10). In phospholipids with monoclinic interchain packing such as that in crystalline DMPC dihydrate (72), this crystal field band near 1418 cm^{-1} appears as a weak shoulder (see Figure 4). Only at very low temperature (117) and high pressure (106) does this crystal field component of aqueous phospholipid dispersions become a well-defined band.

The methylene twisting mode $\tau(CH_2)$ near 1300 cm^{-1} appears as a band that is well-isolated from other features in the Raman spectra of phospholipids (see Figure 2), and thus its peak position and width can be measured accurately. It is a highly depolarized band; therefore, the width of this band can be used to monitor the chain mobility efficiently. Moreover, the frequency of this band can be used to determine the conformational state, since the frequency of this band is one of the most sensitive Raman frequencies of phospholipids to the chain conformation. $\tau(CH_2)$ is at $\sim$1294 cm^{-1} for extended chains, whereas it shifts to $\sim$1305 cm^{-1} for disordered chains in the liquid crystalline phase of phospholipids (92, 103, 113). This band has been used as an internal standard of intensity (117). However, it has been shown recently that the intensity of this band is dependent on temperature and chain length (18, 94, 103).

For a phospholipid with unsaturated hydrocarbon chains, there are several additional bands in the Raman spectra that are due to the CH stretching of HC=CH residues, ν(=CH), the C=C stretching, ν(C=C), the in-plane =CH deformation, $\delta(=CH)_i$, and the out-of-plane =CH deformation, $\delta(=CH)_0$ (18, 28, 54, 59). The frequencies of these bands are dependent on the conformation of the C=C double bond (89). For cis double bonds, $\nu(C=C)$ is at 1650–1665 cm^{-1} (54), $\delta(=CH)_i$ is at 1261–1267 cm^{-1} (18, 59), and $\delta(=CH)_0$ is at 967–970 cm^{-1} (54, 59), whereas for trans double bonds, $\nu(C=C)$ is at 1670–1680 cm^{-1} (54), $\delta(=CH)_i$ cannot be resolved from the $\tau(CH_2)$ band (117), and $\delta(=CH)_0$ is at 810 cm^{-1} (54). The $\nu(=CH)$ band is in the frequency region of 3000 cm^{-1} (18, 88). The intensity ratio between $\delta(=CH)_i$ and $\tau(CH_2)$ may be used as a measure of the degree of unsaturation in the chain (18), since the intensity of $\delta(=CH)_i$ is proportional to the number of double bonds while that of $\tau(CH_2)$ is proportional to the number of CH_2 groups in the methylene chain segments. Therefore, the Raman intensity ratio between $\delta(=CH)_i$ and $\tau(CH_2)$ increases as the degree of unsaturation in the chains increases. Moreover, it has been shown (111) that the intensity of $\nu(C=C)$ is very sensitive to the conformation of the methylene chain segments. When the methylene chain segments are in a disordered state with a large number of gauche conformers, the intensity of $\nu(C=C)$ increases dramatically.

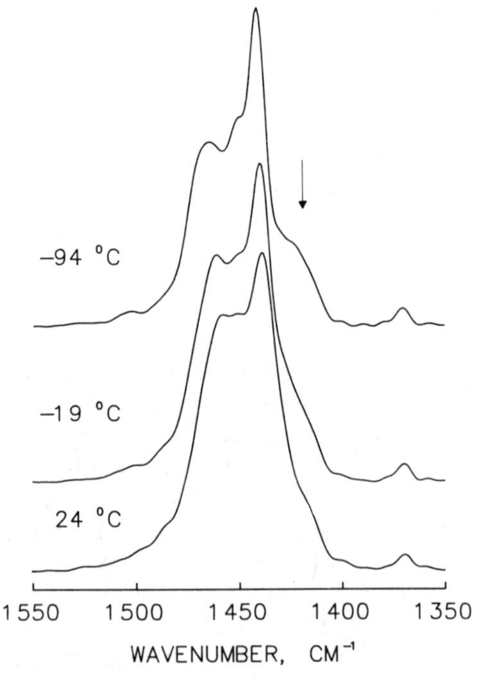

Figure 4 Raman spectra of polycrystalline DMPC dihydrate in the CH_2 scissoring region at several temperatures.

Until recently (2, 7–9, 16, 17, 64, 65), the contribution of Raman spectroscopy to the study of the structures of the head group and the acyl groups in the interface region in phospholipids has been limited because of the weak intensity of the Raman bands that are characteristic of these groups.

The carbonyl stretching modes ν(C=O) in the 1700–1750 cm^{-1} region have been used recently to distinguish between conformations about the interface $C_3C_2C_1$(=O)O–ester groups in phospholipids (7, 65). Two bands have been observed in this frequency region, and it has been demonstrated that the ν(C=O) of an acyl chain with a bent conformation at the C_2 position is at lower frequency (~1716–1728 cm^{-1}), while that of a nearly straight acyl chain is at higher frequency (~1727–1744 cm^{-1}).

A normal frequency calculation for the symmetric, ν_s(O–P–O), and antisymmetric, ν_{as}(O–P–O), O–P–O stretching modes, the symmetric, $\nu_s(PO_2^-)$, and antisymmetric, $\nu_{as}(PO_2^-)$, PO_2^-, stretching modes, and the neighboring C–C stretching modes ν(C–C) of the C–C–O–P(O_2)–O–C–C skeleton in the head groups of phospholipids as a function of rotations about the P–O and the C–O bonds has demonstrated that these normal modes are very sensitive to the conformation of the head groups (9). A rotation about a P–O bond from the usual gauche conformer to a trans conformer results in an increase in the frequency difference between $\nu_s(PO_2^-)$ and $\nu_{as}(PO_2^-)$ and a decrease in the frequency difference between ν_s(O–PO) and ν_{as}(O–P–O). A rotation about a C–O bond from the usual trans conformer towards a gauche conformer results in a decrease in the frequencies of both the ν_s(O–P–O) and ν(C–C) of the C–C group adjacent to the rotated C–O bond.

With the aid of the Raman spectra of a series of choline halides with known conformational states, it has been concluded (2) that the frequency of the total symmetric CN stretching mode ν_s(CN) of the head group is dependent on the conformation of the O–C–C–N^+ backbone in the head group. For a gauche conformation, ν_s(CN) is at ~720 cm^{-1}, whereas for a trans conformation, ν_s(CN) is at ~770 cm^{-1}. Most of the choline backbones of phospholipids in both the gel and the liquid crystalline phases are in the gauche conformational state; the trans conformational state is only present in the crystalline phase of some phospholipids.

The frequencies of methylene vibrational modes of a deuterated methylene chain are different from those of one that is nondeuterated, although the structures are presumably isomorphous. The dynamically and structurally sensitive CH stretching bands shift to the ~2100 cm^{-1} region for a deuterated methylene group where there is no Raman band arising from the nondeuterated one (32). Therefore, in principle, appropriately deuterated phospholipids may be used in Raman spectroscopic studies of the structural properties and phase behaviour of each component of a phospholipid mixture and of individual methylene groups in the acyl chains and in the head group.

The frequency of the symmetric and asymmetric CD_2 stretching modes near 2100 cm^{-1} is strongly dependent on the position of the methylene group in acyl chains that results from the charge distribution of the carbonyl group (5). A sharp decrease in frequency is observed as the position of the methylene group moves away from the polar carbonyl group, and the frequency becomes essentially constant beyond the sixth methylene group from the carbonyl group.

The band width of the symmetric CD_2 stretching mode in deuterated phospholipids changes significantly during the phase transition between the gel and the liquid crystalline phase and has been used to monitor this transition in binary phospholipid mixtures (32, 60–62). This band width has also been used to monitor the conformational change in the deuterated methylene chains (60, 61). However, this band near 2100 cm^{-1} is not a single band of the symmetric CD_2 stretching mode; it contains shoulders arising from the CD_3 symmetric stretching component and the combination between the 1128 cm^{-1} gauche C–C stretching mode and the 985 cm^{-1} CD_2 rocking fundamental. Consequently, the conformation of the chain is only related to the width of one of these shoulders (14).

Thermotropic and High-Pressure Phases in Aqueous Phospholipids

One of many interesting physical properties of aqueous phospholipid dispersions is the thermotropic phase transition. It has been suggested that the function of biological membranes is partially controlled by the phase behavior of the lipids in the bilayers (52, 58, 70, 71, 74). The most thoroughly investigated phase transition of the fully hydrated phospholipid bilayers is the thermotropic phase transition between the liquid crystalline phase LC (L_α)[2] and the lamelar GI gel phase ($P_{\beta'}$), the main transition (22, 23, 44, 46, 55, 80). The main transition temperature (T_m) is dependent on the length, the degree of branching, and the extent of unsaturation of the acyl chains, the degree of hydration, the polar head group structure, and the incorporation of additional components into the bilayers (12, 22, 27, 40, 44, 46, 47, 56, 80, 105).

Below T_m, there is a thermotropic phase transition between the GI ($P_{\beta'}$) and the GII ($L_{\beta'}$) gel phases, which is generally referred to as the pretransition (21, 41, 44). The pretransition temperature (T_p) is also dependent on the nature of the acyl chains and the head group, the degree of hydration, and other pertubations (12, 34, 41, 44, 105). Both T_m and T_p of

[2] Luzzati's notation in the parentheses.

fully hydrated phosphatidylcholine dispersions progressively increase with increasing hydrocarbon chain length, and T_p appears to converge with $T_m \cdot T_p$ and T_m will coincide when the number of carbon atoms in the hydrocarbon chain exceeds 22 (42, 44).

Unsaturated and certain saturated phosphatidylethanolamines (25, 26, 98) in the presence of a very high concentration of mono-valent cations (35) transform to the hexagonal type II phase ($H_{II}\alpha$) at elevated temperature above T_m. This transition temperature (T_{BH}) is dependent on the degree of acyl chain unsaturation. T_{BH} is lowered by higher acyl chain unsaturation (98).

Both the main transition and the pretransition exhibit first-order characteristics (3, 21, 67). As predicted by the thermodynamic Clausius-Clapeyron relationship,

$$\frac{d\mathrm{T}}{d\mathrm{P}} = \frac{\mathrm{T}\Delta\mathrm{V}}{\Delta\mathrm{H}}$$

where T, P, ΔV and ΔH are transition temperature, pressure, volume change, and enthalpy change, respectively, an increase in pressure should raise the phase transition temperature of both the main transition and the pretransition.

The pressure effect on the main transition temperature of a series of aqueous phospholipid dispersions has been investigated, and the $d\mathrm{T}/d\mathrm{P}$ values are in the range of 20–40 K/kbar (1 kbar = 986.9 atm) (36, 50, 99, 113, 114).

A recent Raman spectroscopic study (113) showed that the pretransition temperature of a 1,2-dimyristoyl 3-*sn*-phosphatidylcholine (DMPC) aqueous dispersion can be raised from 14°C at atmospheric pressure to 30°C by increasing the hydrostatic pressure to 1 kbar and that a new phase transition from the GII gel phase to the GIII gel phase was observed at 2.6 kbar and 30°C. The corresponding pressure-induced GII/GIII transitions at 30°C in 1,2-dipalmitoyl-3-*sn*-phosphatidylcholine (DPPC) and 1,2-distearoyl-3-*sn*-phosphatidylcholine (DSPC) were found at 1.7 and 1.1 kbar, respectively (108).

Since this pressure-induced transition between the GII and the GIII phases exhibits first-order characteristics (108, 113), on the basis of thermodynamics, this transition is expected at low temperature and atmospheric pressure. Indeed the corresponding GII to GIII phase transitions at atmospheric pressure for DMPC, DPPC, and DSPC were found at −60, −30, and −6°C, respectively (107, 109; P. T. T. Wong, unpublished).

Furthermore, at higher pressure another gel-gel phase transition (GIII to GIV gel phase) was observed for DSPC, DPPC, and DMPC (106, 112).

Therefore, a total of four gel phases can be found in fully hydrated phospholipid dispersions.

Another recently discovered phase transition, the subtransition, in hydrated 1,2-diacyl-L-phosphatidylcholine bilayers with alkyl chains of 16, 17, and 18 carbon atoms takes place between the GII gel phase and a new low-temperature subphase. This phase transition occurs only after extensive incubation of the sample at temperatures around 0°C (24). The subtransition temperature and enthalpy increase with increasing incubation time and a limiting value for enthalpy is reached after approximate 2 days incubation. The formation of the subphase is accompanied by a decrease in hydration and an increase in inter- and intrachain order (20, 29, 76). The infrared results show that the interchain packing in the subphase resembles that found in triclinically packed acyl systems (20), whereas the X-ray diffraction results indicate that the packing in this phase is that of an orthorhombic sub-cell (76).

Changes in Raman Spectral Features Associated With Phase Transitions

The majority of Raman spectroscopic studies on aqueous phospholipid dispersions concern the phase transition between the liquid crystalline phase and the GI gel phase, i.e. the main transition, and only a limited number of synthetic phosphatidylcholine dispersions have been systematically studied (39).

The spectral regions that show the most spectacular changes at the main transition are the CH stretching and the skeletal C–C stretching regions. At the phase transition from the GI gel phase to the LC liquid crystalline phase, the narrow $\nu_{as}(CH_2)$ band at $\sim$2880 cm^{-1} becomes extremely broad and merges with the broad Fermi resonance band. It appears as if the $\nu_{as}(CH_2)$ band disappeared completely from the spectrum of a LC phase (see Figure 1). The intensities of the ν_s(C–C) and the ν_{as}(C–C) bands decrease abruptly, and the broad gauche C–C stretching band near 1090–1085 cm^{-1} dominates the spectrum in the skeletal C–C stretching region in the LC phase (30, 31, 53, 94, 103, 104) (see Figure 2). Apparently, the number of gauche conformers on the acyl chains increases dramatically and the orientation of the methylene groups on the chains becomes highly disordered in the LC phase. The band shape of the symmetric CH stretching band at 2850 cm^{-1}, however, remains more or less unchanged at the transition, which implies that the interchain packing in these two phases is about the same and is in a hexagonal lattice (45, 85). This is consistent with the results from X-ray diffraction studies (76, 95).

The frequencies of all the normal modes of the methylene chains show discontinuities at the main transition, and those of $\nu_{as}(CH_2)$, $\nu_s(CH_2)$, and

$\tau(CH_2)$ have the largest shift at this transition (31, 113). All the frequencies except for that of ν_s(C–C) increase abruptly at the transition from the GI phase to the LC phase (4, 86, 92, 103, 113). The shift to higher frequencies of these modes in the LC phase is due to the increase in the steric repulsion potential between neighboring chains, which is a result of an increase in the hard core radii of the acyl chains arising from the presence of a large number of gauche conformers in the chains (66). On the other hand, the frequency of ν_s(C–C) at $\sim 1120\ cm^{-1}$ decreases in the LC phase. It has been shown that the frequency of this mode is dependent on the all-trans methylene chain length (115). As the length of all-trans methylene segments is significantly reduced because of the presence of gauche C–C bonds, a lower frequency of this mode is expected for the LC phase. Since the frequency of ν_s(C–C) decreases while that of ν_{as}(C–C) increases abruptly at the main transition (31, 113), the discontinuity of the frequency difference between these two modes at the main transition is much larger than those of the individual frequency shift of these two modes. Consequently, the change in the frequency difference between these modes has been used to monitor the main transition (91).

The widths of all the depolarized and weakly polarized bands of the methylene modes [$\nu_{as}(CH_2)$, $\delta(CH_2)$, $\tau(CH_2)$, ν_{as}(C–C), ν_s(C–C)] increase dramatically in the LC phase (15, 63, 94, 113), which indicates that the rate of the chain reorientational fluctuations is extremely high and the interchain orientation is highly disordered in the LC phase (6, 75). On the other hand, the band width of the highly polarized $\nu_s(CH_2)$ band only increases slightly in the LC phase (14, 113). The width of a highly polarized Raman band is not directly affected by the rate of reorientational fluctuation but is strongly affected by interchain coupling interactions (1, 6). The interchain interaction in the LC phase is very weak as shown by the abrupt decrease in the peak height ratio between $\nu_{as}(CH_2)$ and $\nu_s(CH_2)$ in the LC phase (13, 100, 101, 113, 117). Consequently, the width of $\nu_s(CH_2)$ should decrease in the LC phase. On the other hand, this width should increase in the LC phase because of the distribution of frequencies of this mode that arise from the difference in the local environment of various methylene groups in the highly disordered acyl chains. As a result of the compensation between these two effects, the width of the highly polarized $\nu_s(CH_2)$ band does not change significantly at the main transition.

The Raman spectral changes at the pretransition (i.e. the transition between the GI and GII gel phases) below the main transition temperature or above the main transition pressure (30, 113) are less dramatic. The discontinuity of the frequency shifts of all the Raman bands at the pretransition is extremely small (30, 113). In fact, only the discontinuities for $\nu_s(CH_2)$ and $\nu_{as}(CH_2)$ can be barely observed (113).

In the CH stretching region, both the $\nu_s(CH_2)$ and $\nu_{as}(CH_2)$ bands decrease in intensity, and the intensity of $\nu_{as}(CH_2)$ decreases about twice more than that of $\nu_s(CH_2)$ at the pretransition from the GII to the GI phase (31). The peak height of the $\nu_{as}(CH_2)$ band at $\sim$2880 cm^{-1}, taking that of the CN stretching band at $\sim$720 cm^{-1} as reference, shows a discontinuity at the pretransition (30). The band shape of $\nu_s(CH_2)$ of the GII gel phase is, however, similar to that of the GI phase (31), which indicates that the interchain packing is similar between these two gel phases with a hexagonal lattice (45, 85). X-ray diffraction results by Tardieu et al (95) and Janiak et al (41, 42) show that the chains are packed in a distorted hexagonal lattice in the GII phase, whereas Ruocco & Shipley (76) have described the interchain structure of this phase with an enlarged, disordered orthorhombic hybrid subcell. The Raman spectral features of this phase in the CH stretching region resemble those of *n*-alkanes with hexagonal interchain structure (10, 45, 85).

An abrupt loss of intensity of the all-trans C–C stretching bands and a significant increase in the intensity of the 1090 cm^{-1} gauche C–C stretching band have been observed at the pretransition from the GII phase to the GI phase (30, 32, 48, 116), which leads to the conclusion that the pretransition is characterized by an increase in gauche conformers in the GI phase, although the number of gauche conformers is much smaller than that in the LC phase (30).

The CH_2 twisting mode $\tau(CH_2)$ near 1300 cm^{-1} decreases considerably in intensity at the pretransition from the GII phase to the GI phase, while its frequency remains constant.

The peak height ratio H2880/H2850 shows little change in the GI phase, whereas it shows a dramatic increases in the GII phase with increasing pressure (113) or decreasing temperature (117). A change of slope of this ratio as a function of pressure has been clearly observed at the pretransition (113). These results indicate that the interchain interaction increases only slightly in the temperature or pressure range of the GI phase but that it increases dramatically with decreasing temperature or increasing pressure in the GII phase.

The widths of the depolarized and weakly polarized bands decrease continuously from the GI to the GII phase with a slightly different rate in the GII phase, and an inflection point of the width change has been observed at the pretransition (113). Therefore, the reorientational mobility of the chains in the GI phase is relatively higher than that in the GII phase and decreases continuously in going from the GI to the GII phase.

Raman spectral changes at the structural phase transition between the GII and the GIII gel phases are much larger than those at the pretransition. Frequency discontinuities of the methylene vibrational modes have been observed at both the temperature-induced (107, 109) and pressure-induced (106, 108, 113) GII/GIII phase transition.

The widths of the depolarized and weakly polarized bands decrease further as the temperature or pressure approaches the GII/GIII transition and become much less sensitive to temperature or pressure in the GIII phase (106–109, 113). Therefore, the chain reorientation is further damped by decreasing temperature or increasing pressure in the GII phase, and the orientation of the acyl chains becomes highly ordered in the GIII phase.

In the CH stretching region the $\nu_s(CH_2)$ band becomes broader while the $\nu_{as}(CH_2)$ band becomes narrower and more intense in the GIII phase (Figure 5). These spectral features in the GIII phase resemble those observed in *n*-alkanes that have an orthorhombic subcell structure (45, 85). The correlation field band near 1410 cm^{-1}, which is a characteristic band observed in the Raman spectra of *n*-alkanes with a monoclinic or an orthorhombic structure (10), appears in the GIII phase immediately below the transition temperature (107, 109) or above the transition pressure (Figure 6). In the GIII phase, this correlation field band is not well separated from the $\delta(CH_2)$ band and appears as a shoulder similar to that observed in

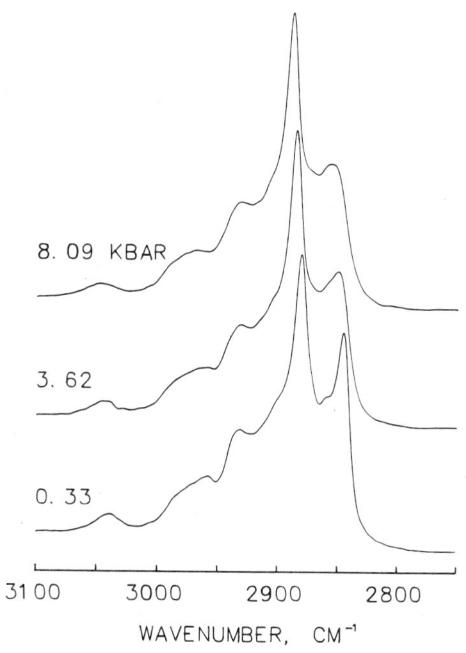

Figure 5 Raman spectra of aqueous DPPC dispersions in the CH stretching region at several pressures.

the Raman spectrum of monoclinic DMPC dihydrate crystal (72) (Figure 4). The appearance of this correlation field component of $\delta(CH_2)$, along with the change in the spectral features in the CH stretching region, indicates that the GII/GIII phase transition is associated with the change in the packing pattern of the acyl chains and suggests that it involves a structural change from a distorted hexagonal packing to a monoclinic packing.

In principle, the correlation field component of the $\delta(CH_2)$ band should appear in the Raman spectra of the GII phase of phospholipids, especially at the temperatures and pressures near the GII/GIII transition, since the damping of the reorientational fluctuation of the acyl chains is increased with decreasing temperature or increasing pressure in the GII phase and, thus, statistically non-equivalent orientations of the chains exist in this phase. However, the Raman intensity of this correlation field band is very weak. Consequently it can be observed only when the reorientational fluctuation is nearly completely damped as that in the GIII phase. In fact, the infrared band of this correlation field component, which is a relatively strong band, has been observed in the GII phase of phospholipids (19).

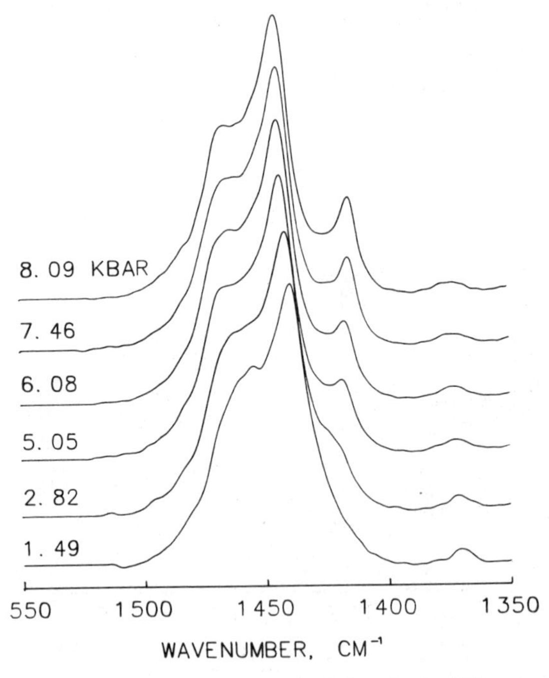

Figure 6 Raman spectra of aqueous DPPC dispersions in the CH_2 scissoring region at several pressures.

The peak height ratio H2880/H2850 increases in the GII phase with decreasing temperature or increasing pressure, and a change of slope occurs at the GI/GIII transition. Therefore, a further increase in interchain interaction in the GII phase with decreasing temperature or increasing pressure is evident, which is undoubtedly a result of the damping of the reorientational fluctuation. This peak height ratio is almost unaffected by temperature or pressure in the GIII phase (106–109, 113). Thus the interchain interaction in the GIII phase becomes less sensitive to temperature and pressure because of the highly ordered structure of the acyl chains in this phase.

The occurrence of a phase transition at higher pressure to a subsequent gel phase (GIV) in the aqueous DMPC, DPPC, and DSPC dispersions is quite evident from the changes in the Raman spectra of these systems (106, 112). The transition pressures between the GIII and GIV phases are ~8, 4.8, and 3.1 kbar for DMPC, DPPC, and DSPC, respectively. Discontinuities in pressure dependences of frequencies and width have been observed at the GIII/GIV transition. Some of these discontinuities are more pronounced than the corresponding ones at the GII/GIII transition (106).

The $\nu_s(CH_2)$ band of the GIV phase is further broadened. A representative spectrum in the CH stretching region of the GIV phase of DPPC at 8.09 kbar is compared with those of the GIII and the GII phases of DPPC in Figure 5. The correlation field component of the $\delta(CH_2)$ band in the GIV phase becomes a well-defined band similar to those observed in orthorhombic phases of *n*-alkanes and polyethylene (10). Representative spectra in the $\delta(CH_2)$ region of the GIV phase of DPPC above 5 kbar are compared with those of the GIII and GII phases of DPPC in Figure 6. These spectral features imply that the acyl chain packing in the GIV phase corresponds to that of *n*-alkanes in an orthorhombic lattice.

A preliminary Raman spectroscopic study of the subtransition of DPPC has recently been made (57). The band shape in the CH stretching region indicates that the structure of the subphase corresponds to a fairly rigid quasi-triclinic packing, which is consistent with the infrared result (20). The integrated intensity of the 1130 cm^{-1} band shows that the acyl chains in the subphase are fully extended. Longitudinal acoustic modes are clearly observed at 160 cm^{-1} and 173 cm^{-1}, which correspond to all-trans acyl chain lengths of fifteen and fourteen carbon atoms, respectively. The appearance of two components, a sharp peak at ~1742 cm^{-1} and a much broader band at ~1730 cm^{-1} in the carbonyl stretching region, suggests a partial and non-equivalent dehydration of the two carbonyl groups in the subphase.

Literature Cited

1. Abbott, R. J., Oxtoby, D. W. 1979. *J. Chem. Phys.* 70:4703–7
2. Akutsu, H. 1981. *Biochemistry* 20: 7359–66
3. Albon, A., Sturtevant, J. M. 1978. *Proc. Natl. Acad. Sci. USA* 75:2258–60
4. Asher, I. M., Leven, I. W. 1977. *Biochim. Biophys. Acta* 468:63–72
5. Bansil, R., Day, J., Meadows, M., Rice, D., Oldfield, E. 1980. *Biochemistry* 19:1938–43
6. Bartoli, F. J., Litovitz, T. A. 1972. *J. Chem. Phys.* 56:404–12
7. Bicknell-Brown, E., Brown, K. G., Person, W. B. 1980. *J. Am. Chem. Soc.* 102:5486–91
8. Bicknell-Brown, E., Brown, K. G., Person, W. B. 1981. *J. Raman Spectrosc.* 11:356–62
9. Bicknell-Brown, E., Brown, K. G., Person, W. B. 1982. *J. Raman Spectrosc.* 12:180–89
10. Boerio, F. J., Koenig, J. L. 1970. *J. Chem. Phys.* 52:3425–31
11. Born, M., Karman, T. V. 1912. *Phys. Z.* 13:297–309
12. Brady, G. W., Fein, D. B. 1977. *Biochim. Biophys. Acta* 464:249–59
13. Brown, K. G., Peticolas, W. L., Brown, E. 1973. *Biochem. Biophys. Res. Commun.* 54:358–64
14. Bryant, G. J., Lavialle, R., Levin, I. W. 1982. *J. Raman Spectrosc.* 12:118–21
15. Bunow, M. R., Levin, I. W. 1977. *Biochim. Biophys. Acta* 1977. 489:191–206
16. Bush, S. F., Adams, R. G., Levin, I. W. 1980. *Biochemistry* 19:4429–36
17. Bush, S. F., Levin, H., Levin, I. W. 1980. *Chem. Phys. Lipids* 27:101–11
18. Butler, M., Salem, N. Jr., Hoss, W. 1979. *Chem. Phys. Lipids* 29:99–102
19. Cameron, D. C., Casal, H. L., Mantsch, H. H. 1980. *Biochemistry* 19:3665–72
20. Cameron, D. G., Mantsch, H. H. 1982. *Biophys. J.* 38:175–84
21. Chapman, D. 1968. *Biol. Membr.* 1:125–202
22. Chapman, D. 1976. *Q. Rev. Biophys.* 8:185–235
23. Chapman, D., Williams, R. M., Ladbrooke, B. D. 1967. *Chem. Phys. Lipids* 1:445–75
24. Chen, S. C., Sturtevant, J. M., Gaffney, B. J. 1980. *Proc. Natl. Acad. Sci. USA* 77:5060–63
25. Cullis, P. R., DeKruijff, B. 1976. *Biochim. Biophys. Acta* 436:523–40
26. Cullis, P. R., DeKruijff, B. 1978. *Biochim. Biophys. Acta* 513:31–42
27. Epand, R. M., Epand, R. F. 1980. *Chem. Phys. Lipids* 27:139–50
28. Faiman, R., Larson, K. 1976. *J. Raman Spectrosc.* 4:387–94
29. Füldner, H. H. 1981. *Biochemistry* 20:5707–10
30. Gaber, B. P., Peticolas, W. L. 1977. *Biochim. Biophys. Acta* 465:260–74
31. Gaber, B. P., Yager, P., Peticolas, W. L. 1978. *Biophys. J.* 21:161–76
32. Gaber, B. P., Yager, P., Peticolas, W. L. 1978. *Biophys. J.* 22:191–207
33. Gaber, B. P., Yager, P., Peticolas, W. L. 1979. *NATO Adv. Study Inst. Ser. C.* 43:241–59
34. Harlos, K. 1978. *Biochim. Biophys. Acta* 511:348–55
35. Harlos, K., Eibl, H. 1981. *Biochemistry* 10:2888–92
36. Heremans, K. 1982. *Ann. Rev. Biophys. Bioeng.* 11:1–21
37. Hornig, D. F. 1948. *J. Chem. Phys.* 16:1063–76
38. Hornig, D. F. 1949. *J. Chem. Phys.* 17:1346
39. Huang, C., Jeffrey, R. L., Levin, I. W. 1982. *J. Am. Chem. Soc.* 104:5926–30
40. Jacobson, K., Papahadjopoulos, D. 1975. *Biochemistry* 14:152–61
41. Janiak, M. J., Small, D. M., Shipley, G. G. 1976. *Biochemistry* 15:4575–80
42. Janiak, M. J., Small, D. M., Shipley, G. G. 1979. *J. Biol. Chem.* 25:6068–78
43. Karvaly, B., Loshchilova, E. 1977. *Biochim. Biophys. Acta* 470:492–96
44. Ladbrooke, B. D., Chapman, D. 1969. *Chem. Phys. Lipids* 3:304–67
45. Larson, K. 1973. *Chem. Phys. Lipids* 10:165–76
46. Lee, A. G. 1975. *Prog. Biophys. Mol. Biol.* 29:3–56
47. Lee, A. G. 1977. *Biochim. Biophys. Acta* 472:285–344
48. Levin, I. W., Bush, S. F. 1981. *Biochim. Biophys. Acta* 640:760–66
49. Levin, I. W., Mushayakarara, E., Bittman, R. 1982. *J. Raman Spectrosc.* 13:231–34
50. Lin, N. I., Kay, R. L. 1977. *Biochemistry* 16:3484–86
51. Lin, T. P., Koenig, J. L. 1962. *J. Mol. Spectrosc.* 9:228–43
52. Linden, C. D., Wright, K. L., McConnell, H. M., Fox, C. F. 1973. *Proc. Natl. Acad. Sci. USA* 70:2271–80
53. Lippert, J. L., Peticolas, W. L. 1971. *Proc. Natl. Acad. Sci. USA* 68:1572–76
54. Lippert, J. L., Peticolas, W. L. 1972. *Biochim. Biophys. Acta* 282:8–17

55. Luzzati, V. 1968. *Biol. Membr.* 1:71–123
56. Mabrey, S., Sturtevant, J. M. 1976. *Proc. Natl. Acad. Sci. USA* 73:3862–66
57. Magni, R., Sheridan, J. P. 1982. *Biophys. J.* 37:11a
58. Melchior, D. G., Steim, J. M. 1979. *Prog. Surf. Membr. Sci.* 13:211–96
59. Mendelsohn, R. 1972. *Biochim. Biophys. Acta* 290:15–21
60. Mendelsohn, R., Dluhy, R., Curatolo, W., Sears, B. 1982. *Chem. Phys. Lipids* 30:287–95
61. Mendelsohn, R., Koch, C. C. 1980. *Biochim. Biophys. Acta* 598:260–71
62. Mendelsohn, R., Maisano, J. 1978. *Biochim. Biophys. Acta* 506:192–201
63. Mendelsohn, R., Sunder, S., Bernstein, H. J. 1975. *Biochim. Biophys. Acta* 413:329–40
64. Mushayakarara, E., Albon N., Levin, I. W. 1982. *Biochim. Biophys. Acta* 686:153–59
65. Mushayakarara, E., Levin, I. W. 1982. *J. Phys. Chem.* 86:2324–27
66. Nagle, J. F. 1973. *J. Chem. Phys.* 58:252–64
67. Nagle, J. F., Wilkinson, D. A. 1978. *Biophys. J.* 23:159–75
68. Okabayashi, H., Okuyama, M., Kitagawa, T. 1975. *Bull. Chem. Soc. Jpn.* 48:2264–69
69. Olf, H. G., Peterlin, A., Peticolas, W. L. 1974. *J. Polym. Sci. Polym. Phys. Ed.* 12:359–84
70. Overath, P., Brenner, M., Gulik-Krzywicki, T., Schechter, E., Letellier, L. 1975. *Biochim. Biophys. Acta* 389:358–69
71. Overath, P., Thilo, L. 1978. *Int. Rev. Biochem.* 19:1–44
72. Pearson, R. H., Pascher, I. 1979. *Nature* 281:499–501
73. Piseri, L., Zerbi, G. 1968. *J. Chem. Phys.* 48:3561–72
74. Racker, E., Knowles, A. F., Eytan, E. 1975. *Ann. NY Acad. Sci.* 264:17–33
75. Rakov, A. V. 1959. *Opt. Spectrosc.* 7:128–31
76. Ruocco, M. J., Shipley, G. G. 1982. *Biochim. Biophys. Acta* 684:59–66
77. Schachtschneider, J. H., Snyder, R. G. 1963. *Spectrochim. Acta* 19:117–68
78. Schaufele, R. F., Shimanouchi, T. 1967. *J. Chem. Phys.* 47:3605–10
79. Shimanouchi, T. 1971. *Kagaku no Ryoiki, Tokyo* 25:97–104
80. Shipley, G. G. 1973. *Biological Membranes*, ed. D. Chapman, D. F. H. Wallach, 2:1–89. New York: Academic
81. Shuker, R. Cammon, R. W. 1970. *Phys. Rev. Lett.* 25:222–25
82. Snyder, R. G. 1960. *J. Mol. Spectrosc.* 4:411–34
83. Snyder, R. G. 1967. *J. Mol. Spectrosc.* 23:224–28
84. Snyder, R. G. 1969. *J. Mol. Spectrosc.* 31:464–65
85. Snyder, R. G., Hsu, S. L., Krimm, S. 1978. *Spectrochim. Acta* 34A:395–406
86. Snyder, R. G., Schachtschneider, J. H. 1963. *Spectrochim. Acta* 19:85–116
87. Snyder, R. G., Scherer, J. R. 1979. *J. Chem. Phys.* 71:3221–28
88. Snyder, R. G., Scherer, J. R., Gaber, B. P. 1980. *Biochim. Biophys. Acta* 601:47–53
89. Snyder, R. G., Strauss, H. L., Elliger, C. A. 1982. *J. Phys. Chem.* 86:5145–50
90. Spiker, R. C. Jr., Levin, I. W. 1975. *Biochim. Biophys. Acta* 388:361–73
91. Spiker, R. C. Jr., Levin, I. W. 1976. *Biochim. Biophys. Acta* 433:475–68
92. Spiker, R. C. Jr., Levin, I. W. 1976. *Biochim. Biophys. Acta* 455:560–75
93. Sunder, S., Bernstein, H. J. 1978. *Chem. Phys. Lipids* 22:279–83
94. Susi, H., Byler, D. M., Damert, W. C. 1980. *Chem. Phys. Lipids* 27:337–44
95. Tardieu, A., Luzzati, V., Reman, F. C. 1973. *J. Mol. Biol.* 75:711–33
96. Tasumi, M., Krimm, S. 1967. *J. Chem. Phys.* 46:755–66
97. Tasumi, M., Shimanouchi, T., Miyazawa, T. 1962. *J. Mol. Spectrosc.* 9:261–87
98. Tilcock, C. P. S., Cullis, P. R. 1982. *Biochim. Biophys. Acta* 684:212–18
99. Trudell, J. R., Rayan, D. G., Chin, J. H., Cohen, E. N. 1974. *Biochim. Biophys. Acta* 373:436–43
100. Van Dael, H., Ceuterickx, P., Lafaut, J. P., Van Cauwelaert, F. H. 1982. *Biochem. Biophys. Res. Commun.* 104:173–80
101. Verma, S. P., Wallach, D. F. H. 1976. *Biochim. Biophys. Acta* 436:307–18
102. Verma, S. P., Wallach, D. F. H. 1977. *Biochim. Biophys. Acta* 486:217–27
103. Vogel, H., Jähnig, F. 1981. *Chem. Phys. Lipids* 29:83–101
104. Wallach, D. F. H., Verma, S. P., Fookson, J. 1979. *Biochim. Biophys. Acta* 559:153–208
105. Wittebort, R. J., Blume, A., Huang, T. H., DasGupta, S. K., Griffin, R. G. 1982. *Biochemistry* 21:3487–502
106. Wong, P. T. T. 1983. In preparation.
107. Wong, P. T. T., Mantsch, H. H. 1982. *Can. J. Chem.* 60:2137–40
108. Wong, P. T. T., Mantsch, H. H. 1982.

Proc. 8th Int. Conf. Raman Spectrosc., Bordeaux, ed. J. Lascombe, P. V. Huong, pp. 757–58. Chichester: Wiley
109. Wong, P. T. T., Mantsch, H. H. 1983. *Biochim. Biophys. Acta* 732:92–98
110. Wong, P. T. T., Mantsch, H. H. 1983. *J. Chem. Phys.* 78:7362–67
111. Wong, P. T. T., Mantsch, H. H. 1983. *J. Phys. Chem.* 87:2436–43
112. Wong, P. T. T., Moffatt, D. J. 1983. *Appl. Spectrosc.* 37:85–87
113. Wong, P. T. T., Murphy, W. F., Mantsch, H. H. 1982. *J. Chem. Phys.* 76:5230–37
114. Yager, P., Peticolas, W. L. 1980. *Biophys. J.* 31:359–70
115. Yellin, N., Levin, I. W. 1977. *Biochemistry* 16:642–47
116. Yellin, N., Levin, I. W. 1977. *Biochim. Biophys. Acta* 468:490–94
117. Yellin, N., Levin, I. W. 1977. *Biochim. Biophys. Acta* 489:117–90

Ann. Rev. Biophys. Bioeng. 1984. 13:25–49

CHARACTERIZATION OF TRANSIENT ENZYME-SUBSTRATE BONDS BY RESONANCE RAMAN SPECTROSCOPY[1]

P. R. Carey and A. C. Storer

Division of Biological Sciences, National Research Council, Ottawa, Canada K1A 0R6

INTRODUCTION AND PERSPECTIVE

A full description of an enzyme catalyzed reaction requires a knowledge of the positions of the atoms and electrons of the substrate and enzyme (including the encompassing solvent) for each species along the reaction pathway. Such information, together with a knowledge of the individual rate constants associated with the enzyme-substrate intermediates, provides a basis for understanding enzyme reactivity. However, such a definition is not complete, since it is essentially a set of static pictures and it ignores the dynamical properties of proteins (66). Even for a chemically well-defined enzyme-substrate intermediate, there are structural fluctuations or rapid interchanges between energetically close conformational states that play a vital role in enzyme action (4, 32, 67). Moreover, the translational energy of the solvent molecules "bombarding" the enzyme-substrate complex and the vibrational motions of the enzyme and substrate atoms have to be taken into account in the total description of enzymolysis (4, 48). The importance of the latter motions is recognized by their use to explain equilibrium and kinetic isotope effects (22), since the differences in vibrational properties of initial and transition states affect free energies of activation and catalytic rates.

[1] NRCC No. 22824.

0084–6589/84/0615–0025$02.00

The interplay between "static" structure and dynamic events in enzyme mediated catalysis highlights several advantages of using vibrational spectroscopy to study enzymolysis. Although, by its very nature, vibrational data pertains to atomic motions, the data can be related to the static picture of structure via theories such as normal coordinate analysis. Also, the physics of vibrational spectroscopy works to our advantage, since the intrinsic time scale is in the picosecond range, allowing one, in principle, to detect the separate vibrational signatures of rapidly interconverting species. Although there has been some success in using FTIR to observe enzyme complexes (1), infrared spectroscopy generally lacks the sensitivity, selectivity, and time resolution (in terms of being able to gather the data quickly) to make it of great value in the study of enzyme intermediates. The vibrational spectroscopic technique of choice is resonance Raman (RR) spectroscopy. In essence, the RR effect allows us to observe selectively the vibrational spectrum of a chromophore in a complex biological milieu at concentrations of 10^{-4} to 10^{-5} M in water (7). In order to monitor the bonds undergoing transformation in the active site, we ensure that the substrate is itself chromophoric or that a chromophore is created at the point of covalent linkage between enzyme and substrate. In physical terms the time scale of the RR effect is $<10^{-11}$ sec, and thus the spectrum effectively provides an instantaneous "snapshot" of the molecular population. In addition, recent technological advances allow RR data to be obtained rapidly (from a single pulse of nanoseconds duration in favorable cases), and these considerations of time scale render the technique ideally suited to the characterization of reaction transients.

For understanding catalysis, there are several attractive aspects about the information content of a RR spectrum. The spectral peak positions are a property of the vibrational motions of the atoms solely in the ground electronic state. The RR spectrum can provide definitive evidence for events of catalytic importance in the active site such as H-bonding, charge and dipole interactions, pK changes, the nature and number of substrate conformations, substrate distortion, and nonbonding interactions. Whereas primary evidence relates to the substrate itself, since this is the source of the chromophore, the properties of the protein making up the active site can be probed in as far as they affect the substrate's RR spectrum. RR peak intensities provide a wealth of information on excited electronic states (21, 55, 63). This can be of great value when the chemical transformations under consideration proceed through the excited state, for example for retinal in the visual process (3, 25, 63). However, even for catalysis occurring in the electronic ground state, a knowledge of low-lying excited states may give clues as to how various quantum states mix to effect molecular distortion and eventual chemical transformation. Much of the RR work so far has involved studying acyl enzyme intermediates found during the hydrolysis of esters by enzymes such as chymotrypsin and papain. Earlier reviews detail how some of the effects outlined above have been observed for these systems (5, 6, 14, 18); recent introductions to the RR effect itself are contained in references (3, 5, 7, 18, 63).

While RR spectroscopy is a powerful technique in its own right, its successful application to enzymology often requires the support of other methods. For example, kinetic studies are sometimes needed to set the ground rules for the reaction under consideration and are essential for pursuing the Holy Grail of structure-rate correlations. Also, spectral interpretation can be set on a solid foundation by joint X-ray crystallographic-Raman investigations. None of these "other techniques" should be seen as a distraction; with the RR approach they are all parts of the whole and they interact synergistically to provide a more profound insight into the workings of the enzyme.

EXPERIMENTATION PROMOTED BY TECHNICAL ADVANCES

A RR experiment is conceptually simple. A laser beam is focused into a solution containing the enzyme-substrate intermediate. The scattered light, one constituent of which is the RR spectrum, is usually detected at right angles to the incident beam and is analyzed for frequency and intensity. The solution is moved through the laser beam by some mechanical means to obviate the problems associated with photolability or photodecomposition [Reference (7), Chapter 3]. These simple physical constraints make it easy to adapt methods employed in other biophysical studies of enzymes, e.g. rapid-flow and cryoenzymological technology, to RR experimentation.

Continuing progress in the technology of lasers and photon detectors is having a major impact on the field. The first study of an enzyme-substrate intermediate (16) used a laser operating in the visible region and recorded the RR spectrum by means of a single photomultiplier coupled to a scanning double monochromator. The availability of only visible laser wavelengths meant that the substrate, a substituted cinnamic acid derivative, had to have an intense electronic transition in the visible spectral range. With the advent of reliable laser wavelengths in the near-UV region, the technique could be extended to a much wider range of substrates. However, a major drawback remained in that it took tens of minutes to record complete RR spectra. This problem has now been largely removed by the use of multichannel technology.

The basic facets of single and multichannel detection are shown schematically in Figure 1. The most common method is to use a scanning spectrometer with a narrow exit slit over the exit port, followed by a photomultiplier tube. By slowly turning the grating, using the accurate

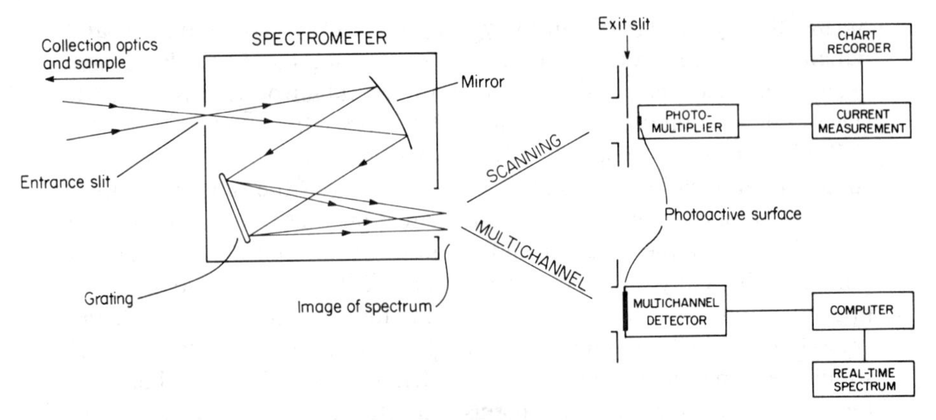

Figure 1 Schematic of a Raman spectrometer showing the options of single channel (scanning) and multichannel detection. Reproduced from (7) with permission.

drive of the spectrometer, the lines of the RR spectrum move in succession across the slit and are detected and recorded as shown in Figure 1. In the second method a multichannel detector is placed at the exit port. This detector is similar to having several hundred minute photomultiplier tubes across the port. All the RR lines are then registered on different elements of the detector all the time. Thus it is possible to observe the entire RR spectrum on a television screen or an oscilloscope in real time. Since each element of the multichannel detector is of comparable sensitivity to a single photomultiplier, the time needed to acquire a complete RR spectrum is reduced markedly. For the systems under consideration here, and using CW lasers, the time is reduced from tens of minutes to a few seconds. Moreover, the multichannel method has considerable advantages in terms of its ability to produce reliable, reproducible results from enzyme reaction mixtures of poor optical quality (15).

EARLY RR STUDIES ON ENZYME SUBSTRATES

Historical Perspective

RR studies in biochemistry have developed in two ways. The first approach uses the chromophores provided by nature, and in the past 15 years there has been a rapid growth in areas such as heme, visual pigment, and metalloprotein RR studies (7). However, this approach cannot be used to probe biological sites that, although they may be of great importance, do not carry a suitable chromophore. Thus, the second approach, the RR labelling technique, was introduced to overcome this limitation (18, 19). A RR label, which usually mimics a natural component, is introduced as a reporter or probe group. Ingenuity is required to ensure that the label causes minimal perturbation to the system being monitored, but with the labelling method any site, in principle, can be probed.

The first labelling study, used methyl orange bound to bovine serum albumin (19) and showed that it was indeed possible to observe selectively and specifically the vibrational spectrum of a chromophore bound to a protein. Since that time, detailed investigations of drug-protein (11, 34, 44, 47, 51), drug-DNA (20), hapten-antibody (10, 28, 35), enzyme-inhibitor (24, 42), and irreversibly labeled enzyme systems (53) have appeared. Each of these systems involves a stable 1 : 1 complex, but it was apparent that by using chromophoric substrates it might be possible to monitor the RR spectrum of a substrate in an unstable enzyme-substrate complex. This expectation was realized in 1974 when the RR spectrum of 4-amino,3-nitro-cinnamoyl-chymotrypsin was reported (16). In addition to showing that RR studies on enzyme substrates were feasible, this study and its extensions (17, 49) provided insight into an old problem—the origin of shifts in λ_{max}, the electronic absorption maximum of the chromophore, upon binding to an enzyme.

What Do Shifts in Substrate λ_{max} Tell Us?

The origin of the shifts in a chromophore's absorption maximum upon binding to protein has been controversial since it was realized many years ago that this effect was responsible for anomalies when pH indicator dyes were used in protein solutions. The power of the RR approach to this problem is that, since RR peak positions are a property solely of the ground electronic state, it is possible to distentangle ground and excited state effects. Thus, Kim et al (33) could show that the red shifts in the λ_{max} of methyl orange upon binding to human serum albumin is not associated with protonation of the dye (as it is for the free dye in solution), since there are no shifts in RR peak positions between the unbound dye at basic pH and the protein-bound dye. Therefore, the red shift is mainly caused by excited state effects.

For enzyme-substrate complexes, changes in the absorption maximum of the chromophoric substrate have been widely used as evidence for changes in the electronic ground state properties of the substrate (1a), e.g. shifts in λ_{max} have been taken as evidence that a conjugated amide carbonyl goes from a sp^2 to a sp^3 state by forming a tetrahedral intermediate in the active site (27). From the above, it is apparent that such conclusions, based on λ_{max} alone, are fraught with danger. Indeed, the first RR study on an enzyme-substrate intermediate showed that, although the absorption bands of the substrate, 4-amino-3-nitrocinnamic acid methyl ester, red shift upon forming the acyl chymotrypsin (e.g. the main $\pi \rightarrow \pi^*$ transition shifts from

323 to 340 nm), the RR peak positions remained the same within experimental error. Thus there was no evidence that the ester function in the active site was perturbed, and any inference, based on absorption data, that a substantial change in chemistry had occurred would be erroneous. Subsequent studies on other acyl enzymes (13) support the conclusion that the change in λ_{max} is a poor indicator of chemical change in the active site.

It will be seen in the next sections that π electron polarization is the main source of shifts in λ_{max}. However, we can already answer the question posed in the heading to the present section: changes in λ_{max} are a good indication that a change in environment or a change in ground electronic state chemistry (or both) has occurred for a chromophoric substrate. Usually, it is difficult or impossible to decide between these possibilities or to define them more accurately on the basis of λ_{max} alone. Shifts in λ_{max} may reflect mainly excited state properties and thus provide no evidence on ground state chemistry: for the latter we need to turn to the RR spectrum.

Acyl Papains: π-Electron Polarization Explains Absorption and RR Shifts

A series of reactions involving the enzyme papain yielded results which initially were quite unexpected (8, 9). Papain is an enzyme of ~23,000 mol wt which hydrolyzes amide or ester substrates via a transient acyl enzyme. Thus, in the hydrolysis of 4-dimethylamino,3-nitrocinnamoyl imidazole catalyzed by papain, the cinnamoyl is transiently linked to enzyme by a cysteine-SH side chain in the active site (Figure 2). The acyl enzyme has a λ_{max} of 411 nm and can be purified from the reaction mixture by adjusting the pH and subsequent column chromatography. The unexpected nature of the RR results is clear from Figure 2. The RR spectrum of the enzyme-substrate intermediate is distinct from that of the substrate or the product. The active site obviously produces drastic changes in the properties of the acyl residue. As can be seen in Figure 2, the spectrum of the product is dominated by bands attributable to ethylenic and ring modes in the 1600 cm^{-1} region and a nitro feature near 1350 cm^{-1}. In contrast, the spectrum of the acyl enzyme shows a very intense peak at 1570 cm^{-1}. There is little evidence for peaks from the product in the spectrum of the intermediate or vice versa.

The dramatic change in the RR spectra of the cinnamoyl chromophore shown in Figure 2 is thought to be attributable to the polarization of π-electrons in the bound cinnamoyl group. A series of model compounds, based on the imidazole esters of cinnamic acid, mimics the absorption and RR properties of the acyl papain studied. The imidazole ester of *p*-dimethylamino cinnamic acid is the principal model. It has a strong electron-attracting group (imidazole) attached to the carbonyl and a very strong electron-donating group (p-dimethylamino) at the other extremity of the cinnamoyl skeleton. Acting in concert through the chemical bonds, these groups set up a highly polarized π-electron system. It has been proposed (9) that essentially the same sort of electron polarization occurs in the acyl group bound at the active site. However, in the active site the polarization probably occurs indirectly by interaction of the acyl residue with protein groups.

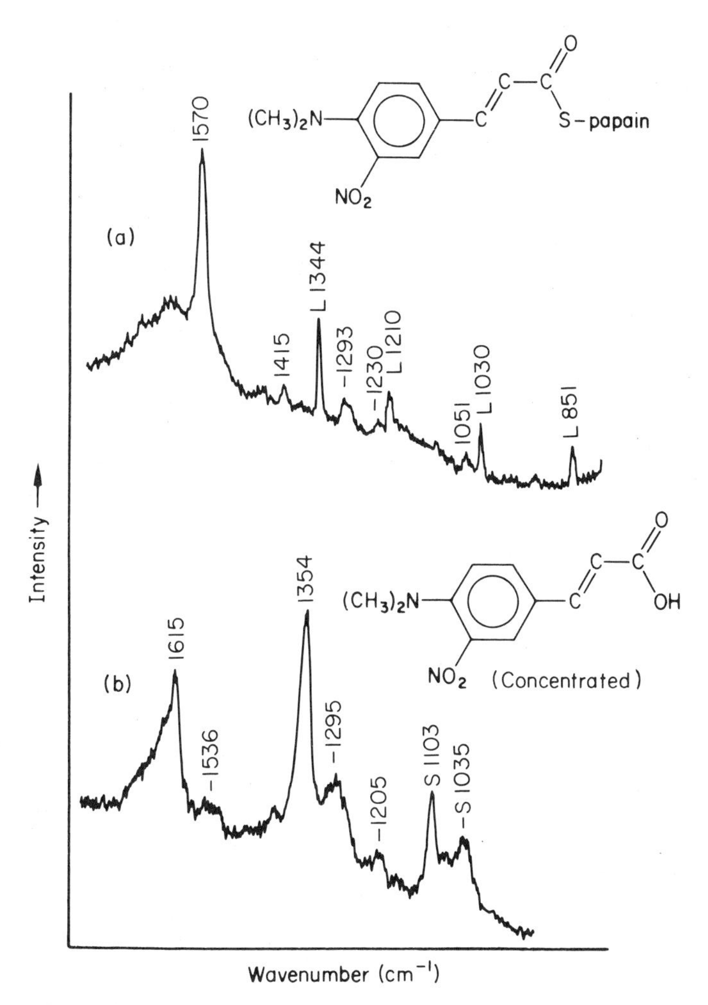

Figure 2 Comparison of the RR spectra of an acyl papain intermediate and the product. (*a*) RR spectrum of 4-dimethylamino-3-nitrocinnamoyl papain. (*b*) Pre-RR spectrum of 4-dimethylamino-3-nitrocinnamic acid. L and S denote laser plasma and solvent peaks, respectively. Adapted with permission from (9).

The development of a series of model compounds presents a good opportunity to test our ideas on electron polarization and, at the same time, to arrive at a semiquantitative estimate of the change in bond lengths occurring in the bound acyl group of the substrate. The approach is to carry out precise structural determinations of the key model compounds by X-ray crystallography. In simple valence-bond terms, the prediction that 4-dimethylamino-cinnamoyl imidazole, for example, has a polarized π-electron system is illustrated by saying that structures of the type *1* make

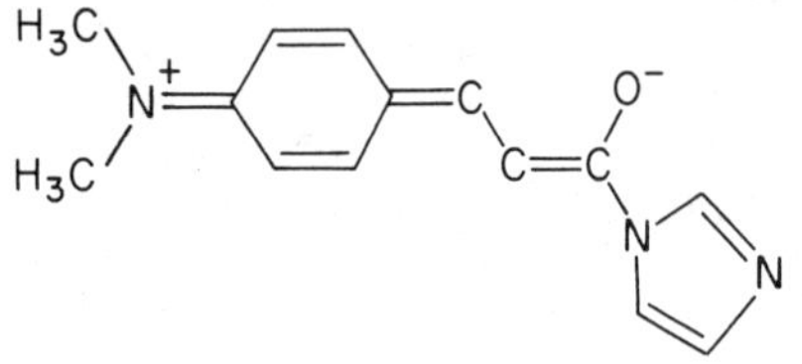

Structure 1

an important contribution to the true structure. This prediction is borne out by X-ray crystallographic studies. The structure of 4-dimethylamino-cinnamoyl imidazole has been compared to that of the product 4-dimethylamino-3-nitrocinnamic acid (C. P. Huber, D. J. Phelps, P. R. Carey, unpublished) and to those of other cinnamic acid derivatives. The comparison shows that the 4-dimethylamino derivative does indeed have a tendency towards quinoid character and that the ethylenic single and double bonds are shortened and lengthened in accord with the simple valence bond picture. By obtaining the structure of the free product and the structure for a model compound that mimics the spectral properties of the chromophore in papain's active site, we can, therefore, arrive at a semiquantitative estimate of the structural changes occurring in the substrate upon binding to the enzyme.

In general, the degree of polarization appears to be much larger for papain bound substrates than for the chymotrypsin bound substrates discussed in the next section. There are two possible causes for this disparity. One is the fact that papain intermediates are thiolesters, whereas acyl chymotrypsins are dioxygen esters. Sulfur attached to an extended π-electron system does bring about a modest red shift, the degree of which may depend on the stereochemistry (60). However, it is unlikely that this is the sole cause of the large shifts seen for acyl-papains. The second cause may be connected with the α-helix dipole present in papain's active site. Cysteine 25, whose –SH side chain binds the acyl group during papain catalysis, is at the end of a portion of α-helical polypeptide chain. α-Helices possess large dipole moments (65), and this could account for the specially high degree of electron polarization (with the resultant spectral changes) observed for papain-bound acyl groups. In contrast, there are no α-helical fragments in the chymotrypsin active site, and the lack of major spectral changes for acyl chymotrypsins may be attributable to this fact.

The origin of the red shifts seen upon binding a substrate to the active site has, in every case examined by RR spectroscopy, been associated with π-electron polarization. Large red shifts, of the order of 30 to 60 nm, are usually accompanied by major π-electron reorganization in the ground state. Such major ground state effects bring about gross changes in the appearance of the RR spectrum as evidenced above and in the binding of *p*-dimethylaminobenzaldehyde to liver alcohol dehydrogenase (31). Small red shifts, in the 5–20 nm range, are usually accompanied by small or negligible shifts in RR peak positions. In the latter case the forces which bring about the shift in λ_{max} operate mainly on the excited electronic state.

Acyl Chymotrypsins

Compared to the acyl papains, the absorption and RR spectral changes observed for acyl chymotrypsins are small. Nevertheless, the changes can still be interpreted in terms of the charge polarization model. RR data have been obtained for about ten acyl chymotrypsins, based on furylacryloyl, thienylacryloyl, or cinnamoyl acyl groups, but many of the generalizations can be illustrated by reference to 5-methylthienylacryloyl chymotrypsin shown in Figure 3. The absorption spectrum of this species at pH 3.0 has a λ_{max} at 339 nm, which is attributable to the acyl group; by excitation with 350-nm radiation the RR spectrum shown in Figure 3 was obtained. The features in the RR spectrum are all due to acryloyl or thienyl modes. Using spectra of this kind, the following generalizations emerged for the thienyl- and furyl-acryloyl based substrates.

MULTIPLE CONFORMATIONS IN THE BOUND AND UNBOUND SUBSTRATES For the native acyl enzymes, the RR spectral profiles in the carbonyl region suggest that the acyl groups bound to Serine 195 in the active site adopt two conformations. These are characterized by having either strong hydrogen bonds to the C=O (giving rise to the broad shoulder near 1700 cm^{-1} in Figure 3) or a nonbonding hydrophobic environment about the C=O group (giving rise to the sharper band near 1725 cm^{-1} in Figure 3) (39).

Additionally, in solution, the ester and acid analogs of the acyl group probably adopt more than one conformation about the acryloyl linkages. Therefore, the measured spectral parameters, such as the ethylenic double-bond stretching frequency $\nu_{C=C}$ in the RR spectra (corresponding to the 1612.5-cm^{-1} feature in Figure 3), should be considered as a weighted mean, $\langle \nu_{C=C} \rangle$, (39).

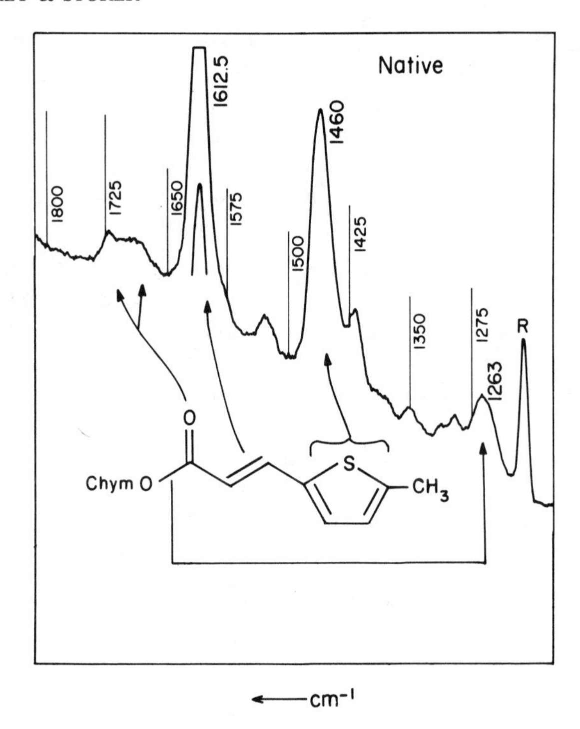

Figure 3 The 350.6-nm excited RR spectrum of 5-methylthienyl acryloyl chymotrypsin at pH 3.0 showing conformationally informative features. Adapted with permission from (39).

π-ELECTRON POLARIZATION AND THE $\langle \nu_{C=C} \rangle$ VS $\langle \lambda_{max} \rangle$ CORRELATION For a series of compounds based on a given acyl group, a correlation exists between $\langle \nu_{C=C} \rangle$ and the measured absorption maximum $\langle \lambda_{max} \rangle$. Changes in these spectral parameters are thought to be attributable to changes in electron polarization. One or more factors can change the polarization (e.g. H bonding at the C=O, electrical charges or dipoles about the acyl chromophore, changes in bulk dielectric, and changes in conformation about the acryloyl's C–C bonds) (39).

NOVEL RR FEATURES Some acyl enzymes show a RR band near 1260 cm^{-1} (39). This is seen clearly in the spectrum in Figure 3 but is absent in the spectra of denatured intermediates and in the spectra of any model compounds, e.g. the acid or ester forms of the acyl group. Recent isotope studies (13) using ^{13}C-labelled substrates have shown that the 1260 cm^{-1} feature is probably associated with $\nu_{C\text{–}O}$, i.e. the bond being cleaved in the active site. However, the appearance of the 1260-cm^{-1} band seems to depend on subtle stereochemical factors and to be unrelated to catalytic activity.

ENZYME ACTIVITY By changing pH it is possible to study the RR spectra as a function of enzyme activity. At pH 3.0 the acyl enzymes are perfectly stable because the enzyme groups required for catalytic activity are protonated and consequently ineffectual. However, at pH 7.0–8.0 the enzyme has optimum activity, and the acyl enzymes have half-lives of seconds. These unstable intermediates have been studied in flow systems, and a steady state of unstable complexes has been generated by mixing stable acyl enzyme at pH 3.0 with buffer at a pH near 7.0. On forming an active acyl enzyme, there is a small increase in $\langle \nu_{C=C} \rangle$ of $\sim 3\ cm^{-1}$ and a blue shift in $\langle \lambda_{max} \rangle$ of 5–10 nm. These changes are ascribed to a reduction in π-electron polarization in the chromophore prior to deacylation, which results from placing a negative or partial negative charge near the acryloyl carbonyl at active pH (49).

Acyl Gylceraldehyde-3-phosphate Dehydrogenases

Chymotrypsin and papain are single-subunit enzymes of ~25,000 mol wt with one active site. In contrast, glyceraldehyde-3-phosphate dehydrogenase is a tetramer composed of four equivalent subunits, each of 40,000 mol wt and each with its own active site. Such enzymes have important control functions because substrate binding in one active site brings about a change in the chemical properties of the remaining three sites with a consequent effect on their activity and/or binding properties.

For the rabbit muscle enzyme, it is possible, by using furylacryloyl phosphate as a substrate, to bind nearly two furylacryloyl molecules per tetrameter (41). The RR spectrum of this species appears to have two $\nu_{C=C}$ peaks; this, taken with absorption and kinetic data, led to the conclusion that the acyl enzyme exists as a mixed population of at least two forms (60). However, the acyl enzyme prepared from sturgeon glyceraldehyde-3-phosphate dehydrogenase appears to consist of a single population (54, 60). The RR and absorption spectral properties were explained again in terms of the electron-polarization model. The furylacryloyl binds to a cysteine S in the enzyme active site, and a series of model derivatives of the type seen in *Structure 2* where X is H, N, O, or S were examined to aid spectral

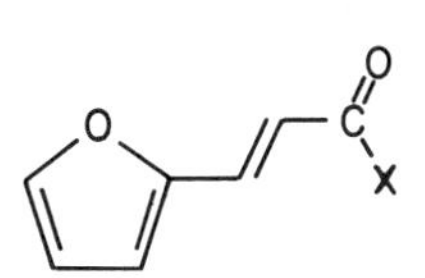

Structure 2

interpretation (60). When X is H, N, or O, $\nu_{C=C}$ shows a clear correlation with λ_{max}. However, when X is S, as in the acyl enzymes, the correlation breaks down. It has been proposed that this anomaly is attributable to empty sulfur $3d\pi$ orbitals, which, in certain conformations, may overlap with, and consequently change the electron distribution within, the ethylenic $p\pi$ orbitals.

CRYOENZYMOLOGY

Considerable progress has been made in recent years in trapping enzyme-substrate intermediates by cryoenzymological means (23, 26). Although there have been several low temperature RR studies involving visual pigments (3) and heme proteins (43), no reports have appeared involving characterization by RR spectroscopy of enzyme-substrate complexes formed at subzero temperatures. Recently, however, this laboratory, in collaboration with Drs. S. Hoffmann and E. T. Kaiser at the University of Chicago, has examined the species trapped in the -25 to $-40°C$ range during the hydrolysis of *O*-(*trans*-*p*-chlorocinnamoyl)- or *O*-(*trans*-*p*-dimethylaminocinnamoyl)-L-β-phenyllactate by carboxypeptidase (28a). The rationale behind these experiments was an attempt to detect a major population of an anhydride acyl enzyme, which has been proposed as a transitory intermediate involving the carbonyl of the substrate and the γ-carboxylate of the enzyme's Glu-270 (36, 40). However, the RR spectra of the enzyme-substrate complexes at subzero temperature were the same in the presence and absence of the inhibitor benzylsuccinate. This fact, together with model studies, showed that no evidence could be found for a buildup of the putative anhydride. The experiments involved the use of a multichannel diode array detector and demonstrated that good quality RR spectra could be obtained under cryoenzymological conditions with a data acquisition time of 20 sec.

USING DITHIOESTERS TO OBSERVE THE POINT OF CATALYTIC ATTACK

Resonance Raman studies of the acyl enzymes discussed above provide new insights into enzyme catalysis and the molecular forces that operate in proteins. However, there are drawbacks associated with these systems since the substrates used are not the natural substrates, which potentially compromises any mechanistic inferences based on the RR data. A further limitation is that intense RR features generally are not observed from the ester bonds composing the acyl enzyme linkage. Because these are the bonds undergoing catalytic transformation, it would be of great interest to monitor their vibrational modes during enzymolysis. These problems have been successfully overcome by the use of the dithioester chromophore, which yields the RR spectrum attributable to those bonds (and their close neighbors) undergoing transformation in a more "natural" enzyme-substrate complex (58).

As mentioned previously, the catalytic hydrolysis of ester substrates by papain proceeds through the formation of an acyl enzyme in which a covalent linkage is formed to an active-site cysteine residue. Absorption spectroscopy enabled Lowe & Williams (38) to monitor this process by using methyl thionohippurate as a substrate. They observed the transient appearance of a chromophoric intermediate with a λ_{max} at 315 nm. Using kinetic evidence and absorption spectral comparisons with model compounds, Lowe & Williams inferred that this intermediate was a dithioacyl enzyme (Figure 4). The intermediate has a λ_{max} at 315 nm, whereas the substrate and product both absorb below 250 nm. The RR data confirm that the intermediate is a dithioacyl enzyme (58).

As can be seen in Figure 5, by excitation with the 324-nm Kr^+ line, the intermediate gives rise to many bands in the 500–1200 cm^{-1} range. The

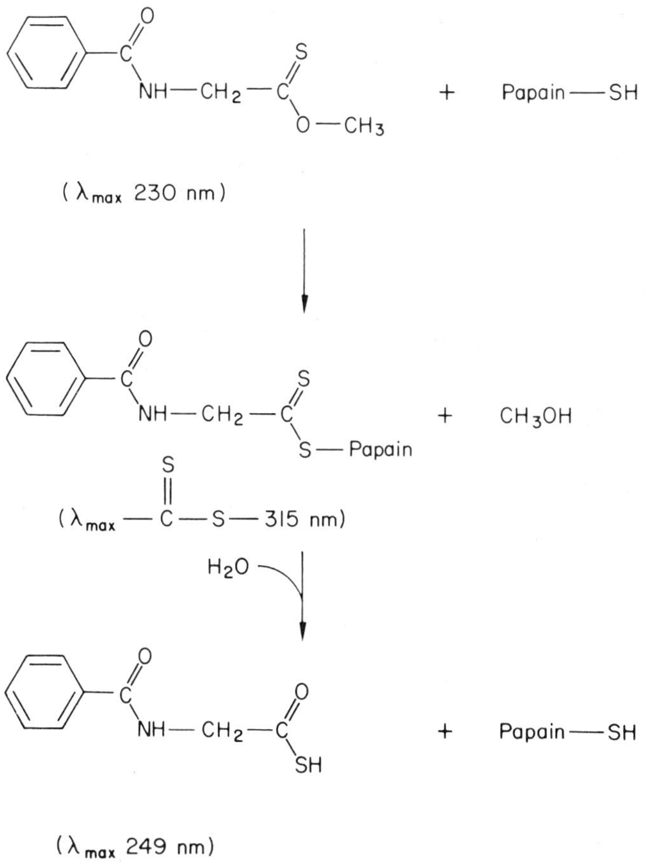

C_6H_5–C(=O)–NH–CH_2–C(=S)–O–CH_3 + Papain—SH

(λ_{max} 230 nm)

↓

C_6H_5–C(=O)–NH–CH_2–C(=S)–S—Papain + CH_3OH

(λ_{max} —C(=S)—S— 315 nm)

↓ H_2O

C_6H_5–C(=O)–NH–CH_2–C(=O)–SH + Papain—SH

(λ_{max} 249 nm)

Figure 4 Reaction involving the formation of a transient dithioacyl papain.

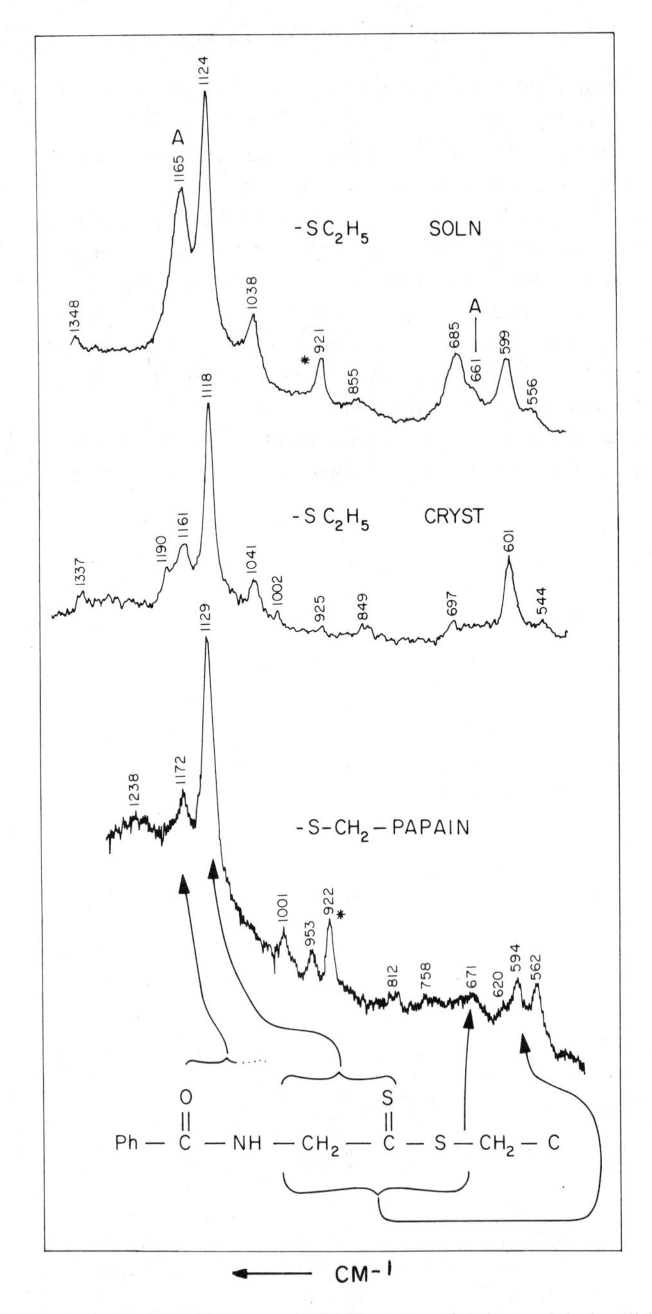

Figure 5 Comparison of the 324-nm excited RR spectra of *N*-benzoylglycine dithioesters: the ethyl ester in $H_2O:CH_3CN$, 98:2 v/v (*top*); the ethyl ester in its polycrystalline form containing only conformer B (*middle*); the dithioacyl papain (*bottom*). Solvent peak is marked with an asterisk. Multichannel detection, typically 20 sec were required for data acquisition. A schematic description of the conformationally sensitive peaks is given.

peaks in Figure 5 contain contributions from the stretching and other motions of the C=S and C–S–C bonds, and thus intense modes from the catalytically crucial bonds are observed in the RR spectrum. Moreover, the reaction scheme can be generalized to use, for example, a polypeptide sequence as substrate and any enzyme forming a transient ester linkage involving an active-site thiol. The "natural" reaction involves the formation of the –C(=O)S–group, so the RR label now involves a single atom replacement-sulfur for oxygen. The spectroscopic data are obtained by examining a reaction mixture (in contrast to the examples discussed above where the acyl enzymes had to be separated from substrate and product) that contains an excess of substrate over enzyme; since for ester substrates acylation is faster than deacylation, there is a buildup of a quasi-steady-state population of intermediate. The RR spectra shows the same time dependence as the absorption band at 315 nm; both disappear as the substrate is used up. At the time of the first RR studies on dithioacyl papains, the RR spectroscopy and much of the chemistry and enzymology of dithioesters was virgin territory. It has since taken several years to build a foundation of knowledge sufficient to interpret the RR spectra of the intermediates with confidence. At the same time, the technical advances outlined above have improved the quality of the RR data and enabled experiments to be undertaken that would have been impossible five years ago.

Interpretation Of Dithioester RR Spectra

Even at the most basic level, vibrational and theoretical investigations had to be undertaken on simple dithioesters, such as $CH_3C(=S)SCH_3$, in order to build a reliable force field for dithioesters, which could then be used as a basis for our understanding of more complex dithioester systems. The work on simple dithioesters included Raman, RR, and infrared analysis coupled with a normal coordinate treatment of methyl and ethyl dithioacetate (61) and several ^{13}C- and D-substituted analogs. The approach was extended to molecules such as $CH_3CH_2 \nmid C(=S)SCH_3$ and $CH_3(C=S)S \nmid CH_2CH_3$, which provided information on the sensitivity of the RR spectrum to rotation about the bonds indicated (46). Ascending the scale of complexity, the spectroscopic and physical chemical properties of dithioesters of the type $RC(=O)NHCH_2C(=S)SC_2H_5$ were studied next (59). These *N*-acylglycine derivatives are important because they form the basis of the series of thionoester substrates and corresponding dithioacyl papains. The series of compounds are related as shown in Figure 6.

The model *N*-acylglycine ethyl dithioesters give rise to intense RR bands in the 500–700 and 1000–1200 cm^{-1} regions of the spectrum. In solution, the relative intensities of the bands were found to be very sensitive to temperature and solvent. These facts, taken with other considerations,

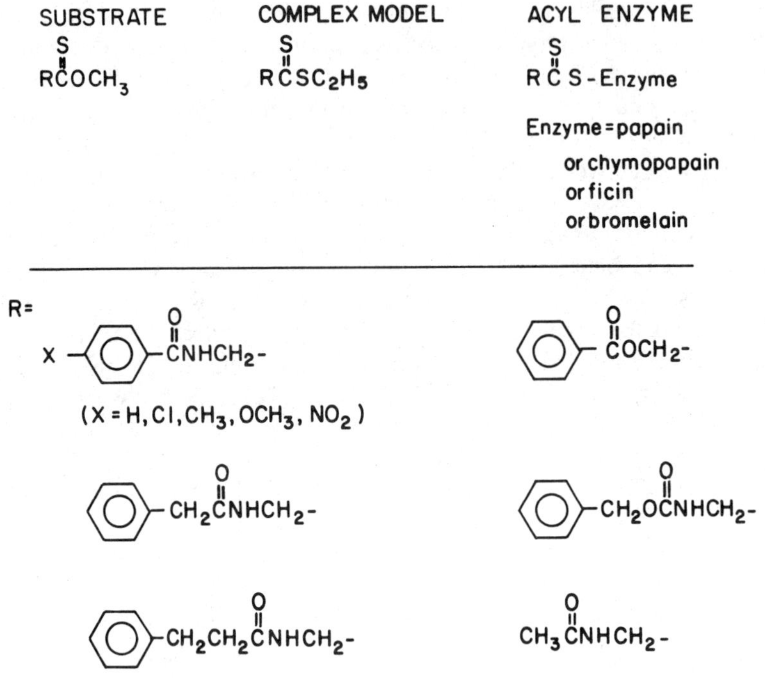

SUBSTRATE	COMPLEX MODEL	ACYL ENZYME
$RC(=S)OCH_3$	$RC(=S)SC_2H_5$	$RC(=S)S$-Enzyme
		Enzyme = papain or chymopapain or ficin or bromelain

R =

$X-C_6H_4-C(=O)NHCH_2-$ ($X = H, Cl, CH_3, OCH_3, NO_2$)

$C_6H_5-C(=O)OCH_2-$

$C_6H_5-CH_2C(=O)NHCH_2-$

$C_6H_5-CH_2OC(=O)NHCH_2-$

$C_6H_5-CH_2CH_2C(=O)NHCH_2-$

$CH_3C(=O)NHCH_2-$

Figure 6 The substrates used to generate dithioacyl enzymes and their corresponding dithioester model compounds.

indicated the presence of more than one conformer. It was found that in aqueous or acetonitrile solutions there are two major conformational states, designated conformers A and B (shown in Figure 7), and that each conformer has a characteristic and separate RR spectrum in both the 600 and 1100 cm^{-1} regions. In order to form an exact description of conformers A and B, combined X-ray crystallographic-Raman analyses (29, 62) were undertaken on single crystals of *N*-acylglycine ethyl dithioesters. The combined X-ray crystallographic-RR approach could, in turn, be used to understand the solution RR spectra of the *N*-acylglycine dithioesters and the RR spectra of the dithioacyl papains. In keeping with the higher thermodynamic stability of conformer B, most of the *N*-acylglycine dithioesters crystallize in this form.

Figure 5 compares the RR spectrum of crystalline

$PhC(=O)NHCH_2C(=S)SC_2H_5$

with the RR spectrum of this molecule in H_2O. In solution (*top* spectrum), the spectral signatures of forms A and B are present, but for the crystals

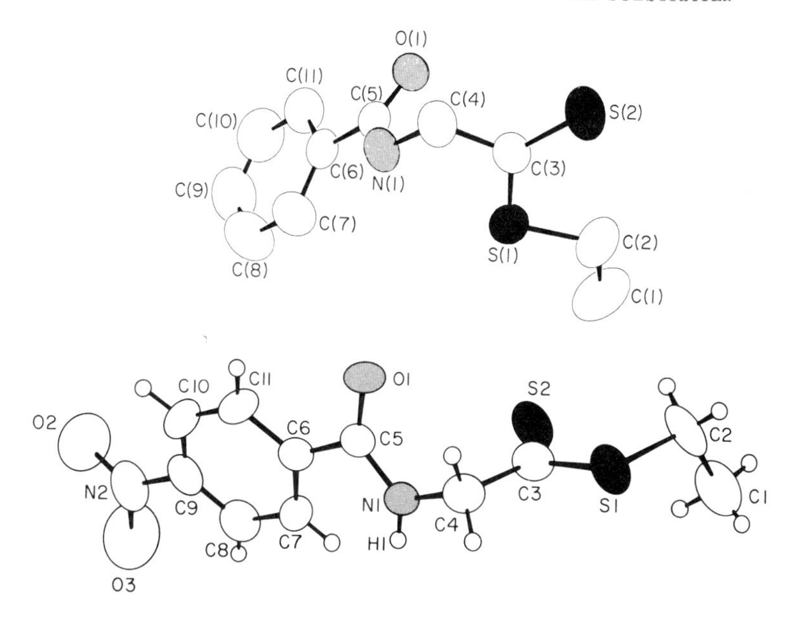

Figure 7 Conformer B (*top*) and conformer A (*bottom*). The structures of crystalline *N*-benzoylglycine ethyl dithioester and *N*-(*p*-nitrobenzoyl)glycine ethyl dithioester, respectively.

only the signature of conformer B is found, e.g. in the upper trace in Figure 5 most of the intensity of the 1165 cm^{-1} peak is due to a conformer A mode that is absent in the spectrum of the crystalline material. An X-ray diffraction analysis on the crystal provides an accurate structure of conformer B, and this is shown in Figure 7. To date, only

$$p\mathrm{NO_2PhC({=}O)NHCH_2C({=}S)SC_2H_5}$$

has crystallized in a form giving rise to a conformer A type RR signature [a fact that is related to the effect the $p\mathrm{NO_2}$ substituent has on the basicity of the NH group (12)]; its structure is compared to conformer B in Figure 7. The major conformational difference between conformers A and B is a rotation of $\sim 150°$ about the C(3)–C(4) bond. The rotation about the C–C linkage changes the vibrational coupling in and about the dithioester moiety and accounts, at least in part, for the different Raman spectral signatures of conformers A and B.

A characteristic of conformer B is that the amide and dithioester groups are in close contact; in fact the N (glycinic nitrogen) $\cdots$ S (thiol) distance is 2.9 Å, somewhat less than the sum of the van der Waals radii, which is 3.35 Å. The amide and ester planes are roughly orthogonal and thus there is no intramolecular H-bonding involving the –NH group. The close contact between the amide nitrogen and the thiol sulfur atom in conformer B can be analyzed in terms of the criteria set out by Rosenfield et al (50) concerning the directional preferences of nonbonded atomic contacts with sulfur. By reexamining crystal data these authors recognized two separate preferred lines of approach, one for nucleophiles and one for electrophiles, to sulfur in a Y–S–X bonding situation. The approach of N to S in conformer B of Figure 7 meets the Rosenfield criterion for the N atom acting as a nucleophile and the S atom acting as an electrophile. In fact, the nitrogen lone pair is pointing towards the sulfur, and the contact can be understood in terms of a HOMO(N)–LUMO(S) attraction (62). From these considerations we might predict that conformer B will have increased contributions from the canonical form seen in *Structure 4*. Support for this comes from a

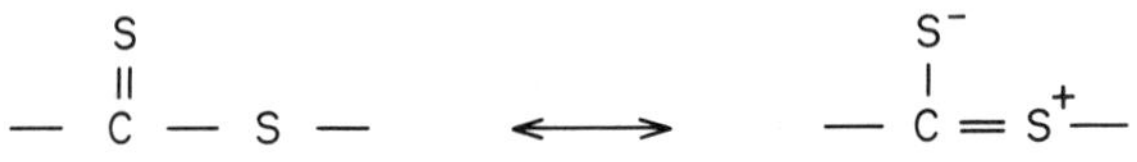

$$—\overset{\overset{\displaystyle S}{\|}}{C}—S— \quad \longleftrightarrow \quad —\overset{\overset{\displaystyle S^-}{|}}{C}=\overset{+}{S}—$$

Structure 3 *Structure 4*

comparison of X-ray crystallographic data for conformers B and A. In conformer B the C=S bond is indeed slightly longer and the C–S bond is shorter compared to conformer A.

A further aid in the analysis of dithioacyl papain RR spectra has been the preparation of isotopically labelled *N*-acylglycine ethyl dithioesters (37). The solution and crystal spectra of *N*-benzoyl and *N*-(β-phenylpropionyl)-glycine ethyl dithioesters substituted with ND, ^{15}N, or ^{13}C=S have provided insight into the normal mode structures within conformer B. The RR spectra of the labelled compounds provide important "benchmarks" with which to compare the RR spectra of the corresponding labelled dithioacyl enzymes; this is discussed in the next section.

Active Site Selects a Relaxed Conformer B

The X-ray analyses provide the exact structures of conformers A and B and, taken with the Raman data, a set of precise structure-spectra correlations. These may now be applied to an analysis of the dithioacyl papain RR spectra. Consideration of these spectra (45, 57), e.g. the RR spectrum of $PhC(=O)NHCH_2C(=S)S$–papain seen in Figure 5, demonstrates unequivocally that between pH 4 and 9 the great majority of acyl groups assume a B-type conformation. In Figure 5 the intense peak at 1129 and the peak at 594 cm^{-1} are modes characteristic of a B conformer—the corresponding peaks are seen in the RR spectrum of the crystalline ethyl dithioester.

The situation is the same for the other acyl papains listed in Figure 6. There is a very close correspondence between the RR spectrum of each dithioacyl papain and that of its corresponding ethyl ester in the B form. All major peaks and most minor peaks in the acyl enzyme spectra are due to conformer B. Although it is possible that some minor peaks are due to a small population of non-B conformers, there is no convincing evidence to support this notion. However, denaturation of the dithioacyl enzymes below pH 3.0 does result in conformer A peaks appearing in the spectrum (45) and, under these conditions, the conformational population of the covalently linked acyl group reverts to that found for the corresponding ethyl ester. Thus, the first conclusion to emerge is that the native active site exerts conformational selection; it binds one of the conformational states available to *N*-acylglycine dithioesters.

The next consideration is how closely the conformation of the bound substrate resembles that of a standard relaxed dithioester. In other words, does the enzyme bring about any conformational distortion? To approach this question the most stable conformation of the *N*-acylglycine ethyl dithioesters in H_2O/CH_3CN solution, conformer B, is taken as a standard state. Comparison with dithioacyl enzyme RR spectra is facilitated by ND, ^{15}N, and $^{13}C{=}S$ substitution in the dithioacyl papains (57) and the corresponding ethyl dithioesters (37) for the acyl groups based on *N*-benzoylglycine and *N*-(β-phenylpropionylglycine). For the latter acyl group in the active site there is no evidence for distortion in the NH–CH_2–C(=S) torsional angles that define conformer B. The same is essentially true for the *N*-benzoyl derivative, although there may be a minor perturbation away from the relaxed standard state (57). Conclusions such as these are formed on the basis of the comparison of RR peak positions, which are known to be conformationally sensitive. In Figure 5, for example, the position of the intense band near 1130 cm^{-1} is very sensitive to torsional angles in the glycinic NH–CH_2–C(=S) linkages: a $\sim 150°$ rotation about CH_2–C(=S) brings about a 40-cm^{-1} shift in this RR peak. Thus, a slight shift from 1124 to 1129 cm^{-1} (Figure 5), going from the standard state to the active site, is taken as evidence for a minor perturbation. Such conclusions are reinforced by the differential sensitivity to $^{13}C{=}S$ substitution in the two states.

DISSECTING THE BOUND SUBSTRATE Figure 5 illustrates (in an approximate manner) how different features in the RR spectra can be used to follow conformational events in different parts of the substrate. We have already discussed the sensitivity of the 1130-cm^{-1} peak to the torsional angles in the NH–CH_2–C(=S) fragment. The less intense peak near 1165 cm^{-1} is sensitive to events in the C(=O)–NH–group, which links the P_1 and P_2 subsites (52), and comparisons show that this amide group is unperturbed for both the *N*-benzoyl- and *N*-(β-phenylpropionyl)-glycine substrates (57). The peak seen in Figure 5 near 670 cm^{-1} is almost a pure group frequency corresponding to the stretching motion of the cysteine S–C bond. Thus, its precise position is sensitive to the chemistry in and about this linkage (37). In contrast, the peak near 600 cm^{-1} is due to a highly coupled mode extending throughout the NH–CH_2–C(=S)–S–C fragment, and its position may therefore be sensitive to events in any part of this fragment.

Detailed comparisons of these features, relying on model spectra and isotopic replacements, have led to the conclusion, already stated, that there is little or no conformational distortion in the C(=O)–NH–CH_2–C(=S) bonds. However, there are minor discrepancies between dithioacyl papain and corresponding ethyl dithioester RR spectra (57). For example, in Figure 5 the shift of the 601-cm^{-1} peak in the crystalline model spectrum to 594 cm^{-1} in the acyl enzyme spectrum is explained in terms of the conformation in cysteine-25's S–C–C bonds and suggests that in the acyl enzyme the C–S–C–C skeleton is significantly nonplanar. Acid denaturation experiments show that most of the spectral discrepancies are a property of the intact active site and lead to the hypothesis that, in the native acyl enzyme, the torsional angles about these cysteine bonds are not those of a relaxed unperturbed state. However, such effects are minor; the enzyme is using the available minimally perturbed conformational space, and the distortions observed are not expected to produce rate acceleration (57).

Some of the disparities between the model and acyl enzyme RR spectra appear to be related to the specificity of the substrate (57). For example, the *N*-(β-phenylpropionyl) glycine dithioacyl group is an analog of phenylalanyl glycine for which papain has a high degree of specificity, and for this acyl group there are few spectral differences detectable. However, for the *N*-benzoyl glycine acyl group, which has a lower degree of complementarity to the active site, there are additional spectral differences—such as the appearance of the 953- and 562-cm^{-1} peaks in the dithioacyl papain spectrum alone (Figure 5). In the case of the *N*-benzoyl glycine the benzene ring is closer to the P_1–P_2 amide group than the enzyme's specificity requires, and the active site may have difficulty in accommodating the ring without some degree of distortion, which may lead to spectral discrepancies not seen for the more specific substrates.

As an interesting contrast to the results for the cinnamoyl papain discussed above, no effects ascribable to charges or dipoles are seen for dithioacyl papains. This may reflect the insensitivity of the dithioester group to charges, compared to the inherently polarizable (and larger) cinnamoyl chromophore. Another possibility is that the charge effects seen for the nonspecific cinnamoyls are silent for the specific substrate in the ground state and are used instead to stabilize the buildup of charge in the transition state (64).

Structure-Rate Correlation

One of the enzymologist's goals is to emulate the success that physical organic chemists have enjoyed, in their studies of relatively small molecules, in correlating kinetic constants with chemical structure. Apart from the inherent massive complexity of enzymes, there is the limitation that usually it has not been possible to disentangle the effect of, say, a subtle change in substrate structure on enzyme-substrate contacts or on a transition state in the reaction pathway. Either of these can bring about a change in rate constant for the reaction step under consideration. However, since the RR spectrum enables us to follow the structure in the active site for a series of related substrates, we should be able to identify or eliminate the effects of differential enzyme-substrate contacts. At the same time, a kinetic analysis has to be undertaken to characterize the rate constants associated with individual steps in the active site. Both pH-stat and stopped-flow kinetic analyses have been carried out on the papain catalyzed hydrolysis of *N*-acylglycine thionoesters (A. C. Storer, P. R. Carey, unpublished). The k_{cat}/K_M values are very similar for oxygen ester substrates and their thionoester counterparts, although the k_{cat}'s for the latter are 20–30 times less than for the oxygen analogs.

The key kinetic parameters for the present discussion involve the rate constants for deacylation, k_3, derived from stopped flow studies on a series of *para*-substituted *N*-benzoylglycine thionoesters. When the *para* substituents are $-OCH_3$, $-CH_3$, $-H$, $-Cl$, $-NO_2$, the k_3's are 0.0297, 0.0612, 0.0818, 0.0928, and 0.1783 sec^{-1} respectively (12). A plot of $\log(k_3)$ vs the Hammett parameter σ is a straight line, indicating a strong correlation between k_3 and the electronic nature of the substituent, with an increase in the electron-donating ability of the substituent leading to a decrease in k_3. The RR data for the *para*-substituted intermediates show that each dithioacyl papain forms a single conformer B population and that the torsional angles about the glycinic $-NH-CH_2-C(=S)$ bonds are invariant (12). This suggests there is no change in enzyme-substrate contacts throughout the series. In addition, the RR spectra of the corresponding ethyl dithioesters in solution provide insight into the nature and strength of the $N \cdots S$ contact throughout the series of the model compounds. Each of the *para*-substituted *N*-benzoylglycine dithioesters in H_2O/CH_3CN solution forms a population of B-conformers having a very similar geometry, and $N \cdots S$ (thiol) interaction, to that selected by the active site. However, the ratio of conformer B to conformer A populations depends on the nature of the *para* substituent. Again, the population ratio correlates with the Hammett constant σ. Increasing the electron-donating ability of the *para* substituent increases the basicity of the glycinic nitrogen, which, in turn, leads to an increase in both the strength of the N··· S (thiol) interaction and the population of conformer B.

Considering the model and dithioacyl enzyme data together, it appears that, in the active site, the *para* substituent modulates the basicity of the glycinic nitrogen and thence the strength of the N··· S (thiol) interaction. The increase in strength is expressed in a reduction in k_3, suggesting that the rate-limiting step in deacylation of the dithioacyl papains involves breaking the N··· S contact. This is consonant with the view that the acyl enzyme is followed in the reaction pathway by a tetrahedral intermediate, since, in the latter, steric considerations show that the N··· S interaction is absent (12).

For natural *thiol*acyl papain intermediates, it has been shown that the rate-limiting step in deacylation does involve the formation of the tetrahedral intermediate. No direct evidence exists to indicate that these acyl enzymes are present in an analogous B conformation. However, X-ray structures of two *N*-acylglycine ethyl thiolesters (30) are both B-conformations with N··· S interactions very similar to those found in dithioesters. Therefore, since the interaction between N and S occurs in both the thiol and dithioesters and since the more specific of the dithioacyl enzymes form the N··· S interaction without any detectable distortion, we suggest that the N··· S interaction is also present in the natural thiolacyl papains. For amide substrates, acylation is the rate-determining step, and it is therefore possible that the mechanistic significance of the N··· S interaction is expressed in this stage of the "natural" reaction. One way in which this could occur would be for the attractive N··· S contact in the acyl enzyme to function as part of a thermodynamic trap, which facilitates the breakdown of the tetrahedral intermediate for acylation to the acyl enzyme rather than allowing the tetrahedral species to revert to the Michaelis complex.

Summary

The present status of our knowledge of dithioacyl papains may be summarized as follows.

1. We can observe the vibrational motions of the chemical group, and its near neighbors, undergoing transformation in the active site.
2. Precise information on the number and nature of the conformers taken up by these groups is obtained. For *N*-acylglycine dithioacyl papains the enzyme selects a single conformer, conformer B, in which the glycinic N is in close contact with the thiol S. The N··· S contact can be understood in terms of a HOMO(N)··· LUMO(S) attraction.
3. For the substrates studied thus far, based on glycine derivatives, there is

no evidence that the acyl enzymes are particularly reactive intermediates. Additionally, torsional angles in the bound substrate's C(=O)–NHCH$_2$–C(=S)–S bonds are those for a relaxed unperturbed conformer B.

4. Changes in an individual rate constant, k_3, can be correlated with changes in strength of a single (N $\cdots$ S) interaction in the active site. From this, it is seen that a conformational change, which breaks the N$\cdots$S contact, is required for deacylation to occur.

PROSPECTS

Although we began this review emphasizing the potential of RR spectroscopy for delineating the structural and dynamical events occurring during enzymolysis, most of the discussion and conclusions have centered on the structural aspects. For the dithioacyl enzymes, in the immediate future, it is likely that the structural bias will remain, e.g. in studies of cysteine proteases other than papain, of substrates which have residues other than glycine in the P_1 site and which fill papain's extended active site, and of temporally homogeneous dithioacyl papains formed in rapid flow systems. However, in the long run, it will be important to see how much dynamical information can be gleaned from the data. For simple molecules, dynamical information on the picosecond time scale can be gained by analyzing Raman band shapes (2), and this may be carried over to the RR band profiles of macromolecular systems—perhaps by investigating the systems down to very low temperatures (43). Other questions in which structure and dynamics are entwined, such as the potential of changes in substrate normal mode structure upon binding to the active site to map distortions along the reaction pathway and the role of entropically important low frequency modes, will have to be addressed as experimental and theoretical capabilities continue to improve.

Literature Cited

1. Belasco, J. G., Knowles, J. R. 1983. *Biochemistry* 22:122–29

1a. Bernhard, S. A., Lau, S.-J. 1972. *Cold Spring Harbor Symp. Quant. Biol.* 36:75–83

2. Brooker, M. H., Papatheodorou, G. N. 1983. In *Advances in Molten Salt Chemistry*, Vol. 5, ed. G. Mamantov, J. Braunstein. Amsterdam: Elsevier, In press.

3. Callender, R., Honig, B. 1977. *Ann. Rev. Biophys. Bioeng.* 6:33–55

4. Careri, G., Fasella, P., Gratton, E. 1979. *Ann. Rev. Biophys. Bioeng.* 8:69–97

5. Carey, P. R. 1978. *Q. Rev. Biophys.* 11:309–70

6. Carey, P. R. 1981. *Can. J. Spectrosc.* 26:134–42

7. Carey, P. R. 1982. *Biochemical Applications of Raman and Resonance Raman Spectroscopies.* New York: Academic

8. Carey, P. R., Carriere, R. G., Lynn, K. R., Schneider, H. 1976. *Biochemistry* 15:2387–93

9. Carey, P. R., Carriere, R. G., Phelps, D. J., Schneider, H. 1978. *Biochemistry* 17: 1081–87
10. Carey, P. R., Froese, A., Schneider, H. 1973. *Biochemistry* 12:2198–2208
11. Carey, P. R., King, R. W. 1979. *Biochemistry* 13:2834–38
12. Carey, P. R., Lee, H., Ozaki, Y., Storer, A. C. 1983. In preparation
13. Carey, P. R., Phelps, D. J. 1983. *Can. J. Chem.* In press
14. Carey, P. R., Salares, V. R. 1980. *Adv. Infrared Raman Spectrosc.* 7:1–58
15. Carey, P. R., Sans Cartier, L. R. 1983. *J. Raman Spectrosc.* 14:271–75
16. Carey, P. R., Schneider, H. 1974. *Biochem. Biophys. Res. Commun.* 57:831–37
17. Carey, P. R., Schneider, H. 1976. *J. Mol. Biol.* 102:679–93
18. Carey, P. R., Schneider, H. 1978. *Acc. Chem. Res.* 11:122–28
19. Carey, P. R., Schneider, H., Bernstein, H. J. 1972. *Biochem. Biophys. Res. Commun.* 47:588–95
20. Chinsky, L., Turpin, P. Y., Duquesne, M., Brahms, J. 1975. *Biochem. Biophys. Res. Commun.* 65:1440–46
21. Clark, R. J. H., Stewart, B. 1979. *Struct. Bonding, Berlin* 36:1–80
22. Cleland, W. W., O'Leary, M. H., Northrop, D. B., eds. 1977. *Isotope Effects on Enzyme Catalysed Reactions.* Baltimore: Univ. Park Press. 303 pp
23. Douzou, P. 1977. *Cryobiochemistry, an Introduction.* London/New York: Academic. 286 pp
24. Dupaix, A., Bechet, J.-J., Yon, J., Merlin, J.-C., Delhaye, M., Hill, M. 1975. *Proc. Natl. Acad. Sci.* USA 72:4223–27
25. Eyring, G., Curry, B., Brock, A., Lugtenburg, J., Mathies, R. 1982. *Biochemistry* 21:384–93
26. Fink, A. L., Cartwright, S. J. 1981. *CRC Crit. Rev. Biochem.* 11:145–207
27. Fink, A. L., Meehan, P. 1979. *Proc. Natl. Acad. Sci. USA* 76:1566–69; Hunkapiller, M. W., Forgac, M. D., Richards, J. H. 1976. *Biochemistry* 15:5581–88; Petkov, D. D. 1978. *Biochim. Biophys. Acta* 523:538–41
28. Gettins, P., Dwek, R. A., Perutz, R. N. 1981. *Biochem. J.* 197:119–25
28a. Hoffman, S. J., Chu, S. S., Lee, H., Kaiser, E. T., Carey, P. R. 1983. *J. Am. Chem. Soc.* In press
29. Huber, C. P., Ozaki, Y., Pliura, D. H., Storer, A. C., Carey, P. R. 1982. *Biochemistry* 21:3109–15
30. Huber, C. P., Storer, A. C., Carey, P. R. 1983. In preparation
31. Jagodzinski, P. W., Funk, G. F., Peticolas, W. L. 1982. *Biochemistry* 21:2193–2202
32. Karplus, M., McCammon, J. A. 1981. *CRC Crit. Rev. Biochem.* 9:293–349
33. Kim, B.-K., Kagayama, A., Saito, Y., Machida, K., Uno, T. 1975. *Bull. Chem. Soc. Jpn.* 48:1394–96
34. Kumar, K., King, R. W., Carey, P. R. 1976. *Biochemistry* 15:2195–2202
35. Kumar, K., Phelps, D. J., Carey, P. R., Young, N. M. 1978. *Biochem. J.* 175: 727–35
36. Kuo, L. C., Fukuyama, J. M., Makinen, M. W. 1983. *J. Mol. Biol.* 163:63–105
37. Lee, H., Storer, A. C., Carey, P. R. 1983. *Biochemistry* 22:4781–89
38. Lowe, G., Williams, A. 1965. *Biochem. J.* 96:189–93
39. MacClement, B. A. E., Carriere, R. G., Phelps, D. J., Carey, P. R. 1981. *Biochemistry* 20:3438–47
40. Makinen, M. W., Yamamura, K., Kaiser, E. T. 1976. *Proc. Natl. Acad. Sci. USA* 73:3882–86
41. Malhotra, O. P., Bernhard, S. A. 1973. *Proc. Natl. Acad. Sci. USA* 70:2077–81
42. McFarland, J. T., Watters, K. L., Peterson, R. L. 1975. *Biochemistry* 14: 624–30
43. Ondrias, M. R., Rousseau, D. L., Simon, S. R. 1983. *J. Biol. Chem.* 258:5638–42
44. Ozaki, Y., King, R. W., Carey, P. R. 1981. *Biochemistry* 20: 3219–25
45. Ozaki, Y., Pliura, D. H., Carey, P. R., Storer, A. C. 1982. *Biochemistry* 21: 3102–8
46. Ozaki, Y., Storer, A. C., Carey, P. R. 1982. *Can. J. Chem.* 60:190–98
47. Petersen, R. L., Li, T.-Y., McFarland, J. T., Watters, K. L. 1977. *Biochemistry* 16:726–31
48. Peticolas, W. L. 1979. *Methods Enzymol.* 61:425–58
49. Phelps, D. J., Schneider, H., Carey, P. R. 1981. *Biochemistry* 20:3447–54
50. Rosenfield, R. E., Parthasarathy, R., Dunitz, J. D. 1977. *J. Am. Chem. Soc.* 99:4860–62
51. Saperstein, D. D., Rein, A. J., Poe, M., Leahy, M. F. 1978. *J. Am. Chem. Soc.* 100:4296–4300
52. Schecter, I., Berger, A. 1967. *Biochem. Biophys. Res. Commun.* 27:157–62
53. Scheule, R. K., Han, S. L., Van Wart, H. E., Vallee, B. L., Scheraga, H. A. 1981. *Biochemistry* 20:1778–84
54. Schmidt, J., Benecky, M., Kafina, M., Watters, K. L., McFarland, J. T. 1978. *FEBS Lett.* 96:263–68
55. Siebrand, W., Zgierski, M. Z. 1979. In *Excited States*, ed. E. C. Lim, Vol. 4, Chap. 1. New York: Academic
56. Deleted in press
57. Storer, A. C., Lee, H., Carey, P. R. 1983. *Biochemistry* 22:4789–96

58. Storer, A. C., Murphy, W. F., Carey, P. R. 1979. *J. Biol. Chem.* 254:3163–65
59. Storer, A. C., Ozaki, Y., Carey, P. R. 1982. *Can. J. Chem.* 60:199–209
60. Storer, A. C., Phelps, D. J., Carey, P. R. 1981. *Biochemistry* 20:3454–61
61. Teixeira-Dias, J. J. C., Jardim-Barreto, V. M., Ozaki, Y., Storer, A. C., Carey, P. R. 1982. *Can. J. Chem.* 60:174–89
62. Varughese, K., Storer, A. C., Carey, P. R. 1983. In preparation
63. Warshel, A. 1977. *Ann. Rev. Biophys. Bioeng.* 6:273–300
64. Warshel, A. 1981. *Acc. Chem. Res.* 14:284–90
65. Warwicker, J., Watson, H. C. 1982. *J. Mol. Biol.* 157:671–79, and references therein
66. Weber, G. 1975. *Adv. Protein Chem.* 29:1
67. Welch, G. R., Somogyi, B., Damjanovich, S. 1982. *Prog. Biophys. Mol. Biol.* 39:109–46

Ann. Rev. Biophys. Bioeng. 1984. 13:51–83

BACTERIAL MOTILITY AND THE BACTERIAL FLAGELLAR MOTOR

Robert M. Macnab and Shin-Ichi Aizawa

Department of Molecular Biophysics and Biochemistry,
Yale University, New Haven, Connecticut 06511

INTRODUCTION

Bacteria are exposed to a wide range of environments, some hostile, some benign. Because of their small size and simple structure, their ability to modify their environment is quite limited. The dominant strategy that has evolved, therefore, is selective motion toward environments that enhance the prospects for survival. This phenomenon, taxis, involves sensory reception, signal transduction, and modulation of the motor apparatus. There are a number of recent reviews of the overall process of taxis, and especially of chemotaxis (e.g. 12, 29, 54, 64, 67, 84, 107). In this review, the emphasis is on the motor apparatus, its mechanism of action, and its modulation.

There are several distinct types of bacterial motility, but by far the most extensively studied is motility generated by external organelles called flagella. We shall concentrate on flagellar motility and discuss the other types briefly. The relationship between bacterial flagella and other mechanoenzymes such as eukaryotic cilia/flagella and muscle is also considered.

Bacterial motility and the flagellar apparatus are examined from a variety of points of view: structure and assembly of the flagella (including the underlying genetics); hydrodynamics and mechanics of the external motion; energy conversion within the motor; and motor switching. Neither the review nor the literature citations are intended to be exhaustive. Frequently, more recent papers are cited, and the interested reader can use these to trace earlier contributions; this is especially true of the section on

0084–6589/84/0615–0051$02.00

flagellar genetics. The reader is also referred to other recent reviews of bacterial flagella and motility (8, 9, 37, 65, 66, 100, 101).

BACTERIAL MOTILITY: MAJOR TYPES AND SALIENT CHARACTERISTICS

Bacterial motility occurs in a variety of contexts, such as translation through three-dimensional aqueous media or along surfaces; it may be individual or social; and it may or may not involve clearly recognizable external organelles (flagella). Before going into molecular aspects, the salient features of the various types of motility are described (see also 65).

Flagellar Motility

Flagellated bacteria are capable of "swimming," i.e. translation through a homogeneous liquid medium; they do not require an interface in order to exert thrust, although some species (e.g. *Proteus mirabilis*) also display communal surface motility ("swarming"; 30) on thin aqueous layers such as a moist agar surface. Speeds of swimming bacteria range from 20 to as much as 60 $\mu m\ s^{-1}$. This can represent speeds of up to 30 body lengths per second—an impressive performance!

The flagella are organized in a variety of ways, depending on species. The simplest arrangement is a single flagellum at one pole of the cell. Multiple flagella may be organized in tightly arrayed polar tufts containing as many as 75 flagella, or they may originate at random points around the cell surface. The latter arrangement (peritrichous flagellation) is characteristic of several commonly studied species such as *Escherichia coli*, *Salmonella*, and *Bacillus subtilis* (Figure 1*a*). In one truly remarkable symbiotic situation, large numbers of bacterial cells—imbedded column upon column in the cell surface of a host protozoan—possess lateral arrays of flagella that are used to propel the protozoan through the medium (106).

The bacterial flagellum differs fundamentally from the eukaryotic organelle of the same name, in that the external filament, which is the only part visible by either light or electron microscopic examination of whole cells, is not itself the mechanoenzyme but merely a passive transmitter of mechanical force generated at the base. The prokaryotic flagellar filament is a simple naked structure of the same level of complexity as a single microtubule. In contrast, the eukaryotic flagellum (91) is bounded by an extension of the cell membrane and has a complex architecture of microtubule pairs; driven by ATP hydrolysis, these pairs slide past each other under the constraints of various links and arms.

The primary motion of the prokaryotic flagellum is rotation, which in order to generate thrust as well as torque, employs helical geometry. It is

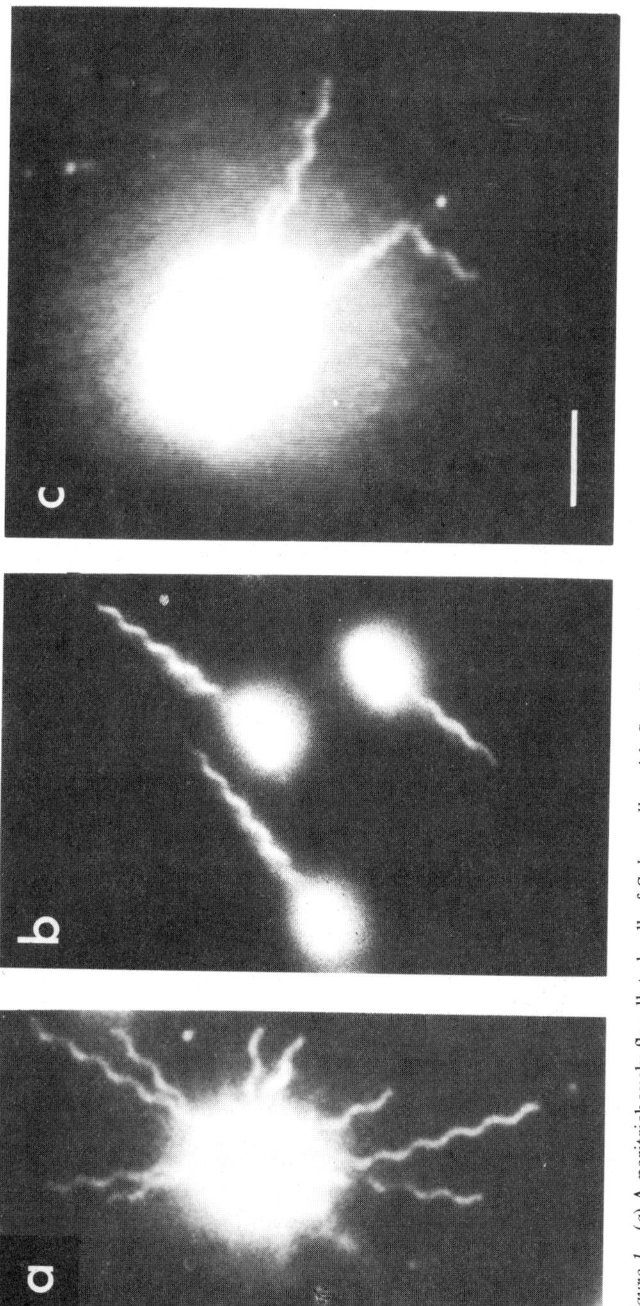

Figure 1 (*a*) A peritrichously flagellated cell of *Salmonella*, with flagella dispersed. (*b*) Swimming *Salmonella* cells with rotating flagella operating cooperatively in a bundle. (*c*) *Salmonella* cell, showing flagellar filament in the heteromorphous state generated dynamically during tumbling. All micrographs were obtained by high-intensity dark-field light microscopy. Bar = 5 μm. [Figures 1*a* and *b* from (72) with permission; Figure 1*c* from (48).]

worth describing briefly the nature of the evidence, since many workers had been anticipating a conformational mechanism of helical wave propagation. It is of two types. (*a*) Advantage is taken of morphological asymmetries in cell shape. Cells can be "tethered" to a glass surface by a single filament (using antiflagellar antibody), and the effect of motor operation on the cell body observed. The appearance of these asymmetries should be quite different under precessional and rotational motion and has been observed to conform to the latter. (*b*) Body motion can be detected under conditions where the filament has no macroscopic helicity ("straight" mutants, see below), provided that the filament has been anchored to an object with appreciable frictional resistance, such as another cell or a glass surface. Observations of both sorts, which have been made in a variety of ways and confirmed in a number of laboratories, provide a convincing case for the rotational mechanism (3, 60, 97).

The frictional resistance of the cell body and that of the filaments are comparable, and so either can be effective in generating thrust. In many species the helicity resides exclusively in the filaments, but in others [notably the spirilla (55); but see (83)] the cell body is helical as well.

The motion of bacteria (or any small organism in liquid media) occurs in a domain where only viscous forces are important, and where inertial forces and turbulence are negligible. This greatly simplifies analysis, and the force balance of an organism with rotating helical filaments or cell bodies has been worked out in detail (e.g. 15, 111). Such analyses have been useful in enabling a lower bound [of a few hundred/rev (3)] to be placed on the stoichiometry of utilization of protons, the energy source for the flagellum.

The translational motion generated by flagellar rotation does not continue indefinitely. Intermittent changes of direction occur, whose details vary with the flagellation pattern of the species but whose net effect in all cases is the random exploration of three-dimensional space. The crucial feature of tactic behavior is the modulation of these directional changes by sensory information from the environment, with the net result that a time bias is injected into the trajectory; a directional change is less likely to occur when the cell is receiving temporal gradient information that is favorable (7, 70). The underlying cause of the motility changes is a reversal of the motor from counterclockwise (CCW) to clockwise (CW) rotation (60). The capacity of the motor to switch rotational sense, and of the sensory system to modulate the switching probabilities, are fundamentally important features of bacterial motility.

The motility pattern of peritrichous organisms (such as *Salmonella* or *E. coli*) is now described in more detail, partly because it has been most studied and partly because it raises some interesting questions that can only be answered by considering the molecular architecture of the flagellar filament. The dominant sense of motor rotation in wild-type cells is CCW (60). CCW rotation of the flagella, which are left-handed helices (71, 92), results in a positive wave velocity (72), i.e. in a proximal-to-distal propagation of the helical waveform, which therefore exerts a pushing force on the cell. Under these conditions, the individual rotating flagella are observed to operate as a well-defined in-phase bundle (Figure 1*b*) oriented more or less parallel to the long axis of the cell. Formation of the bundle can be explained by passive hydrodynamic and mechanical interactions (1, 63); there is no reason to postulate a coordinating signal. Obviously, however, all of the flagella in the bundle must be rotating in the same sense; this synchrony of sense is pertinent to the important question of the mechanism for switching between the two senses of rotation, to be discussed later. The swimming motion is periodically (at random intervals of order 1 s) interrupted by tumbling, a chaotic angular motion (lasting for times of order 0.1 s) that reorients the cell before another period of swimming ensues (7). Although there is definite evidence (60) from the tethering of behavorial mutants that CW rotation causes tumbling in free cells, it is not at all obvious why this should be. In fact it has been shown (63), by theoretical modelling and by building a working model, that the negative wave velocity caused by CW rotation of a left-handed helix would be expected to produce a jamming situation in the bundle and not the observed violent tumbling motion. An explanation of this apparent paradox will emerge when we consider filament structure.

Spirochetal Motility

Spirochetes (such as *Treponema pallidum*, the causative agent of syphilis) resemble spirilla in that they have helical bodies that screw their way through the medium, but in this case there are no externally visible organelles (14)—a spirochete is an apparently "horseless carriage." Electron microscopic examination of the cells, however, reveals (e.g. 33) that spirochetes are indeed flagellated, but the flagellar filaments (also given the neutral terms "axial fibrils" or "periplasmic fibrils," since their mode of action has not been directly demonstrated), which originate close to the two poles of the cell, never emerge through the most external superficial layer, the outer membrane. Instead, they lie longitudinally in the annular cylindrical space between the outer membrane and the "protoplasmic cylinder" (i.e. the thin peptidoglycan layer, the cell membrane, and the cytoplasm within). Morphologically, they closely resemble the organelles of externally flagellated bacteria and have been shown (87) to be essential for motility.

With the reasonable assumption that spirochetal flagella rotate, and with the postulate that the outer membrane is nowhere attached to the protoplasmic cylinder, Berg (4) has described a model for spirochetal motility in which the rotating flagella slip with respect to the protoplasmic cylinder and generate shear forces on the outer membrane that (in the limit) cause a pure rolling motion. Thus the torque generated (via the flagella) between the helically shaped cell and the outer cylinder causes them mutually to rotate. The outer membrane thereby propagates a conformational helical waveform that drives the cell. There is a group of spirochetes (leptospira) for which a somewhat different mechanism may apply (6). Here, prominent shape changes are observed to occur at the poles, and the flagella (one per pole) only extend a relatively short distance toward the middle. The remainder of the cell body is helical in shape. If the flagella are artificially freed from the external sheath, they show a pronounced tendency to coil like a watch spring (80). Berg et al (6) propose that the polar regions of the cell body are forced to distort in a gyrating motion, which generates the torque for rotation of the long, helical mid-stretch of the cell body.

Gliding Motility

Gliding motility, which occurs only in thin aqueous layers, is the third major class of bacterial motility and is still poorly understood (30). No external organelles are evident. Gliding motility can be subdivided into two classes based on speed and possibly on mechanism.

The more rapid form (speeds of order 1 to 10 $\mu m\ s^{-1}$) occurs in many of the cyanobacteria (also called blue-green algae, even though they are prokaryotes) and in soil species such as cytophaga. In spite of reports of a variety of structures such as fibrils (89) and rings (88), there has been no definitive identification of the motor apparatus. In other attempts, no motor structures were detected (58). Pate & Chang (88) and Lapidus & Berg (58) both report that latex beads, artificially attached to cells of cytophaga, move back and forth in a longitudinal direction. Different beads could be seen moving in opposite directions on the same cell. Beads of different sizes and shapes all travelled at about the same rate (about 2 $\mu m\ s^{-1}$), which was also the rate observed for the gliding of the cells themselves. The beads maintained a fixed orientation as they travelled. The constancy of speed and orientation argues strongly for a rigid rather than a viscous coupling. Pate & Chang (88) also observed examples of beads rotating instead of translating; pivoting of cells about one pole was noted in both studies. Rotation of cells and of beads suggests that the underlying motor mechanism may be the same as that of flagellated species. The conversion of rotational to linear force could be accomplished by tilting the motor so that only one edge was in contact with the substrate, much as a tilted rotary mop travels across a floor. If this description is correct, the flagellar motors in these gliding bacteria would be wheels in the full sense that we use the term for wheels on manmade vehicles.

The slower type of gliding (of order 1 $\mu m\ min^{-1}$) is characteristic of the fruiting myxobacteria. No motor structures have been detected, nor has the attachment or movement of latex beads been observed. Evidence has been presented that is consistent with a model for differential rates of excretion of surfactants at the two poles of the cell (25, 44), but this model cannot be regarded as proven at the present time.

STRUCTURE OF BACTERIAL FLAGELLA

The bacterial flagellum may be divided into three major regions: the flagellar filament, the hook, and the motor (Figure 2).

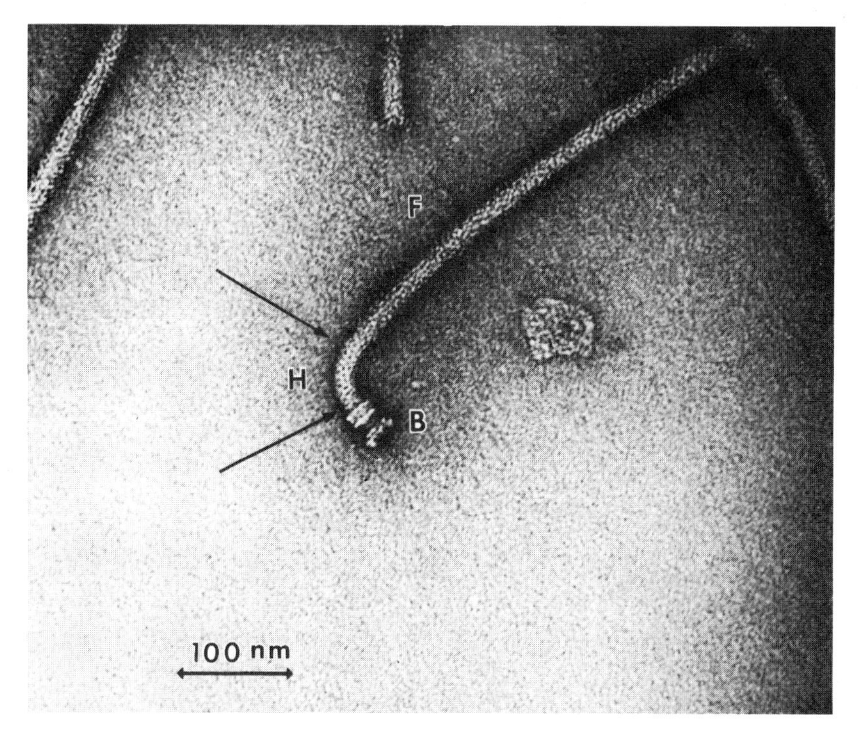

Figure 2 Isolated "intact flagellum" or, more correctly, filament/hook/basal body complex from *Salmonella*. The helical filament (F) extends to the upper right and is connected by the tightly helical hook (H) to the basal body (B), which has four rings and a central rod. (S.-I. Aizawa and H. Kagawa, unpublished micrograph.)

The Flagellar Filament

The flagellar filament is the most prominent part of the structure, only 20 nm in diameter but extending for distances of as much as 10 μm from the cell surface and comprising the vast majority of the protein mass of the total organelle. It is a relatively simple structure, constructed in many species from subunits of a single protein, flagellin; it plays only a passive role in motility, in the sense that it has no capacity to convert metabolic energy into work. Nonetheless, it has some remarkable structural and mechanical features.

The flagellar filament is a totally self-assembling biological structure. There are only quaternary interactions among the subunits, and no enzymatic machinery is needed to ensure that the interactions are properly made. A preparation of filaments may readily be dissociated to monomer and reassociated by appropriate choice of physicochemical conditions (2). It has well-defined repeating, geometrical properties, and so we may properly speak of it as a crystal. However, although the protein subunit (with a molecular weight of 55 K) has dimensions of around 5 nm, the macroscopically evident periodicity—i.e. the helical shape of the filament—is typically around 2 μm. Analysis of electron micrographs (53, 81) reveals that the subunits are arranged around a pseudohexagonal, pseudocylindrical lattice; pseudohexagonal, because (Figure 3) there is both a distortion and tilting of the lattice with respect to the filament axis; pseudocylindrical, because subunits in each of the 11 component longitudinal "fibrils" that comprise the closed lattice are not in truly identical environments. The macroscopic helix is generated because fibrils at inner radii are in a more compressed state than those at outer radii. In the terminology of Klug (49), the subunits, though of identical primary structure, are only quasi-equivalent in their quaternary interactions.

Evidence has accumulated over the years for a variety of distinct crystal forms (polymorphs) that flagellar filaments may adopt, each with its own well-defined geometrical parameters of curvature and torsion (e.g. see 2). There are straight forms (zero curvature), highly coiled forms (low torsion), and forms in between; some are left-handed, some right-handed. Asakura and co-workers have beautifully described "phase diagrams"—as a function of parameters such as pH and ionic strength—for reversible interconversions among these forms (e.g. see 43). Macnab & Ornston (72) and Hotani (35) have demonstrated in vivo and in vitro, respectively, that interconversion can be accomplished by mechanical force (see below). Different polymorphs may also occur as a result of amino acid changes in the primary sequence.

What structural features could be responsible for this structural versatility? Unfortunately, the most interesting feature of flagellin, namely its pronounced tendency to form a crystal that is semi-infinite in one dimension only, has thus far prevented high-resolution structural information from being obtained by X-ray crystallography. Three-dimensional image reconstruction of electron micrographs has provided a structure at ca 1.5-nm resolution (94). The putative flagellin subunit appears to be elongated, to have three main density domains, and to point inwards and at a downward angle to the filament axis (a feature that may confer the ability to withstand lateral stress during motility). One cannot, at the resolution available, make any statement regarding the detailed subunit interactions, but a remarkably convincing set of theoretical arguments has been put forward by Asakura and co-workers and in even more detail by Calladine. There is only space here to mention a few salient points; the reader is referred to two excellent chapters (13, 43) in a symposium volume for further details. It is postulated that, in a highly cooperative fashion in the

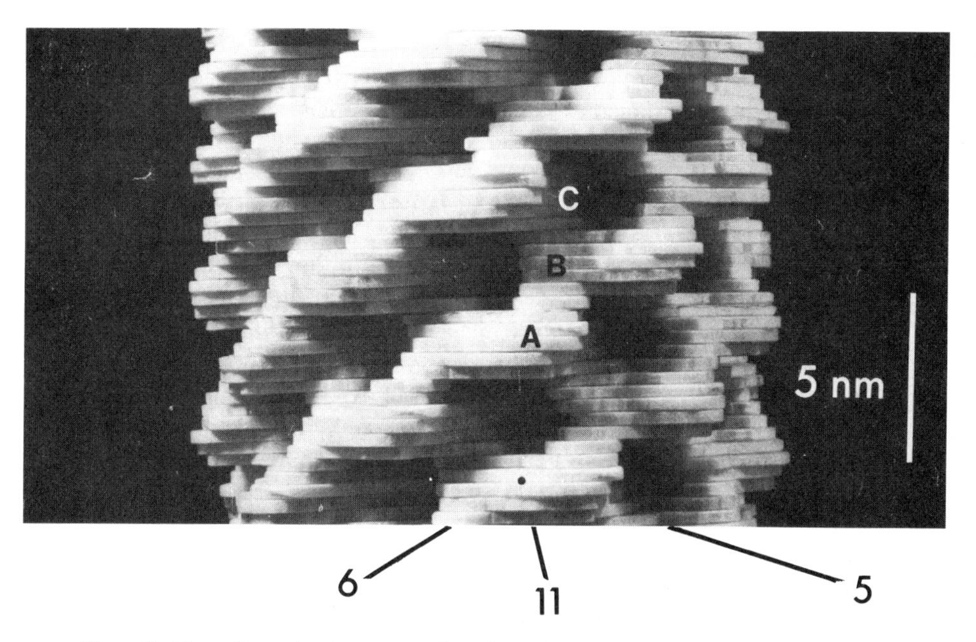

Figure 3 Three-dimensional reconstruction of a portion of the hook structure of *Salmonella*, illustrating the pseudohexagonal lattice that is characteristic of both the hook and the filament. The major helical directions contributing to this lattice are indicated by arrows. Note the deep corrugation in a direction normal to the 6-start direction and thus almost parallel to the longitudinal (11-start) direction; this may be important in permitting the extreme curvature and torsion that is characteristic of the hook. The three major density domains of the putative subunit are labelled A, B, and C. [Based on (109), with permission.]

longitudinal (11-start) direction, subunits are bi-stable in their interactions. Suppose, for example, that in a particular polymorph five of the 11-start fibrils are in one state, which we shall term L (for long), and the remaining six are in the other state S (for short). To minimize mismatch, the L fibrils are all adjacent, as are the S fibrils. Then, with elastic distortion to further minimize mismatch at the interface between L and S, we have the following situation:

$$[S\ S_{+\delta}\ S_{+2\delta}\ L_{-2\delta}\ L_{-\delta}\ L\ L_{-\delta}\ L_{-2\delta}\ S_{+2\delta}\ S_{+\delta}\ S],$$

with the length variation of the fibrils generating the curvature, and their tilt angle generating the torsion. Modelling of this sort predicts 12 polymorphs (11S, 10S/1L, . . . , 11L) with constant increments of torsion and sinusoidally varying increments of curvature. Nine polymorphs have been observed to date, including the two extreme forms (which of course are straight); all conform well to this description. The entire phenomenon bears a strong resemblance to all-or-none conformational changes in oligomeric enzymes (78) except that, in the present case, absolute cooperativity only applies in the 11-start direction and not in the other two directions of the pseudo-hexagonal lattice. The question arises of how a given set of physicochemical conditions can simultaneously give rise to two different structural states (L and S)—surely only one should be preferred? The answer must be that, under the given conditions, neither structure alone is geometrically compatible with the closed cylindrical lattice. Calladine uses the simple analogy of a masonry voussoir (an element of an annulus, used in constructing arches); if it subtends too large or too small an angle it could not be used to construct a closed ring, whereas a mixture of such units might generate angles summing to 360°. Consider, for example, the use of 31° and 34° voussoirs. Eleven 31° units only subtend 341°, eleven 34° units subtend 374°, whereas a combination of five 31° units and six 34° units forms a virtually closed ring (359°). Thus, even though in isolation the 31° conformer (say) might be more stable, the use of some 34° units would permit the final interaction to be made, and the stability achieved thereby could compensate for the use of the 34° units.

It may be noted that the two extreme, straight polymorphs that were mentioned earlier are pure examples of each of the two possible types of tertiary and quaternary interaction. They therefore offer the possibility, should higher resolution structures become available, of clarifying the nature of the bi-stability that is responsible for flagellar helicity in wild-type cells.

The existence of quasi-equivalence and polymorphism in flagellar filaments is all very interesting from a structural point of view, but what does it have to do with bacterial motility? The most important point in this regard is that, without quasi-equivalence, there could be no helicity to the filament and hence no way of converting torque into thrust. Note, incidentally, that the phase of the structure is arbitrary; one could permute the state of the 11 fibrils without altering the overall structure. This was at one time proposed as a model for bacterial motility: phase permutation enforced at the base and driving a helical wave against the resistance of the medium (2, 49, 62). The demonstration of a true rotary mechanism for bacterial motility not only invalidates this model but requires that phase permutation have such a high activation energy barrier that it is essentially prohibited. Otherwise the filaments would "back-permute" when rotated and the net phase velocity would be drastically reduced (13; cf discussion of hook structure and function below).

Dynamic conversion between polymorphs, however, does appear to be important in motility. Recall that tumbling could not be explained in simple terms of CW rotation of the left-handed filaments within a bundle. Extensive study, by high-intensity dark-field light microscopy (71), of the behavior of filaments during tumbling revealed that they were undergoing polymorphic transitions (72). These could be understood very simply in terms of the effect of the viscous load on a rotating filament. The filament is placed under a torsional load (left-handed during CCW rotation, right-handed during CW rotation) that is maximal at the point of attachment at the cell surface. We have already remarked that the polymorphic family includes examples of both right- and left-handed helices; the intrinsically stable polymorph in wild-type strains under physiological conditions is left-handed (71). CW rotation, however, introduces a new factor into the stability considerations, since the torsional load will tend to stabilize right-handed structures. If the effect is large enough, a polymorphic transition will be induced, starting at the base and propagating outward as a dynamic crystal fault. This indeed is what is observed (Figure 1*c*). The phenomenon obviates the jamming problem referred to earlier, because a right-handed helix rotating CW, like a left-handed helix rotating CCW, has a positive wave velocity, the waveform propagating from proximal to distal. The contour shape of a filament in transition is observed to have a "dog-leg" at the region of the crystal fault, which can be explained quantitatively in terms of a phase matching of the fibrils at the junction (34). The geometry of such a heteromorphous filament results in a high resistance to rotation and therefore to an enhanced angular motion of the cell body—the desired end result of a tumble.

A quite remarkable demonstration by Hotani (35) of in vitro polymorphic interconversion fully supports the interpretation of the in vivo situation. Any helical structure, regardless of its handedness, will be destabilized if it is subjected to axial flow. Isolated flagellar filaments, attached to glass by one end, were found to undergo endless cycles of RH → LH → RH transitions under these conditions.

The Flagellar Hook

In all flagellated species that have been examined, the filament is connected to the basal structure via a short, curved structure called the hook (Figure 2). Examination of indefinite structures ("polyhooks") from regulatory mutants has shown that the curve is actually three-dimensional, corresponding to a helix of very high curvature and torsion (about 22 and 36 rad μm^{-1}, respectively; 57). These values are much higher than those for the flagellar filament (1.6 and 2.1 rad μm^{-1}, respectively). This structure has recently been shown (43a) to be a right-handed member of a family of polymorphs that can be interconverted by manipulating conditions (41), as was described above for flagellar filaments. The hook is assembled from a single protein which, though distinct from flagellin, is organized in a very similar lattice (109, 110). Three-dimensional reconstruction indicates that even the overall shape and angle of insertion of the subunits are similar. It also, however, reveals one difference that may be significant in terms of the properties and function of the hook (Figure 3). Whereas the flagellar filament shows only slight modulation of surface density in the 5- and 6-start direction of the lattice, the hook is highly corrugated, with alternating ridges and grooves running parallel to the 6-start direction. Because the lattice is considerably distorted from hexagonal symmetry, the corrugation (perpendicular to the 6-start direction) is almost parallel to the 11-start direction. For technical reasons, the reconstructions were carried out on a straight polymorph; but if we now imagine introducing curvature in the 11-start direction (by invoking the bistable feature discussed earlier for the filament), it can be seen that this curvature could be quite extensive without causing steric hindrance at large radii. This reasoning agrees well with the high curvature that is actually observed.

What is the function of the hook? The explanation that it is a "connector" is not very satisfactory, as it leads to a regressive pattern of thought: what connects the connector? A popular theory (5, 64, 100) for hook function is that it is a flexible coupling that permits torque on the motor axis to be redirected as needed, for example along the long body axis of a peritrichously flagellated cell. This may be true, but the situation is not as simple as it sounds. First, what do we mean by a "flexible" coupling? Recall that, in electron micrographs, the hook displays a pronounced and well-defined curvature—sufficient in fact that its 80-nm length generates an approximately 90° elbow joint that would be very appropriate for a peritrichous cell, since a flagellum originating on the cylindrical surface of the rod-shaped body could be directed back along the body axis. However, after the motor had rotated through 180°, the joint would be pointing in the opposite direction—an undesirable geometry, which could be corrected in either of two ways. The hook could continuously and elastically flex as needed, or it could inelastically permute the phasing of its quasi-equivalent fibrils so that its waveform remained stationary. The latter seems much more likely, but it has further implications. (*a*) The activation energy for permutation, which must be high in the filament to prevent cancellation of wave propagation, must be low in the hook. (*b*) In contrast to the rule (34) obeyed by heteromorphous flagellar filaments—that the fibrils must be in phase at the junction—we are here demanding that the phase of the filament rotates while the phase of the hook stands still, thus introducing a cyclic mismatch at the junction. The hook, like the filament, probably undergoes polymorphic conversion when the motor changes sense of rotation. If, during CW rotation, it went from the highly curved right-handed conformation to the extreme right-handed straight one, this would contribute greatly to bundle dispersion and the vigor of the tumble. Wagenkecht et al (110) have postulated that such events might contribute the asymmetry between forward and reverse swimming that is desirable in a polar monoflagellate bacterium such as *Caulobacter* in order to provide a randomization of orientation, equivalent to tumbling in peritrichous bacteria.

The Basal Body

Dissolution of the peptidoglycan layer and the inner and outer membranes of a variety of bacterial species results in the release of what has been termed an "intact flagellum" (Figure 2), consisting of the filament-hook complex joined to a structure termed the basal body (16, 21, 24). The basal body varies in detail from species to species, but in all cases it has a cylindrically symmetrical stack of rings with a central rod. In the gram-positive bacterium *Bacillus subtilis* there are only two rings (24), in the gram-negative bacteria *E. coli* and *Salmonella* (Figure 4) there are an additional two outer rings for a total of four (22), whereas in the polarly monoflagellated gram-negative species *Caulobacter crescentus* there is a fifth ring at the innermost position (40).

We may assume that the outer two rings of *E. coli* (which are co-planar with the outer membrane and peptidoglycan layer, respectively; 23) serve as bearings but do not participate in chemomechanical energy conversion. This assumption is further supported by their absence in *B. subtilis*, which has a quite different outer wall. The significance of the fifth ring in *Caulobacter* is less clear. *Caulobacter* sheds its flagellum at a particular point in the cell cycle; conceivably the fifth ring, which is very thick, could be an ejection apparatus analogous to the contractile tail sheath of many

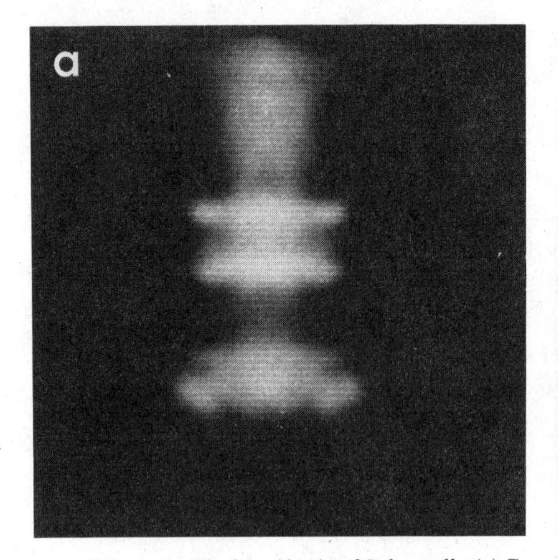

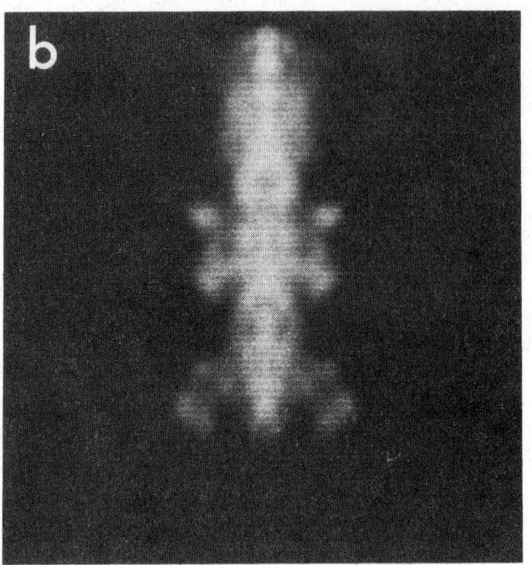

Figure 4 The basal body of *Salmonella*. (*a*) Computer average of 12 separate images, digitized, aligned, averaged, and mirror symmetrized using SPIDER, a program written by J. Frank. (*b*) A central axial section of the computed three-dimensional reconstruction (assuming cylindrical symmetry) from (*a*). (B. Stallmeyer, S.-I. Aizawa, D. DeRosier, and R. M. Macnab, unpublished data.)

bacteriophages (112). It may, on the other hand, be an essential component of the motor that is more tenaciously attached in this species but, in others, is lost during the isolation protocol.

Before attempting to understand the motor in more detail, it will be helpful to review current knowledge of the genetics of flagellar assembly and function.

Flagellar and Motility Genes and Their Regulation

The bacterial flagellum presumably is one of the simplest motor organelles. How many genes are necessary for its assembly and operation? The answer is surprisingly high: in the case of *Salmonella* or *E. coli*, around 35 genes for the organelle itself, another 9 for transduction of signals to it, another 3 or possibly more coding for primary receptors that are not transducers, and of course all of the genes for the system for generation of protonmotive force (PMF). The flagellar genetics of other species (e.g. *Caulobacter*; see ref. 39) promises to be of comparable complexity. The relevant aspects of the *Salmonella* and *E. coli* genomes are given in Table 1; there is almost total homology between the two (20, 56). The genes fall into three classes: (*a*) *fla*, indicating involvement in flagellar assembly; (*b*) *mot*, involvement in flagellar rotation; (*c*) *che*, involvement in switch regulation and hence chemotaxis. Note that although most of the genes carry only a single

Table 1 Flagellar and motility genes and gene products in *Salmonella* and *E. coli*[1]

Gene	Product MW	Product location	Function of gene product	References[2]
flaFI (*flaU*)	?	?	Necessary for addition of P ring. Repressor of late operons?	50, e
flaFII (*flbA*)	?	?	Necessary for any detectable structure. Genetic evidence suggests motor component.	e
flaFIII (*flaW*)	?	?	Necessary for any detectable structure. Genetic evidence suggests motor component.	e
flaFIV (*flaV*)	? (11 K)	HBB[3]	Necessary for addition of L ring and outer cylinder.	52
flaFV (*flaK*)	42 K (42 K)	HBB	Hook protein. Universal joint (?) connecting filament and basal body.	52, 57
flaFVI (*flaX*)	32 K (30 K)	HBB	Necessary for any detectable structure.	a, e
flaFVII (*flaL*)	30 K (27 K)	HBB	Necessary for any detectable structure.	52, a
flaFVIII (*flaY*)	?	?	Necessary for addition of L ring and outer cylinder	
flaFIX (*flaM*)	38 K (38 K)	HBB	Necessary for addition of P ring	52, a
flaFX (*flaZ*)	?	?	Necessary for any detectable structure. Regulates *hag* expression?	50, e, f
flaW (*flaS*)	60 K (60 K)	HBB	Hook/filament junction protein. Mutants have structure up through hook.	50, 52, d
flaU (*flaT*)	? (35 K)	HBB	Hook/filament junction protein? Mutants have structure up through hook.	50, 52, d

[*continued overleaf*]

Table 1 *(continued)*

Gene	Product MW	Product location	Function of gene product	References[2]
flaC (*flaH*)	?	?	Necessary for any detectable structure. Interaction with cell wall?	51
flaM (*flaG*)	?	?	Necessary for any detectable structure.	
[Chemotaxis genes: *cheZ*, *cheY*, *cheB*, *cheR*, *tap*, *tar*, *cheW*, *cheA*]				
motB	? (39 K)	Inner membrane	Necessary for motor rotation.	31, 95
motA	? (31 K)	Inner membrane	Necessary for motor rotation.	31, 95
flaE (*flaI*)	? (22 K)	?	Necessary for any detectable structure. Necessary for expression of all other operons.	50, 98, g
flaK (*flbB*)	? (13 K)	?	Necessary for any detectable structure. Necessary for expression of all other operons. Operon is under cAMP/CAP control.	50, 98, g
*fla**[4]	?	?	Unknown	a
flaL (*flaD*)	?	?	Activator of late operon expression? Mutants have structure up through hook.	e, f
H1[5] (*hag*)	51–57 K (60 K)	?	Flagellin, i.e. filament protein.	Numerous
flaV (*flbC*)	53 K? (?)	HBB?	Hook-filament junction protein? Control of filament assembly. Mutants have structure up through hook.	42, 56, d
flaAI (*flaN*)	?	?	Necessary for any detectable structure.	
flaAII.1 (*fla**)[4]	?	?	Necessary for any detectable structure.	g, h
flaAII.2 (*flaB*)	? (38 K)	Inner membrane?	Necessary for any detectable structure. *motC*, *cheV*. Switch component of motor? Interacts with *cheY*, *cheZ*, *flaQ*.	18, 86, a, c, g
(*fla**)[4]	(28 K)	?	Unknown	g

flaAIII (*flaC*)	? (56 K)	?	Necessary for any detectable structure.	g
flaS (*flaO*)	? (17 K)	?	Necessary for any detectable structure.	g
flaR (*flaE*)	? (54 K)	Cytoplasm	Control of hook length.	90, 103, g
flaQ (*flaA*)	? (38 K)	Inner membrane?	Necessary for any detectable structure. *motE*, *cheC*. Switch component of motor? Interacts with *cheY*, *cheZ*, *flaAII.2*, *flaN*.	85, 86, a, b, c, g
flaN (*motD*)	? (47 K)	?	*motD*, *che**.[4] Switch component of motor? Interacts with *flaQ*.	50, a, g
flaP (*flbD*)	?	?	Necessary for any detectable structure.	
flaB (*flaR*)	?	?	Necessary for any detectable structure.	
flaD (*flaQ*)	?	?	Necessary for any detectable structure.	
flaX (*flaP*)	?	?	Necessary for any detectable structure.	f, h
HBB proteins of unknown genetic origin				
?	65K	HBB	Inner face of M ring.	a
?	27 K	HBB	Detected in outer ring preparation.	52, a

[1] Genes are in genome order. The *Salmonella* gene is given first; the homologous *E. coli* gene, when named differently, follows below in parentheses. The same format is used for product molecular weights.

[2] Only directly relevant references are given. In addition, References (37, 56, 100, 104, 105) apply to most entries. a: S. I. Aizawa, S. Yamaguchi, and R. Macnab, unpublished results; b: D. Clegg and D. E. Koshland, Jr., personal communication; c: M. Eisenbach, personal communication; d: M. Homma and T. Iino, personal communication; e: Y. Komeda, personal communication; f: K. Kutsukake, personal communication; g: P. Matsumura, personal communication; h: S. Yamaguchi, personal communication.

[3] HBB: hook basal body.

[4] No gene symbol has been assigned to these genes or phenotypic alleles.

[5] An alternative gene (*H2*) for flagellin exists in a distant region of the genome.

designation, there are a few with multiple designations; the significance of this will become apparent shortly.

The genes are organized into a number of operons, and it has been demonstrated in *E. coli* (by the use of *lac* fusions) that they are interregulated in a complex fashion (50). A "master operon" (*flbB*, homologous to the *Salmonella flaK* operon) is itself under positive regulation by cyclic AMP (98) via the same "catabolite gene activator protein" (CAP protein) that participates in a number of other aspects of cell function (19). *FlbB* operon gene products then appear to act as direct positive regulators of most of the other operons, with the exception of some that are involved in the final stage of flagellar assembly or function, including the operon containing the flagellin gene and the *mocha* operon that contains the *motA*, *motB*, *cheA*, and *cheW* genes (99). These late operons appear to be under secondary control by products of *flbB*-activated operons; the absence of any of about 22 proteins prevents late operon expression. It seems unlikely that such a large number of proteins could all be acting directly as DNA-binding proteins, and recently Y. Komeda (personal communication) has obtained evidence suggesting that only two gene products may be acting directly as regulators of the late operons, one as an activator and the other as a repressor. Positive control by other gene products is then postulated to be indirect, by incorporating the repressor protein into the structure of the motor.

Regulation is especially critical with regard to the flagellin structural gene (*H1* or *H2* in *Salmonella*; *hag* in *E. coli*). A flagellar filament of typical length contains about 20,000 subunits, and there are 5–10 flagella per cell; synthesis of proteins in such large amounts, if they cannot be incorporated because of a basal body defect, would be very wasteful. It has been established that there is negligible flagellin mRNA under these conditions (102). It also makes sense that *mot* and *che* genes (for example, those of the *mocha* operon) should be under late control, since there is no point in enabling the rotational mechanism, or the sensory transduction apparatus, unless the rest of the organelle is properly in place. Although the other chemotaxis operons have not been examined in this regard, it seems likely that they, too, will be under late control.

Table 1 summarizes current knowledge regarding the roles of the motility and flagellar genes. Many of their products are known or presumed to be structural components of the flagellum and are discussed below. Others are known to play a genetic regulatory role, as has been discussed. Another interesting possibility is that some of the *fla* genes may code for enzymes (in addition, of course, to the "super-enzyme," i.e. the motor itself). It seems unlikely that the entire organelle can be assembled without any enzymatic intervention, especially since it has to be incorporated into the triple-layered cell surface (outer membrane, peptidoglycan layer, and cell membrane). One might postulate, as one example, a lysozyme-like activity in order to penetrate the peptidoglycan layer. There may also be a requirement for nonenzymatic ("scaffolding") accessory proteins (cf phage assembly; 112).

Protein Composition of the Flagellum

One would like to establish the correspondence between each structural gene, its product, a morphological feature of the motor, and a functional role. Only in the case of the filament and the hook is the correspondence complete and, in the case of the hook, the functional role is not fully understood. For all remaining structural components, the evidence is still fragmentary (Table 1). It derives from a variety of experimental approaches such as specific radiolabeling of proteins by programmed gene expression, and biochemical and electron microscopic analysis of structures from mutant strains. The product molecular weights are known for about 19 *fla* and *mot* genes; of these about 11 are known or thought to be structural. Only one polypeptide/morphological correspondence (other than those for the hook and filament) is indicated. Monoclonal antibody against a 65-K protein component of the basal body causes decoration of the inner face of the M ring (S.-I. Aizawa et al, in preparation).

Morphological assembly maps constructed by Suzuki and co-workers (Figure 5; 104, 105) on the basis of electron microscopy of partial structures assembled by *fla* mutants have provided some insight regarding the correspondence between genes and morphology but do not permit any definite assignments. Also, it was not possible to purify and further characterize these structures. A complementary approach that promises to yield useful information is the analysis of purified complete structures assembled by conditional mutants under permissive conditions (S.-I. Aizawa, S. Yamaguchi, and R. M. Macnab, unpublished results).

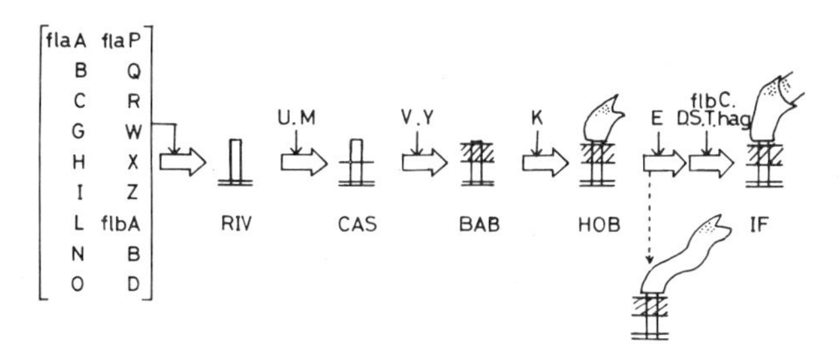

Figure 5 Scheme for flagellar assembly in *E. coli* (a similar scheme exists for *Salmonella*). The *fla* genes necessary to reach each successive stage of assembly are indicated. [From (105), with permission.]

The Process of Flagellar Assembly

Much of the order of flagellar assembly can be inferred from the partial structural studies referred to above (Figure 5). However, the early events of assembly are obscure, since the first detectable structure in *E. coli* required 18 genes!

The actual mechanism of assembly is not understood at all. Even the simplest part of the structure, the filament, presents enigmas. It has been demonstrated beyond doubt (26, 36) that the filament is assembled distally; by analogy, one may suppose this applies to the hook also. Although both the hook and the filament appear to have a hollow core, these are probably not large enough to accommodate filament subunits in their native state (94, 109). The problem is even more acute if one assumes that the monomers must pass up through the center of the rod, which is narrower still, yet has to be strong enough to withstand the full torsional load when the motor is running. If there is a continuous path for flagellin extrusion, what prevents it from constituting an ion leak? Mutants lacking the hook and filament would seem to be especially vulnerable in this regard. It seems hard to imagine that three distinct structures—rod, hook, and filament—could all accommodate flagellin monomers in a "watertight" seal; perhaps there is a special structure for this purpose at the departure port on the cytoplasmic face of the motor. This structure might also act as a "conformase" that converts flagellin monomers to an extended conformation to permit the export process. One wonders also whether the whole process can be driven simply by the concentration gradient of monomer or whether there is an active transport system.

Whereas the length of the filament is arbitrary, the length of the hook is rather closely defined. The mechanism of length determination is controlled by an unknown mechanism involving the *flaR* gene (96, 103). Length determination of biological structures (e.g. phage tails; 112) is generally a poorly understood phenomenon. One commonly advanced hypothesis, cumulative strain, would seem particularly inappropriate for the hook, which has to withstand massive mechanical stress during motility.

The Basal Body vs the "Total Motor"

Although, for practical reasons, attempts to understand motor function have focused on the basal body, there is good reason to suppose that this structure, though clearly very important, may not constitute the entire motor.

For example, the fact that no *mot* or *che* mutations have been found to map in basal body genes suggests that none of their products may be involved in energization or switching; this is not surprising for components that lie beyond the cytoplasmic membrane, but it makes questionable the suggestion that the M and S rings are the rotor and stator (although they may well be rotating and stationary, respectively).

Also, several proteins known to be important to motor function are not present in the basal body. These include the *motA* and *motB* proteins (31), without which the assembled flagellum cannot rotate, and the *flaQ* (and probably the *flaN* and *flaAII.2*) proteins that are necessary for flagellar assembly, rotation, and switching (see below). There is also a problem in reconciling the geometries of the basal body and the cytoplasmic membrane. A careful study by DePamphilis & Adler (23) established that the L ring is more or less flush with the outer, or Lipopolysaccharide membrane. The P ring therefore exists about 4 nm inward, placing it more or less in the plane of the Peptidoglycan layer, the extremely thin covalent mesh that provides mechanical strength against lysis under hypotonic conditions. DePamphilis & Adler also demonstrated association of the inner (cytoplasmic) membrane with the cytoplasmic-proximal pair of rings; in most images, the S ring could be seen riding just above the membrane (Supramembrane). Because of the small center-to-center separation (ca 3 nm) between S and M rings, DePamphilis & Adler concluded that the M ring is imbedded in the outer face of the cytoplasmic Membrane but is not thick enough to span it. This suggests that the unaccounted-for motility proteins might occupy the remaining volume.

The motA and motB Proteins

The *motA* and *motB* proteins are found in the cytoplasmic membrane fraction (31) but have not been localized further. Both genes of *E. coli* have been sequenced recently (G. E. Dean, J. Stader, P. Matsumura, and R. M. Macnab, in preparation). The predicted protein sequences have some overall similarities but little, if any, detailed sequence homology. The *motA* protein sequence is quite hydrophobic, with two possible helical membrane spanning regions near the N-terminus and another two about a third of the way from the C-terminus. In the region of the putative aqueous domain immediately after the second predicted helix, and again immediately before the third one, are regions of extremely high (and uncompensated) charge density; 7 basic residues in a 21 residue segment, and 9 acidic residues in an 18 residue segment, respectively. Close to the N-terminus of the *motB* sequence is a quite hydrophobic region that indicates two spanning helices. As with the *motA* protein, there are two regions of high charge density, one at the N-terminus and the other near the C-terminus. The presence of basic and acidic segments in both the *motA* and *motB* proteins suggests strongly that they will somehow be electrostatically paired, possibly intra-chain, inter-chain, or both. For both proteins, there is a prediction that there will be large cytoplasmic-proximal domains that might occupy the unaccounted-for volume in the motor.

How is it possible to assemble the flagellum in the absence of the *motA* and *motB* proteins? Let us assume that they are structural components of the motor (although it should be emphasized that this has not been established). Perhaps motor assembly is modular, with the central portion starting with the switch complex (see below) and proceeding independently: switch complex → switch complex/basal body → switch complex/basal body/hook → switch complex/basal body/hook/filament, and then the *motA* and *motB* proteins being added on peripherally. The *motB* protein can in fact enable rotation of a preexisting core, and it does so in a progressive fashion that argues against an absolutely required *motB*/core stoichiometry (8). Likewise, it is significant that the *motA* and *motB* proteins do not appear to be involved in switching (since no mutants of Che$^-$ phenotype have been found). An intriguing observation is that, although the core does not require the *motB* protein for assembly, it may be stabilized or regulated by its presence: *motB* mutants produce smaller numbers of flagella than do other *mot* mutants (H. Kagawa and Y. Komeda, personal communication).

The Switch Complex

We turn next to the products of a remarkable set of genes, to date three (*flaAII.2*, *flaQ*, and *flaN*, see Table 1), that appear to be centrally important to the motor's function, since they are necessary for its assembly, rotation, and switching. The simplest explanation would be that these gene products are in fact structural components of the motor and constitute a switch complex, which is sufficiently central in the geometrical sense that the basal body cannot be assembled in their absence (no partial structures were detected in *flaAII.2* or *flaQ* mutants; 104, 105; *flaN* mutants have not been examined in this regard), and sufficiently central in the mechanistic sense that slight tampering with their structure affects switching or blocks rotation. In the case of *flaAII.2*, it has been demonstrated beyond question that the gene product is mechanistically involved, since a conditional mutant lost function in <0.5 s in a temperature-jump experiment (18).

The most likely location for these gene products is at the cytoplasmic-proximal face of the motor. There are several reasons for supposing this. (*a*) They have not been detected in basal body preparations (D. Clegg, and D. E. Koshland, Jr., personal communication; S.-I. Aizawa and R. M. Macnab, unpublished observations). (*b*) They are probably at an early stage of the assembly process, since no partial structures were detected in mutants from these gene classes (104, 105). (*c*) The involvement in energy transduction and switching suggests a membrane-associated location, since the energy source is the transmembrane PMF. (*d*) They are the only genes where *che* alleles can give rise to either a CW or a CCW rotational bias (18, 48, 84, 108), which indicates that they are at the focus of signals from the sensory system. They are in this sense bi-functional, having the possibility of defects that make the protein less capable of creating (or perhaps itself adopting) the CW state or the CCW state. In contrast, cytoplasmic components of the chemotaxis system appear to mediate only one type of signal, and hence mutants are invariably of the same bias. (*e*) In contrast to other CW *che* mutants, CW *flaAII.2* and *flaQ* mutants retain their phenotype in cell envelopes devoid of cytoplasmic content (M. Eisenbach, personal communication). (*f*) Intergenic suppression analysis of *che* mutants (85, 86) suggests that at least two of these three genes in *E. coli* (*flaA* and *flaB*, homologous to *flaQ* and *flaAII.2* respectively) code for proteins that are in physical interaction with cytoplasmic components (*cheY* and *cheZ*). (*g*) Intergenic suppression analysis of *mot* alleles of these genes suggests that all three products may form a complex (S.-I. Aizawa, S. Yamaguchi, and R. M. Macnab, unpublished observations).

A Speculative Model for the Structure of the Motor

Synthesizing the various lines of evidence into a speculative model (Figure 6), we envisage the basal body as a passive structure, with one half of the

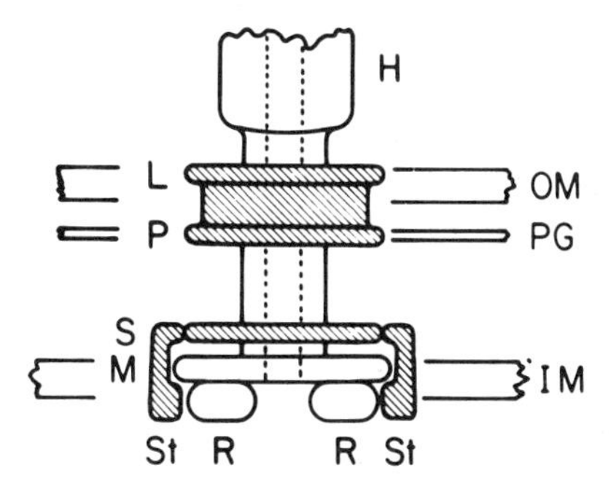

Figure 6 A speculative model for the flagellar motor. The model utilizes existing structural evidence for the basal body (cf Figure 4), correlations with surface structure, primary sequence information for the *motA* and *motB* proteins, and a variety of genetic and physiological evidence regarding nonbasal-body motility proteins. Putative stationary features are shown cross-hatched. It is proposed that the energy-transducing machinery consists of (*a*) a stator complex or set of complexes (St), constructed from the *motA* and *motB* proteins and anchored to the S ring and to some feature of the cell surface, possibly the inner membrane (IM) or the peptidoglycan layer (PG); (*b*) a bistable (switchable) rotor complex R that is attached to the M ring and is constructed from the *flaAII.2*, *flaN*, and *flaQ* proteins (see text). The L and P rings constitute a stationary bushing anchored to, and coplanar with, the outer membrane (OM) and peptidoglycan layer, respectively. [The L, P, S, M terminology for the rings was introduced by DePamphilis & Adler (22).] The dotted lines represent the hollow core through which hook and filament proteins are presumed to be exported for assembly.

energy-transducing complex (the rotor) composed of the *flaAII.2*, *flaN*, and *flaQ* proteins bound to the cytoplasmic-proximal face of the M ring. The stator half of the energy-transducing complex is a membrane-spanning structure, built from the *motA* and *motB* proteins and interacting both with the rotor complex (in a dynamic fashion powered by proton flux through the interface between the two) and with an anchored structure, possibly the S ring. Whereas the *motA*/*motB* stator complex is of fixed symmetry, the *flaAII.2*/*flaN*/*flaQ* rotor complex is bistable—a switch that can adopt two essentially antisymmetric conformations, with probabilities that are determined in part by its own primary structures and in part by allosteric regulation by *cheY* and *cheZ* proteins (and possibly by small molecule effectors).

DYNAMIC PROPERTIES OF THE FLAGELLAR MOTOR

Energetics and Rotation Mechanism

Unlike virtually all other known mechanoenzymes, the flagellar motor is not an ATPase; bacteria swim perfectly well even when their ATP pools have been drastically reduced (59). The energy source for rotation is either PMF itself or possibly some unidentified energy form derived from it (27, 47, 74, 75, 76). Based on ion substitution studies, it does not seem likely that any other ion (with the possible exception of OH^-) is involved (46, 74). Further support for these statements comes from recent work (M. Eisenbach, personal communication) in a model system of cell envelopes devoid of cytoplasmic content. Their flagella can be made to rotate in simple buffer, regardless of its ionic composition, by imposing a pH gradient across the envelope membrane.

The statement that the motor is proton-driven applies to bacteria that occupy temperate environments, but not to certain bacteria whose alkaline environment is such that PMF is not a suitable energy source. Such organisms rely on sodium potential—generated from PMF, presumably with a high H^+/Na^+ stoichiometry—for transport (28, 32) and also for motility. The implications of this will emerge when we consider specific mechanisms.

In proton-driven motility, chemical potential (transmembrane pH difference) and transmembrane electrical potential are found to be equally effective (47, 73, 93). Since measurement of motility is necessarily a measurement of the rate of an energy conversion process (there is no technique available for measuring stalling torque), the equivalence is more than a thermodynamic statement and suggests that protons generate torque by an identical mechanism regardless of whether the original energy form was electrical or chemical. The conversion of electrical potential to chemical potential can be accomplished by the use of a proton well, as described by Mitchell (77). The motor can be driven artificially by a PMF of reverse polarity (73). Over an appreciable range of potential values, the work output is approximately proportional to the potential (47, 73, 93). Further, the torque is load-independent (11, 73) at all except very light loads (as, for example, in a filament-less mutant; 9). The simplest conclusion from these observations is that the device works with a fixed stoichiometry of protons per revolution (i.e. it is tightly coupled in the stoichiometric sense) and with a high efficiency, the energy conversion events proceeding close to equilibrium. In this class of mechanism, protons flow through the device by a dynamically evolving path generated by motor rotation; the activation energy for transfer without rotation is presumed to be infinite.

Several more or less specific models for the coupling mechanism have been presented. (*a*) Lauger (61) invokes the creation of a cation-binding site by the juxtaposition of two anionic groups, one from the stator and one from the rotor. The electrostatic stabilization provided by either group alone is presumed to be insufficient to remove the ion from the bulk medium, so that only the joint site can be occupied. These sites exist in linear arrays at an angle to each other: one radial, the other at a tilt. Migration of the cation can therefore proceed only if accompanied by mutual rotation of the arrays. (*b*) Macnab (66) also employs the idea of mutually inclined arrays (in this case axial rather than radial, though this is incidental to the mechanism), but now consisting of hydrogen bonding chains (cf. 79), with each elementary migration event requiring the formation of a new bridging hydrogen bond between rotor and stator. (*c*) Glagolev & Skulachev (27) describe a proton association/dissociation (acid/base) mechanism in which exit on the cytoplasmic face requires an angular displacement. It is not specified whether the motor could rotationally diffuse in the absence of a proton flux. (*d*) Berg & Khan (8) also invoke an acid/base mechanism but explicitly require that a proton be bound in order to sufficiently lower the activation energy barrier for an elementary rotational step of the motor.

Although it is not possible at this stage to decide which (if any) of these models is correct, the existence of sodium-driven motility argues against the hydrogen bonding model (there is no evidence that sodium can participate in a specific divalent bonding arrangement). The ion-binding model of Lauger (61) remains intact, while the acid/base models can readily be modified as salt/base models. The finding that the *motA* and *motB* proteins have regions of very high charge of either acidic or basic residues (G. E. Dean, J. Stader, P. Matsumura, and R. M. Macnab, in preparation) may be a clue indicating an acid/base mechanism.

It is worth reemphasizing that all of the above models assume tight coupling. In contrast, Oosawa & Masai (82) have described a loosely coupled model in which thermal radial motion of an arm on the stator, containing a basic protonable site S, is electrostatically coupled to diagonally-placed, negatively-charged sites on the rotor. At one extreme of the swing, protonation/deprotonation can only occur from the outside; at the other, from the cytoplasm. Assume a high proton activity on the outside and a low activity on the inside. Then the stator statistically will swing in the inward direction in its protonated form SH^+, and outwardly in its unprotonated form S, exerting net torque on the rotor each cycle. However, the stator arm can cycle (and permit a proton flux) even though the rotor is held fixed. Thus, at infinite load this device has zero efficiency. As the authors point out, a crucial test of such a model is whether the flagellar motor has an appreciable conductance at infinite load. In order to accommodate the finding that torque is load-independent, Oosawa & Masai assume that the flux depends only on the PMF and not on the load. Although this argument is logically sound, it implies exceedingly loose coupling and hence an efficiency that would seem to be intolerably low (cf 68).

If the making or breaking of bonds between protons and other molecular structures is rate limiting, there should be appreciable isotope and temperature effects on motor kinetics. Khan & Berg (45), however, found that the motor behavior for cells suspended in H_2O- and D_2O-based media was identical. They also found that at a defined PMF (and after proper viscosity correction) there was no temperature effect on motility. Khan & Berg claim, we believe erroneously, that the absence of isotope and temperature effects discriminates in favor of an acid/base model. The presence of such effects would have discriminated strongly against any tightly coupled mechanism operating at high efficiency in favor of any mechanism (including those that are tightly coupled in the stoichiometric sense, see below) operating at low efficiency. [The argument is straightforward. Suppose that by raising the temperature we succeed in doubling the speed of motor rotation against a specified load. Since the opposing force is proportional to speed, the device is now performing twice the work per revolution (four times the work per unit time), using the same number of protons per revolution at the same potential as before. The efficiency at the lower temperature can therefore have been no higher than 50%.] The fact that no temperature effects were observed makes a loosely coupled mechanism such as Oosawa & Masai's seem unlikely; one would expect an increased flux as a result of more rapid cycling of the stator arm. Depending on the nature of the kinetic limitation, one would also expect (stoichiometrically) tightly coupled models operating at low efficiency to show thermal and possibly isotope effects, as will become apparent.

Combining the observations (*a*) that the device can be driven by ΔpH, (*b*) that over a considerable operational range its torque is proportional to PMF and independent of load and (*c*) that it is free of temperature and isotope effects, we are faced with the description of a device that converts thermal energy to work (which could in principle be the lifting of a weight, although in a swimming bacterium it is the regeneration of thermal energy) with close to 100% efficiency. Since the thermal energy source derives from a concentration difference and not a temperature difference, this description is not in violation of the second law of thermodynamics.

A device that can utilize a purely entropic energy form (ΔpH) is fundamentally different from a device like muscle which, to a first approximation, utilizes a purely enthalpic energy term (ΔH of ATP hydrolysis). It is also distinct because, in the elementary set of events that constitute a step, there is no reason to suppose that there is a loss of coupling, such as occurs in muscle at the point where the myosin head dissociates from the actin filament. Thus, a flagellar motor ought to be able to come to true equilibrium against an appropriate constraint; muscle necessarily expends energy even in isometric contraction.

Use of pure thermal energy to perform systematic work (rotation against a load) with high efficiency requires that the load automatically develop as a constraint that is opposite and nearly equal to the driving force. How can this be accomplished? A simple one-step-at-a-time mechanism will not work, since the single proton does not "know" what its potential is, and therefore how much work it "ought" to perform. However, if the consequence of proton use can be stored, in much the same way as a man-made ion concentration cell develops an electrode potential, then the PMF and the constraint can be brought close to equilibrium with each other. Under these conditions, forward and reverse elementary steps have nearly equal probabilities and occur at a rate that is high compared to the net rate of stepping. This situation can only exist when the opposing force is conservative; one cannot use friction as the constraint. This is the fundamental reason the one-step-at-a-time model will not work.

Macnab (68) has discussed a model in which the constraint is developed in the form of high-energy (strained) proton-motor bonds, so that only protons above a critical energy can be utilized.

Berg & Khan (8) have described an alternative model in which the motor itself is in a "high energy" state, with elastic energy in the stator as the stored energy form. Imagine the rotor being held fixed. A few net protons would be "wasted" as the stator complex made a statistical walk to the mean position where it was sufficiently elastically constrained that the trans-motor PMF was equal to the next elementary increase of elastic energy. At that point there would be no further (net) proton flux, and the device would have reached equilibrium. The rotor (tightly coupled to the current stator position) would be under a torque derived from the stored elastic energy and, if released, would accelerate to an angular velocity such that the work done against external friction was nearly equal to the energy loss of the protons transferred (provided the protons and the elastic state of the stator were still close to equilibrium, i.e. that the motor rotation was the rate-limiting step). At low loads, the elastic energy stored would be insufficient and the device would operate at low efficiency. In Berg & Khan's model it is the link between the stator and the cell wall that constitutes the elastic storage medium. However, the rotating portion of the organelle, which includes the entire external filament, should be capable of acting in the same fashion by storing torsional energy.

It is possible to have obligatory coupling between proton transfer and motor rotation and still operate at low efficiency. We have already seen one example of that in the limit of low external load. Loss of efficiency is also encountered at high PMF; both in *E. coli* and *B. subtilis*, saturation was observed (47, 93). Berg and Khan (8) suggest this may be a result of a finite range in the elastic constraint, e.g. a parabolic energy well bounded by infinite walls. The reason may be more fundamental than that. If the mechanism is one of diffusion, gated by protons, then the upper limit to the rate—regardless of the available PMF—is the unidirectional diffusion rate for the prevailing frictional geometry. This in fact enables one to make a rough calculation of the maximum speed in any given situation. Suppose n protons take the motor through one revolution in n elementary steps and are driving a filament of length l and radius a at a helical radius r through a medium of viscosity η (constituting the major frictional load). Then the elementary diffusional step is $2\pi r/n$, the diffusion coefficient D is ca $kT/[4\pi\eta l/(\ln l/a+0.193)]$ (17), the mean diffusion time t is $[2\pi r/n]^2/2D$ and the rotation rate is $1/nt$. For $n = 300$, $l = 10\ \mu m$, $r = 0.25\ \mu m$, $a = 10$ nm, $\eta = 1$ cp, the limiting speed is calculated to be about 60 Hz, in the same range as is observed experimentally on fully energized cells. Viewed in this way, there is no essential difference between the saturation of the motor at high PMF and moderate external load, and the saturation at moderate PMF and low external load. Berg & Khan (8), carrying out a similar calculation under the latter conditions, arrived at a limiting speed of 300 Hz, in the range that was observed experimentally for mutants lacking the flagellar filament (9). For any given PMF, the device can be made to run close to equilibrium by securing an appropriately high frictional geometry.

Prevailing pH values may be important to motor operation, especially if the mechanism involves acid-base dissociation. Again, the effect is a kinetic rather than a thermodynamic one. If, for example, an elementary motion is only possible if the outside site is protonated and the inside is not (or vice versa), then at very high pH the device is proton starved on both sides, whereas at very low pH it is saturated on both sides; in either case the system is kinetically limited. Another way of stating this is that only where the PMF is close to equilibrium with all possible states of the system can it be utilized with close to 100% efficiency.

Another aspect of the dynamics of motor rotation may be mentioned. Rotation does not proceed at PMF values less than about 30 mV in either *E. coli* or *B. subtilis* (47, 93) (this phenomenon was not, however, noted in *Streptococcus*; 73). Also, nonenergized cells do not even display rotational diffusion (9). At first the existence of a threshold for rotation seems paradoxical, since it is known that the device can be driven by thermal energy (ΔpH). The paradox can be resolved if one assumes that the concerted activity of several protons is necessary to overcome an activation energy barrier to rotation.

We turn now to the question of stepping. A macroscopic turbine does not step because the driving medium is continuous. Protons, however, are quanta and in a tightly coupled model deliver quanta of work; even in a loosely coupled model such as that of Oosawa & Masai, one might expect to see some remnant of quantization in the output. Consideration of the scale of the motor and the assumption that it is constructed from medium-sized protein subunits indicates that individual components of the motor are likely to have axial symmetries of the order of 10-fold. [DePamphilis & Adler (22) in fact suggested on the basis of image averaging that the M-ring may have 16-fold symmetry.] It is therefore certain that there are going to be at least 10 or so steps per revolution. Multiple sites per subunit or a vernier arrangement of subunits between rotor and stator would decrease the size of the elementary step substantially, and in fact hydrodynamic energy calculations show that the number of protons required to complete one revolution is of the order of a few hundred (3).

The ability to detect an elementary step is dependent on the precision with which such steps are quantized at the output. A loosely coupled model would be expected to exhibit a broad distribution of work output per proton, depending on the geometric relationship of the weakly coupled sites. Even a tightly coupled system would show thermal fluctuation in the stored constraint in the motor and in the coupling of the motor to the external environment. In extensive studies, Berg and co-workers have failed to detect any elementary stepping events (9).

Motor Switching

We consider finally the phenomenon of motor switching, which results in a PMF of invariant polarity generating mechanical work in either of two polarities, CCW or CW rotation. Scale and symmetry considerations indicate that there are almost certainly many (say at least 10) parallel paths for proton flux within a motor. Also, in multiflagellate species there are many motors. In both contexts, the question of coordination of switching arises.

At the level of the individual motor, it is observed that rotation changes abruptly (the deceleration time being too fast to detect) from one sense to the other, and the speed of rotation in the two senses is approximately the same (9). This implies that all modular elements within a motor switch simultaneously, an event that could derive either from a high degree of cooperativity within the motor structure itself or from a common signal from the cell. However, if there is a cell-wide signal, it ought to introduce some degree of correlation of switching events among the flagella on a multiflagellated cell. This idea is at first very appealing. In bipolar multiflagellated cells, for example, it is observed microscopically that both flagellar tufts switch simultaneously (55); peritrichous multiflagellates swim with a coordinated bundle (Figure 1*b*) and then undergo a tumble. However, in both of these situations a high degree of mechanical and hydrodynamic coupling is present, raising the question of whether the synchronization would still exist if the individual motors were not so coupled. The experimental result is that individual motors switch totally asynchronously. The observations have been made in two ways, on dispersed flagella observed by high-intensity dark-field light microscopy [M. Kihara and R. M. Macnab, unpublished observations, cited in (46); 69], and on killed cells used as markers—attached by specific antibodies—of the functioning of motors on elongated live cells (38). Since there is no global switch signal to the motors, it seems most reasonable to assume that the abrupt switching of an individual motor stems from a tightly cooperative switch complex (cf Figure 6). The model of individual stator elements (8), each with their own "switching particle," therefore seems unlikely.

What is the physical cause of switching? Although switching is modulated by environmental stimuli, it occurs even in an isotropic and temporally invariant environment. Under constant conditions, the probabilities per unit time (in either direction) of switching are constant; CCW and CW intervals conform to a Poissonian distribution (10). Another significant point is that the switching rates in the two directions appear to be related. The effect of a tactic stimulus, or of a *che* mutation, is to alter both the CCW → CW and CW → CCW probabilities in a reciprocal fashion (46).

It is inferred from genetic studies that certain proteins (*cheY* and *cheZ* products) bind to presumed components of the switch complex (*flaAII.2* and *flaQ* products) (84, 85). From the phenotype of *cheY* and *cheZ* mutants, one infers that the corresponding wild-type proteins favor the CW and

CCW states, respectively. It may be that they bind competitively and act as allosteric regulators. Allosteric regulation implies that the motor itself has two conformations; if so, it should in principle be able to interconvert (at some frequency) even in the absence of the effectors. Pursuing this line of thought a little further, it might be that the switching events originate internally, as a thermal isomerization process (46), and that the role of the *cheY* and *cheZ* proteins (and ultimately the entire sensory transduction system) is to modulate the probabilities but not to obligatorily cause switching.

The reciprocity remains a puzzling feature: in almost any process one can think of, it ought to be possible to modulate the rate in one direction without altering the reverse rate, and it ought to be possible to alter both rates in the same sense (as, for example, the V_{max} of an enzyme).

CONCLUSION

The subject of bacterial motility and its control by sensory inputs continues to stimulate and to challenge. Though simple by the standards of behavior in higher organisms, it is still a remarkable example of the integration of a large number of genes and their products into a single functioning system. The flagellar motor, apparently a highly efficient device for converting thermal energy into work, represents a novel class of mechanoenzyme. The elucidation of the molecular mechanism is likely to be some distance in the future but will be of fundamental importance to our understanding of energy transduction mechanisms in biological systems.

ACKNOWLEDGMENTS

We are grateful to a number of colleagues for providing unpublished information that has been incorporated into this review: M. Eisenbach, H. Kagawa, Y. Komeda, D. E. Koshland, Jr., P. Matsumura, J. S. Parkinson, J. Stader, B. Stallmeyer, and S. Yamaguchi. M. Eisenbach and B. Taylor provided helpful criticism. The work from this laboratory has been supported by USPHS Grant AI12202.

Literature Cited

1. Anderson, R. A. 1975. In *Swimming and Flying in Nature*, ed. T. Y. Wu, C. J. Brokaw, C. J. Brennen, 1:45–56. New York: Plenum
2. Asakura, S. 1970. *Adv. Biophys.* 1:99–155
3. Berg, H. C. 1974. *Nature* 249:77–79
4. Berg, H. C. 1976. *J. Theor. Biol.* 56:269–73
5. Berg, H. C., Anderson, R. A. 1973. *Nature* 245:380–82
6. Berg, H. C., Bromley, D. B., Charon, N. W. 1978. *Symp. Soc. Gen. Microbiol.* 28:285–95
7. Berg, H. C., Brown, D.A. 1972. *Nature* 239:500–4
8. Berg, H. C., Khan, S. 1983. In *Mobility and Recognition in Cell Biology*, ed. H.

Sund, C. Veeger, pp. 485–97. Berlin/New York: de Gruyter
9. Berg, H. C., Manson, M. D., Conley, M. P. 1982. In *Prokaryotic and Eukaryotic Flagella*, ed. W. B. Amos, J. G. Duckett, pp. 1–31. Cambridge, England: Cambridge Univ. Press
10. Berg, H. C., Tedesco, P. M. 1975. *Proc. Natl. Acad. Sci. USA* 72:3235–39
11. Berg, H. C., Turner, L. 1979. *Nature* 278:349–51
12. Boyd, A., Simon, M. 1982. *Ann. Rev. Physiol.* 44:501–17
13. Calladine, C. R. 1982. See Ref. 9, pp. 33–51
14. Canale-Parola, E. 1978. *Ann. Rev. Microbiol.* 32:69–99
15. Chwang, A. T., Wu, T. Y. 1971. *Proc. R. Soc. London Ser. B* 178:327–46
16. Cohen-Bazire, G., London, J. 1967. *J. Bacteriol.* 94:458–65
17. Cox, R. G. 1970. *J. Fluid Mech.* 44:791–810
18. Dean, G. E., Aizawa, S.-I., Macnab, R. M. 1983. *J. Bacteriol.* 154:84–91
19. de Crombrugghe, B., Busby, S., Buc, H. 1983. In *Biological Regulation and Development*, ed. K. Yamamoto, Vol. 3B. New York: Plenum. In press
20. DeFranco, A. L., Parkinson, J. S., Koshland, D. E. Jr. 1979. *J. Bacteriol.* 139:107–14
21. DePamphilis, M. L., Adler, J. 1971. *J. Bacteriol.* 105:376–83
22. DePamphilis, M. L., Adler, J. 1971. *J. Bacteriol.* 105:384–95
23. DePamphilis, M. L., Adler, J. 1971. *J. Bacteriol.* 105:396–407
24. Dimmitt, K., Simon, M. 1971. *J. Bacteriol.* 105:369–75
25. Dworkin, M., Keller, K. H., Weisberg, D. 1983. *J. Bacteriol.* 155:1367–71
26. Emerson, S. U., Tokuyasu, K., Simon, M. I. 1970. *Science* 169:190–92
27. Glagolev, A. N., Skulachev, V. P. 1978. *Nature* 272:280–82
28. Guffanti, A. A., Cohn, D. E., Kaback, H. R., Krulwich, T. A. 1981. *Proc. Natl. Acad. Sci. USA* 78:1481–84
29. Hazelbauer, G. L., Harayama, S. 1983. *Int. Rev. Cytol.* 81:33–70
30. Henrichsen, J. 1972. *Bacteriol. Rev.* 36:478–503
31. Hilmen, M., Simon, M. 1976. In *Cell Motility*, ed. R. Goldman, T. Pollard, J. Rosenbaum, pp. 35–45. New York: Cold Spring Harbor Lab.
32. Hirota, N., Kitada, M., Imae, Y. 1981. *FEBS Lett.* 132:278–80
33. Holt, S. C. 1978. *Microbiol. Rev.* 42:114–60
34. Hotani, H. 1976. *J. Mol. Biol.* 106:151–66
35. Hotani, H. 1982. *J. Mol. Biol.* 156:791–806
36. Iino, T. 1969. *J. Gen. Microbiol.* 56:227–39
37. Iino, T. 1977. *Ann. Rev. Genet.* 11:161–82
38. Ishihara, A., Segall, J. E., Block, S. M., Berg, H. C. 1983. *J. Bacteriol.* 155:228–37
39. Johnson, R. C., Ferber, D. M., Ely, B. 1983. *J. Bacteriol.* 154:1137–44
40. Johnson, R. C., Walsh, M. P., Ely, B., Shapiro, L. 1979. *J. Bacteriol.* 138:984–89
41. Kagawa, H., Aizawa, S.-I., Asakura, S. 1979. *J. Mol. Biol.* 129:333–36
42. Kagawa, H., Morishita, H., Enomoto, M. 1981. *J. Mol. Biol.* 153:465–70
43. Kamiya, R., Hotani, H., Asakura, S. 1982. See Ref. 9, pp. 53–76
43a. Kato, S., Okamoto, M., Asakura, S. 1983. Submitted for publication
44. Keller, K. H., Grady, M., Dworkin, M. 1983. *J. Bacteriol.* 155:1358–66
45. Khan, S., Berg, H. C. 1983. *Cell* 32:913–19
46. Khan, S., Macnab, R. M. 1980. *J. Mol. Biol.* 138:563–97
47. Khan, S., Macnab, R. M. 1980. *J. Mol. Biol.* 138:599–614
48. Khan, S., Macnab, R. M., DeFranco, A. L., Koshland, D. E. Jr. 1978. *Proc. Natl. Acad. Sci. USA* 75:4150–54
49. Klug, A. 1967. *Symp. Int. Soc. Cell Biol.* 6:1–18
50. Komeda, Y. 1982. *J. Bacteriol.* 150:16–26
51. Komeda, Y., Icho, T., Iino, T. 1977. *J. Bacteriol.* 129:908–15
52. Komeda, Y., Silverman, M., Matsumura, P., Simon, M. 1978. *J. Bacteriol.* 134:655–67
53. Kondoh, H., Yanagida, M. 1975. *J. Mol. Biol.* 96:641–52
54. Koshland, D. E. Jr. 1981. *Ann. Rev. Biochem.* 50:765–82
55. Krieg, N. R., Tomelty, J. P., Wells, J. S. Jr. 1967. *J. Bacteriol.* 94:1431–36
56. Kutsukake, K., Iino, T., Komeda, Y., Yamaguchi, S. 1980. *Mol. Gen. Genet.* 178:59–67
57. Kutsukake, K., Suzuki, T., Yamaguchi, S., Iino, T. 1979. *J. Bacteriol.* 140:267–75
58. Lapidus, I. R., Berg, H. C. 1982. *J. Bacteriol.* 151:384–98
59. Larsen, S. H., Adler, J., Gargus, J. J., Hogg, R. W. 1974. *Proc. Natl. Acad. Sci. USA* 71:1239–43
60. Larsen, S. H., Reader, R. W., Kort, E. N., Tso, W.-W., Adler, J. 1974. *Nature* 249:74–77
61. Lauger, P. 1977. *Nature* 268:360–62

62. Lowy, J., Spencer, M. 1968. *Symp. Soc. Exp. Biol.* 22:215–36
63. Macnab, R. M. 1977. *Proc. Natl. Acad. Sci. USA* 74:221–25
64. Macnab, R. M. 1978. *CRC Crit. Rev. Biochem.* 5:291–341
65. Macnab, R. M. 1979. In *Encyclopedia of Plant Physiology, New Series*, ed. W. Haupt, M. E. Feinleib, 7:207–23. Berlin/Heidelberg: Springer-Verlag
66. Macnab, R. M. 1979. *Trends Biochem. Sci.* 4:N10–N13
67. Macnab, R. M. 1980. In *Biological Regulation and Development*, ed. R. F. Goldberger, 2:377–412. New York: Plenum
68. Macnab, R. M. 1983. In *Biological Structures and Coupled Flows*, ed. A. Oplatka, pp. 147–60. Philadelphia/ Rehovot: Balaban Int. Sci. Serv.
69. Macnab, R. M., Han, D. P. 1983. *Cell* 32:109–17
70. Macnab, R. M., Koshland, D. E. Jr. 1972. *Proc. Natl. Acad. Sci. USA* 69:2509–12
71. Macnab, R. M., Koshland, D. E. Jr. 1974. *J. Mol. Biol.* 84:399–406
72. Macnab, R. M., Ornston, M. K. 1977. *J. Mol. Biol.* 112:1–30
73. Manson, M. D., Tedesco, P. M., Berg, H. C. 1980. *J. Mol. Biol.* 138:541–61
74. Manson, M. D., Tedesco, P. M., Berg, H. C., Harold, F. M., van der Drift, C. 1977. *Proc. Natl. Acad. Sci. USA* 74:3060–64
75. Matsuura, S., Shioi, J.-I., Imae, Y. 1977. *FEBS Lett.* 82:187–90
76. Matsuura, S., Shioi, J.-I., Imae, Y., Iida, S. 1979. *J. Bacteriol.* 140:28–36
77. Mitchell, P. 1969. *Theor. Exp. Biophys.* 2:159–216
78. Monod, J., Wyman, J., Changeux, J.-P. 1965. *J. Mol. Biol.* 12:88–118
79. Nagle, J. F., Morowitz, H. J. 1978. *Proc. Natl. Acad. Sci. USA* 75:298–302
80. Nauman, R. K., Holt, S. C., Cox, C. D. 1969. *J. Bacteriol.* 98:264–80
81. O'Brien, E. J., Bennett, P. M. 1972. *J. Mol. Biol.* 70:133–52
82. Oosawa, F., Masai, J. 1982. *J. Phys. Soc. Japan* 51:631–41
83. Padgett, P. J., Friedman, M. W., Krieg, N. R. 1983. *J. Bacteriol.* 153:1543–44
84. Parkinson, J. S. 1981. In *Genetics as a Tool in Microbiology*, ed. S. W. Glover, D. A. Hopwood, pp. 265–90. Cambridge, England: Cambridge Univ. Press
85. Parkinson, J. S., Parker, S. R. 1979. *Proc. Natl. Acad. Sci. USA* 76:2390–94
86. Parkinson, J. S., Parker, S. R., Talbert, P. B., Houts, S. E. 1983. *J. Bacteriol.* 155:265–74
87. Paster, B. J., Canale-Parola, E. 1980. *J. Bacteriol.* 141:359–64
88. Pate, J. L., Chang, L.-Y. E. 1979. *Curr. Microbiol.* 2:59–64
89. Pate, J. L., Ordal, E. J. 1967. *J. Cell Biol.* 35:37–51
90. Patterson-Delafield, J., Martinez, R. J., Stocker, B. A. D., Yamaguchi, S. 1973. *Arch. Mikrobiol.* 90:107–20
91. Satir, P., Ojakian, G. K. 1979. See Ref 65, pp. 224–49
92. Shimada, K., Kamiya, R., Asakura, S. 1975. *Nature* 254:332–34
93. Shioi, J.-I., Matsuura, S., Imae, Y. 1980. *J. Bacteriol.* 144:891–97
94. Shirakihara, Y., Wakabayashi, T. 1979. *J. Mol. Biol.* 131:485–507
95. Silverman, M., Matsumura, P., Simon, M. 1976. *Proc. Natl. Acad. Sci. USA* 73:3126–30
96. Silverman, M., Simon, M. 1972. *J. Bacteriol.* 112:986–93
97. Silverman, M., Simon, M. 1974. *Nature* 249:73–74
98. Silverman, M., Simon, M. 1974. *J. Bacteriol.* 120:1196–1203
99. Silverman, M., Simon, M. 1976. *Nature* 264:577–80
100. Silverman, M., Simon, M. 1977. *Ann. Rev. Microbiol.* 31:397–419
101. Simon, M., Silverman, M., Matsumura, P., Ridgway, H., Komeda, Y., Hilmen, M. 1978. *Symp. Soc. Gen. Microbiol.* 28:271–86
102. Suzuki, H., Iino, T. 1975. *J. Mol. Biol.* 95:549–56
103. Suzuki, T., Iino, T. 1981. *J. Bacteriol.* 148:973–79
104. Suzuki, T., Iino, T., Horiguchi, T., Yamaguchi, S. 1978. *J. Bacteriol.* 133:904–15
105. Suzuki, T., Komeda, Y. 1981. *J. Bacteriol.* 145:1036–41
106. Tamm, S. L. 1982. *J. Cell Biol.* 94:697–709
107. Taylor, B. L., Laszlo, D. J. 1981. In *The Perception of Behavioral Chemicals*, ed. D. N. Norris, pp. 1–27. Amsterdam: Elsevier/North Holland Biomed.
108. Tsui-Collins, A. L., Stocker, B. A. D. 1976. *J. Bacteriol.* 128:754–65
109. Wagenknecht, T., DeRosier, D., Aizawa, S.-I., Macnab, R. M. 1982. *J. Mol. Biol.* 162:69–87
110. Wagenknecht, T., DeRosier, D., Shapiro, L., Weissborn, A. 1981. *J. Mol. Biol.* 151:439–65
111. Winet, H., Keller, S. R. 1976. *J. Exp. Biol.* 65:577–602
112. Wood, W. B., King, J. 1979. In *Comprehensive Virology*, ed. H. Frankel-Conrat, R. R. Wagner, 13:581–633. New York: Plenum

Ann. Rev. Biophys. Bioeng. 1984. 13: 85–103

MAGNETIC GUIDANCE OF ORGANISMS

Richard B. Frankel

Francis Bitter National Magnet Laboratory, Massachusetts Institute of Technology, Cambridge, Massachusetts 02139

INTRODUCTION

Human beings have used magnetic compasses as navigation aids since at least the eleventh century AD (55, 63). The magnetic compass facilitated long-distance navigation on the high seas and was one of the technological advances that led to the great European voyages of discovery in the thirteenth and fourteenth centuries. The invention of the magnetic compass is usually credited to the Chinese (55), who discovered over 2000 years ago that certain pieces of naturally occurring magnetite (Fe_3O_4), a common mineral, would orient in astronomically significant directions when allowed to rotate by flotation or pivoting. How the magnetic compass worked was not explained until 1600, when William Gilbert of Colchester published his theory, based on his study of magnetized needles and small spheres of magnetite, that the earth itself is a giant magnetic dipole (27).

Evidence has accumulated that the earth's magnetic dipolar field may also play a role in the orientation, navigation, and homing of a wide variety of organisms including bacteria (7, 8), algae (46), snails (13), planaria (14), honey bees (28, 50), salmon (60), salamanders (57), homing pigeons (28, 68), robins (74), mice (51), and possibly humans (2, 3, 30). In addition, training experiments on pigeons (12), skates (36), and tuna (M. M. Walker and J. Kirschvink, private communication) have demonstrated the ability of these organisms to sense magnetic fields. Two interaction mechanisms have been elucidated: (*a*) detection by the organism of the electric field induced by the Faraday effect as the organism moves through the magnetic field; (*b*) interaction of the magnetic field with magnetic material in the organism.

0084–6589/84/0615–0085$02.00

Magnetic Induction

The magnetic induction, or Faraday effect, mechanism is apparently operative in marine sharks, skates, and rays, which are sensitive to electric fields as low as 0.005 μV/cm in seawater (35). The animals detect electric fields through the ampullae of Lorenzini, which are long conductive channels that connect electrically sensitive cells in the snout with pores on the skin. The flow of ocean currents and the motion of the animal through the geomagnetic field induce voltage gradients with sign and magnitude that depend on orientation and that are in general above the animal's sensitivity threshold. Kalmijn (37) demonstrated that skates could be trained to use magnetic fields of the order of the geomagnetic field as an orientational cue. Brown et al (15) used electrophysiological measurements to show that the ampullae of Lorenzini can detect variations in the geomagnetic field. Jungerman & Rosenblum (34) have considered the possibility of the magnetic induction mechanism for an animal moving in air. They concluded that a circular, electrically conducting loop millimeters in size would be required to overcome thermal noise, with voltages induced by changes in magnetic flux in the loop as the animal changes its heading.

Magnetic Material in Organisms

Evidence for orientation by the second mechanism was obtained for homing pigeons in the classical experiment of Keeton (39). He glued small bar magnets to the backs of the heads of a group of homing pigeons and compared their homing ability with that of a group of control birds carrying brass weights. Under sunny skies both groups oriented and homed equally well when released from unfamiliar sites many miles from the home loft; but under overcast skies when the birds could not see the sun, the orientation of the birds carrying magnets was disrupted whereas control birds oriented normally. Subsequently, Walcott & Green (70) used Helmholtz coils attached to pigeons' heads to change the orientation of the birds under overcast conditions. The orientation depended on the direction of the magnetic field, as determined by the direction of current in coils. Pigeon orientation is also affected by magnetic anomalies and magnetic storms (40, 67). These observations suggest that in addition to a magnetic compass a homing pigeon may have a magnetic "map" (29, 68). The experimental situation has been reviewed by Walcott (68), Gould (28, 29), and Able (1). Magnetic effects in orientation of migratory birds have been reviewed by Able (1) and Wiltschko (73). Although attempts to observe magnetic sensitivity of pigeons by cardiac response have not been successful, Bookman (12) was able to train pigeons to detect the presence of magnetic fields in a flight cage.

Birds possess magnetic material that could act as a magnetic sensor. Walcott et al (69) dissected pigeons with nonmagnetic tools and found inducible magnetic remanence in head and neck sections. Magnetic material was localized in a piece of tissue between the dura and the skull. Each pigeon had inducible remanence of 10^{-5} to 10^{-6} emu, which disappeared at 575°C, indicating the presence of Fe_3O_4. Presti & Pettigrew (58) found magnetic material in the neck musculature of pigeons and migratory white crowned sparrows but did not find localized magnetic materials in the heads. It is likely, but not yet proven, that there is a connection between the magnetic material and magnetic sensitivity. Elucidation of anatomical structure is clearly required. Yorke (75), Kirschvink & Gould (44), and Presti & Pettigrew (58) have speculated on the role of Fe_3O_4 in a magnetic sensor. Yorke points out that, if a pigeon can somehow measure the total magnetization of its ensemble of magnetic particles, there is enough magnetic material present to indicate the field direction with high accuracy.

A possible connection between Fe_3O_4 and magnetic field effects on behavior is also found in honey bees. The behavioral effects have been reviewed by Martin & Lindauer (50) and Gould (28). Honey bee workers communicate the location of a food source to other workers in a hive by means of a "waggle dance" on a vertical honeycomb. The angle between the direction of the dance and the vertical direction indicates the angle between the food source and the sun. There are consistent errors (*missweissungen*) in the dance angle, which vanish when the magnetic field in the hive is nulled by means of external coils. In anomalous situations where bees are made to dance on horizontal surfaces, after an initial period of disorientation they dance along the eight magnetic compass directions (N, NE, E, SE, etc) (28, 42, 50). If the field in the hive is nulled, the dances become disoriented again. There is also evidence that bees can use the diurnal variations in the geomagnetic field to set their circadian rhythms (50).

Gould et al (31) have found that honey bees also contain Fe_3O_4. They measured an average induced magnetization of about 2×10^{-6} emu per bee, distributed between single-domain and superparamagnetic sized particles, mostly localized to the abdomen. Recently, Kuterbach et al (45) have found bands of cells around the abdominal segments that contain numerous iron-rich granules. The granules are primarily a hydrous iron oxide, which can be a precursor in the precipitation of Fe_3O_4 (25, 65).

Fe_3O_4 appears to be widely distributed in the biological world (48) in addition to its presence in pigeons and honey bees. Magnetic inclusions have been reported in organisms as diverse as dolphins (76), butterflies (33), tuna (71), green turtles (56), marine crustacea (16), bacteria (24), and humans (4, 43). The first identification of Fe_3O_4 in an organism was by Lowenstam

(47), who found it in the tooth denticles on the radulae of a group of mollusks called chitons. Fe_3O_4 is very hard as well as magnetic, making it useful as tooth mineral for chitons, which scrape algae off rocks. This illustrates the fact that the presence of Fe_3O_4 does not necessarily mean that the organism has a magnetic detector. In addition to being magnetic and hard, Fe_3O_4 is apparently one of the densest materials that can be mineralized by living organisms. This property might also play a role in certain cases.

MAGNETOTACTIC BACTERIA

The best documention to date for the connection between magnetically sensitive behavior and the presence of Fe_3O_4 is in aquatic bacteria that orient and swim along magnetic field lines (7–9). This behavior is termed magnetotaxis. Magnetotactic bacteria were discovered serendipitously in the early 1970s by Blakemore (7), who found bacteria, from both freshwater and marine muds, that accumulated at the North side of drops of water placed on a microscope slide. These bacteria swam toward and away from the south pole and north pole of a bar magnet, respectively.[1] Subsequently, Kalmijn & Blakemore (38) used homogeneous magnetic fields produced by Helmholtz coils to show that New England bacteria swim along magnetic field lines in the field direction, that is, in the direction indicated by the North-seeking pole of a magnetic compass needle. When the field produced by the coils is reversed by reversing the direction of current flow, the bacteria respond immediately by executing "U-turns" and continuing to swim in the field direction. Killed cells orient along the field lines and rotate when the field direction is reversed, but they do not move along the field lines. Thus, magnetotactic bacteria from New England behave as self-propelled magnetic dipoles and are predominantly North-seeking (38, 53).

Magnetosomes

Magnetotactic bacteria are found in the sediments of many aquatic environments (8, 20, 23, 53). In addition to their worldwide distribution, the diversity of morphological types suggests that magnetotaxis is a feature of a number of bacterial species. Two characteristics unify these species. Apparently, they are all anaerobic or microaerophilic (8) and they all contain magnetosomes (5), which are unique intracytoplasmic structures consisting of membrane-bounded magnetite (Fe_3O_4) (24, 52, 66) (Figure 1). One species, *Aquaspirillum magnetotacticum*, has been isolated and grown

[1] A bacterium that swims toward the south pole of a bar magnet will swim Northward in the geomagnetic field.

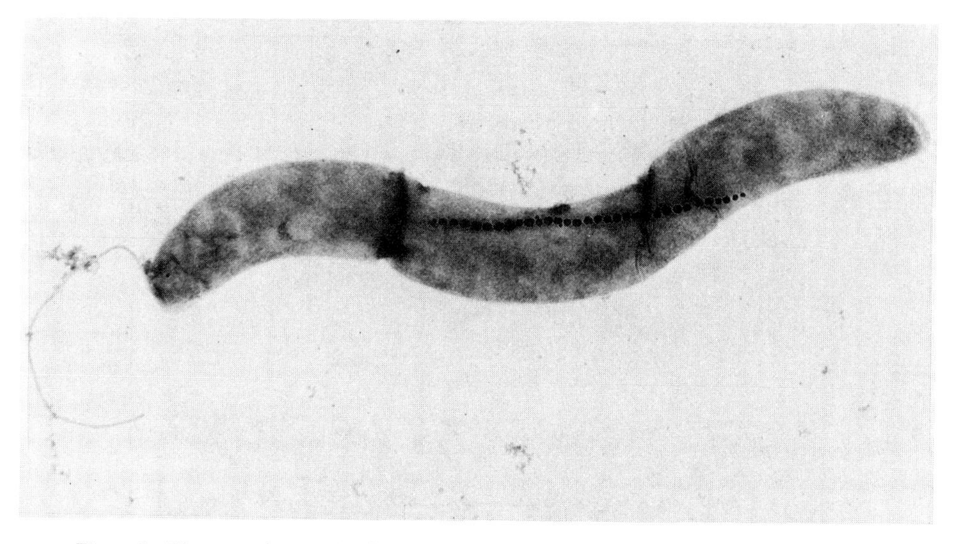

Figure 1 Electron micrograph of a magnetotactic spirillum. The chain of electron-opaque particles are the magnetosomes, the enveloped 500-Å particles of Fe_3O_4. The cell length is approximately 3 μm.

in pure culture in a chemically defined medium (11). Iron accounts for 2% of the dry weight of this species. Most of the iron (80–90%) is present in the form of the cytoplasmic membrane-bounded, 40–50-nm wide particles of Fe_3O_4 (24) (Figure 2). Cells also contain ferrous iron and hydrous-ferric-oxide (ferrihydrite) (25) (Figure 3). The enveloped Fe_3O_4 particles (magnetosomes) are arranged in a chain that longitudinally traverses the cell (Figure 1) in close proximity to the inner surface of the cytoplasmic membrane (5). The number of magnetosomes in the chain is variable, depending upon the culture conditions, but typically averages 20. Magnetosomes are enveloped by electron-transparent and electron-dense layers, and each is separated from those adjacent to it by 10-nm regions containing cytoplasmic material free of ribosomes or other particulate elements (5). The chemical composition of the distinctive region surrounding the bacterial magnetite grains is unknown but may be important in their formation. Magnetite particles extracted from cells by brief sonication retain an envelope, although their interparticle separation is less than 50% of that separating particles in chains within intact cells (5). Magnetosomes containing Fe_3O_4 have been identified in several other bacterial species as well (52, 66). In two cases, high resolution electron microscopy has shown that the Fe_3O_4 particles are single crystals (52; S. Mann and R. J. P. Williams, unpublished information).

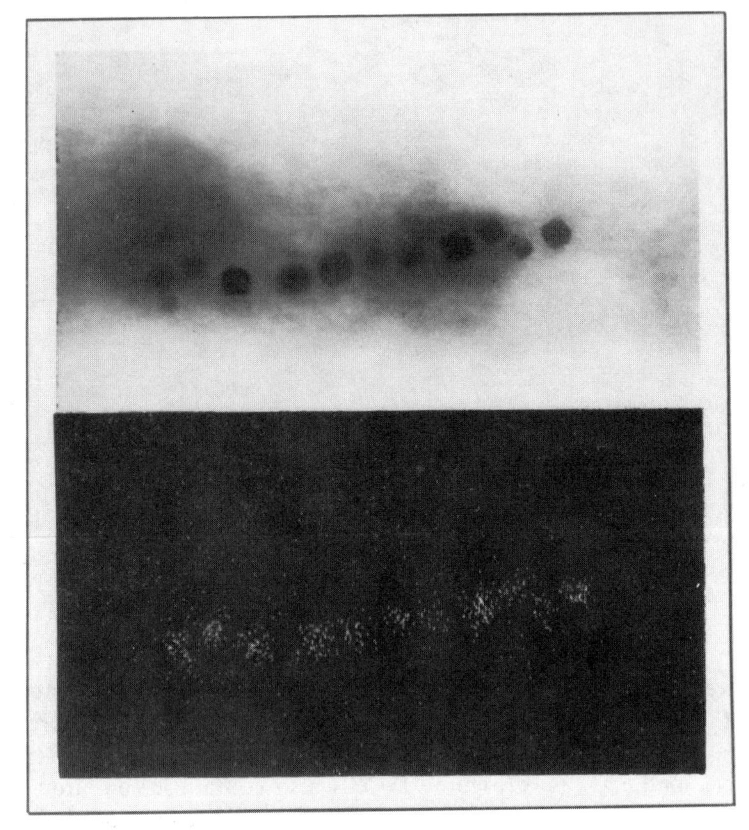

Figure 2 Transmission electron micrograph of *A. magnetotacticum* (*top*) and iron X-ray pulse map (*bottom*) showing that iron in the cell is localized in the magnetosomes [after (25)].

Mechanism of Magnetotaxis

The magnetosome chain in *A. magnetotacticum* imparts a magnetic dipole moment to the cell, parallel to the axis of motility, sufficient to orient the cell in the geomagnetic field (22). First, consider the size of the Fe_3O_4 particles. Large particles of Fe_3O_4 form magnetic domains that reduce the remanent magnetic moment and hence the magnetostatic energy. The domains are separated by transition regions or domain walls. When the particle length d is less than the width of a domain wall, it cannot form domains and will be a single magnetic domain. The upper limit for single magnetic domains d_{sd} is thus approximately the width of a domain wall d_w, which is a function of the exchange and anisotropy energy of the material

$$d_w = \left(\frac{kT_c}{Ka^3}\right)^{1/2} a, \qquad 1.$$

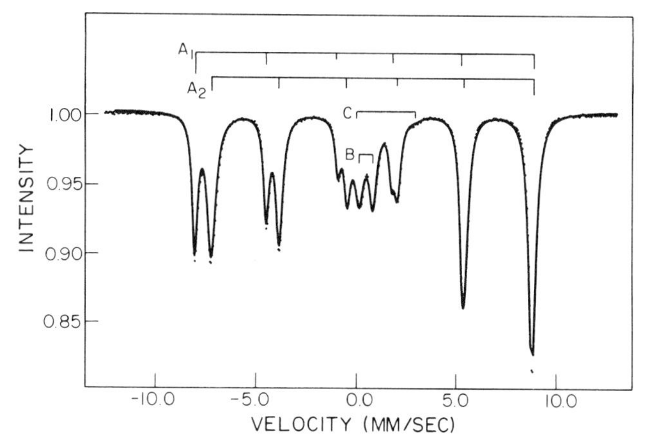

Figure 3 Mössbauer spectrum of frozen cells of *A. magnetotacticum*. Subspectra A_1 and A_2 are due to Fe_3O_4. Subspectrum B is a ferric doublet; subspectrum C is a ferrous doublet [after (25)]. The ferric and ferrous materials are precursors to Fe_3O_4 precipitation.

where k is Boltzmann's constant, T_c the Curie temperature, K the anisotropy energy, and a the atomic spacing. Substituting values for Fe_3O_4 (54) yields $d_w = 500$ Å. More precise calculations by Butler & Banerjee (17) for equidimensional particles yield $d_{sd} = 760$ Å. d_{sd} increases with the axial ratio (length/width). On the other hand, if the particle dimension is less than a certain value d_s, it will be superparamagnetic at room temperature; that is, thermal energy will cause transitions of the single domain magnetic moment between equivalent easy magnetic axes of the particle with a consequent loss of the time-averaged remanent moment (6). The transition frequency f is a function of the anisotropy energy, volume, and the thermal energy:

$$f \sim f_0 \exp\left[\frac{-KV}{2kT}\right], \qquad 2.$$

where f_0 is a constant of the order of $10^9\ s^{-1}$ and $V(=d^3)$ is the particle volume. Particles of dimensions greater than 350 Å are stable for times greater than 10^6 years; hence $d_s < 350$ Å. Thus, particles of Fe_3O_4 with dimensions 350 Å $< d <$ 760 Å are permanent, single magnetic domains with remanent moments of 480 G/cm^3. So we can assume that each 500-Å particle produced by a bacterium has a magnetic moment of 6.0×10^{-14} emu.

Second, consider the chain structure. When the single domain particles are organized in a chain as they are in *A. magnetotacticum*, the interactions between the particle moments will cause them to be oriented parallel to each other along the chain direction (32). Thus, the moment of the entire chain will be equal to the sum of the individual particle moments. For chains of twenty-two particles, this gives a total remanent moment $M = 1.3 \times 10^{-12}$ emu. Since the particles are fixed in the bacterium by the magnetosome envelope, the bacterium is, in effect, a swimming magnetic dipole.

The simplest hypothesis for magnetotaxis is passive orientation of the swimming bacterium along the magnetic field lines by the torque exerted by the field on the magnetic moment (22). Thermal energy, on the other hand, will tend to disorient the bacterium during swimming. In a magnetic field **B**, the energy of the bacterial moment

$$E_{\mathrm{m}} = -\mathbf{M} \cdot \mathbf{B} = -MB \cos \theta, \qquad 3.$$

where θ is the angle between **M** and **B**. The thermally averaged orientation of an ensemble of moments, or equivalently, the time averaged orientation of a single moment

$$\langle \cos \theta \rangle = \frac{\int \cos \theta \, e^{-E_{\mathrm{m}}/kT} \, dV}{\int e^{-E_{\mathrm{m}}/dt} \, dV} = L(\alpha). \qquad 4.$$

$L(\alpha)$ is the Langevin function,

$$L(\alpha) = \coth(\alpha) - \frac{1}{\alpha}; \quad \alpha = MB/kT, \qquad 5.$$

and is plotted in Figure 4. If we consider *A. magnetotacticum* in the earth's magnetic field of 0.5 G at room temperature, then $\alpha \sim 16$ and $\langle \cos \theta \rangle > 0.9$. Because the Langevin function asymptotically approaches one as α increases, the orientation would not improve significantly if there were more particles and the moment per bacterium were larger. Thus each bacterium is in effect a biomagnetic compass optimized to the geomagnetic field at room temperature.

For passively oriented bacteria, the migration velocity along the magnetic field lines

$$v_{\mathrm{B}} = v_0 \langle \cos \theta \rangle, \qquad 6.$$

where v_0 is the forward velocity of the swimming bacterium and θ is the angle between the axis of motility and the magnetic field. If v_0 is independent of B and the magnetic moment is parallel to the axis of

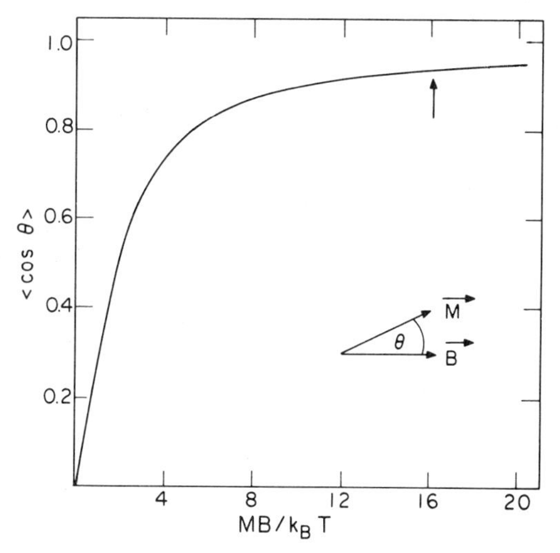

Figure 4 Langevin function plotted against MB/kT. The vertical arrow indicates the average orientation of bacteria with moments $M = 1.3 \times 10^{-12}$ emu in an 0.5-G field at 300 K.

motility,

$$v_B = v_0 L(\alpha), \tag{7}$$

providing that the velocity is averaged over a time that is long compared to the rotational diffusion time

$$\tau = \frac{8\pi r^3 \eta}{kT} \simeq 1 \text{ s}, \tag{8}$$

where r, the effective hydrodynamic radius of bacterium $= 0.5\ \mu$m, and the viscosity of water $= 0.01$ poise. Equation 7 is the basis of a method for measuring the magnetic moments of individual bacteria (36).

For bacteria with moments $>10^{-12}$ emu, the migration speed along magnetic field lines is thus $>90\%$ of their forward speed. This is in contrast to the relatively slow migration of chemotactic bacteria that vary the duration of swimming according to the orientation of the chemical concentration gradient.

The width of a "U-turn" executed by a bacterium following an instantaneous field reversal can also be calculated (C. P. Bean, private communication). In the world of the bacterium, inertial effects are negligible (59), and hence the torque exerted on the moment by the field is proportional to angular velocity:

$$\mathbf{M} \times \mathbf{B} = MB \sin\theta = 8\pi r^3 \eta \frac{d\theta}{dt}. \tag{9}$$

We define

$$\tau_0 = \frac{8\pi r^3 \eta}{MB}; \quad 10.$$

then the velocity perpendicular to the field

$$v_\perp = v_0 \sin\theta$$

$$= v_0 \tau_0 \frac{d\theta}{dt}. \quad 11.$$

The width of the "U-turn"

$$W = \int v_\perp \, dt$$

$$= v_0 \tau_0 \pi. \quad 12.$$

From the definition of τ_0, W is inversely proportional to MB and a measurement of W can in principle yield M. An actual measurement, however, would be experimentally complicated by the requirement that the bacterium confines itself to a horizontal plane during the "U-turn." The method has been applied to microorganisms from Rio de Janeiro (20). An unidentified coccus with 6-μm diameter had a magnetic moment of 4.7×10^{-12} emu.

The average magnetic moment per cell of bacteria in culture has been determined by elastic light scattering (62) and magnetically induced birefringence (61). The latter method depends on the anisotropic optical polarizability of the bacteria, which results in a net birefringence Δn in the sample when the bacteria are oriented by an applied magnetic field. The orientation is axial, not polar, and is proportional to the ensemble average of the second Legendre polynomial

$$\Delta n \propto \langle P_2(\cos\phi) \rangle. \quad 13.$$

The angular distribution function of the bacteria about the field direction is given by

$$f(\phi, \alpha) = \frac{\alpha \exp[\alpha \cos\phi] \sin\phi}{\alpha \sinh\alpha}; \quad 14.$$

hence

$$\langle P_2(\cos\phi) \rangle = \int P_2(\cos\phi) f(\phi, \alpha) \, d\phi$$

$$= 1 - \frac{3 \coth\alpha}{\alpha} + \frac{3}{\alpha^2}. \quad 15.$$

Measurements of Δn as a function of B can be fitted with Equation 15 to yield the average moment (61). Bacteria cultured under particular conditions had an average magnetic moment of 10^{-13} emu.

Aquaspirillum magnetotacticum is bipolarly flagellated, that is, it has a flagellum at each end of the cell and can swim in either direction along the magnetic field lines. However, most other magnetotactic bacterial species in sediments are asymmetrically flagellated and have unidirectional motility. As noted above, these bacteria from New England swim along magnetic field lines in the field direction. Based on the passive orientation hypothesis, this occurs if the bacterial moment is oriented in the cell forward with respect to the flagellum (Figure 5). Then the bacterium will propel itself in the field direction when the moment is oriented in the field, and will be North-seeking in the geomagnetic field. If the bacterial moment were oriented in the cell rearward with respect to the flagellum, the cell would propel itself opposite to the field direction when the moment was oriented in the field, and hence would be South-seeking in the geomagnetic field.

South-seeking bacteria have been produced in the laboratory by subjecting North-seeking cells to magnetic pulses (38) or ac magnetic fields (10) that are strong enough to overcome the magnetic interaction forces between the particles in the chain and cause their moments to rotate and reorient along the chain the opposite direction. Pulse strengths of several hundred gauss are required, consistent with magnetic measurements on freeze-dried cells (19) and in agreement with estimates based on the "chain of spheres" model of Jacobs & Bean (32), who considered the magnetic properties of a chain of single domain particles in a different context before the discovery of magnetotactic bacteria.

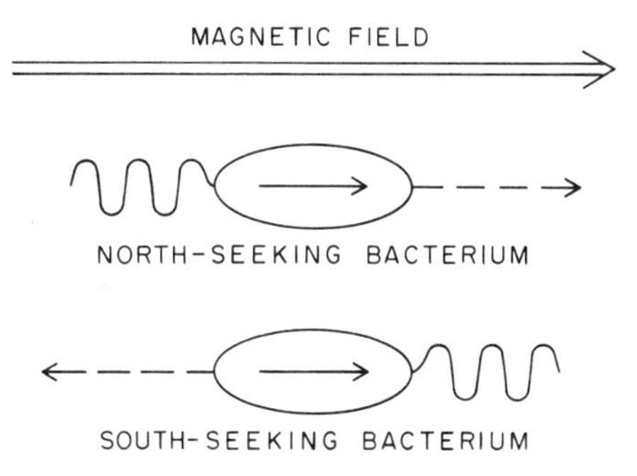

Figure 5 Illustration of North-seeking and South-seeking bacteria in a magnetic field. The solid arrows indicate the magnetic dipole moments of the cells that are due to the Fe_3O_4 particles. The arrowhead is the North-seeking pole. When the North-seeking pole is forward with respect to the flagellum, the bacterium will migrate in the field direction, or Northward in the geomagnetic field, i.e. the cell is North-seeking. If the South-seeking pole is forward, the bacterium will migrate opposite to the field direction, or Southward in the geomagnetic field, i.e. the cell is South-seeking.

Biological Advantage

The predominance of North-seeking bacteria in the Northern Hemisphere is due to the inclination of the geomagnetic field (8, 9). Since many sediment-dwelling bacteria are anaerobic or microaerophilic, it is advantageous for them to have mechanisms that prevent them from swimming up toward the toxic, higher oxygen concentration at the water surface and that keep them in the sediments. Since the geomagnetic field is approximately dipolar, the magnetic field lines at the earth's surface are inclined at an angle that increases with latitude. The total flux density at latitude θ is approximately

$$B_G = 0.3(\sin^2 \theta + 1)^{1/2} G, \qquad 16.$$

and the inclination I of the field from the horizontal is given by

$$\tan I = 2 \tan \theta. \qquad 17.$$

In the Northern Hemisphere the field is inclined downwards, pointing straight down at the North magnetic pole. In the Southern Hemisphere the field is inclined upwards, at an angle increasing with latitude, pointing straight up at the South magnetic pole. At the geomagnetic equator the field is horizontal.

Because of the inclination of the field lines, North-seeking bacteria migrate downward in the Northern Hemisphere and upward in the Southern Hemisphere (Figure 6). South-seeking bacteria migrate upward in the Northern Hemisphere and downward in the Southern Hemisphere. At the equator, both polarity types migrate horizontally. Because downward directed motion is advantageous, North-seeking bacteria should be favored

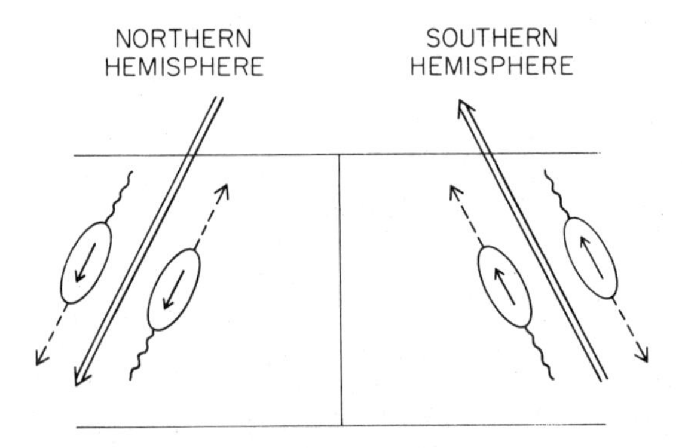

Figure 6 Because of the inclination of the geomagnetic field, North-seeking and South-seeking bacteria migrate downward and upward, respectively, in the Northern Hemisphere and upward and downward, respectively, in the Southern Hemisphere. At the geomagnetic equator, both polarities migrate horizontally.

in the Northern Hemisphere and South-seeking bacteria should be favored in the Southern Hemisphere. At the equator neither polarity would be favored.

Examination of bacteria in sediments from various places in the world confirms this hypothesis. In contrast to New England (inclination 70°N) and other Northern Hemisphere locales, magnetotactic bacteria in freshwater and marine sediments in Australia and New Zealand (inclination 70°S) are almost exclusively South-seeking (10, 41), as are bacteria in Rio de Janerio (inclination 25°S) (20, 23). These bacteria have chains of particles and can be remagnetized to North-seeking polarity (10). At the geomagnetic equator in Brazil (inclination 0°) both North-seeking and South-seeking bacteria are present in roughly equal numbers (23). Thus the vertical component of the geomagnetic field selects the predominant cell polarity in natural environments, with downward directed motion advantageous for, and upward directed motion detrimental to, survival of the organisms. At the geomagnetic equator where motion is directed horizontally, both polarities benefit because horizontally directed motion presumably reduces harmful upward migration. Examination of sediments on the coast of Brazil just South of the geomagnetic equator shows that South-seeking bacteria predominate when the magnetic inclination is greater than about 8° (F. F. Torres de Araujo et al, unpublished information).

The role of the vertical magnetic field component has also been confirmed in laboratory experiments (10, 23). When a sediment sample from New England, initially containing North-seeking bacteria, was placed in a coil that produced a field of twice the magnitude and opposite sign to the ambient vertical field, the polarity of the bacteria in the sample inverted over several weeks, that is over many bacterial generations. In a sample placed in a coil that canceled the vertical component of the ambient magnetic field, the population in the sample tended toward equal numbers of both polarities, again over many generations. Equal numbers of both polarities also resulted when samples initially containing all North- or all South-seeking bacteria were placed in an enclosure that canceled the ambient magnetic field. Further experiments in null field by Blakemore (8) confirmed the role of oxygen. When samples with tight stoppers were placed in the zero field enclosure, bacteria of both polarities were ultimately found in the sediment and in the water column up to the surface. When the sample bottles were loosely stoppered, allowing diffusion of air, bacteria were found in the sediments but not in the water column.

While the ability to synthesize Fe_3O_4 and construct magnetosomes is certainly genetically encoded, the polarity of the magnetosome chain cannot be encoded. If a bacterium that lacks magnetosomes starts to synthesize them *de novo*, there is equal probability that when the particles grow to permanent single domain size, the chain will magnetize with North-seeking pole forward as with South-seeking pole forward; a population of these bacteria will consist of 1:1 North-seekers and South-seekers. If however, the daughter cells inherit some of the parental magnetosomes during cell division, they will inherit the parental polarity. As they synthesize new magnetosomes at the ends of their inherited chains, the magnetic field produced by the existing particles will magnetize the new particles in the same orientation. Thus, North-seeking bacteria can produce North-seeking progeny and South-seeking bacteria can produce South-seeking progeny. This has been cited as a rudimentary example of "gene-culture transmission" (C. J. Lumsden, to be published). However, there are mechanisms by which some progeny with the opposite polarity can be produced in each generation (21). For example, if in the cell division process some of the daughter cells inherit no parental magnetosomes, these cells will synthesize them *de novo* and about one half those cells will end up with the polarity opposite to that of the parental generation. So in New England where North-seeking bacteria are found and predominate, some South-seekers are produced in each population division. Under normal circumstances, these South-seekers are unfavored by being directed upwards towards the surface, when they are separated from the sediments, and their total population remains low compared to the North-seeking population. However, when the vertical magnetic field is inverted, as in the experiment described above, these South-seekers are suddenly favored and their progeny eventually predominate as the previously favored North-seeking population declines in their newly unfavorable circumstances. When the vertical component is set equal to zero, neither polarity is favored and the North-seeking and South-seeking populations eventually equalize.

We can envision a similar process occurring in natural environments during reversals or excursions of the geomagnetic field. During these processes the vertical component changes sign over thousands of years. This would be accompanied by a change in the predominant polarity of the magnetotactic bacterial population in that locale. Models for the equilibrium polarity ratio as a function of geomagnetic field inclination have been proposed (N. Germano et al, unpublished information). In these models, even a small differential survival probability strongly favors one polarity over the other. This is consistent with the observation that in natural environments a geomagnetic inclination of 8° is sufficient to select one polarity (F. F. Torres de Araujo et al unpublished information).

Other possible advantages of rapid straight-line motion to magnetotactic bacteria might include population dispersal, escape, and outrunning chemical diffusion. There are also consequences of magnetotaxis and

Fe_3O_4 synthesis that may or may not be advantageous. Magnetic bacteria that are within 4 μm of each other will experience magnetic forces greater than the forces of Brownian motion. Fe_3O_4 synthesis also increases the density of the bacteria, helping them to stay down in the sediments even when they are not swimming, and may serve some metabolic functions as well (8).

The high density of Fe_3O_4 ($\rho = 5.1$ gms/cm^3) could conceivably play a role in geotaxis, whereby microorganisms would be oriented in the gravitational field so that they could swim upward or downward (E. Purcell, private communication). Imagine a cylindrical bacterium of length $2l = 4$ μm with a segment of Fe_3O_4 at one end. When suspended in water this hypothetical bacterium would tend to orient with the Fe_3O_4 segment downward. If the flagellum were opposite the Fe_3O_4 segment, the bacterium would migrate downward. The mass of Fe_3O_4 required can be estimated as follows. The orientational energy of the bacterium in the gravitational field

$$E_g = -V(\rho - \rho_0)lg \cos\theta \tag{18.}$$

where V is the volume and ρ is the density of the Fe_3O_4 segment, ρ_0 is the density of water, l is one half the length of the cell (approximately the distance between the center of the cell and the center of the Fe_3O_4 segment), g is the acceleration due to gravity and θ is the angle between the bacterium and the vertical direction. We assume that the density of the bacterium exclusive of the Fe_3O_4 segment is the same as water. Then the thermally averaged orientation

$$\langle \cos\theta \rangle = L(\beta) = \coth(\beta) - \frac{1}{\beta}, \tag{19.}$$

where $L(\beta)$ is the Langevin function and

$$\beta = V(\rho - \rho_0)lg/kT. \tag{20.}$$

Arbitrarily defining "good" average orientation as $\langle \cos\theta \rangle \geq 0.8$, means $\beta \geq 10$. With $\rho - \rho_0 = 4.1$, $T = 300$ K, and $l = 2$ μm, this requires $V \sim 5 \times 10^{-13}$ cm^3, or 2.5×10^{-12} g of Fe_3O_4. A smaller value of l would require an even larger mass of Fe_3O_4. For comparison, *A. magnetotacticum* typically contains 1.25×10^{-14} g of Fe_3O_4, which is sufficient for orientation ($\langle \cos\theta \rangle \geq 0.9$) in the geomagnetic field. Hence in *A. magnetotacticum* neither is there enough Fe_3O_4, nor is the Fe_3O_4 present in the cell appropriately distributed to function in geotaxis. This calculation illustrates that Fe_3O_4 is typically five hundred time more efficient for magnetotaxis than for geotaxis in microorganisms.

Biomineralization of Fe_3O_4

Progress has been made in elucidating the Fe_3O_4 biomineralization process. On the basis of extensive Mössbauer spectroscopic analysis, it has been proposed that *A. magnetotacticum* precipitates Fe_3O_4 in the sequence Fe^{3+} quinate → Fe^{2+} → low density hydrous ferric oxide → high density hydrous ferric oxide (ferrihydrite) → Fe_3O_4 (25). In nonmagnetic variants the process stops with high density hydrous ferric oxide (ferrihydrite) or with low density hydrous ferric oxide. In the proposed process, iron enters the cell as Fe^{3+} chelated by quinic acid. Reduction to Fe^{2+} releases iron from the chelator. Fe^{2+} is reoxidized and accumulated as the low density hydrous iron oxide. By analogy with the deposition of iron in the micellar cores of the protein ferritin (18), this oxidation step might involve molecular oxygen, which is required for Fe_3O_4 precipitation in *A. magnetotacticum* (D. Bazylinski and R. P. Blakemore, unpublished information). Dehydration of the low density hydrous ferric oxide results in ferrihydrite. Finally, partial reduction of ferrihydrite and further dehydration yields Fe_3O_4.

Mössbauer spectroscopy measurements of whole cells above freezing temperatures show that diffusive motions of the Fe_3O_4 particles are small, that is, the particles are effectively fixed in the cell. The ferrihydrite is associated with the magnetosome chain, but Fe^{2+} is located elsewhere in the cell, possibly in the cell wall (S. Ofer et al, to be published).

Fe_3O_4 is thermodynamically stable with respect to hematite and ferrihydrite at low E_H and high pH (26). It is known that magnetite can be precipitated by the controlled addition of O_2 to mixed ferrous and ferric hydroxides (green rust) (49) or by the addition of Fe^{2+} ions to lepidocrocite (γ-FeOOH) (64). The latter process is a surface reaction involving dissolution of the γ-FeOOH and reprecipitation of Fe_3O_4. Lepidocrocite has not been detected in *A. magnetotacticum*, but a similar reaction might occur with ferrihydrite, especially on poorly crystallized or amorphous particles. In any case, the fact that the precipitation process requires control of the size and location of the Fe_3O_4 as well as probable spatial segregation of regions of differing E_H and possibly pH in the cell suggests that the precipitation process is "matrix mediated" (48, 72). The magnetosome envelope is probably an integral element in the precipitation process and may function as a locus for enzymatic activities, including control of E_H and pH, as well as a structural element.

Reduction of a ferrihydrite precursor to Fe_3O_4 also occurs in the marine chiton, a mollusk of the genus *Polyplacophora* (65). In this organism the radular teeth undergo a sequential mineralization process that results in a surface coating of Fe_3O_4. Iron is transported to the superior epithelial cells of the radula as ferritin. Then iron is transferred to a preformed organic matrix on the tooth surface as ferrihydrite. Finally, the ferrihydrite is reduced to Fe_3O_4. Thus the Fe_3O_4 precipitation processes in chitons and in magnetotactic bacteria appear to be similar.

CONCLUSION

In conclusion, magnetotactic bacteria precipitate Fe_3O_4. The precipitation process is controlled to produce the magnetosome chain, a highly structured magnetic element in the cell that results in the migration of the bacterium along geomagnetic field lines. The vertical component of the geomagnetic field is the important element in determining the predominant polarity of bacteria in natural environments, with downward directed motion favored over upward directed motion. Downward directed motion guides the bacteria to, and keeps them in, their preferred habitat, the relatively anoxic sediments. In fact, the low O_2 environment appears to be a requirement for Fe_3O_4 precipitation.

The fact that the magnetic sensitivity of magnetotactic bacteria is based on Fe_3O_4 suggests that the magnetic sensitivity of some complex organisms could also be connected with Fe_3O_4 precipitates. Of course, the passive orientation mechanism is suitable only for motile microorganisms. In larger organisms Fe_3O_4 would be part of a sensory system. The detailed anatomical and physiological description of such systems remains for future research.

ACKNOWLEDGMENTS

I gratefully acknowledge the continued collaboration of Richard Blakemore in the study of magnetotactic bacteria. I am also grateful to many colleagues and friends for contributions and discussions. G. Lynch provided editorial assistance. This work was partially supported by the Office of Naval Research. The Francis Bitter National Magnet Laboratory is supported by the National Science Foundation.

Literature Cited

1. Able, K. P. 1980. In *Animal Migration, Orientation and Navigation*, ed. S. A. Gauthreaux, pp. 283–373. New York: Academic
2. Baker, R. R. 1980. *Science* 210:555–57
3. Baker, R. R. 1981. *Human Navigation and the Sixth Sense*. London: Hodder & Stoughton. 310 pp.
4. Baker, R. R., Mather, J. G., Kennaugh, J. H. 1983. *Nature* 301:78–80
5. Balkwill, D. L., Maratea, D., Blakemore, R. P. 1980. *J. Bacteriol.* 141:1399–1408
6. Bean, C. P., Livingston, J. D. 1959. *J. Appl. Phys.* 30:120S–29S
7. Blakemore, R. P. 1975. *Science* 190:377–79
8. Blakemore, R. P. 1982. *Ann. Rev. Microbiol.* 36:217–38
9. Blakemore, R. P., Frankel, R. B. 1981. *Sci. Am.* 245(6):58–65
10. Blakemore, R. P., Frankel, R. B., Kalmijn, A. J. 1981. *Nature* 286:384–85

11. Blakemore, R. P., Maratea, D., Wolfe, R. S. 1979. *J. Bacteriol.* 140:720–29
12. Bookman, M. A. 1977. *Nature* 267:340–42
13. Brown, F. A. Jr., Barnwell, F. H., Webb, H. M. 1964. *Biol. Bull.* 127:221–31
14. Brown, F. A. Jr., Park, Y. H. 1965. *Biol. Bull.* 128:347–55
15. Brown, H. R., Ilyinsky, O. B., Muravejko, V. M., Corshkov, E. S., Fonarev, G. A. 1979. *Nature* 277:648–49
16. Buskirk, R. E. 1981. *Trans. Am. Geophys. Union* 62:850
17. Butler, R. F., Banerjee, S. K. 1975. *J. Geophys. Res.* 80:4049–58
18. Clegg, G. A., Fitton, J. E., Harrison, P. M., Treffry, A. 1980. *Mol. Biol.* 36:56–80
19. Denham, C. R., Blakemore, R. P., Frankel, R. B. 1980. *IEEE Trans. Magn.* MAG-16:1006–7
20. Esquivel, D. M. S., Lins de Barros, H. G. P., Farina, M., Aragao, P. H. A., Danon, J. 1983. *Biol. Cell.* 47:227–34
21. Frankel, R. B. 1982. *Comments Mol. Cell. Biophys.* 1:293–310
22. Frankel, R. B., Blakemore, R. P. 1980. *J. Magn. Magn. Mater.* 15–18:1562–44
23. Frankel, R. B., Blakemore, R. P., Torres de Araujo, F. F., Esquivel, D. M. S., Danon, J. 1981. *Science* 212:1269–70
24. Frankel, R. B., Blakemore, R. P., Wolfe, R. S. 1979. *Science* 203:1355–56
25. Frankel, R. B., Papaefthymiou, G. C., Blakemore, R. P., O'Brien, W. D. 1983. *Biochim. Biophys. Acta* 763:147–59
26. Garrels, R. M., Christ, C. L. 1965. *Solution, Minerals and Equilibria.* New York: Harper & Row. 450 pp.
27. Gilbert, W. 1600. *De Magnete Magneticisque Corporibus et de Magne Magnete Tellure.* London: Petrus Short. 240 pp.
28. Gould, J. L. 1980. *Am. Sci.* 68:256–67
29. Gould, J. L. 1981. *Nature* 296:205–11
30. Gould, J. L., Able, K. P. 1981. *Science* 212:1061–63
31. Gould, J. L., Kirschvink, J. L., Deffeyes, K. S. 1978. *Science* 201:1026–28
32. Jacobs, I. S., Bean, C. P. 1955. *Phys. Rev.* 100:1060–67
33. Jones, D. S., MacFadden, B. J. 1982. *J. Exp. Biol.* 96:1–9
34. Jungerman, R. L., Rosenblum, B. 1980. *J. Theor. Biol.* 87:25–32
35. Kalmijn, A. J. 1974. In *Handbook of Sensory Physiology*, ed. A. Fessard, 3(3):147–200. Berlin: Springer-Verlag
36. Kalmijn, A. J. 1981. *IEEE Trans. Magn.* MAG-17:1113–23
37. Kalmijn, A. J. 1982. *Science* 218:916–18
38. Kalmijn, A. J., Blakemore, R. P. 1978. In *Animal Migration, Navigation and Homing*, ed. K. Schmidt-Koenig, W. T. Keeton, pp. 354–55. New York: Springer-Verlag
39. Keeton, W. T. 1971. *Proc. Natl. Acad. Sci. USA* 68:102–6
40. Keeton, W. T., Larkin, T. S., Windsor, D. M. 1974. *J. Comp. Physiol.* 95:95–103
41. Kirschvink, J. L. 1980. *J. Exp. Biol.* 86:345–47
42. Kirschvink, J. L. 1981. *BioSystems* 14:193–203
43. Kirschvink, J. L. 1981. *J. Exp. Biol.* 92:333–35
44. Kirschvink, J. L., Gould, J. L. 1981. *BioSystems* 13:181–201
45. Kuterbach, D. A., Walcott, B., Reeder, R. J., Frankel, R. B. 1982. *Science* 218:695–97
46. Lins de Barros, H. G. P., Esquivel, D. M. S., Danon, J., de Oliveira, L. P. H. 1982. *An. Acad. Bras. Cienc.* 54:258–59
47. Lowenstam, H. A. 1962. *Geol. Soc. Am. Bull.* 73:435–38
48. Lowenstam, H. A. 1981. *Science* 211:1126–31
49. Mackay, A. L. 1960. In *Reactivity of Solids*, ed. J. H. DeBoer, pp. 571–83. Amsterdam: Elsevier
50. Martin, H., Lindauer, M. 1977. *J. Comp. Physiol.* 122:145–88
51. Mather, J., Baker, R. 1981. *Nature* 291:152–54
52. Matsuda, T., Endo, J., Osakabe, N., Tonomura, A., Arii, T. 1983. *Nature* 302:411–12
53. Moench, T. T., Konetzka, W. A. 1978. *Arch. Microbiol.* 119:203–12
54. Morrish, A. H. 1968. *The Physical Principles of Magnetism.* New York: Wiley. 680 pp.
55. Needham, J. 1962. *Science and Civilization in China*, 4(1):231–334. Cambridge: Cambridge Univ. Press
56. Perry, A., Bauer, G. B., Dizon, A. S. 1981. *Trans. Am. Geophys. Union* 62:850
57. Phillips, J. B. 1977. *J. Comp. Physiol. A.* 121:273–88
58. Presti, D., Pettigrew, J. D. 1980. *Nature* 285:99–101
59. Purcell, E. M. 1977. *Am. J. Phys.* 45:3–10
60. Quinn, T. P. 1980. *J. Comp. Physiol.* 137:243–48
61. Rosenblatt, C., Torres de Araujo, F. F., Frankel, R. B. 1982. *Biophys. J.* 40:83–85
62. Rosenblatt, C., Torres de Araujo, F. F., Frankel, R. B. 1982. *J. Appl. Phys.* 53:2727–29
63. Stoner, E. C. 1972. *Encyclopedia Brittanica*, 14:584–86. Chicago: Benton
64. Tamaura, Y., Ho, K., Katsura, T. 1983. *J. Chem. Soc. Dalton Trans.* 189–94
65. Towe, K. M., Lowenstam, H. A. 1967. *J. Ultrastruct. Res.* 17:1–13
66. Towe, K. M., Moench, T. T. 1981. *Earth Planet. Sci. Lett.* 52:213–20

67. Walcott, C. 1978. See Ref. 38, pp. 143–51
68. Walcott, C. 1980. *IEEE Trans. Magn.* MAG-16: 1008–13
69. Walcott, C., Gould, J. L., Kirschvink, J. L. 1979. *Science* 205: 1027–29
70. Walcott, C., Green, R. 1974. *Science* 184: 180–82
71. Walker, M. M., Dizon, A. S. 1981. *Trans. Am. Geophys. Union* 62: 850
72. Webb, J. 1983. In *Biomineralization and Biological Metal Accumulators*, ed. P. Westbroek, E. W. de Jong, pp. 413–22. New York: Reidel
73. Wiltschko, W. 1978. See Ref. 38, pp. 302–10
74. Wiltschko, W., Wiltschko, R. 1972. *Science* 176: 62–64
75. Yorke, E. D. 1981. *J. Theor. Biol.* 89: 533–37
76. Zoeger, J., Dunn, J. R., Fuller, M. 1981. *Science* 213: 892–94

Ann. Rev. Biophys. Bioeng. 1984. 13: 105–24

MULTIFREQUENCY PHASE AND MODULATION FLUOROMETRY[1]

Enrico Gratton

Department of Physics, University of Illinois at Urbana-Champaign, Urbana, Illinois 61801

David M. Jameson

Department of Pharmacology, The University of Texas Health Science Center at Dallas, Dallas, Texas 75235

Robert D. Hall

Laboratory of Molecular Biophysics, National Institute of Environmental Health Sciences, Research Triangle Park, North Carolina 27709

INTRODUCTION

General Aspects

Determination of characteristic fluorescence parameters i.e. spectral properties, quantum yields, polarizations, and lifetimes, is important in the study of excited states of atoms, molecules, and crystals and also for diagnostic purposes in the physical, chemical, biological, and medical sciences. The appeal of fluorescence methodologies lies in their intrinsic sensitivity and also in the characteristic time scale of the emission process. The excited fluorescent state typically persists for nanoseconds or less, a period that corresponds to the time scale of many important biological processes including diffusion or transport of small molecules over a few angstroms, rotational motions of proteins, internal motions of protein residues, proton transfer reactions, and others. Steady state fluorescence data such as spectra, quantum yields, or polarizations are not always capable of revealing the molecular origins of the observed effects. The correct physical interpretation of many fluorescence experiments requires precise knowledge of the excited state lifetime (59). Moreover, emissions from even simple biological systems will almost always have heterogeneous aspects. This heterogeneity may arise as a consequence of the number and types of fluorophores present or from the complexities of the fluorophore's environment. The characterization of heterogeneous emissions and the assignment of their origins is one of the more challenging problems facing the fluorescence practitioner, and time-resolved measurements offer a powerful means of examining and quantitating these emissions. In this review we describe recent advances of a technique frequently employed for the determination of fluorescence lifetimes and some relevant applications to the biophysical field.

A tremendous effort has been expended during the past twenty years on the development of instrumentation and theory for the measurement and analysis of fluorescence lifetimes. Most of this effort has gone into the development of the impulse response technique, which yields direct recordings of the fluorescence intensity versus time after a brief exciting pulse (1, 58). The harmonic response technique, although practiced in one form or another for more than half a century, has only recently become a popular and powerful alternative to pulse methods. In the traditional harmonic response approach, fluorescence is excited by light with an intensity modulated sinusoidally at high frequencies, typically in the megahertz range. In this case, the fluorescence signal will also be modulated sinusoidally, but the finite persistence of the excited state will lead to a phase delay and demodulation of the fluorescence relative to the excitation (see Figure 1). Measurement of the phase delay and modulation ratio provides independent determinations of the fluorescence lifetime (52). The very recent appearance of true multifrequency phase fluorometry has enormously extended the scope and power of the harmonic response technique. In this review we discuss the theoretical basis of multifrequency phase fluorometry, the practical realization of the method, phase and modulation data, and the application of the method to spectroscopic and biochemical problems.

We must clarify at the outset what we mean by multifrequency phase fluorometry. A true multifrequency instrument permits facile selection of arbitrary modulation frequencies over a wide frequency range, e.g. from hundreds of kilohertz to hundreds of megahertz. The aim, however, is to utilize the optimal frequencies for the particular problem in hand. The

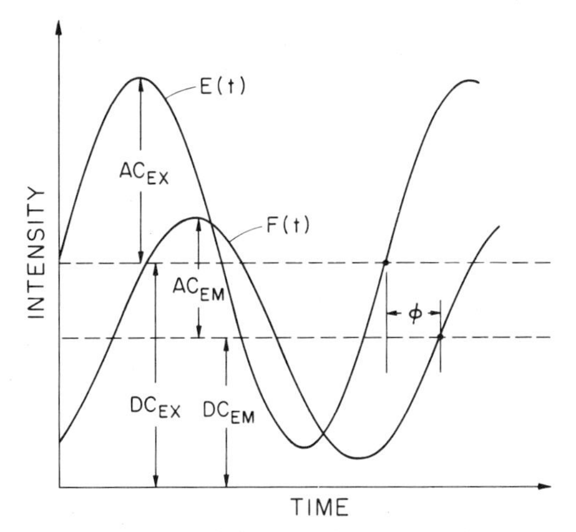

Figure 1 Schematic representation of the excitation E(*t*) and fluorescence F(*t*) waveforms. Fluorescence is delayed by an angle ϕ and demodulated with respect to excitation.

characteristics of the fluorescence system, i.e. the time scale of the emission process, and not instrumental limitations, should dictate the number and choice of frequencies utilized. The criteria for frequency selection is discussed in detail in a later section.

BASIC PRINCIPLES

The theory of the phase fluorometer was developed by Dushinsky (11). He demonstrated that a fluorophore characterized by a single exponential decay time τ will, upon excitation by light with an intensity modulated sinusoidally at an angular frequency, ω, emit light sinusoidally modulated at the same frequency but delayed in phase and demodulated with respect to the excitation (Figure 1). The demodulation is the ratio of the signal amplitude at frequency ω to the average signal (the AC/DC ratio). The relations between phase shift and modulation ratio and the characteristic time τ are

$$\phi = \tan\,(\omega\tau^{\mathrm{P}}), \quad M = (1+(\omega\tau^{\mathrm{M}})^2)^{-1/2}, \qquad 1.$$

which provide the basis of two independent determinations of the fluorescence lifetime, i.e. phase (τ^{P}) and modulation (τ^{M}) lifetimes.

If the system described above is replaced by a system of noninteracting fluorophores, each giving rise to a single exponential decay, then a composite sinusoidal emission waveform will result, with a frequency ω and a phase delay and demodulation given by

$$\phi = \tan^{-1}\,(S/G), \quad M = (S^2+G^2)^{-1/2}, \qquad 2.$$

where

$$S = \sum_i f_i M_i \sin \phi_i, \quad G = \sum_i f_i M_i \cos \phi_i. \tag{3.}$$

The values of ϕ_i and M_i for each component are given by Equation 1; f_i is the fractional intensity of the ith component ($\sum_i f_i = 1$). In such a system $\tau^P \neq \tau^M$ and τ^P and τ^M are frequency dependent. The functions S and G are the sine and cosine transforms of the impulse response (60). If $I(t)$ represents the free decay after excitation then

$$S = \int_0^\infty I(t) \sin(\omega\tau)dt \Big/ \int_0^\infty I(t)dt,$$

$$G = \int_0^\infty I(t) \cos(\omega\tau)dt \Big/ \int_0^\infty I(t)dt. \tag{4.}$$

These equations provide the basis for the mathematical equivalence between the harmonic and impulse response methods and are useful from the theoretical viewpoint. Certain kinetic decay schemes that are difficult to describe directly in the frequency domain are amenable to time domain formulation. Equation 4 then provides the means to obtain the frequency response of such kinetic schemes. Direct transformation of data between the frequency and time domains is not possible in practice, however, since the actual frequency and time ranges utilized do not extend from zero to infinity, which precludes exact evaluation of the integrals. A critical comparison of practical differences between the two approaches as well as a discussion of the validity of Equation 4 are presented in Reference (28).

INSTRUMENTATION

Brief History

The development of phase fluorometry has been reviewed in varying detail and from different perspectives several times (e.g. see 6, 28, 53). Phase fluorometers trace their ancestry back to Gaviola's instrument (13), which utilized the Kerr effect to modulate the exciting light and relied upon visual detection of the fluorescence. Some of the important developments in phase fluorometry in the last half century include ultrasonic diffraction grating (39, 52), the use of photomultipliers (46), and measurement of the modulation ratio (4, 21, 52). Many of the early instruments featured direct measurements of the phase delay at the modulation frequency using devices such as oscilloscopes, RC phase compensators, cable delays, and light path compensators (2, 3, 5, 45). Heterodyne techniques permitted the transposition of the signal to the low frequency domain, with subsequent improvement in sensitivity and precision. The advantages of cross-correlation techniques (discussed in detail later) were demonstrated by Spencer & Weber (52). Early cross-correlation techniques, however, were limited to a few modulation frequencies (49, 52). The modern approach, utilizing a wide and continuously variable frequency range, has realized the full potential of the harmonic method. An excellent account of the development of phase and modulation instrumentation, including the heterodyne and cross-correlation techniques, has been given by Teale (53).

In recent years a renewed interest in phase fluorometry has led to the appearance of a number of instruments (15, 17–19, 25, 38, 47). The more recent instruments differ in their approach to the light modulation principle and the strategy for signal detection and processing, but almost all utilize a CW laser for sample excitation. As we shall see later, synchrotron radiation is also now being utilized for phase and modulation fluorometry. The appearance of commercially available phase and modulation fluorometers, such as the SLM 4800 series, has served to popularize the technique in the biochemical and biophysical fields.

Modern Multifrequency Instrumentation

A multifrequency phase fluorometer using the method of cross-correlation has been recently described (15). Figure 2 shows the layout of this instrument, which was used to obtain much of the data presented in this review. This particular instrument operates over a frequency range of 1 to 160 MHz and gives a time resolution of several picoseconds. The light source is an argon-ion laser and the light modulator is a Pockel's cell. A frequency synthesizer provides the modulated signal to drive the Pockel's

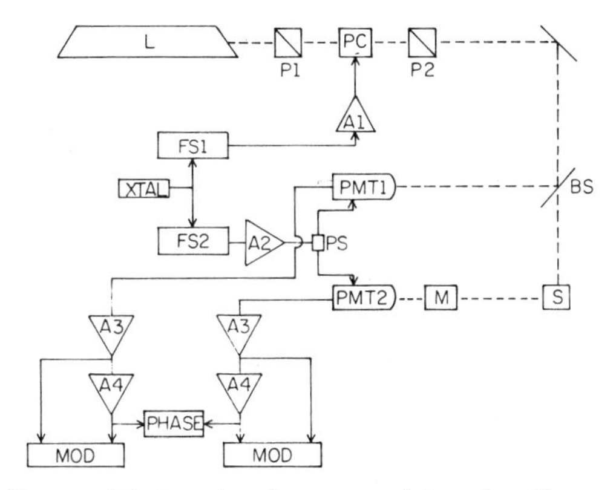

Figure 2 Block diagram of the laser-based cross-correlation phase fluorometer of Reference (15). L, argon ion laser; A1, 10 W rf power amplifier; A2, 2W rf power amplifier; XTAL, 10 MHz quartz oscillator; FS1 and FS2, frequency synthesizers; PS, power splitter; PMT1 and PMT2, photomultipliers; M, monochromator; BS, beam splitter; S, sample; A3, pre-amplifier; A4, ac tuned amplifier; PHASE, phase meter; MOD, modulation meter.

cell. A second frequency synthesizer, locked in phase with the first, provides the signal that modulates the photomultiplier's response by varying the voltage at the last dynode. The two synthesizers differ in frequency by a small increment, 31 Hz, which corresponds to the cross-correlation frequency. The optical components are standard; all polarizers are calcite prisms.

In a typical measurement the phase delay and modulation ratio for scattered light (from glycogen or a suspension of latex particles) is determined relative to the signal generated by a reference photomultiplier (see Figure 2) or an internal electronic reference signal. The phase delay and modulation ratio of the fluorescence is then determined relative to the scattered signal (Figure 1). These measurements are repeated at selected modulation frequencies; typical results for fluorophores characterized by single exponential decay and the usual mode of data presentation are shown in Figure 3. Figure 4 shows data for a double exponentially decaying system and the result of a two-component analysis; in this presentation the lifetime values are utilized rather than phase and modulation values. Precision of phase and modulation data is on the order of 0.1° and 2 parts per thousand respectively. Phase and modulation lifetimes are calculated according to Equation 1. Curves corresponding to various decay schemes can be plotted over the data and used to evaluate the fit of the data to a particular scheme.

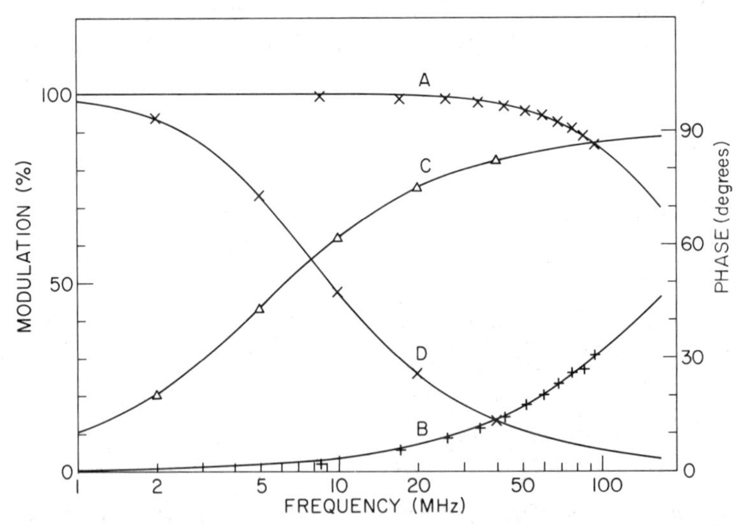

Figure 3 Multifrequency phase (+, Δ) and modulation (×) data for DENS (2,5-diethylaminonaphthalene sulfonate) in water (curves C and D) and *p*-terphenyl (curves A and B) in cyclohexane. Solid lines correspond to single exponential decays of 29.28 nsec for DENS and 980 psec for *p*-terphenyl.

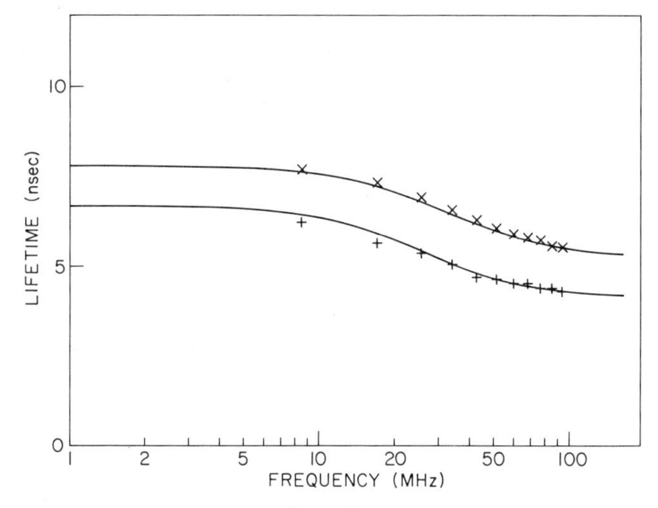

Figure 4 Multifrequency lifetime data [$\tau^{P}(+), \tau^{M}(\times)$] for tryptophan at 20°C, pH 9.25. Solid lines correspond to the best fit using two exponential components $\tau_1 = 3.193 \pm 0.026$ nsec, $\tau_2 = 9.00$ nsec, and $f_1 = 0.396 \pm 0.005$. As noted in the text $\tau^{P} < \tau^{M}$ for heterogeneous emission of independent species. Also τ^{P} and τ^{M} reach constant values at the extremities of the frequency range [see Reference (28) for a discussion of this effect].

ANALYSIS OF FLUORESCENCE DECAYS

One of the most important features of multifrequency phase and modulation fluorometry is the ability to analyze emission processes other than single exponential decays. Several methods have been proposed for analysis of heterogeneous emissions in terms of the sum of exponential components. Weber obtained the exact solution for the derivation of *N* exponential lifetime components and their fractional intensities, given phase and modulation data at *N* modulation frequencies (61). Methods based upon nonlinear least-squares statistical analysis have also been described (27, 28). A critical comparison of the exact and the statistical approaches has revealed that the exact solution requires very high precision in the data and that the uncertainty in the derived parameters increases with the number of measurement frequencies, in contrast to the statistical approach (27). We should emphasize, however, that the exact solution was developed for applications wherein only a few modulation frequencies are available.

Analysis of phase and modulation data has the following two important aspects. We must first recognize which decay schemes are permissible (single exponential, multiexponential, or nonexponential) using well-established statistical methods, based on analysis of the chi-square. Once we adhere to a particular decay scheme, we must evaluate the uncertainty of the derived parameters and the correlation between these parameters. Also in this case, statistical methods furnish a solution based on analysis of the covariance matrix of the errors associated with the parameters (7).

Coincidence of phase and modulation lifetimes at all modulation frequencies is a necessary and sufficient condition for correct description of a decay as single exponential. Since measured quantities have an associated error, however, and since the frequency range is limited in practice, one cannot rigorously demonstrate the mathematical equivalence of phase and modulation lifetimes but rather must consider this equivalence within the precision of the measurement.

When phase and modulation lifetimes differ, the data must be analyzed according to a more complex scheme, e.g. multiexponential decay. Emitting systems that can properly be described by a double exponential decay include systems composed of two independent emitting species, systems that demonstrate excited state reactions between well-defined states, systems exhibiting energy transfer between two species, isotropic rotators observed through appropriately oriented polarizers, and others. Except for the systems composed of two independently emitting species, the parameters associated with the two-component decay cannot be assigned to particular molecular entities.

Analysis of a double exponential emission gives three independent parameters: two lifetime values and one fractional intensity contribution. We may inquire as to the resolvability of lifetime pairs, given a particular experimental precision and range of modulation frequencies. Our criterion for resolvability is that the derived lifetimes be distinct within their associated errors. Different criteria for resolvability will, of course, lead to different results. The uncertainty in the values of the derived parameters can be determined from the covariance matrix of the errors (discussion of the covariance matrix can be found in any good statistics text; e.g. see 7). In our analysis we assume that the experimental errors in the measured phase and modulation values are independent of modulation frequency [see Reference (15) for a discussion of statistical and systematic errors in phase fluorometry]. In such a case, each term in the covariance matrix has an associated error that becomes a common factor upon evaluation of the matrix. The covariance matrix is then a linear function of the errors associated with the measured phase and modulation values and depends upon the values of the parameters (τ_1, τ_2, and f_1) and the frequency set utilized. Furthermore, if the angular modulation frequency range utilized is sufficient to encompass the inverse of the lifetimes of the two components, then the covariance matrix will also be largely independent of the frequency set. The important variables are then the number of frequencies utilized and the values of the parameters. In the case of a double exponential decay with each component contributing equally to the integrated intensity ($f_1 = f_2$), we can calculate the minimum ratio of the lifetimes that permit resolution of the components. In Figure 5 we report the results of this analysis for a set of eight frequencies (1, 2, 4, 8, 16, 32, 64, and 128 MHz) with errors of $\pm 0.2°$ and ± 0.004 associated with phase and modulation data, respectively. An interesting feature of this analysis is that the width of the nonresolvability zone remains approximately constant from 1 to 500 nsec. Two components (with $f_1 = f_2$) in this lifetime range can be resolved if the ratio of the lifetimes is approximately 1.6. As we have already suggested, the resolvability ratio is proportional to the precision of the measurements.

The method used for the resolvability analysis has general applicability, and other cases with $f_1 \neq f_2$ can be readily evaluated. In Figure 6 we analyze the dependence of the resolvability ratio on the number of frequencies utilized. The errors on the parameters are proportional to the reciprocal of a fractional power (ranging from 1/2 to 1/3 in all cases studied) of the number of frequencies. This analysis demonstrates that decreasing the errors in the parameters by a factor of two (to subsequently improve the resolvability ratio twofold) requires that the number of modulation frequencies be increased by a factor ranging from 4 to 8. The use of a large number of modulation frequencies is impractical and unnecessary and, except for special applications, 3 to 10 frequencies scaled logarithmically in the range of 1 to 200 MHz suffice for practical purposes. A frequency set such as 1, 2, 4, 8, 16, 32, 64, 128, and 256 MHz is ideal.

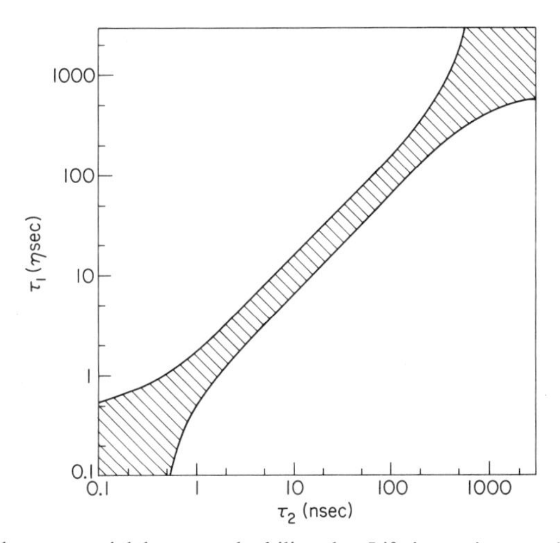

Figure 5 Double exponential decay resolvability plot. Lifetime pairs τ_1 and $\tau_2 (f_1 = f_2)$ in the shaded region cannot be resolved using the frequency set 1, 2, 4, 8, 16, 32, 64, and 128 MHz if the precision of phase and modulation measurements is $>0.2°$ and >0.004 respectively.

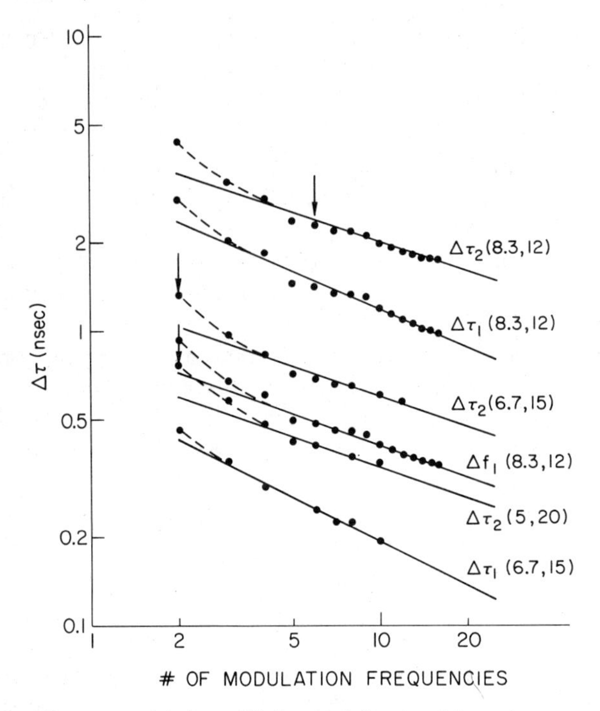

Figure 6 Double exponential decay lifetime and fractional intensity uncertainties as a function of the number of frequencies utilized for the measurements. In parentheses selected pairs of lifetime in nanoseconds ($f_1 = f_2$). Arrows indicate the minimum number of frequencies needed. Frequencies are chosen equally spaced on a log scale in the range 1 to 256 MHz.

In some cases of two independently emitting species, the lifetimes of the two components may be known already and the fractional contributions may be the unknown quantities. In such cases, analysis of the covariance matrix of the errors, similar to that performed for the previous case, demonstrates the possibility of obtaining the fractional contributions with high precision (0.7–0.8% error), using a single modulation frequency. The optimum modulation frequency is that which corresponds to the average (in the log scale) of the frequencies given by the inverse of the two lifetimes. Figure 7 shows the error on the fractional intensity as a function of the modulation frequency for a given lifetime pair. This analysis is pertinent to the consideration of the optimum modulation frequency for use with phase-sensitive detection techniques (discussed later). The small error on the fractional intensities justifies the use of a single modulation frequency to obtain phase-resolved spectra of individual components in a mixture when the two lifetimes are given.

Having outlined the general rules for resolvability of a double exponential decay, we may now consider the application of multifrequency phase and modulation data to other emission decay schemes.

1. The emission from a system undergoing an excited state reaction between two well-defined states will be double exponential, but the two lifetime values cannot be assigned to either of the reacting species. Such a system can be described by six parameters: two radiative decay rates for each molecular species, the forward and reverse reaction rates, the ratio of extinction coefficients of each species at excitation wavelength, and the ratio of fluorescence intensities of each species at emission wavelength. Since analysis of a double exponential decay yields only three independent parameters, the system cannot be analyzed in the absence of simplifying considerations. Simplification is often possible, however, through judicious choice of excitation and emission wavelengths to minimize the contribution of one species. Also, in many cases some of the rates are quite large compared to others. Hence, the number of parameters can often be reduced and the system fully determined. The form of the equations, applicable in phase fluorometry, for excited state reactions has been given (34) and can also be obtained from the impulse response using Equation 4. Systems that fall in the general category of excited state reactions include collisional quenching of the fluorescence, energy transfer reactions, excimer formation, and proton transfer reactions.

2. Measurements on the decay of emission anisotropy will yield a double exponential decay for the case of an isotropic rotator and for some other rotational modes as well (60). The harmonic response of these systems

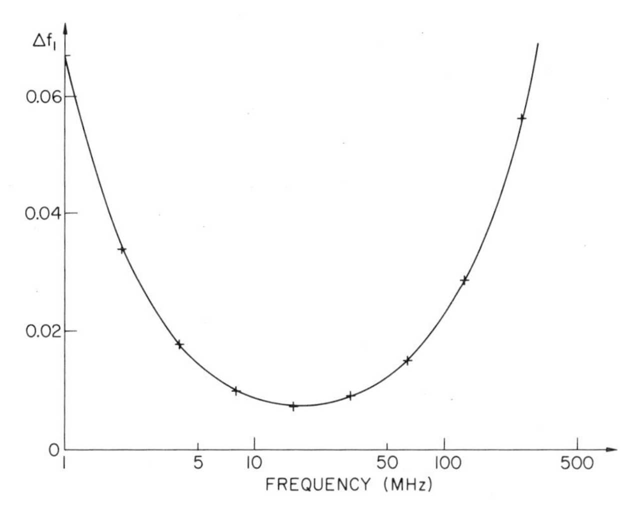

Figure 7 Frequency dependence of fractional intensity uncertainty for a double exponential decay ($\tau_1 = 6.7$ nsec, $\tau_2 = 15$ nsec, and $f_1 = 0.5$).

can be derived from the impulse response through the application of Equation 4.

Differential Methods

Phase and modulation fluorometry is inherently a differential technique, since the phase delay and modulation ratio of the emission is measured relative to a reference. In some cases a convenient choice of the reference can simplify the analysis and improve the precision of the measurement. We have pointed out that in the case of an interconverting system the decay rate is a property of the system rather than corresponding to particular molecular entities. If one can isolate the emission from each species of interest, for example by proper selection of the excitation or emission wavelengths or appropriately oriented polarizers, then the phase delay between the two species, and their relative modulation ratios, can be directly determined. Weber (60) has derived general expressions for the differential tangent and modulation ratio corresponding to two emissions, each of which is represented by a sum of exponentials:

$$\Delta = \tan(\phi_2 - \phi_1), \quad Y = M_2/M_1, \tag{5.}$$

where ϕ_1, M_1, and ϕ_2, M_2 are the phase delay and modulation ratio of the emission from molecular species 1 and 2 respectively and are given by expressions such as Equation 2.

1. For the case of two independent species one finds

$$\Delta = \omega(\tau_2 - \tau_1)/(1 + \omega^2\tau_1\tau_2), \quad Y^2 = [1 + (\omega\tau_1)^2]/[1 + (\omega\tau_2)^2]. \tag{6.}$$

These expressions are often used when the phase and modulation of a sample are measured relative to a fluorescence reference signal in place of a scatter solution in order to minimize certain systematic errors such as wavelength-related time response of the photomultiplier. The shape of the differential tangent versus frequency curve corresponds to a Lorentzian with a maximum:

$$\omega^2_{max} = 1/(\tau_1\tau_2). \tag{7.}$$

2. For the case of the parallel and perpendicular components of the emission from an isotropic rotator, excited with parallel polarized light, one obtains

$$\Delta = (3\omega rR)/[(k^2 + \omega^2)(1 + r - 2r^2) + R(R + 2k + kr)],$$
$$Y^2 = \{[k + 6R/(1 - r)]^2 + \omega^2\}/\{[k + 6R/(1 + 2r)]^2 + \omega^2\}, \tag{8.}$$

where r is the limiting anisotropy, R the rotational rate, and k the radiative decay rate.

In this case also, the differential tangent versus frequency curve is a Lorentzian with a maximum given by

$$\omega_{max}^2 = k^2 + R(R + 2k + kr)/(1 + r - 2r^2). \qquad 9.$$

For anisotropic rotators, the frequency dependence of the differential tangent curve has maxima corresponding to each individual rotation rate, and the absolute value of each maximum is less than that corresponding to an isotropic rotator, a difference termed the tangent defect (60). Examples of isotropic and more complex rotations are given in the applications section.

3. For the case of an interconverting system (e.g. excited state deprotonation or energy transfer) in which the back reaction rate is negligible (and only the initial species is directly excited), the differential tangent and modulation ratio have a simple expression:

$$\Delta = \omega\tau_2, \quad Y = [1 + (\omega\tau_2)^2]^{-1/2}. \qquad 10.$$

In this particular case, differential methods permit us to obtain the forward reaction rate independently of the radiative decay rate (34).

APPLICATIONS

Noninteracting Systems Characterized By Double Exponential Decays

The analysis of double exponential decays for the component lifetimes and their fractional contributions may proceed with two general goals in mind. The first goal may be a characterization of the nature and extent of the heterogeneity, with the aim, for example, of studying the molecular photophysics of a fluorophore, determining distribution of fluorophores in different environments, or achieving a compositional description of a system. The second goal may be utilization of the heterogeneity as a tool in the analysis of the dynamics of equilibria of complex systems. An example of the latter case would be a titration experiment to determine the extent of binding in a protein-ligand system.

Phase and modulation lifetime heterogeneity studies have been carried out on a number of systems including membranes (30, 41) and various protein systems (10, 12, 22, 24, 44, 48, 50, 51, 57). A study of the pH-dependent heterogeneity of tryptophan (29) was carried out with the aim of evaluating Weber's exact solution. A recent review discusses a number of these studies (28).

Excited State Reactions

Several types of excited state reactions have been investigated, using phase fluorometry, with the aim of obtaining information on the reaction rates as well as spectral properties. For example, quenching of fluorescence by small molecules such as oxygen and acrylamide has been extensively studied. In particular, quenching of the tryptophan fluorescence of a number of globular proteins was studied by phase fluorometry (37). In these studies, however, only a single modulation frequency was utilized and only phase data were obtained. Lifetime results were important in establishing the dynamic character of quenching but could not be used to analyze the details of the quenching process. A detailed multifrequency phase and modulation study on oxygen quenching of the prophyrin emission of iron-free myoglobin and hemoglobin was recently reported (26). The results were interpreted in terms of a general model for dynamic quenching of the fluorescence of globular proteins, which, in the limit of low oxygen concentrations, yields double exponential decay. The multifrequency data permitted assignment of the rate of acquisition of quencher by the protein, the exit rate of quencher from the protein, and the migration rate of quencher in the protein interior.

Dipolar Relaxations

A complete theory to account for the dynamics of dipolar relaxations in pure solvents and complex environments such as protein interiors has not yet appeared. The information we seek from dipolar relaxation studies concerns the modalities of the relaxation of the initial Frank-Condon state. A multistate or continuous model is required to describe the phenomenon accurately. However, theoretical difficulties in elaborating a continuous model have led many researchers to adopt the simpler phenomenological approach of considering only an initial unrelaxed and a final relaxed state (33, 62). In this case the general framework of a double exponential decay can be applied. In Figure 8 we report the result of this type of analysis for dansylaziridine derivative [S-(dansyl aminoethyl)-2-thioethanol] in propylene glycol at $-42°C$. The results of a double exponential analysis indicate that the decay rates change in a continuous fashion across the emission spectrum and that the fractional intensity of one component increases to a value greater than unity towards the red (note that for a relaxing system the value of a fractional intensity may exceed unity). For a two-state system, however, the values of the fractional intensities must become constant at the two extremities of the spectrum, and the decay rates of the two components must be wavelength independent. The dansylaziridine derivative results thus demonstrate the inadequacy of the simple two-state approach.

Spectral relaxation studies are often presented in the form of time-resolved spectra, i.e. the emission spectra at selected times after the exciting light pulse. Multifrequency phase fluorometry can also be utilized to obtain time-resolved spectra as shown in Figure 9 for TNS (*p*-2-toluidinyl-6-

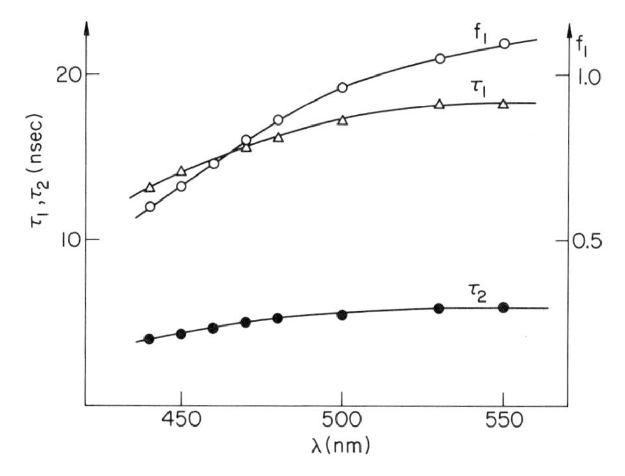

Figure 8 Double exponential decay analysis of dansylaziridine derivative in propylene glycol at −42°C.

naphthalene sulfonic acid) in glycerol (14). Although this example shows the capabilities of multifrequency phase fluorometry to accurately record the characteristics of the emission of relaxing systems, a more complete theoretical treatment of solvent relaxation processes is required for rigorous data analysis.

Rotations

Expressions for the differential tangent and ratio of modulations for isotropic and anisotropic rotators were derived by Weber (60). Mantulin & Weber (40) performed a number of differential phase measurements at two

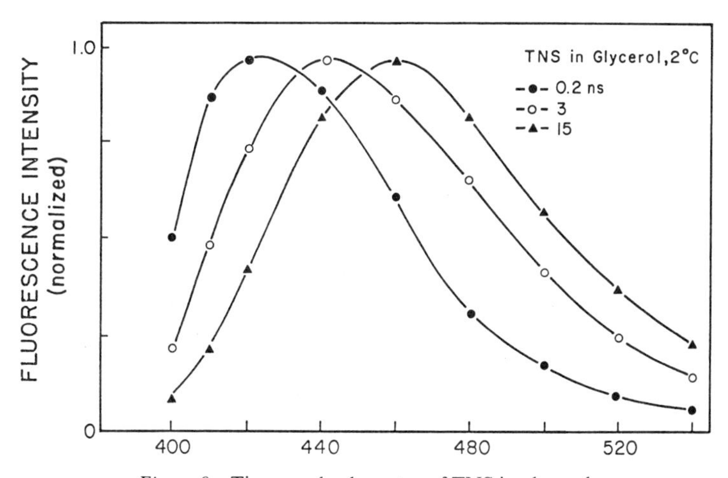

Figure 9 Time-resolved spectra of TNS in glycerol.

modulation frequencies on a series of unsubstituted aromatic hydrocarbons (anthracene, perylene, and chrysene) as well as fluorophores capable of forming hydrogen bonds with the solvent, propylene glycol. In these studies the rotational rates were altered by changing the temperature, and hence the viscosity, of the solvent. The results indicated strongly anisotropic rotations for the unsubstituted fluorophores and isotropic rotations for those fluorophores capable of forming two or more hydrogen bonds. These investigations inspired a number of differential phase studies on fluorophore rotations in model and natural membrane systems (8, 9, 36).

Multifrequency phase and modulation fluorometry was used extensively by Hauser and co-workers for investigations of rotation of small molecules in pure solvents (20, 31, 32). Rotational diffusion times for oxypyrene trisulfonate, rhodamine 6G, and perylene in water were measured; only isotropic rotations were discernible. The phase fluorometer utilized was operational at frequencies up to 400 MHz, and rotational diffusion rates of a few tenths of picoseconds were measurable representing the shortest rotational times directly observed using phase fluorometry.

Figure 10 presents some recent results (Eccleston, Jameson, and Gratton, unpublished observations) on the differential tangent of a fluorescent derivative of GDP (2′-amino-2′-deoxy GDP derivatized on the 2′ amino group with fluorescamine) free in solution and bound to elongation factor

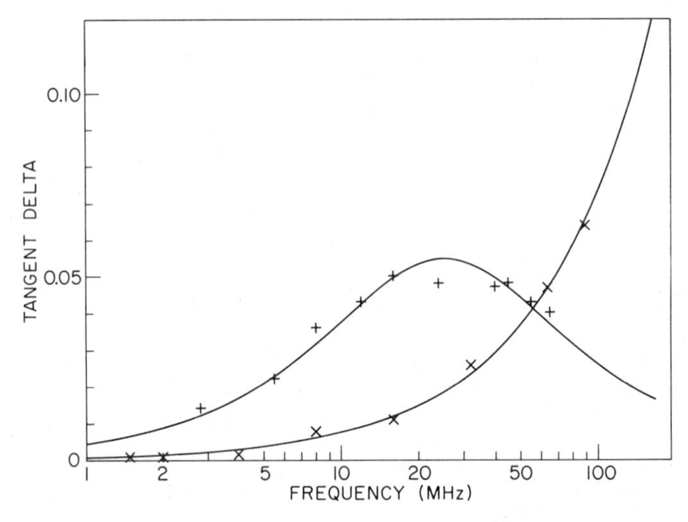

Figure 10 Polarized differential tangent plot for fluoram-GDP free (×) and bound (+) to elongation factor TU from *E. coli*. Excitation wavelength 351 nm, emission wavelength >460 nm. Solid lines correspond to rotational rate of 68 nsec and limiting anisotropy 0.08 for the bound species and rotational rate of 0.15 nsec and limiting anisotropy of 0.35 for the free species.

Tu from *Escherichia coli*. The phase fluorometer described in Reference (15) was used to obtain data over the frequency range of 1 to 100 MHz. The bound fluorophore gave the characteristic bell-shaped curve peaking at intermediate frequencies while the curve corresponding to free fluorophore (resulting from displacement of the probe from the protein by the addition of excess GDP) was shifted to higher frequencies. We should note that a multifrequency analysis on the emission of the free and bound probe indicated a single exponential decay in each case with lifetimes of 7.70 and 11.03 nsec respectively. This application demonstrates that multifrequency measurements at a single temperature and viscosity can yield rotational rates and limiting anisotropies; the information content is thus the same as that from decay of anisotropy experiments carried out using pulse techniques.

Phase-Sensitive Detection

Phase-sensitive detection has been a standard methodology in fluorometry for many years (6, 28, 53). Veselova et al, however, first applied the technique to resolve individual intensity components of heterogeneous emissions from simple fluorescence solutions (54–56). The basic idea of phase-sensitive detection is that emissions corresponding to two species can be individually recorded upon illumination of the system with light modulated at an appropriate frequency if the species differ in lifetime and if the emission is viewed with a detector sensitive to the phase delay. Two conditions render phase-sensitive detection feasible: first, the total intensity of a sinusoidally modulated fluorescence signal is the sum of the intensity components of the individual emitting species; second, the amplitude of the phase-detected signal depends upon both the phase angle of detection and the phase angles of the individual components relative to a reference signal of the same frequency. These considerations are expressed in the relationship (28)

$$PSD = \textstyle\sum_i I_i(\lambda) f_i M_i \cos(\phi_i - \phi_D), \qquad 11.$$

where $I_i(\lambda)$ is the relative intensity as a function of wavelength, f_i is the fractional intensity contribution, M_i is the modulation, ϕ_i the phase angle of the ith species, and ϕ_D the phase angle of the detector. Clearly, setting $\phi_D = \phi_i \pm 90°$ eliminates the fluorescence of the ith species. Selection of the appropriate modulation frequency for phase sensitive detection has been discussed in the analysis section.

Phase-sensitive detection has been applied to several systems relevant to the spectroscopy of proteins such as mixtures of tyrosine and tryptophan (35, 42) and human serum albumin in which the tyrosine contribution was isolated. The resolution of anthracene-diethylaniline exciplex emission (35) illustrated the acquisition of spectra at different detector phase angles. The effect of dipolar relaxation upon phase-resolved spectra has been investigated using *n*-acetyl-L-tryptophanamide in propylene glycol (33); a simple method for estimating spectral relaxation times from phase sensitive intensities was given. The effect of modulation frequency upon the resolving power of the phase-sensitive technique was recently illustrated (42).

CONCLUSIONS AND FUTURE PROSPECTS

In the last few years, multifrequency phase and modulation fluorometry has developed into a powerful technique for the examination of excited state processes. Processes amenable to analysis include heterogeneous emissions, excited state reactions, dipolar relaxations, and rotational motions. Multifrequency instruments generally utilize sinusoidal modulation of continuous light sources. Intrinsically modulated sources such as synchrotron radiation and mode-locked lasers may, however, be utilized to collect phase and modulation data (16, 44). For example, the data in Figure 3 on *p*-terphenyl and in Figure 4 on tryptophan were obtained with a multifrequency apparatus utilizing synchrotron radiation at the ADONE storage ring in Frascati, Italy. The instrumentation used in these studies is virtually identical to that described in Reference (15), with the exception of the modulated light source; the frequencies utilized represent the harmonics of the fundamental ring frequency (8.568 MHz at ADONE). Note that one cannot distinguish between measurements done with a pulsed source and those utilizing sinusoidal light modulation. The extensive wavelength range available from synchrotron radiation is a particularly advantageous feature, as is the fact that all wavelengths are rigorously simultaneous. High repetition rate pulsed sources are ideal for multifrequency phase and modulation measurements, since the harmonic content of such sources extends to very high frequencies. The availability of very high modulation frequencies (in the gigahertz range) would permit complete characterization of rotational modes of small molecules in fluid solvents. Such rotational studies would also be extremely useful in investigation of rotations of tyrosine and tryptophan residues in proteins; molecular dynamics calculations predict very fast rotations, on the picosecond time scale, for some of these residues (23, 43).

Since data acquisition and analysis in phase fluorometry is rapid (compared to the more commonly utilized pulse techniques), one may in principle perform kinetic lifetime studies on systems with rate constants in the range of seconds or less. One may also envision the use of multifrequency techniques in analytical applications such as quantitation of sample purity.

Finally, we may speculate that multifrequency phase and modulation fluorometry could eventually prove to be an important diagnostic technique in cell biology and clinical chemistry. Lifetime measurements offer distinct advantages over intensity and even polarization data in that the contributions from various components (scattered light, for example) can often be unequivocally assigned. The requisite instrumentation is no longer a laboratory curiosity but is, in fact, easy to operate and readily accessible to researchers in diverse fields.

ACKNOWLEDGMENTS

Sincere and grateful appreciation is expressed to G. Weber for his support and encouragement of the development of multifrequency phase fluorometry. We are indebted to M. Limkeman for statistical analysis. Research in our lab (E.G.) is supported by Grants PCM79-18646 and ICR-Physics 1-2-22190.

Literature Cited

1. Badea, M. G., Brand, L. 1979. *Methods Enzymol.* 61:378–425
2. Bailey, E. A., Rollefson, G. K. 1953. *J. Chem. Phys.* 21:1315–22
3. Bauer, R., Rozwadowski, M. 1959. *Bull. Acad. Pol. Sci. Ser. Sci. Math. Astron. Phys.* 7:365–68
4. Birks, J. B., Dyson, D. J. 1961. *J. Sci. Instrum.* 38:282–85
5. Birks, J. B., Little, W. A. 1953. *Proc. Phys. Soc. London Sect. A* 66:921–28
6. Birks, J. B., Munro, I. H. 1967. *Prog. React. Kinet.* 4:239–303
7. Brandt, S. 1976. *Statistical and Computational Methods in Data Analysis.* Amsterdam: North-Holland. 2nd ed.
8. Chong, P., Cossins, A. R. 1983. *Biochemistry* 22:409–15
9. Cossins, A. R., Kent, J., Proffer, C. L. 1908. *Biochim. Biophys. Acta* 599:341–58
10. Dalbey, R. E., Weiel, J., Yount, R. G. 1983. *Biochemistry* 22:4696–4706
11. Dushinsky, F. 1933. *Z. Phys.* 81:7–21
12. Eftink, M. R., Jameson, D. M. 1982. *Biochemistry* 21:4443–49
13. Gaviola, E. 1926. *Ann. Phys. Leipzig* 81:681–710
14. Gratton, E., Lakowicz, J. R. 1983. Presented at Ann. Meet. Am. Soc. Photobiol., 11th, Madison, Wis.
15. Gratton, E., Limkeman, M. 1983. *Biophys. J.* In press
16. Gratton, E., Lopez-Delgado, R. 1980. *Nuovo Cimento B* 56:110–24
17. Gugger, H., Calzaferri, G. 1980. *J. Photochem.* 13:21–33
18. Gugger, H., Calzaferri, G. 1980. *J. Photochem.* 13:295–307
19. Haar, H. P., Hauser, M. 1978. *Rev. Sci. Instrum.* 49:632–33
20. Haar, H. P., Klein, U. K. A., Hafner, F. W., Hauser, M. 1977. *Chem. Phys. Lett.* 49:563–67
21. Hamilton, T. D. S. 1957. *Proc. Phys. Soc. London Sect. B* 70:144–45
22. Herron, J., Voss, E. W. 1981. *J. Biochem. Biophys. Methods* 5:1–17
23. Ichiye, T., Karplus, M. 1983. *Biochemistry* 22:2884–93
24. Ide, G., Engelborghs, Y. 1983. *J. Biol. Chem.* 256:11684–87
25. Ide, G. Engelborghs, Y., Persoons, A. 1983. *Rev. Sci. Instrum.* 54:841–44
26. Jameson, D. M., Alpert, B., Gratton, E., Weber, G. 1983. *Biophys. J.* In press
27. Jameson, D. M., Gratton, E. 1983. In *New Directions in Molecular Luminescence*, ed. D. Eastwood, L. Cline-Love, pp. 67–81. Philadelphia: ASTM.
28. Jameson, D. M., Gratton, E., Hall, R. D. 1983. *Appl. Spectrosc. Rev.* 20: In press
29. Jameson, D. M., Weber, G. 1981. *J. Phys. Chem.* 85:953–58
30. Klausner, R. D., Kleinfeld, A. M., Hoover, R. L., Karnovsky, M. J. 1980. *J. Biol. Chem.* 255:1286–95
31. Klein, U. K. A., Haar, H. P. 1978. *Chem. Phys. Lett.* 58:531–35

32. Klein, U. K. A., Haar, H. P. 1979. *Chem. Phys. Lett.* 63:40–42
33. Lakowicz, J. R., Balter, A. 1982. *Photochem. Photobiol.* 36:125–32
34. Lakowicz, J. R., Balter, A. 1982. *Biophys. Chem.* 16:99–115
35. Lakowicz, J. R., Cherek, H. 1981. *J. Biochem. Biophys. Methods* 5:19–35
36. Lakowicz, J. R., Prendergast, F. G., Hogen, D. 1979. *Biochemistry* 18:508–19
37. Lakowicz, J. R., Weber, G. 1973. *Biochemistry* 12:4171–79
38. Lytle, F. E., Pelletier, M. J., Harris, T. D. 1979. *Appl. Spectrosc.* 33:28–32
39. Maercks, O. 1938. *Z. Phys.* 109:685–99
40. Mantulin, W. W., Weber, G. 1977. *J. Chem. Phys.* 66:4092–99
41. Matayoshi, E. D., Kleinfeld, A. M. 1981. *Biophys. J.* 35:215–35
42. Matheis, J. R., Mitchell, G. W., Spencer, R. D. 1983. See Ref. 27. In press
43. McCammon, J. A., Wolynes, P. G., Karplus, M. 1979. *Biochemistry* 18:927–42
44. Moya, I., Garcia, R. 1983. *Biochim. Biophys. Acta* 722:480–91
45. Muller, A., Lumry, R., Kokubun, H. 1965. *Rev. Sci. Instrum.* 36:1214–26
46. Ravilious, C. F., Farrar, R. T., Liebson, S. H. 1954. *J. Opt. Soc. Am.* 44:238–41
47. Saleem, I., Rimai, L. 1977. *Biophys. J.* 20:335–42
48. Sarkar, H. K., Song, P.-S., Leong, T.-Y., Briggs, W. R. 1982. *Photochem. Photobiol.* 35:593–95
49. Schmillen, A. 1953. *Z. Phys.* 135:294–308
50. Schuldiner, S., Spencer, R. D., Weber, G., Weil, R., Kaback, H. R. 1975. *J. Biol. Chem.* 250:8893–98
51. Sebban, P., Moya, I. 1983. *Biochim. Biophys. Acta* 722:436–42
52. Spencer, R. D., Weber, G. 1969. *Ann. NY Acad. Sci.* 158:361–76
53. Teale, F. W. J. 1983. In *Time Resolved Fluorescence Spectroscopy in Biochemistry and Biology*, ed. E. B. Cundall, R. E. Dale, *NATO ASI*, *Life Sci. Ser. A* 69:59–80. London: Plenum
54. Veselova, T. V., Cherkasov, A. S., Shirokov, V. I. 1970. *Opt. Spectrosc. USSR* 29:617–18
55. Veselova, T. V., Limareva, L. A., Cherkasov, A. S. 1965. *Izv. Akad. Nauk. SSSR, Bull. Phys. Ser.* 29:1345–54
56. Veselova, T. V., Shirokov, V. I. 1972. *Izv. Akad. Nauk. SSSR, Bull. Phys. Ser.* 36:925–28
57. Visser, A. J. W. G., Grande, H. J., Veeger, C. 1980. *Biophys. Chem.* 12:35–49
58. Ware, W. R. 1971. In *Creation and Detection of the Excited State*, ed. A. A. Lamola, 1A:213–302. New York: Dekker
59. Weber, G. 1976. *Horizon Biochem. Biophys.* 2:163–98
60. Weber, G. 1977. *J. Chem. Phys.* 66:4081–91
61. Weber, G. 1981. *J. Phys. Chem.* 85:949–53
62. Weber, G., Mitchell, G. W. 1976. In *Excited States of Biological Molecules*, ed. J. B. Birks. New York: Wiley

Ann. Rev. Biophys. Bioeng. 1984. 13: 125–44

SOLID STATE NMR STUDIES OF PROTEIN INTERNAL DYNAMICS[1]

Dennis A. Torchia

National Institute of Dental Research, National Institutes of Health, Bethesda, Maryland 20205

INTRODUCTION

NMR spectroscopy has been applied to study the dynamics of small molecules and polymers in solids for over thirty years (1, 35, 36, 51, 55); however, complex macromolecules like proteins have been studied in the solid state only in the past decade. Although a variety of elegant experiments for achieving high resolution and sensitivity in solid state spectra were worked out about twenty years ago (2, 9, 23, 33, 46, 54), the full potential of these experiments was realized only with the advent of modern pulse Fourier transform technology (15, 22, 36, 41, 47, 51, 55).

In this article the term "high resolution" is not limited to spectra that contain only resolved narrow lines but it includes spectra that are the superposition of many lines, each corresponding to a particular orientation of the sample in the external field. Such spectra contain much more information than those consisting of isotropic lineshapes, since one can often measure and analyze NMR parameters as a function of orientation.

At about the same time that advances in NMR methodology were taking place, interest in protein dynamics was rapidly increasing (21, 28). Although the detailed three-dimensional structures of crystalline proteins had been solved using diffraction methods, it was recognized that proteins do not generally have rigid structures. Solution NMR studies, for example, provided particularly clear evidence for motion of aromatic rings in a variety of proteins (21). One limitation of solution NMR is that the information contained in the angular dependence of NMR interactions is averaged by the rapid overall motion of the protein, and many types of NMR measurements are insensitive to internal motions that are significantly slower than the overall motion. In solids, the orientation-dependent NMR parameters are sensitive to internal motions covering a wide range of time scales and provide a much more rigorous test of motional models than is possible using solution data.

This article first gives a brief account of solid state NMR spectroscopy and then reviews NMR studies of the internal dynamics of model compounds and three proteins. Several recent, more general reviews cover related areas of solid state NMR not discussed herein (34, 38, 63, 65).

NMR OF SOLIDS

As modern solid state NMR experiments have been discussed in detail in recent monographs (22, 36, 51, 55), I will limit myself to a brief summary of experiments that have proven particularly useful in the study of dynamics of proteins and model compounds.

^2H Lineshapes

In contrast to spectra of liquids, spectra of solids contain signals whose positions are orientation dependent. In the case of a spin-one nucleus like ^{2}H where the signal frequency is determined by the quadrupole interaction, the NMR frequency is given by

$$\omega = \omega_0 \pm \omega_Q f(\theta', \phi') \qquad 1.$$

where,

$$f(\theta', \phi') = (3 \cos^2 \theta' - 1 + \eta \sin^2 \theta' \cos 2\phi')/2, \qquad 2.$$

$\omega_Q = 3e^2qQ/4h$, η is the asymmetry parameter, and the polar angles θ' and ϕ' define the orientation of the magnetic field in the principal axis system of the ^{2}H field gradient tensor (55). In this report we focus our attention upon deuterons that are directly bonded to carbon atoms. In this case, the field gradient tensor is very nearly axially symmetric, i.e. $\eta = 0$ is an excellent approximation, and the unique (symmetry) axis of the field gradient tensor is along the C–^{2}H bond axis, $\hat{\mu}$ (Figure 1*a*). The NMR frequency therefore depends only upon θ', the angle made by the C–^{2}H bond axis and the magnetic field B_0.

According to Equation 1, a single crystal containing a single type of ^{2}H has a spectrum consisting of two signals that are located at $\pm\omega_Q f(\theta', \phi')$. As the orientation of the magnetic field changes with respect to the principal axes of the field gradient tensor, the separation between the ^{2}H signals varies (Figure 2*a–c*). The spectrum of a polycrystalline powder consists of a superposition of doublets that correspond to all possible orientations of the

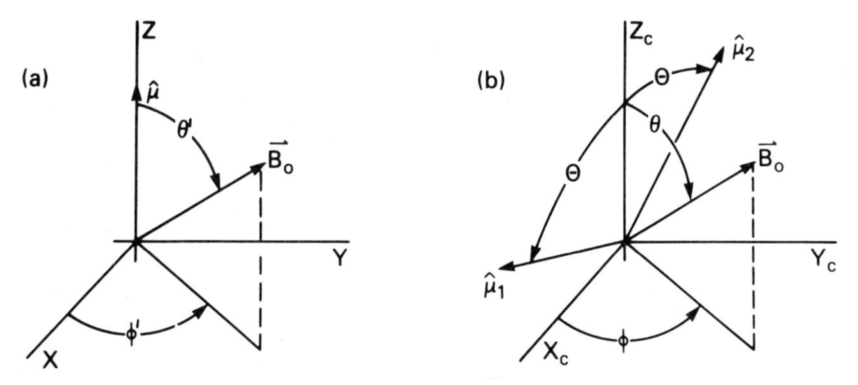

Figure 1 (*a*) Polar angles (θ', ϕ') that specify the orientation of the external magnetic field, B_0, in the principal axis system of the electric field gradient tensor (X, Y, Z). The C–^{2}H bond axis, $\hat{\mu}$, is along Z. (*b*) The polar angles (θ, ϕ) and (Θ, Φ) that specify the orientations of B_0 and the C–^{2}H bond axis, $\hat{\mu}$, when the latter executes motion in a crystal fixed system (X_c, Y_c, Z_c). In the two-site jump model illustrated, $\hat{\mu}_1 = (60°, 0°)$, $\hat{\mu}_2 = (60°, 180°)$.

magnetic field. The spectrum of a crystalline powder is therefore a symmetric pattern about ω_0, and examples of axially symmetric ($\eta = 0$) and asymmetric ($\eta > 0$) lineshapes are shown in Figures 2*d* and 2*e*, respectively. In the absence of motion, the observed ^{2}H spectrum is usually axially symmetric (Figure 2*d*), and the separation between the two maxima is

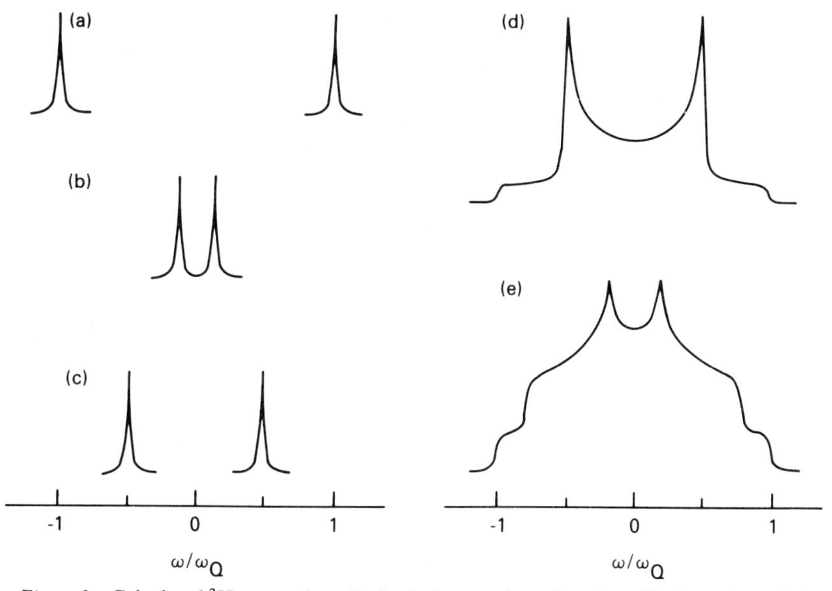

Figure 2 Calculated ^{2}H spectra ($\eta = 0$) of a single crystal as a function of θ', the angle made by the C–^{2}H bond axis and B_0; (*a*) $\theta' = 0°$; (*b*) $\theta' = 60°$; (*c*) $\theta' = 90°$. Calculated spectrum for a polycrystalline sample: (*d*) $\eta = 0$; (*e*) $\eta = 0.6$.

$\omega_Q/2\pi$ in Hz. Typically, $\omega_Q/2\pi$ is ca 125 kHz for a static C–^{2}H bond so that the entire powder pattern covers about 0.25 MHz. The instrumentation and experimental procedures required to obtain accurate lineshapes over this wide frequency range have been discussed recently (20).

In order to see how molecular motion affects the ^{2}H lineshape we consider the case of two-site exchange. In this model the C–^{2}H bond axis jumps between two orientations that have equal equilibrium probability. The orientations of $\hat{\mu}$ and B_0 in the crystal fixed axis system (Figure 1*b*) are specified by the respective polar angles (Θ, Φ) and (θ, ϕ). If the correlation time for the two-site jump is large, $\tau \gg 1/\omega_Q$, separate signals corresponding to the two orientations are observed (56). As τ decreases, these signals broaden and eventually coalesce when $\tau \sim 1/\omega_Q$ (intermediate exchange). Finally, a narrow signal at the averaged frequency of the two orientations is observed when $\tau \ll 1/\omega_Q$ (fast exchange).

The left side of Figure 3 illustrates the effect of two-site jumps on ^{2}H powder spectra. Figure 3*a* is a typical slow exchange spectrum having $\eta = 0$, $\omega_Q/2\pi = 125$ kHz. At the opposite extreme, Figure 3*d* shows that rapid

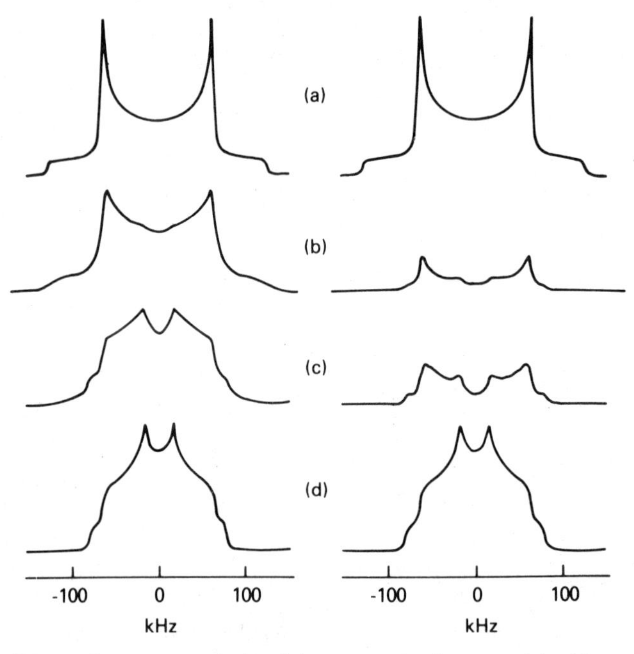

Figure 3 ^{2}H powder spectra calculated for a two-site jump model with $\mu = (60°, 0°)$; $\mu_2 = (60°, 180°)$; and (*a*) $\tau = 10^{-3}$ s; (*b*) $\tau = 10^{-5}$ s; (*c*) $\tau = 10^{-6}$ s; (*d*) $\tau = 10^{-9}$ s. Spectra on the left neglect echo distortion. Spectra on the right show the distortion that results from a quadrupole echo pulse sequence $(\pi/2)_x - t - (\pi/2)_y - t$ when $t = 30$ μs and $\omega_Q/2\pi = 125$ kHz.

two-site exchange produces an axially asymmetric narrowed spectrum, $\eta = 0.6$, $\omega_Q/2\pi = 78.1$ kHz. Since fast exchange spectra are a superposition of narrow lines at average frequencies, they are readily calculated by diagonalizing the averaged field gradient tensor (53). In contrast, intermediate exchange spectra (Figure 3*b*, *c*) require a computer program that calculates the lineshape at each orientation of B_0 (36, 55). Furthermore, in intermediate exchange, certain orientations of B_0 give rise to signals having very short T_2 values (i.e. large linewidths), and intensity of these signals is lost during the delay period of the quadrupole echo pulse sequence (57). The spectra on the right side of Figure 3 show the extent of intensity loss as a function of correlation time. In the slow and fast exchange limits (Figure 3*a*, *d*), intensity loss is negligible because T_2 is large. In contrast, the intermediate exchange spectra are significantly attenuated. More importantly, the lineshapes are severely distorted because T_2 is anisotropic. Therefore, echo distortion must be included in an intermediate exchange lineshape calculation.

Another source of distortion in ^{2}H spectra arises because of finite pulse power. This distortion is usually small and is readily included in a lineshape calculation (10).

^{2}H Spin-Lattice Relaxation

We have just seen that ^{2}H lineshapes depend on τ only in the intermediate exchange domain, 10^{-8} s $< \tau < 10^{-4}$ s. The fast limit lineshape provides information about the extent of molecular motion (i.e. the order parameter) but only puts an upper limit on the correlation time. Internal motions of protein sidechains often have correlation times in the range 10^{-7} to 10^{-12} s, and spin-lattice relaxation time measurements are a useful means of investigating such motions. Equations for T_1 (60) derived for a variety of motional models demonstrate two important features of T_1 behavior in solids: (*a*) T_1 is anisotropic and (*b*) the anisotropy depends upon τ. These features are observed in the two sets of inversion-recovery spectra (Figure 4), calculated for the two-site exchange model of Figure 1*b*. In contrast with liquid state spectra, the different lineshapes of the partially relaxed spectra in the left and right panels make it possible to determine a unique correlation time from a single set of relaxation spectra. Furthermore, T_1 anisotropy in solid state spectra is a means of discriminating among possible models of molecular motion (60).

Lineshapes and Relaxation of Spin $I = 1/2$ Nuclei

Spin $I = 1/2$ nuclei lack quadrupole moments; therefore, their solid state spectra are dominated by *I–S* dipole-dipole coupling ($I = {}^{13}$C, ^{15}N; $S = {}^1$H) and chemical shift anisotropy. These interactions are negligible in

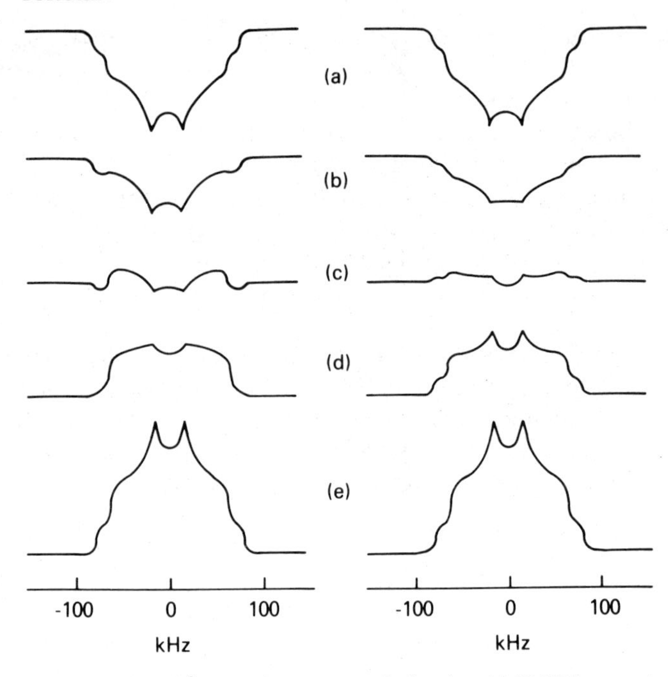

Figure 4 Partially relaxed ^{2}H powder spectra calculated at 38.45 MHz assuming a two-site jump (see Figure 1*b*) and applying the pulse sequence $(\pi)-t_1-(\pi/2)_x-t-(\pi/2)_y-t$ with (*a*) $t_1 = 1.5$ ms; (*b*) $t_1 = 5$ ms; (*c*) $t_1 = 10$ ms; (*d*) $t_1 = 20$ ms; (*e*) $t_1 = 500$ ms. The spectra on the left and right have respective τ values of 0.35 and 20 ns and show the different T_1 anisotropy predicted for small and large correlation times.

^{2}H spectra. In spite of their distinct physical origins, these different angular dependent interactions can all be represented as second rank tensors (22, 36, 55), i.e. dipolar, chemical shift, and electric field gradient tensors. Therefore, many features of ^{2}H spectra are similar to spectra of spin $I = 1/2$ nuclei. One important difference between ^{2}H and spin $I = 1/2$ spectra has just been noted; namely, that two interactions, I–S coupling and chemical shift anisotropy, determine NMR observables in the spin $I = 1/2$ case. Since the former interaction produces very broad featureless lineshapes, it is normally eliminated during signal detection by applying a strong radio frequency field at the proton resonance frequency (41). In this circumstance the NMR lineshape is usually determined by chemical shift anisotropy, and the NMR frequency is given by

$$\omega = \omega_0(1-\sigma_i)+\omega_{cs}f(\theta', \phi')$$

where σ_i is the isotropic chemical shift, $\omega_{cs} = (\sigma_{ZZ}-\sigma_i)\omega_0$, $f(\theta', \phi')$ is given in Equation 2, the polar angles define B_0 in the principal axis system of the chemical shift tensor, and $\eta = (\sigma_{XX} - \sigma_{YY})/(\sigma_{ZZ} - \sigma_i)$. Because the angular dependence of the NMR frequency is the same for the chemical shift and quadrupole interactions, molecular motion affects chemical shift and ^2H powder patterns in essentially the same way. Since ω_{cs} is 10–100 times smaller than ω_Q, chemical shift lineshapes are much narrower than ^2H lineshapes and are therefore sensitive to slower motions.

Although the contribution of the I–S dipolar interaction to the lineshape is eliminated by high power proton decoupling, this interaction normally dominates spin-lattice relaxation of spin $I = 1/2$ nuclei that are directly bonded to protons. Since the orientation of the I–S dipolar tensor is known, detailed information about motion is obtained from the anisotropy of the dipolar T_1, provided that orientation of the chemical shift tensor is known (19, 60). Since the orientation of this tensor is often not known for carbons bonded to protons, one is limited to measurement of the T_1, averaged over the chemical shift powder lineshape, and to a measurement of the average nuclear Overhauser enhancement (NOE). These measurements are usually sufficient to determine an approximate value (accurate to within a factor of two) of τ.

When T_1 values of the I nucleus are very large, the I–S interaction is used to transfer polarization from ^1H nuclei to the spin $I = 1/2$ nuclei (cross-polarization (41). The rate at which polarization is transferred from S to I is a measure of the strength of the residual I–S coupling (16), and in a heterogeneous system one can often enhance signals from ordered regions of a sample relative to disordered (mobile) regions on the basis of differences in I–S coupling (58). Finally I–S coupling is normally responsible for relaxation in the rotating frame, and analysis of these measurements has provided information about slow motions in solid polymers (48). Such information has also been obtained from ^2H spin alignment measurements (56).

INTERNAL DYNAMICS IN CRYSTALLINE AMINO ACIDS AND PEPTIDES

NMR studies of molecular dynamics in model compounds should be a valuable guide to the internal dynamics of proteins, as X-ray studies of amino acids and peptides provided the basis for structural determinations of proteins. At first sight, crystalline amino acids and peptides may appear to be unlikely materials in which to find significant molecular motions because these crystals have high melting temperatures and apparently rigid lattices. However, proton spin-lattice relaxation measurements (3–5) showed that, in addition to the expected rotations of NH_3 and CH_3 groups, several amino acids had significant sidechain motions with $\tau \lesssim 10^{-8}$ s.

Because proton T_1 values in solids are strongly affected by spin diffusion, quantitative information about these motions was not obtained. I now review more recent studies that provide more detailed information about dynamics in these model systems.

Aromatic Sidechains

The spectrum of L-[d_5]phenylalanine (Figure 5*a*) is the superposition of two powder patterns, a slow limit nearly axially symmetric pattern with $\omega_Q/2\pi$ ca 128 kHz and a fast limit pattern having $\omega_Q/2\pi = 79$ kHz and $\eta = 0.6$ (Figure 5*b*). Because the slow limit spectrum has large T_1 values, it is suppressed when the quadrupole echo sequence is repeated rapidly (Figure 5*c*). The resulting $\eta = 0.6$ spectrum is in excellent agreement with the spectrum calculated assuming fast ($\tau < 10^{-8}$ s) 180° ring flips (Figure 5*d*). Computer simulation of inversion-recovery spectra of the fast flipping L-[d_5]phenylalanine lineshape shows that τ is 4.7×10^{-10} s at 20°C.

The calculated spectrum in Figure 5*b* shows that about half of the phenylalanine rings in the sample flip rapidly. Other investigators have

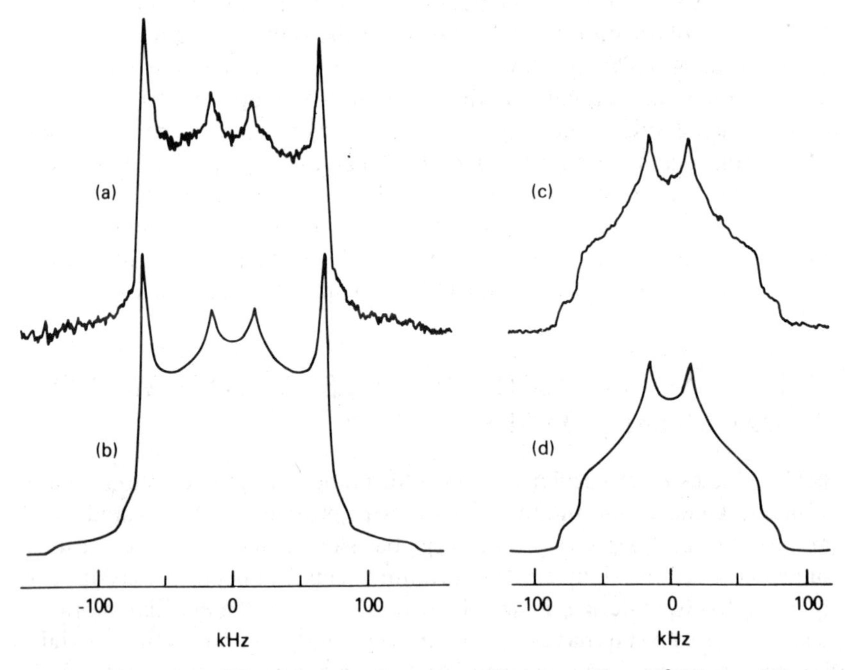

Figure 5 38.45 MHz ^2H quadrupole echo spectra of polycrystalline L-[$^2H_5$]phenylalanine obtained with acquisition delays of (*a*) 25 s; (*c*) 75 ms. Calculated spectra, (*b*) and (*d*), were obtained as described in the text.

found varying amounts of the fast flipping form in their samples (18, 31, 32, 42). This result suggests that L-[d_5]phenylalanine contains two different crystalline forms (42), and the percentage of the fast flipping component depends upon conditions of crystallization. The crystal structure of L-[d_5]phenylalanine is not known. This is unfortunate, since fast flip patterns have now been reported for Phe residues in bacteriorhodopsin (32, 37, 42) and the fd coat protein (17). It is therefore of great interest to know the crystal structure of L-phenylalanine that permits fast ring flips to take place. The crystal structure of L-phenylalanine . HCl is known, but ring flips in this compound are slow at 20°C on the ^{2}H NMR time scale (43).

In contrast with L-phenylalanine, available evidence indicates that the rings of L-tyrosine, L-tryptophan, and L-histidine do not flip on the NMR time scale. ^{2}H NMR spectra of polycrystalline samples of these amino acids deuterated at various ring positions are all nearly axially symmetric patterns, $\eta < 0.1$, $\omega_Q/2\pi$ ca 135 kHz (32, 44, 49).

One example of fast flips of a tyrosine ring in a crystalline peptide, [tyrosine-3,5-^{2}H]-[Leu^5]enkephalin, has been reported. The ^{2}H NMR spectrum changes from a slow limit axially symmetric pattern at −20°C to an axially asymmetric pattern (η ca 0.6) at 101°C. Good simulations of the experimental spectra are obtained assuming a two-site jump model of tyrosine ring motion, provided that echo distortions are included in the lineshape calculation. This study shows that the disorder reported in the X-ray structure (52) probably results from motion of the tyrosine ring.

Proline

Since the N–C$^\alpha$ bond is an integral part of the pyrrolidine ring, the proline sidechain cannot execute large angle internal motions like aromatic amino acids. In spite of this constraint on sidechain mobility, a great deal of evidence indicates that the proline ring is flexible in solution. The degree of flexibility varies with position in the ring according to $C^\gamma > C^\beta > C^\delta > C^\alpha$ (21). ^{2}H NMR spectra of L-[d_7]proline provide direct evidence of ring flexibility in the polycrystalline powder. At −65°C a single axially symmetric show limit pattern having $\omega_Q/2\pi = 128$ kHz is observed (Figure 6*a*). In contrast, at 20°C the L-proline lineshape (Figure 6*b*) is a superposition of four different patterns arising from the differences in flexibility of the deuterons in the ring. An excellent simulation of this experimental pattern is obtained (Figure 6*c*) assuming that the α–C–^{2}H bond is static, and that the β–, γ–, and δ–C–^{2}H bond axes reorient rapidly ($\tau < 10^{-8}$ s) through angles having respective rms values of 17°, 20°, and 12°. This example illustrates the usefulness of solid state NMR for studying proline ring dynamics, and studies of crystalline proline enriched at specific ring positions should provide a detailed picture of proline ring dynamics.

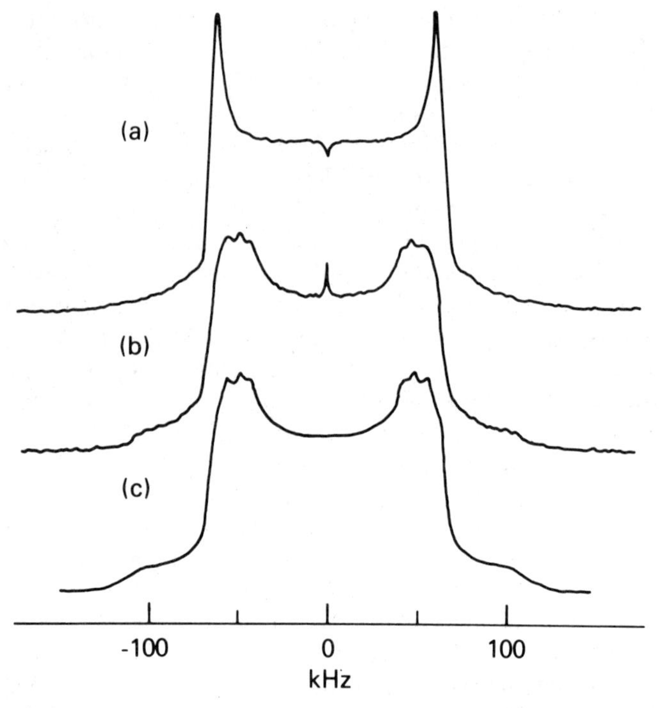

Figure 6 ^2H quadrupole echo spectra of polycrystalline L-[$^2H_7$]proline at (*a*) $-65°C$; (*b*) 20°C. The calculated spectrum was obtained by assuming that the β–, γ–, and δ–C–^{2}H bond axes rapidly reorient through rms angles of 17°, 20°, and 12°, respectively.

Methyl Groups

Because the methyl group has a compact, nearly axially symmetric structure, one might expect methyl reorientation to be insensitive to local environment. A striking demonstration that this is not the case is provided by L-alanine. In the polycrystalline amino acid at 25°C, ^1H and ^2H T_1 measurements show that the methyl group executes three-site jumps with a correlation time of 1.2 ns and an activation energy of ca 22 kJ/mol (3, 7, 29, 31). In contrast, alanyl methyl groups in two peptides, *t*-Boc-L-Ala-L-[3,3,3-$^2H_3$]Ala-OMe and Cyclo-(L-Ala-L-[3,3,3-$^2H_3$]Ala) have correlation times of 0.085 ns at 22°C and respective activation energies of 11.4 and 10.8 kJ/mol (7). The unusually large activation energy obtained for L-alanine is correlated with the unusually tight packing of the methyl groups in the crystal lattice of L-alanine (50). Abnormally close packing is not observed (40) in the crystal structure of cyclo-(L-Ala-L-Ala), which suggests that crystal packing has a marked effect on the rate of methyl reorientation.

Activation energies of 14.0, 17.6, and 15.5 kJ/mol have been reported respectively (29, 31) for methyl reorientation in polycrystalline samples of D,L-[methyl-$^2H_6$]valine, D,L-[3,4,4,4-$^2H_4$]threonine, and L-[methyl-$^2H_3$]leucine. These values were obtained by analyzing ^2H NMR spin-lattice relaxation measurements and are about 4 kJ/mol higher than obtained from ^1H relaxation data (3). The difference in activation energies has been ascribed to problems with quantitative analysis of the proton relaxation data (29).

Although quantitative analysis of ^2H spin-lattice relaxation is reasonably straightforward, one must proceed with care if multiple motions are present. For instance, if methyl reorientation has a correlation time several orders of magnitude different from ω_0^{-1}, another motion having a correlation time closer to ω_0^{-1} may be primarily responsible for spin-lattice relaxation. This can be checked by comparing the T_1 anisotropy predicted for a specific model with that measured. An interesting example that illustrates this point is polycrystalline L-[methyl-$^2H_3$]methionine. At temperatures below 15°C the powder pattern is very nearly that of a methyl group undergoing rapid threefold jumps (30). However, the inversion-recovery spectra observed at 8°C (30) show little if any anisotropy in T_1, in contrast with the anisotropy predicted (60) for a three-site jump motion and observed (7, 30) for L-[methyl-$^2H_3$]methionine at lower temperatures. This suggests that motions other than three-site jumps are responsible for spin-lattice relaxation at 8°C, a conclusion that is confirmed by the nonlinear Arrhenius plot observed at temperatures above −20°C (30). Direct evidence for motion of the S–methyl bond is the appearance of axially asymmetric powder patterns at temperatures above 20°C (30). This evidence of a significant reorientation of the S–methyl bond axis is consistent with the reported crystal structure of L-[methyl-$^2H_3$]methionine (62), which shows that the methyl groups are disordered. It should be possible to determine the detailed dynamics of methionine using crystalline samples labeled at various specific sites.

Oldfield and colleagues have made the interesting observation that the activation energy for methyl reorientation is the same for L- and D,L-methionine (30). However spectra of D,L-[methyl-$^2H_3$]methionine show no evidence of reorientation of the S–methyl bond axis at temperatures up to 40°C.

INTERNAL DYNAMICS OF PROTEINS

Labeled amino acids must be incorporated into proteins to achieve the resolution needed to obtain detailed information about their dynamics. The internal dynamics of about ten labeled proteins have been investigated using pulsed solid state NMR spectroscopy (38). Because of space limitations we discuss only the three proteins that have been studied in greatest detail: collagen, the coat protein of bacteriophage fd, and bacteriorhodopsin in the purple membrane of *Halobacterium halobium*.

Collagen

Collagen is the major protein component in connective tissues where collagen molecules form cross-linked ordered fibers having high tensile strength (11). The collagen molecule consists of three polypeptide chains wound about a common axis. The resulting structure is a helix with the dimensions 1.5 × 300 nm. Glycine residues are located in the interior of the triple helical structure. In contrast, the sidechains of the other amino acid residues are located on the surface of the helix, and their mutual interactions are thought to direct the assembly of collagen fibers and stabilize their structure. Solid state NMR spectroscopy has been used to elucidate these interactions by studying the dynamics of specifically labeled amino acids in collagen.

The first studies of collagen molecular dynamics were carried out on reconstituted fibers of lathyritic chick calvaria collagen (8, 24–27, 61). Although the molecules in these fibrils are not cross-linked, available evidence indicates that they are a good model for the dynamics of cross-linked soft tissue collagen (26).

Collagen fibrils, enriched with [2-^{13}C]glycine, [1-^{13}C]glycine (26, 61), and [3,3,3-$^{2}H_3$]alanine (24, 25), were studied to obtain information about the dynamics of the collagen backbone. Because collagen molecules align with their long axes parallel to the fiber axis, it was assumed that backbone reorientation occurred primarily as a consequence of rotation about the long axis of the molecule. Analysis of measured relaxation times and lineshapes provided an estimate of the range of azimuthal angle, γ_{rms}, through which the molecule reorients. If the molecule assumes a distribution of azimuthal angles, $p(\gamma)$, γ_{rms} is defined by

$$\gamma_{rms}^2 = \int_{-\infty}^{\infty} p(\gamma)\gamma^2 d\gamma.$$

If γ_{rms} is less than about 0.7 radian, lineshapes depend only upon γ_{rms} and not on $p(\gamma)$ in the fast limit (45). Therefore, under these conditions one can use the simple equal population two-site exchange equations to analyze the NMR lineshapes and obtain γ_{rms}, since $\gamma_{rms} = (\gamma_1 - \gamma_2)/2$ where γ_1 and γ_2 are the values of gamma in sites 1 and 2 respectively. Analysis of the [2-^{13}C]glycine-labeled collagen T_1 and of NOE data using the two-site exchange equations yields $\gamma_{rms} = 12°$. A similar value of γ_{rms}, 15°, was obtained by analysis of the lineshape of [3,3,3-d_3]alanine-labeled collagen.

A recent analysis of the lineshape of [1-^{13}C]glycine-labeled collagen yielded a value of $\gamma_{rms} = 41°$, a significantly larger value of γ_{rms} than found previously. This discrepancy arises in part because of the uncertainties in the values of γ_{rms}, which are probably as large as a factor of two. However, a physical argument suggests that γ_{rms} obtained from a ^{13}C lineshape analysis should be larger than γ_{rms} obtained from relaxation measurements. Molecular motions with correlation times of ca 10^{-4} s will effectively narrow the chemical shift lineshape, whereas much faster, presumably smaller amplitude motions, $\tau \sim 10^{-8}$ s, are responsible for spin-lattice relaxation. The ^{2}H lineshape is averaged by motions having $\tau \lesssim 10^{-5}$ s, so that γ_{rms} obtained from ^{2}H lineshape analysis should be in between those obtained by the other two methods.

So far we have discussed estimates of γ_{rms} obtained for fibrils of lathyritic collagen. Recently, [1-^{13}C]glycine has been incorporated into collagen in intact rat tissues (45), and values of γ_{rms} of 14°, 32°, and 31° have been obtained for collagen in mineralized rat calvaria, demineralized rat calvaria, and rat tail tendon. These results, when compared with the value of 41° obtained for lathyritic collagen, are the first quantitative measurement of the extent to which cross-links and mineral restrict backbone motion in collagen fibers.

Spectra of collagen samples labeled at the terminal carbons of glutamic acid, lysine, and methionine (27) show lineshapes that are narrowed by nearly isotropic motion on the $\tau < 10^{-3}$ s time scale. Relaxation time and NOE measurements show that more restricted motions take place on the 10^{-9} s time scale.

More detailed models for motion of collagen sidechains are obtained by analysis of ^{2}H lineshapes. The spectrum of [d_{10}]leucine-labeled collagen fibrils at −85°C (Figure 7*a*), is an axially symmetric fast limit methyl pattern ($\eta = 0$, $\omega_Q/2\pi = 38$ kHz). However, as temperature increases, the pattern transforms to an $\eta \sim 1$ pattern at 30° (Figure 7*c*) (8). The spectra of leucine-labeled collagen fibrils is simulated over the entire temperature range from −85°C to +30°C using a two-site exchange model, suggested by X-ray structures of various crystalline peptides containing leucine (8). It is found that τ for the two-site exchange decreases from $< 10^{-4}$ s at −85° to ca 8×10^{-7} s at 30°C. Although the lineshape calculated using the two-site exchange model is in good agreement with the observed lineshape, the width of the observed pattern is only about 75% of the calculated pattern. This discrepancy was ascribed to the neglect of backbone reorientation and sidechain librations in the calculation. Inclusion of these motions would reduce the width of the calculated pattern, as would the inclusion of additional sidechain conformations in the analysis.

The powder pattern observed for reconstituted collagen labeled with

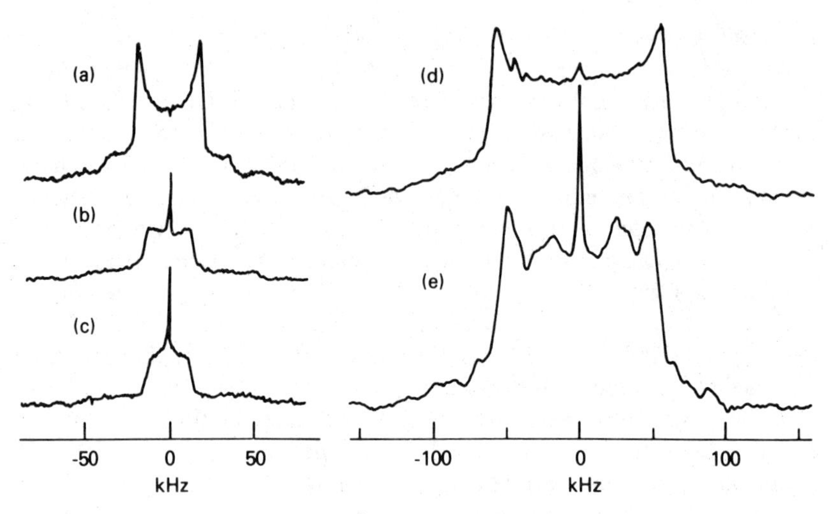

Figure 7 Temperature dependence of ^2H spectra of Leu and Pro sidechains in reconstituted fibers of chick calvaria collagen. [$^2H_{10}$]Leu label (*a*)–(*c*), [$^2H_7$]Pro label (*d*)–(*e*). (*a*) $-85°C$; (*b*) $-6°C$; (*c*) 30°C; (*d*) $-20°C$; (*e*) 20°C.

[d_7]proline at $-20°C$ (Figure 7*d*) is nearly an $\eta = 0$ slow limit pattern, $\omega_Q/2\pi = 110$ kHz, indicating that proline ring motions are small in amplitude or slow on the NMR time scale at this temperature. In contrast, the spectrum obtained at $+20°C$ (Figure 7*e*) shows splittings of ca 40, 60, and 95 kHz. The smaller splittings are tentatively assigned to β and γ deuterons, for reasons discussed earlier, and correspond to motion through rms angles of 25–30°.

The Coat Protein of Bacteriophage fd

The filamentous virus fd consists of 2700 copies of a major coat protein (88 wt %), 5 copies of a minor coat protein (2 wt %), and a circle of single-stranded DNA (10 wt %). The assembled virus is a 900 nm × 6 nm particle in which the DNA is encapsulated in a shell of helical coat protein (6, 59). Because of the size and organized structure of fd, solid state NMR techniques are required to study internal dynamics of fd even in solution. Spin-lattice relaxation times and NOE values have been measured for fd uniformly labeled at peptide nitrogen sites and at selected α-carbon sites (12). The NOE values indicate that the backbone motions have $\tau \sim \omega_0^{-1}$, and the large T_1 values suggest that the amplitude of the backbone motion is small. Model calculations show that T_1 and NOE values obtained by assuming that the N–H and C^α–H bond axes reorient in a cone of half angle $\theta = 5°$ are consistent with the observed T_1 and NOE data (12). These small values of θ are also in accord with the observation that the powder patterns of labeled fd are not significantly different from patterns observed for crystalline amino acids.

In contrast with the small amplitude motions observed for the backbone atoms of fd, ^{13}C and ^{2}H spectra of fd containing enriched Phe and Tyr showed that the aromatic rings of these residues execute large amplitude motions (18). The $\eta = 0.6$ fast limit lineshapes observed for the Phe and Tyr sidechains in fd are almost identical to the fast 180° flip pattern in Figures 5*c* and 5*d*. This observation is strong evidence that the aromatic rings in fd undergo rapid ($\tau < 10^{-8}$ s) jumps through angles of 180°. The observed widths of the ^{2}H patterns are slightly smaller than predicted by the two-site jump equations, and the chemical shift powder pattern of fd labeled with ^{13}C Tyr is significantly narrower than the pattern calculated for two-site jumps. These observations are evidence that Tyr sidechain mobility is not limited to 180° jumps.

In view of the proximity of the single Trp residue to the Phe and Tyr residues of the coat protein, one might expect the Trp ring to execute large amplitude motion. However, it has been concluded (18) that this is not the case. While the fd spectra are consistent with this conclusion, other interpretations of the data are possible. The spectrum of [$^{2}H_{5}$]Trp-labeled fd is a highly distorted quadrupole echo pattern with low signal-to-noise ratio. These spectral features are precisely those expected if motions occur on the microsecond time scale. The amplitude of the motions could be small but need not be. For example, if the Trp ring executes a two-site flip of 180° about the C^{β}–C^{γ} bond, the equilibrium populations in the two sites would be expected to be different because the C^{β}–C^{γ} bond axis is not a twofold symmetry axis. A small difference in free energy of 1 kcal/mol in the two sites produces equilibrium populations in the ratio 5.6 : 1. Because of the unequal equilibrium populations and the small change in orientation of C–^{2}H bond axes produced by the 180° ring flip, the fast limit powder pattern for four out of the five ring deuterons is indistinguishable from the static pattern. For the fifth deuteron η is ca 0.3, but its pattern is difficult to observe in the presence of signals from the four other deuterons, unless signal to noise is excellent and echo distortion is small.

The chemical shift powder patterns observed for [$^{13}C^{\gamma}$]Trp and [^{15}N]Trp-labeled fd (18) do not resolve the question of motion of the Trp sidechain. The width of the [$^{13}C^{\gamma}$]Trp powder pattern calculated from spinning sideband intensities is about 85% that of the polycrystalline amino acid. The width of the powder pattern obtained from a difference spectrum is about 95% of that of the amino acid; but this spectrum has a low signal-to-noise ratio. The [^{15}N]Trp-labeled fd spectrum is very similar to the spectrum of the polycrystalline amino acid and thus suggests a static sidechain in the protein. However, comparisons of shift powder patterns to assess the extent of Trp sidechain motion in proteins will be unreliable until the orientations of the relevant shift tensors are determined.

I have, in part, dwelt on the topic of internal dynamics in fd because uncertainties in internal dynamics limit the information that can be obtained from NMR structural studies of the magnetically oriented virus (13, 14). Also, the question of overall virus dynamics must be carefully examined because the virus is oriented in dilute solution (59). The correlation times calculated at 25°C for an ellipsoid of revolution (64) having the volume and axial ratio of the fd virus are τ_A ca 10^{-1} s, $\tau_B = 2 \times 10^{-5}$ s, $\tau_C = 5 \times 10^{-6}$ s. If reorientation of fd is as fast as this calculation suggests, then rotation about the long axis of the virus will average anisotropic NMR interactions in those virus particles that do not orient perfectly parallel to B_0.

Bacteriorhodopsin in the Purple Membrane

The purple membrane of *H. halobium* contains a single protein, bacteriorhodopsin, which constitutes nearly 60 wt % of the membrane. The protein contains a single polypeptide chain of 248 amino acid residues. The amino acid sequence of the protein is known, and its three dimensional structure and dynamics are being actively studied (39).

A variety of ^{2}H-labeled amino acids have been incorporated into the *H. halobium* bacteriorhodopsin, and the internal dynamics of the protein has been investigated at a variety of backbone and sidechain sites (29, 31, 32, 37, 42).

Spectra of [2,2-$^2H_2$]Gly- and [2-$^2H_1$]Val-labeled bacteriorhodopsin (31, 37) suggest that the dynamical behavior of the protein backbone is heterogeneous. There is a strong isotropic component in each spectrum, suggesting that a portion of the protein backbone is highly mobile. In addition, the major component in each spectrum is axially symmetric with $\omega_Q/2\pi$ ca 125 kHz, which indicates a nearly rigid backbone. However, the lineshape of this component in each spectrum differs significantly from a slow limit $\eta = 0$ pattern, that suggests some backbone residues have significant motions on the microsecond time scale. If motions on the microsecond time scale are present, T_2 will be small and signal intensity reductions will be significant because of quadrupole echo distortions. However, absolute intensity measurements have not yet been reported for backbone labeled bacteriorhodopsin, and it is not now possible to ascertain the extent to which motions on the microsecond time scale are present.

Valine, leucine, and threonine deuterated at the methyl positions have been incorporated into bacteriorhodopsin (29, 31), and spin-lattice relaxation times have been measured over a wide range of temperatures. Correlation times obtained from these measurements have been used to derive activation energies. The activation energies obtained for threonine and leucine are the same in the protein and crystalline powders (29). In contrast, the activation energy obtained for valine is about 40% larger in the crystalline powder than in the protein, and it has been suggested that the valine methyls are more tightly packed in the crystalline powder than in the protein.

The lineshapes of [methyl-$^2H_6$]Val-labeled bacteriorhodopsin provide further evidence that the sidechain is less constrained in the protein than in the crystal. At $-30°C$ the crystal and protein lineshapes are the axially symmetric ($\eta = 0$, $\omega_Q/2\pi = 39$ kHz) patterns expected for methyl rotation. In contrast, at temperatures of 25°C and above, the lineshapes of the labeled protein are complex. They are axially asymmetric with a strong central (possibly isotropic) component. These lineshapes have not been analyzed in detail. They could result from small amplitude reorientation about the C^α–C^β bond axis, but they could also result from three-site (ca 120°) jumps about the C^α–C^β axis. The populations of the three sites will differ since the rotation axis is not a symmetry axis.

The lineshape of the [methyl-$^2H_3$]Leu-labeled bacteriorhodopsin is also axially asymmetric at 37°C and could also result from 120° jumps among two or more sidechain conformations that have different equilibrium populations. Careful lineshape analysis is needed to elucidate the detailed dynamics of the Val and Leu sidechains in bacteriorhodopsin.

The dynamics of sidechains containing aliphatic hydroxyl groups was studied by obtaining spectra of [3,3-^{2}H]Ser- and [3,4,4,4-$^2H_4$]Thr-labeled bacteriorhodopsin (29, 31). The spectrum of polycrystalline D,L-[3,3-$^2H_2$]serine is a nearly axially symmetric pattern with $\omega_Q/2\pi = 122$ kHz, which indicates that the sidechain is rigid. In the purple membrane, the [3,3-$^2H_2$]Ser-labeled spectrum is complicated by the fact that about 20% of the label is transferred to alanine. However, it appears that the powder pattern of most Ser residues in the protein resembles that observed in the crystalline state. Again, in the protein spectrum, a large zero frequency component is observed, some of which may be due to isotropically mobile residues at the protein surface or in the C-terminal region of the polypeptide chain.

The spectra of [3,4,4,4-$^2H_4$]Thr in bacteriorhodopsin and as a crystalline powder show a superposition of a fast limit methyl pattern and a slow limit methine pattern. Therefore, if motions other than a three-site jump of the methyl group occur in the sidechains of Ser and Thr, they do not measurably affect the ^{2}H powder lineshapes.

Spectra of [$^2H_5$]Phe- and [$^2H_2$]Tyr-labeled bacteriorhodopsin at 25°C have the appearance expected for rings undergoing rapid 180° jumps (29, 31, 32, 42). Spectra of [$^2H_5$]Phe-labeled bacteriorhodopsin have been measured from $-30°$ to 25°C and have been simulated using a two-site jump model with a temperature-dependent correlation time (Figure 8) (42). Even at $-30°C$, the spectra (Figure 8*c*) provide evidence that $\tau < 10^{-4}$ s for one or two Phe residues. At 12°C, the experimental spectrum is simulated assuming that all rings flip on the microsecond time scale, and at 25°C the correlation time for flips shortens to $< 10^{-8}$ s. It was noted (42) that at 12° and 25°C it is possible that one or two Phe rings have correlation times different from those used in the simulations.

Although a detailed lineshape analysis of the spectra of Tyr-labeled bacteriorhodopsin has not been made, the available data (31, 32, 37) provide strong evidence that most Tyr rings flip rapidly at 25°C. In contrast, spectra of $[2,2\text{-}^2H_2]$Tyr-labeled bacteriorhodopsin show no evidence for reorientation about the Tyr C^{α}–C^{β} bond (37).

The spectrum of $[^2H_5]$Trp-labeled bacteriorhodopsin resembles the slow limit pattern observed in the crystalline amino acid and suggests that the Trp sidechain is not mobile (31, 32, 37). However, this conclusion should be regarded as tentative since, for the reasons given earlier, the ^2H lineshape may be insensitive to the motion of this sidechain.

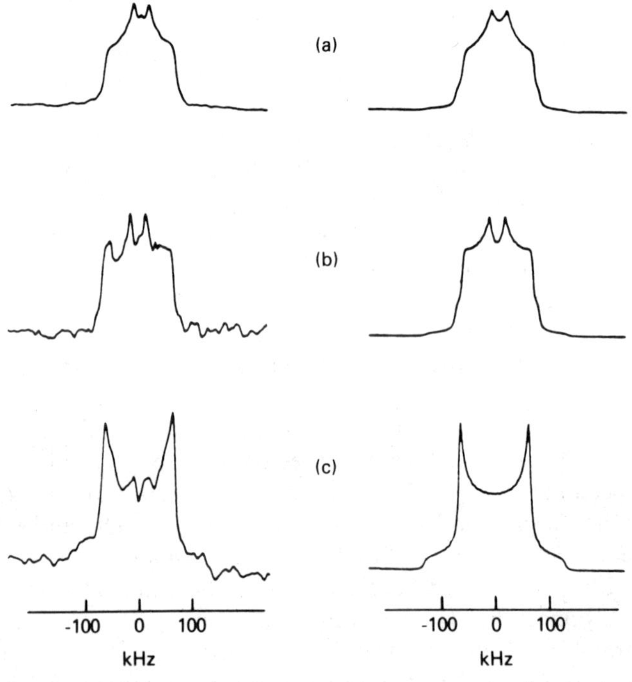

Figure 8 Comparison of experimental (*left*) and calculated (*right*) ^2H quadrupole echo spectra of $[^2H_5]$Phe-labeled bacteriorhodopsin: (*a*) 25°C; (*b*) 12°C; (*c*) $-30°C$. Reproduced with permission from R. G. Griffin.

CONCLUSIONS AND PROSPECTS

Solid state NMR spectroscopy is a powerful method for studying internal dynamics on the 10^{-2} to 10^{-12} s time scale. Correlation times and activation energies for specific models of motion have been determined using this technique. However, the models employed to date (e.g. N-site jumps) are idealizations, which must be replaced by more realistic models. The elementary models have shown that internal dynamics are very sensitive to packing in crystalline amino acids, and it is likely that a combination of NMR, X-ray diffraction, and molecular dynamics calculations will elucidate the connection between packing and internal dynamics of amino acid sidechains. Such studies will also greatly refine our knowledge of intermolecular potential functions and of possible internal motions within proteins.

The models of protein motions that have been obtained from analysis of NMR data will undoubtedly be modified and refined as new data are acquired and data analysis becomes more sophisticated. Perhaps the major limitation of studies of protein dynamics is that one almost always observes signals of a labeled amino acid that occurs repeatedly in the protein sequence. As a result, it is possible to distinguish only pronounced heterogeneity in the dynamics of the amino acid residue of interest. This problem can be overcome by using oriented samples. Their narrow signals are dispersed throughout the spectrum because the labeled residues have different orientations with respect to B_0.

Studies of magnetically oriented myoglobin and phage fd are underway (13, 14, 30), and major advances in our understanding of protein dynamics will undoubtedly result from studies of oriented samples.

Literature Cited

1. Abragam, A. 1961. *The Principles of Nuclear Magnetism*. London: Oxford. 599 pp.
2. Andrew, E. R., Bradbury, A., Eades, R. G. 1958. *Nature* 182:1659
3. Andrew, E. R., Hinshaw, W. S., Hutchins, M. G., Sjoblom, R. O. I. 1975. *Mol. Phys.* 31:1479–88
4. Andrew, E. R., Hinshaw, W. S., Hutchins, M. G., Sjoblom, R. O. I. 1976. *Mol. Phys.* 32:795–806
5. Andrew, E. R., Hinshaw, W. S., Hutchins, M. G., Sjoblom, R. O. I. 1977. *Mol. Phys.* 34:1695–1706
6. Banner, D. W., Nave, C., Marvin, D. A. 1981. *Nature* 289:814–16
7. Batchelder, L. S., Niu, C. H., Torchia, D. A. 1983. *J. Am. Chem. Soc.* 105:2228–31
8. Batchelder, L. S., Sullivan, C. E., Jelinski, L. W., Torchia, D. A. 1982. *Proc. Natl. Acad. Sci. USA* 79:386–89
9. Bloch, F. 1958. *Phys. Rev.* 111:841–53
10. Bloom, M., Davis, J. H., Valic, M. I. 1980. *Can. J. Phys.* 58:1510–17
11. Bornstein, P., Traub, W. 1979. In *The Proteins*, ed. H. Neurath, R. L. Hill, 4:411–632. New York: Academic. 679 pp. 3rd ed.
12. Cross, T. A., Opella, S. J. 1982. *J. Mol. Biol.* 159:543–49
13. Cross, T. A., Opella, S. J. 1983. *J. Am. Chem. Soc.* 105:1505–6
14. Cross, T. A., Tsang, P., Opella, S. J. 1983. *Biochemistry* 22:721–26

15. Davis, J. H., Jeffrey, K. R., Bloom, M., Valic, M. I., Higgs, T. P. 1976. *Chem. Phys. Lett.* 42:390–94
16. Demco, D. E., Tegenfeldt, J., Waugh, J. S. 1975. *Phys. Rev. B* 11:4133–51
17. Gall, C. M., Cross, T. A., DiVerdi, J. A., Opella, S. J. 1982. *Proc. Natl. Acad. Sci. USA* 79:101–5
18. Gall, C. M., DiVerdi, J. A., Opella, S. J. 1981. *J. Am. Chem. Soc.* 103:5039–43
19. Gibby, M. G., Pines, A., Waugh, J. S. 1972. *Chem. Phys. Lett.* 16:296–99
20. Griffin, R. G. 1981. *Methods Enzymol.* 72:108–74
21. Gurd, F. R. N., Rothgeb, T. M. 1979. *Adv. Protein Chem.* 33:73–165
22. Haeberlen, U. 1976. *Advances in Magnetic Resonance, Suppl. 1.* New York: Academic. 190 pp.
23. Hartmann, S. R., Hahn, E. L. 1962. *Phys. Rev.* 128:2042–53
24. Jelinski, L. W., Sullivan, C. E., Batchelder, L. S., Torchia, D. A. 1980. *Biophys. J.* 10:515–29
25. Jelinski, L. W., Sullivan, C. E., Torchia, D. A. 1980. *Nature* 284:531–34
26. Jelinski, L. W., Torchia, D. A. 1979. *J. Mol. Biol.* 133:45–65
27. Jelinski, L. W., Torchia, D. A. 1980. *J. Mol. Biol.* 138:255–72
28. Karplus, M., McCammon, J. A. 1981. *CRC Crit. Rev. Biochem.* 9:293–349
29. Keniry, M. A., Kintanar, A., Smith, R. L., Gutowsky, H. S., Oldfield, E. 1983. *Biochemistry*. In press
30. Keniry, M. A., Rothgeb, T. M., Smith, R. L., Gutowsky, H. S., Oldfield, E. 1983. *Biochemistry* 22:1917–26
31. Keniry, M. A., Smith, R. L., Gutowsky, H. S., Oldfield, E. 1983. In *Stereodynamics of Molecular Systems*, ed. R. H. Sarma, E. Cleminti, 3:435–50. New York: Adenine
32. Kinsey, R. A., Kintanar, A., Oldfield, E. 1981. *J. Biol. Chem.* 256:9028–36
33. Lowe, I. J. 1959. *Phys. Rev. Lett.* 2:285–87
34. Lyerla, J. R., Yannoni, C. S., Fyfe, C. A. 1982. *Acc. Chem. Res.* 15:208–16
35. McCall, D. W. 1971. *Acc. Chem. Res.* 4:223–32
36. Mehring, M. 1983. *High Resolution NMR Spectroscopy in Solids*. Berlin: Springer. 342 pp. 2nd ed.
37. Oldfield, E., Kinsey, R. A., Kintanar, A. 1982. *Methods Enzymol.* 88:310–25
38. Opella, S. J. 1982. *Ann. Rev. Phys. Chem.* 33:533–62
39. Packer, L., ed. 1982. *Methods Enzymology*, Vol. 88. New York: Academic. 836 pp.
40. Pierce, L., Hayashi, N. 1961. *J. Chem. Phys.* 35:479–85
41. Pines, A., Gibby, M. G., Waugh, J. S. 1973. *J. Chem. Phys.* 59:569–90
42. Rice, D. M., Blume, A., Herzfeld, J., Wittebort, R. J., Huang, T. H., DasGupta, S. K., Griffin, R. G. 1981. In *Stereodynamics of Molecular Systems*, ed. R. H. Sarma, 2:255–70. New York: Adenine
43. Rice, D. M., Gingrich, P., Herzfeld, J., Griffin, R. G. 1982. *Biophys. J.* 37:145a (Abstr.)
44. Rice, D. M., Wittebort, R. J., Griffin, R. G., Meriovitch, E., Meinwald, Y., Freed, J. H., Scheraga, H. A. 1981. *J. Am. Chem. Soc.* 103:7707–10
45. Sarkar, S. K., Sullivan, C. E., Torchia, D. A. 1983. *J. Biol. Chem.* 258:9762–67
46. Sarles, L. R., Cotts, R. M. 1958. *Phys. Rev.* 111:853–59
47. Schaefer, J., Stejskal, E. O. 1976. *J. Am. Chem. Soc.* 98:1031–32
48. Schaefer, J., Stejskal, E. O. 1979. In *Topics in Carbon-13 NMR Spectroscopy*, ed. G. C. Levy, 3:283–324. New York: Wiley. 397 pp.
49. Schramm, S., Oldfield, E. 1983. *Biochemistry* 22:2903–13
50. Simpson, H. J., Marsh, R. E. 1966. *Acta Crystallogr.* 20:550–55
51. Slichter, C. P. 1978. *Principles of Magnetic Resonance.* Berlin: Springer. 300 pp. 2nd ed.
52. Smith, G. D., Griffin, J. F. 1981. *Science* 199:1214–16
53. Soda, G., Chiba, T. 1969. *J. Chem. Phys.* 50:439–55
54. Solomon, I. 1958. *Phys. Rev.* 110:61–65
55. Spiess, H. W. 1978. In *NMR Basic Principles and Progress*, ed. P. Diehl, E. Fluck, R. Kosfeld, 15:55–214. Berlin: Springer. 214 pp.
56. Spiess, H. W. 1983. *Colloid Polymer Sci.* 261:193–209
57. Spiess, H. W., Sillescu, H. 1980. *J. Magn. Reson.* 42:381–89
58. Sutherland, J. W. H., Egan, W., Schechter, A. N., Torchia, D. A. 1979. *Biochemistry* 18:1797–1803
59. Torbet, J., Maret, G. 1981. *Biopolymers* 20:2657–69
60. Torchia, D. A., Szabo, A. 1982. *J. Magn. Reson.* 49:107–121
61. Torchia, D. A., VanderHart, D. L. 1976. *J. Mol. Biol.* 104:315–21
62. Torii, K., Iitaka, Y. 1973. *Acta Crystallogr. B* 29:2799–807
63. Wasylishen, R. E., Fyfe, C. A. 1982. *Ann. Rep. NMR Spectrosc.* 12:1–80
64. Woessner, D. E. 1962. *J. Chem. Phys.* 37:647–54
65. Yannoni, C. S. 1982. *Acc. Chem. Res.* 15:201–8

Ann. Rev. Biophys. Bioeng. 1984. 13: 145–65

AMINO ACID, PEPTIDE, AND PROTEIN VOLUME IN SOLUTION

A. A. Zamyatnin

P. K. Anokhin Institute of Normal Physiology, USSR Academy of Medical Sciences, Hertzen Street 6, Moscow 103009, USSR

INTRODUCTION

Modern physicochemical biology steadily enriches our knowledge of the functional role of proteins and their elements—amino acids and peptides. These substances are involved in all essential molecular processes occurring in the living cell and perform key regulatory functions at the cellular and organismic levels.

Their functional specificity arises from their unique physicochemical properties, specific chemical composition, and the variety of intra- and intermolecular interactions of protein residues with each other and with neighboring molecules. As a result, the protein assumes a particular form and volume. Spatial coordinates of the atoms of all macromolecules completely characterize the macromolecule structure.

It is difficult to define a portion of space filled by a macromolecule, i.e. the volume. To do so, one must estimate the surface surrounding the macromolecule or the volume of all its atoms and all intramolecular cavities. Nevertheless, there are a number of definitions of macromolecule volume. For example, the concept of hydrodynamic volume is used to characterize a space occupied by the macromolecule together with its "impregnating" solvent. Or the volume may be defined by measuring the structures observed with the electron microscope, using X-ray small-angle scattering, etc.

Of special interest are the macromolecule volume values obtained under conditions approaching natural ones, i.e. without electron beams or X rays, without applying great hydrodynamic fields and hydrostatic pressure, and

0084–6589/84/0615–0145$02.00

without drying. In this case, special thermodynamic parameters—partial and apparent volumes—are introduced. Sometimes these values are used as auxiliary ones, e.g. in hydrodynamic (76), ultrasonic (22, 73), and neutron scattering (64, 96) studies. In this work we review the studies of volume that are essential for understanding the physical chemistry of proteins.

THERMODYNAMIC VOLUME

Thermodynamic volume is one of the three thermodynamic parameters involved in the classical equation of state (57)

$$f(P, V, T) = 0, \qquad 1.$$

where P is pressure, V volume, and T absolute temperature. The knowledge of any two of these parameters permits a complete description of a system in an equilibrium state.

Partial Volume

In systems consisting of k components, partial volume values (38, 108) are introduced as

$$\bar{V}_i = \left.\frac{\partial V}{\partial N_i}\right|_{P,T,N_1,\ldots,N_{i-1},N_{i+1},\ldots,N_k}, \qquad 2.$$

where V is total system volume, N_i is the amount of the ith component, and $\bar{V}_i$ is the partial mole (molar or molal) volume, which characterizes the ith component with P, T, and other $(k-1)$ components constant. The total system volume is

$$V = \sum_{i=1}^{k} N_i \bar{V}_i. \qquad 3.$$

In the case of protein solutions we have to deal with two- or more-component systems containing protein, water, salts, etc. However, it is more convenient to reduce such a system to a binary one; one component is water with salts, the other the solute (protein only, for example).

According to Equation 3, for binary solutions comprising N_1 moles of solvent and N_2 moles of solute the volume of such a system at constant P and T is expressed as

$$V = N_1\bar{V}_1 + N_2\bar{V}_2, \qquad 4.$$

where $\bar{V}_1$ and $\bar{V}_2$ are partial mole volumes of solvent and solute, respectively, as defined from Equation 2:

$$\bar{V}_1 = \left.\frac{\partial V}{\partial N_1}\right|_{P,T,N_2}, \qquad 5.$$

$$\bar{V}_2 = \left.\frac{\partial V}{\partial N_2}\right|_{P,T,N_1}. \quad 6.$$

In some rare cases the plot of function

$$V = f(N_2) \quad 7.$$

is linear (curve 1 in Figure 1). Examples are aqueous solutions of glycolamide (76), sucrose (92), and RNase (92). In these cases, the partial volumes are volumes of pure components.

Usually, a plot of Equation 7 differs from a straight line (curves 2 and 3 in Figure 1). Here, the mole volumes of solvent and solute are not equal to the values for pure phases of these substances (38, 39, 89, 108), and in most cases finding the values of $\bar{V}_1$ and $\bar{V}_2$ presents a special and rather laborious problem.

Apparent Volume

For simplicity, we consider the mole volume of solvent in solution as coinciding with that of solvent in the pure phase. Then Equation 4 should be written otherwise, namely,

$$V = N_1 V_1^0 + N_2 \Phi_2, \quad 8.$$

where V_1^0 is volume of solvent in the pure phase, and Φ_2 is the apparent mole volume of solute, which is

$$\Phi_2 = (V - N_1 V_1^0)/N_2. \quad 9.$$

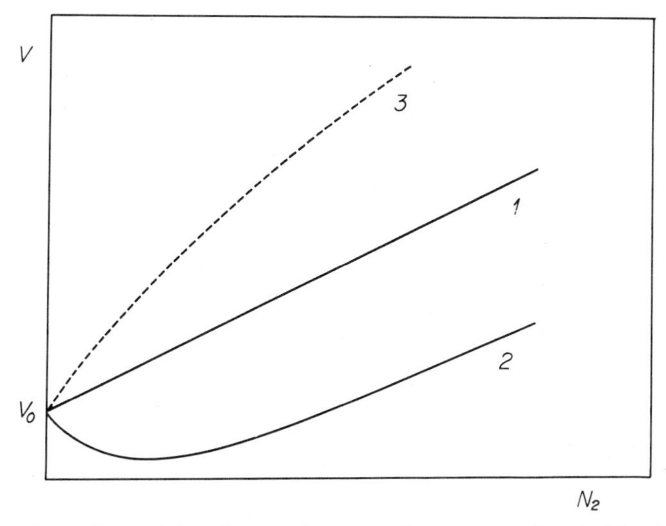

Figure 1 Dependence of the solution volume on solute concentration: (*1*) ideal binary, (*2*) non-ideal (of electrolyte), (*3*) non-ideal (of non-polar solute).

The values of the partial and apparent volumes of solute can be related by differentiating Equation 8 with respect to N_2:

$$\bar{V}_2 = \left.\frac{\partial V}{\partial N_2}\right|_{P,T,N_1} = N_2 \left.\frac{\partial \Phi_2}{\partial N_2}\right|_{P,T,N_1} + \Phi_2. \qquad 10.$$

In most cases, the thermodynamic volume of protein is measured in dilute solutions where the apparent volume is approximately equal to the partial one.

If the amount of solute is expressed in grams, the partial ($\bar{v}_2$) or apparent (ϕ_2) specific volumes are dealt with as follows:

$$\bar{v}_2 = \bar{V}_2/M_2, \qquad 11.$$

$$\phi_2 = \Phi_2/M_2, \qquad 12.$$

where M_2 is the molecular weight of the solute.

Volume Change

Physicochemical processes occurring in a protein system can produce changes in a number of parameters, including the volume. In accordance with Equation 3 the volume change is

$$\Delta V = \sum_{i=1}^{k} N_i \Delta \bar{V}_i. \qquad 13.$$

When considering a binary solution, protein and solvent, we must write Equation 13 as

$$\Delta V = N_1 \Delta \bar{V}_1 + N_2 \Delta \bar{V}_2. \qquad 14.$$

It follows from Equation 14 that the volume change ΔV can be due to both changes in the protein and in the solvent. If the apparent value Φ_2 is considered, then, according to Equations 8 and 13 the volume change is

$$\Delta \Phi = N_2 \Delta \Phi_2 \qquad 15.$$

and is governed by protein volume changes only (107, 108).

EXPERIMENTAL TECHNIQUES

There are two types of techniques for the measurement of the thermodynamic volume. By absolute methods the solution density is measured (densitometry), from which the values $\bar{V}_2$ or Φ_2 are calculated (108). The second type (dilatometry) is represented by measuring $\Delta \Phi$ directly (34, 45, 46, 108).

Densitometry

We shall briefly consider only those techniques that have been used to measure protein solutions.

PYCNOMETRY (108) Pycnometry is based on weighing the solution and solvent in a special vessel (pycnometer) of known volume. The limiting accuracy is $\sim 10^{-5}$ g/cm^3. An increase in accuracy requires a greater quantity of the solute.

FALLING DROP METHOD (52, 71) The falling drop method can be used with extremely small quantities of the solution (less than 1 mm^3). A droplet of the solution is introduced into a special column filled with a mixture of inert liquids (e.g. bromobenzene, kerosene) that establish a controlled density gradient along the column. The density is estimated from the site where the drop stops. The accuracy is about 10^{-6} g/cm^3.

MAGNETIC FLOAT METHOD (43, 44, 99, 100, 109) A float with a small magnet inside is submerged in the solution. The whole system is placed in the magnetic field of the external solenoid, which can equilibrate the float. The solenoid current corresponds to the density of the solution. The accuracy approaches $2 \cdot 10^{-8}$ g/cm^3 (21). However, for high accuracy a large amount of material is required.

TUNING FORK METHOD (41, 42, 67) The tuning fork method yields a sufficiently high degree of accuracy ($5 \cdot 10^{-6}$ g/cm^3) and requires a comparatively small amount of solution (about 0.6 cm^3). The solution is placed in a U-shaped cuvette (tuning fork) that can oscillate at a low amplitude and at a natural frequency. This frequency depends on solution density and is measured with modern electronic techniques.

VOLUME DETERMINATION FROM DENSITY If the molecular weight of the dissolved substance, M_2, is known, the apparent volume can be calculated using Equation 9, which is easily transformed to

$$\Phi_2' = M_2/\rho_0 - 1000(\rho - \rho_0)/c_2\rho_0, \qquad 16.$$

$$\Phi_2'' = M_2/\rho - 1000(\rho - \rho_0)/m_2\rho\rho_0, \qquad 17.$$

where Φ_2' is the apparent molar volume and Φ_2'' the apparent molal volume of solute, ρ is solution density, ρ_0 the density of solvent in the pure phase, c_2 the molarity (the number of moles of solute in 1000 ml of solution), m_2 the molality (the number of moles of solute in 1000 g of solvent) and

$$m_2 = 1000c_2/(1000\rho - c_2M_2). \qquad 18.$$

If M_2 is unknown, the introduction of weight percent, ω_2, and densities ρ

and ρ_0 to Equation 8 gives

$$100/\rho = (100-\omega_2)/\rho_0+\omega_2\phi_2. \qquad 19.$$

Thus the apparent specific volume of solute is

$$\phi_2 = [100/\rho-(100-\omega_2)/\rho_0]/\omega_2. \qquad 20.$$

The error in the experimental determination of apparent specific volume of protein, which depends on the accuracy of density and concentration measurements as well as on the purity of the studied preparation, is usually $\pm 10^{-3}$ cm^3/g.

Dilatometry

The term "dilatometry" seems to have appeared in connection with the study of the expansion of substances upon heating. However, such a type of measurement can also be termed constrictometry since, for example, water within a temperature interval 0 to 4°C does not expand but constricts.

There is a great diversity in the design of dilatometers used to study protein solutions (45, 46, 108). Dilatometers that study two- or more-component systems have been developed in India (81) and Denmark (52) and can be used to study volume changes produced by contact between various substances. Figure 2 shows that a protein solution in a dilatometer is initially separated from the second solution (110). After reaching temperature equilibrium, the partition is removed, the components are mixed, and a reaction followed by a volume change takes place. In the case shown in Figure 2, there is a volume increase. The volume change is judged from the shift of one of the cuvette surfaces using detectors of various types. Capillary detectors are most widely used. In some dilatometers the capillary liquid is layered directly onto the components (52); in others it is separated with an elastic membrane (101, 105, 106). The sensitivity of these dilatometers is about 10^{-5} cm^3.

EXPERIMENTAL-THEORETICAL METHOD

There is virtually an infinite variety of different substances, and it would be impossible to characterize all of them experimentally. A method was required to extrapolate the results obtained for some chemical compounds to others. Such an approach was applied in studies by Kopp (40, 47), who obtained the volume values for a number of atoms contained in molecules of liquid substances and proposed the concept of mole volume as an additive function of atomic volumes. Thus, to estimate the volume of a liquid it proved sufficient to know its chemical composition and have

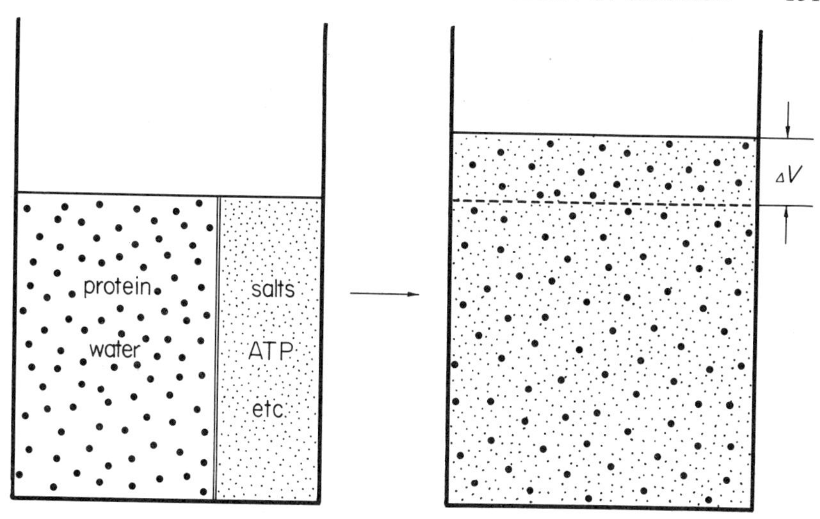

Figure 2 Schematic diagram of a dilatometer for studying volume changes in a two-component system.

available the whole set of atomic volume values of its constituent chemical elements. We shall call such a method experimental-theoretical.

Traube (90) showed that atomic volumes of chemical elements constituting dissolved substances differ from those obtained by Kopp (Table 1), probably because the surroundings of the same atom differ in pure liquid and in solution.

These ideas were used by Cohn & Edsall to determine apparent specific

Table 1 Atomic volumes of the basic chemical elements of proteins

Atom	Atomic volume according to Kopp (cm^3/g-at.)	Apparent atomic volume according to Traube (cm^3/g-at.)
C	11.0	9.9
H	5.5	3.1
N	—	1.5
$O_{Carboxyl}$	7.8	2.3[a] 0.4[b]
$O_{Carbonyl}$	12.2	5.5
S	22.6	15.5

[a] Oxygen volume of the first hydroxyl of a compound.
[b] The volume of oxygen atoms of all the remaining hydroxyls of the same compound.

Table 2 Values of the apparent mole volumes of the main atomic groups of proteins

Atomic group	$-NH_2$	$-CH_2-$	$-COOH$	$-CONH-$	$-OH$
Group volume (cm^3/mol)	7.7	16.3	18.9	20.0	5.4

volumes of amino acids (14, 16, 17). They found the volumes for the main chemical groups that comprise the amino acids (Table 2), calculated apparent specific volumes of amino acids, and extended the additivity principle to obtain the apparent volume of proteins (15) using the Equation

$$\phi_p = \sum_i \phi_i \omega_i \Big/ \sum_i \omega_i, \qquad 21.$$

where ϕ_p is the apparent specific volume of the protein, ϕ_i the apparent specific volume of the ith amino acid residue and ω_i its weight percent. In spite of a number of simplifications (102) this method yields rather accurate results.

This method was also applied to determine protein volumes from protein composition alone (103, 104, 107). It was proposed to use the volumes of amino acid residues obtained experimentally or from

$$\Phi^x_{AR} = \Phi^e_{AR} + \Phi_{gr}, \qquad 22.$$

where Φ^x_{AR} is the desired volume, Φ^e_{AR} the experimental volume of an amino acid residue similar in chemical composition, and Φ_{gr} the volume of chemical group by which the desired amino acid differs from the known one.

This method can probably be used in studying protein complexes, glycoproteins, nucleoproteins, etc.

AMINO ACIDS

At the present time the values of apparent volumes for the 20 usual amino acids that occur in proteins have been obtained experimentally. Earlier (15, 107), such values had been estimated for only 11 amino acids; the rest were calculated using equations such as Equation 22. Among the values presented in Table 3, 17 have been measured anew and only 3 (for asparagine, glutamine, and lysine) are the result of older measurements. Analysis shows that the old values were overestimated, probably because of some methodical errors that have now become smaller (with technical progress, with increasing purity of preparations and accuracy in determi-

Table 3 Values of apparent mole and specific volumes of amino acids at 25°C

Amino acid	M_{AA}† (dalton)	Φ^e_{AA} (cm^3/mol)	Reference	ϕ^e_{AA} (cm^3/g)	(Φ^e_{AA})‡ (cm^3/mol)
Glycine	75.07	43.2	56	0.575	(43.5)
Alanine	89.09	60.4	33	0.679	(60.6)
Valine	117.15	90.8	56	0.775	(92.7)
Leucine	131.18	107.5	37	0.819	(108.4)
Isoleucine	131.18	107.5		0.819	
Phenylalanine	165.19	121.2	37	0.734	(121.3)
Serine	105.09	60.3	37	0.574	(60.8)
Threonine	119.12	76.8	79	0.645	
Methionine	149.21	105.3	56	0.706	(105.1)
Cysteine	121.16	73.4	56	0.606	
Tyrosine	181.19	123.1	79	0.679	
Tryptophane	204.23	143.9	56	0.705	(144.1)
Aspartic acid	132.10	73.8	56	0.559	(74.1)
Glutamic acid	146.12	85.9	56	0.588	
Asparagine	132.12	78.0	15	0.590	(78.0)
Glutamine	146.15	93.9	15	0.642	(93.9)
Lysine	147.20	108.5	15	0.737	(108.5)
Arginine	175.21	127.3	56	0.727	
Histidine	155.16	98.8	56	0.637	(99.3)
Proline	115.13	81.0	37	0.704	(81.0)

† M_{AA} is the molecular weight of amino acid.
‡ The values obtained in 1943 (15).

nation of concentration, etc). That is why we chose the minimal rather than mean (7, 8) values Φ^e_{AA} for glycine, alanine, valine, and serine (1, 33, 37, 56).

Some Φ^e_{AA} values were not obtained for *l*-isomers of amino acids. However, the differences in apparent specific volume (56) such as those for *d*-alanine, *l*-alanine, and *dl*-alanine (60.43, 60.47, and 60.50 cm^3/mol, respectively) are negligibly small for the accuracy required.

The values of Φ^e_{AA} calculated earlier (14, 107) for amino acids with uncharged polar side groups (threonine, cysteine, and tyrosine) appear very close to those measured recently (the differences are less than 1 cm^3/mol), whereas for amino acids charged at neutral pH they differ appreciably from the recently measured values. For glutamic acid, the calculated value was 90.4 instead of 85.9 cm^3/mol, for arginine 111.5 instead of 127.3 cm^3/mol. These differences indicate that the Φ^e_{AA} value is greatly dependent on the site the charged group occupies in the amino acid residue.

Table 3 does not include the volumes of most rare amino acids. These can be easily calculated using Equation 22. But in this review, β-alanine and pyroglutamic acid volumes are needed. For the first, the volume is assumed to be 58.3 cm^3/mol (1, 37, 78). For the second, the Φ^e_{AA} value could not be found in the literature. Based on the structural similarity of this amino acid (75) with proline and using Equation 22,

$$\Phi_{Pyr.A} = \Phi^{e}_{Proline} + \Phi_{CO} - \Phi_{CH_2} \qquad 23.$$

which gives the value 80.1 cm^3/mol.

AMINO ACID RESIDUES

It is impossible to estimate densitometrically the volume of an isolated part of a molecule. Therefore, in the case of amino acid residues, we deal with a hypothetical value. Except for this, two types of volumes of amino acid residues should be considered. The first can be used to estimate the real volume of protein in solution. In this case, the volume of amino acid residue includes the value determined by intramolecular interactions in proteins. The second type is used to calculate the hypothetical protein volume defined by its composition only, i.e. without intramolecular interactions. The first type will be conventionally termed the real volume and the second one the hypothetical volume of an amino acid residue.

Real Volumes

The values of the real volumes of amino acid residues, ϕ_{AR}, widely used to calculate apparent protein volume, ϕ_p, were obtained by Cohn & Edsall in 1943 (Table 4). The transition from the amino acid (see Table 3) to the amino acid residue was performed by subtracting the volume of one water molecule, Φ_{H_2O}, the electrostriction effect, Φ_E, that is due to a pair of charged groups (not involved in side radical) as well as Traube's co-volume, $\Phi_{co\text{-}V}$, inherent in whole molecules, using

$$\Phi^{e}_{AR} = \Phi^{e}_{AA} - \Phi_{H_2O} - \Phi_E - \Phi_{co\text{-}V}. \qquad 24.$$

The magnitudes in this Equation are the averages of numerous model experiments and amount to $\Phi_{H_2O} = 6.6$ cm^3/mol, $\Phi_E = -13.3$ cm^3/mol, $\Phi_{co\text{-}V} = 14.1$ cm^3/mol. As a result,

$$\Phi^{e}_{AR} = \Phi^{e}_{AA} - 7.4 \text{ cm}^3/\text{mol}. \qquad 25.$$

The calculation of apparent protein volume based on these data gives an error of no more than 2% (103, 104, 107).

Hypothetical Volumes

In 1972 we made an attempt to calculate the volume of a protein defined by its composition alone (107). Based on the available data as well as on some assumptions, we obtained ϕ^{*}_{AR} values. These suggested that amino acid residues do not participate in intramolecular interactions in proteins. In

Table 4 Values of apparent mole and specific volumes of amino acid residues at 25°C

Amino acid residue	M_{AR}† (dalton)	(ϕ_{AR})‡ (cm^3/g)	Φ^*_{AR} (cm^3/mol)	ϕ^*_{AR} (cm^3/g)	Φ_R (cm^3/mol)
gly	57.05	(0.64)	34.8	0.610	0
ala	71.07	(0.74)	52.0	0.732	17.2
val	99.13	(0.86)	82.4	0.831	47.6
leu	113.16	(0.90)	99.1	0.876	64.3
ile	113.16	(0.90)	99.1	0.876	64.3
phe	147.18	(0.77)	112.8	0.766	78.0
ser	87.08	(0.63)	51.9	0.596	17.1
thr	101.11	(0.70)	68.4	0.676	33.6
met	131.20	(0.75)	96.9	0.739	62.1
cys	103.14	(0.63)	65.0	0.630	30.2
tyr	163.18	(0.71)	114.7	0.703	79.9
try	186.22	(0.74)	135.5	0.728	100.7
asp	114.08	(0.60)	65.4	0.573	30.6
glu	128.11	(0.66)	77.5	0.605	42.7
asn	114.10	(0.62)	69.6	0.610	34.8
gln	128.13	(0.67)	85.5	0.667	50.7
lys	129.18	(0.82)	100.1	0.775	65.3
arg	157.20	(0.70)	118.9	0.756	84.1
his	137.14	(0.67)	90.4	0.659	55.6
pro	97.12	(0.76)	72.6	0.748	37.8

† M_{AR} is the mole weight of amino acid residue.
‡ Values obtained in 1943 (15) are suitable approximations to estimate the apparent specific volume of protein, ϕ_p.

most cases, the ϕ^*_{AR} values appeared to be smaller than real ϕ_{AR} values, whereas interactions between the amino acid residues can only result in increases in volumes.

New Φ^e_{AA} values allow us to reestimate the ϕ^*_{AR} values. We shall try to determine the volume by which the glycyl differs from glycine without considering intramolecular interactions. Note that glycyl is the smallest amino acid residue and represents a regularly occurring block of the polypeptide backbone in proteins of any composition.

To obtain the most reliable value for the glycyl volume one has to reduce the number of averaged values taken for calculations; to use model compounds with the same distance between the amino and carboxyl groups (50, 51), which participate in the formation of peptide bonds; and to minimize the effect produced by intramolecular interactions.

These conditions are met by comparing the Φ^e_{AA} values of glycine and asparagine. First, the chemical composition of asparagine molecule is absolutely identical with the sum of chemical compositions of glycine and glycyl. Second, the chemical structure of asparagine is composed of glycine (minus one hydrogen atom) and a glycyl isomer (plus one hydrogen atom) that presents a regularly repeating block in the backbone of all polypeptides. The system of equations

$$\Phi^{e}_{Glycine} = \Phi_{gly} + X, \quad 26.$$

$$\Phi^{e}_{Asparagine} = 2\Phi_{gly} + X \quad 27.$$

describes this, where

$$X = \Phi_{AA} - \Phi_{AR} \quad 28.$$

is the difference between the volumes of amino acid and the corresponding residue. The solution of the Equations 26–27 using the data of Table 3 gives $\Phi_{gly} = 34.8$ cm^3/mol and $X = 8.4$ cm^3/mol. Thus the apparent mole volume of an isolated amino acid residue is

$$\Phi_{AR} = \Phi^{e}_{AA} - 8.4 \text{ cm}^3\text{/mol.} \quad 29.$$

Equation 29 was obtained using Φ^{e}_{AA} of asparagine obtained more than 40 years ago (14, 15, 24). However, the new volume values of glycine and serine (33), which, like asparagine, are characterized by uncharged polar side radicals (49), have not undergone considerable changes. Therefore, one should expect that the asparagine volume will either be the same or will decrease somewhat in further measurements. This means that the resulting compositional protein volume (103, 104, 107) will further decrease while the conformational volume will increase.

Using Equation 29 we obtained new values of the apparent molar, Φ^{*}_{AR}, and specific, ϕ^{*}_{AR}, volumes of amino acid residues, which are given in Table 4 and can be used for calculating the compositional volume of a protein.

Table 4 also presents the apparent volume values of amino acid side radicals obtained by

$$\Phi_{R} = \Phi_{AR} - \Phi_{gly} = \Phi_{AR} - 34.8 \text{ cm}^3\text{/mol.} \quad 30.$$

This value denotes the volume of the residue that extends beyond the polypeptide backbone (except for one hydrogen atom).

PEPTIDES

Several terms are used to characterize the molecules composed of amino acid residues that are connected with each other by peptide bonds: oligopeptides, peptides, polypeptides, and proteins. The sequence of these terms corresponds roughly to the increase in molecular weight of these classes of chemical compounds. However, precise identification is difficult. All proteins are peptides, but not all peptides can be called proteins (e.g. regular synthetic polypeptides).

One approach to the quantitative definition of oligopeptides, for example, is based on a notation system. In the case of the decimal system, small numbers are used that do not exceed ten. In our case "the amino acid dictionary" consists of twenty items and, therefore, the molecules whose length is no more than about 20 amino acid residues can be called oligopeptides. This length is characteristic of biologically active peptides.

Unfortunately, experimental data on apparent mole volumes of oligopeptides, Φ^e_{pept}, are very scanty and are concerned with very short compounds (no more than 3–5 amino acid residues; see Table 5).

To calculate the volumes of such oligopeptides it is convenient to use the Φ^e_{pept} values for di-, tri-, tetra- and penta-glycyls (33) as well as for those of the side chains, Φ_R (Table 4). One uses the equation

$$\Phi_{k-pept} = \Phi_{k-gly} + \sum_{i=1}^{k} \Phi_R, \qquad 31.$$

where $k = 1, 2, 3, 4$, or 5. Table 5 shows good agreement of the calculated values with Φ^e_{pept}. Since experimental data for longer peptides are not available, it is reasonable to estimate the ϕ_{pept} from Equation 21 and the values of ϕ_{AR} from Table 4.

The calculated data on the volumes of some biologically active peptides consisting of 2 to 14 amino acid residues are presented in Table 6. It contains the volumes for the peptides with 2–5 residues derived from Equation 31. For longer peptides (6–14 residues), calculated values of compositional volume are given. The choice of biologically active peptides shown in Table 6 is rather random. There is no correlation between the

Table 5 Apparent mole volumes of short peptides at 25°C

Peptide	Φ^e_{pept} (cm^3/mol)	Reference	Φ_{pept}† (cm^3/mol)
$^+H_2$-gly-gly-O^-	76.27	33	
	76.5	18	
$^+H_2$-ala-ala-O^-	110.30	33	110.7
	110.9	18	
$^+H_2$-gly-ala-O^-	92.7	18	93.5
$^+H_2$-ser-ser-O^-	111.8	33	110.5
$^+H_2$-gly-gly-gly-O^-	111.81	33	
$^+H_2$-ala-ala-ala-O^-	163.80	33	163.4
$^+H_2$-ser-ser-ser-O^-	166.0	33	163.1
$^+H_2$-gly-gly-gly-gly-O^-	149.70	33	
$^+H_2$-ala-ala-ala-ala-O^-	220.10	33	218.5
$^+H_2$-gly-gly-gly-gly-gly-O^-	187.10	33	

† Calculated values (see the text).

Table 6 Apparent mole and specific volumes of biologically active peptides

No.†	Peptide	Amino acid sequence in peptide‖	Reference	M_{pept}¶ (dalton)	Φ_{pept}†† (cm^3/mol)	ϕ_{pept}†† (cm^3/g)
2	Carnosine‡	$^+H_2$-β-ala-his-O^-	58	226.2	147.0	0.650
3	Growth factor	$^+H_2$-gly-his-lys-O^-	66	341.4	232.7	0.682
4	Insulin B_{22-25}	$^+H_2$-arg-gly-phe-phe-O^-	94	526.6	389.8	0.740
5	Leu-enkephalin	$^+H_2$-tyr-gly-gly-phe-leu-O^-	65	555.6	409.3	0.737
6	Sex peptide	$^+H_2$-arg-gly-pro-phe-pro-ile-O^-	72	686.9	510.8*	0.744*
7	$ACTH_{4-10}$	$^+H_2$-met-glu-his-phe-arg-try-gly-O^-	95	962.1	666.8*	0.693*
8	Angiotensin II	$^+H_2$-asp-arg-val-tyr-ile-his-pro-phe-O^-	54	1046.2	756.3*	0.723*
9	Bradykinin	$^+H_2$-arg-pro-pro-gly-phe-ser-pro-phe-arg-O^-	54	1062.3	767.9*	0.723*
10	Luliberin	pyr-his-try-ser-tyr-gly-leu-arg-pro-gly-NH_2	75	1183.3	832.3*	0.703*
11	Substance P†††	$^+H_2$-arg-pro-lys-pro-gln-gln-phe-phe-gly-leu-met-NH_2	48	1349.6	999.3*	0.740*
12	Kassinin†††	$^+H_2$-asp-val-pro-lys-ser-asp-gln-phe-val-gly-leu-met-NH_2	98	1334.5	957.0*	0.717*
13	Neurotensin	pyr-leu-tyr-glu-asn-lys-pro-arg-arg-pro-tyr-ile-leu-O^-	80	1675.0	1228.8*	0.734*
14	Bombesin†††	pyr-gln-arg-leu-gly-asn-gln-try-ala-val-gly-his-leu-met-NH_2	2	1620.9	1167.2*	0.720*

† No. is the number of amino acid residues in peptide.
‡ The distance between charged N- and C-end groups is taken as that in $^+H_2$-gly-gly-O^-.
††† The volume of an amide at the C-end of peptide was not taken into account.
‖ pyr is pyroglutamic acid.
¶ M_{pept} is the molecular weight of peptide.
†† The values with the sign (*) are the compositional volumes.

specific volume and molecular weight of the oligopeptides. However, it can be seen that all peptides contain at least one cyclic amino acid side chain. This residue can be either terminal or a neighbor of glycyl (except kassinin) whose side chain has the smallest volume (according to Table 4 it equals zero). The same properties are exhibited by the structure of a fragment of β-casein—morphyceptin (10), xenopsin (48), MSH-release-inhibiting factor (59), met-enkephalin (65), serum thymus factor (68), morphogenetic peptide (74), thyroliberin (75), adipokinetic hormone (84), anorexigenic peptide (91), cholecystokinin$_{26\text{-}33}$ (97), and many others. This property was observed in 90% of endogenous peptides studied (about hundred) and can be evidence that the cyclic side chains often interact with receptors. The position of such a residue at an end or next to glycyl provides favorable sterical conditions for possible ligand-receptor interaction (for example, stacking type).

Since there are no data on apparent volumes of biologically active peptides, it is difficult to judge the values of their conformational volumes. However, it is possible to do so for quasi-regular peptides with 45 residues such as (pro-pro-gly)$_{15}$. The ϕ^{e}_{pept} value of the latter (in 3% water solution of acetic acid) is 0.74 cm^3/g (86). The compositional volume calculated from the data of Table 4 is 0.702 cm^3/g. Thus, the conformational volume determined by intra- and intermolecular interactions in the triple superhelix formed by this peptide is $4 \cdot 10^{-2}$ cm^3/g.

PROTEINS

The total volume of a protein, in principle, ranges from 0.60 to 0.90 cm^3/g. The limits of this range correspond to the ϕ_{AR} values for aspartyl and leucyl (Table 4). However, the range observed for natural proteins is much less (0.70–0.75 cm^3/g). The value of ϕ_p for proteins is often taken to be 0.75 without any measurements or calculations (108).

Kauzmann seems to be the first to have proposed the valid physical description of the total protein volume (35). According to his conception the protein volume is determined by the following factors: (*a*) constitutive volume, i.e. the volume of all protein atoms; (*b*) cavities in the protein structure resulting from imperfect packing of atoms; (*c*) ionized and polar groups that are solvated in solution; (*d*) structural changes near the hydrocarbon regions and other nonpolar groups of the protein molecule.

Evidently, the main part of the protein volume consists of the volumes of the amino acid residues. The concept of compositional volume (104, 107, 108) coincides with Kauzmann's concept of constitutive volume with one refinement: definition of the former suggests that chemical groups of a protein amino acid residue interact only with solvent molecules and chemical groups of its own residue. All other interactions (Kauzmann's factors 2–4) are evidence that there is an additional conformational volume. The refinement of ϕ_{AR}^{e} values and new precise data on ϕ_{p}^{e} did not change this point of view. Thus, the conformational volume is

$$\phi_{p}^{e} - \phi_{p}^{*} = \phi_{p}^{conf} > 0, \qquad 32.$$

where ϕ_{p}^{e} is protein volume estimated experimentally and ϕ_{p}^{*} is compositional volume. According to our new estimates, the ϕ_{p}^{conf} value can be as high as 0.03 cm^3/g.

The validity of this estimate is supported by numerous data on volume changes in protein solutions (Table 7). In this Table the first value exceeds our upper estimate of ϕ_{p}^{conf}. However, the effects of swelling and dissolution were observed when changing from dry protein to the dissolved state; that is why these effects can not be classified as volume changes in solution. The upper estimate of ϕ_{p}^{conf} is also exceeded by the volume change that accompanies the coil–β-form transition. However, during this process both intramolecular and intermolecular interactions are involved. All of the other processes described in Table 7, whether they reflect conformational changes of the protein molecule or formation of additional intermolecular interactions, are characterized by volume changes smaller than our upper estimate for ϕ_{p}^{conf}.

The aggregation following heat denaturation is an example of protein-protein interaction. In this case, the additional volume consists of two contributions, one of which is the result of intra- and the other of intermolecular interactions. The volume changes may be of different signs (Table 7). The dependence of volume on temperature is shown in Figure 3. The curve *3-2-1* illustrates a so-called "direct" gel formed by a temperature decrease, e.g. for gelatin (28). In the case of actomyosin (55), the gel is formed as the temperature increases and is termed a "reverse" gel. Corresponding

Table 7 The limiting values of volume changes observed in solution of proteins and synthetic polypeptides

No.	Process	Volume change (cm^3/g)	Reference
1	Swelling and dissolution	$-5 \cdot 10^{-2}$	11, 25
2	Helix-coil transition	$+10^{-2}$	60–62
3	α–β-transition	$-3 \cdot 10^{-2}$	61
4	Coil–β-form transition	$+4 \cdot 10^{-2}$	53, 61
5	G-F transition	$+7 \cdot 10^{-3}$	23, 31, 32, 83
6	Antigen-antibody reaction	$+7 \cdot 10^{-3}$	63
7	Sol-gel transition	$-6 \div +1 \cdot 10^{-4}$	28, 55
8	Thermal denaturation and aggregation	$-3 \div +9 \cdot 10^{-3}$	6, 9, 13, 27, 82, 87

to this transition is a curve read as *1-2-3*. Such differences in effects may be explained by the fact that the hydrophobicity of actomyosin is, to some extent, greater than that of gelatin (5) as deduced from Tanford's theory (4, 88, 89). The weakening of hydrogen bonds and strengthening of hydrophobic interactions with increasing temperature (77) determines the preferable type of interaction at a given temperature.

Heat denaturation of proteins is also illustrated in Figure 3. The curve *1-2-3* describes the process of heat denaturation of ovalbumin (6) and α-chimotrypsinogen (36), and the curve *1-2-4* shows denaturation of serum albumin (6) and RNase (29).

Numerous studies deal with the effect of various chemical agents (including denaturants) on protein volume. However, we shall limit ourselves to citing the fundamental review by Lapanje (45).

A schematic breakdown of the contributions to the total protein volume is shown in Figure 4 (112, 113). The largest portion is the compositional volume, calculated by a modified method of Cohn & Edsall using the values of ϕ^*_{AR} from Table 4. The modification is the substitution of the averaged values of apparent specific volumes of amino acid residues ϕ_{AR} for the ϕ^*_{AR} values. In any process that is not accompanied by the breakdown of peptide bonds, ϕ^*_p = const. Another important component is the conformational volume determined by Equation 32. Since the conformational component is positive, its value according to Figure 4 depends on the position of ϕ^e_p

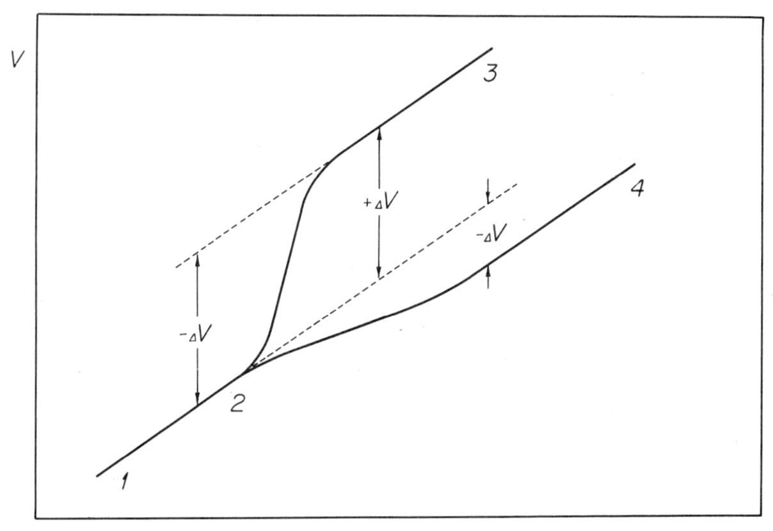

Figure 3 Dependence of protein solution volume on temperature.

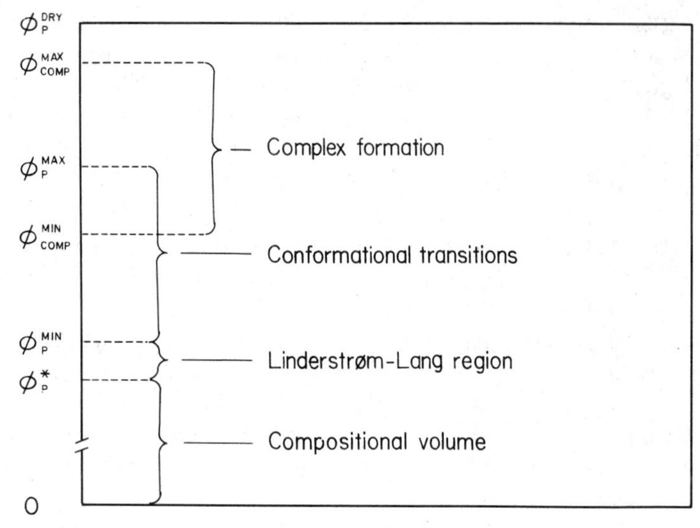

Figure 4 Schematic illustration of the contributions to the volume of a protein in solution.

value in the inequality

$$\phi_p^{min} \leq \phi_p^e \leq \phi_p^{max}, \qquad 33.$$

where ϕ_p^{min} and ϕ_p^{max} are the limiting values of protein volume with regard to all possible conformations.

The difference $\phi_p^{min} - \phi_p^*$ represents a volume described by Linderstrøm-Lang (50, 51). This volume is a small addition to the compositional volume and results from the interaction of side radicals of amino acid residues with their nearest neighbors in the peptide chain. Any conformational rearrangements of protein occurring without breakage of the peptide bonds do not essentially influence this difference.

Figure 4 shows two further regions of conformational transitions and complex formation. When complex formation does not occur, the conformational transitions accompanying the breakage or formation of weak (non-covalent) bonds, i.e. redistribution of hydrophilic and hydrophobic interactions in protein monomers, can result in a volume change within ϕ_p^{max}–ϕ_p^{min}. In the other case, the range of complex formation ϕ_{comp}^{max}–ϕ_{comp}^{min} must overlap that of the conformational transitions, as these two processes can be accompanied by volume changes of different signs (curves 2 and 3 in Figure 1).

Figure 4 shows also that the magnitude of dry protein volume is greater than any of the values for protein in solution. This fact is based on the data on protein dissolution (item 1 in Table 7) as well as on the crystallographic estimation of protein volume (12, 69, 70).

CONCLUDING REMARKS

Both absolute values of thermodynamic protein volume and relative values, or volume changes, have been considered in the present work. Apparent volumes of amino acids, peptides, and proteins were mostly estimated in aqueous solutions. These experimental conditions are only an approximation to conditions in vivo. The accuracy of modern experimental technique makes it possible to carry out studies under conditions approaching physiological ones. However, at present, only a few such studies, e.g. the investigation of muscle proteins and myofibrils, have been reported (3, 19, 20, 30, 111).

The complete thermodynamic characterization of proteins must include data obtained with high pressure (26, 93), calorimetric (85), and ultrasonic (22, 73) techniques, along with densitometry and dilatometry. Only then will it be possible to interpret the "behavior" of biological molecules under various conditions.

Such complex investigations will undoubtedly be a reliable basis for interpreting the functional mechanisms of amino acids, peptides, and proteins in vivo.

ACKNOWLEDGMENTS

I would like to express my appreciation to Prof. K. V. Sudakov, Prof. A. M. Molchanov, Prof. Yu. L. Sokolov, and Dr. V. V. Sherstnev for supporting this work and for fruitful discussions.

Special thanks are due to M. S. Il'ina, N. V. Dushkina, O. E. Golovchenko, and P. I. Bratyshev for their technical assistance.

Literature Cited

1. Ahluwalia, J. C., Ostiguy, C., Perron, G., Desnoyers, J. E. 1977. *Can. J. Chem.* 55:3364–67
2. Anastasi, A., Erspamer, V., Bucci, M. 1972. *Arch. Biochem. Biophys.* 148:443–46
3. Baskin, R. J. 1964. *Biochim. Biophys. Acta* 88:517–27
4. Bigelow, C. C. 1967. *J. Theor. Biol.* 16:187–211
5. Bigelow, C. C., Channon, M. 1976. In *Handbook of Biochemistry and Molecular Biology*, ed. G. D. Fasman, 1: 209–43. Cleveland: CRC. 427 pp.
6. Bull, H. B., Breese, K. 1973. *Biopolymers* 12:2351–58
7. Cabani, S., Conti, G., Matteoli, E., Tiné, M. R. 1981. *J. Chem. Soc. Faraday Trans. 1* 77:2377–84
8. Cabani, S., Conti, G., Matteoli, E., Tiné, M. R. 1981. *J. Chem. Soc. Faraday Trans. 1* 77:2385–94
9. Cassel, J. M., Christensen, R. G. 1967. *Biopolymers* 5:431–37
10. Chang, K.-J., Killian, A., Hazum, E., Cuatrecasas, P. 1981. *Science* 212:75–77
11. Chick, H., Martin, C. J. 1913. *Biochem. J.* 7:92–99
12. Chothia, C. 1975. *Nature* 254:304–8
13. Christensen, R. G., Cassel, J. M. 1967. *Biopolymers* 5:685–89
14. Cohn, E. J., Edsall, J. T. 1943. In *Proteins, Amino Acids and Peptides*, pp. 155–76. New York: Reinhold
15. Cohn, E. J., Edsall, J. T. 1943. See Ref. 14, pp. 370–81
16. Cohn, J. T., McMeekin, T. L., Edsall, J.

T., Blanchard, M. H. 1933. *J. Biol. Chem.* 100: Proc. XXVIII–XXXI
17. Cohn, J. T., McMeekin, T. L., Edsall, J. T., Blanchard, M. H. 1934. *J. Am. Chem. Soc.* 56:784–94
18. Dyke, S. H., Hedwig, G. R., Watson, I. D. 1981. *J. Solution Chem.* 10:321–31
19. Gabelova, N. A., Zamyatnin, A. A. 1970. *Stud. Biophys.* 20:35–42
20. Gabelova, N. A., Zamyatnin, A. A. 1971. *Biofizika* 16:163–71
21. Geffken, W., Beckman, C., Kruis, A. 1933. *Z. Phys. Chem.* 20:398–419
22. Gekko, K., Noguchi, H. 1979. *J. Phys. Chem.* 83:2706–14
23. Gerber, B. R., Noguchi, H. 1967. *J. Mol. Biol.* 26:197–210
24. Greenstein, J. P., Wyman, J. 1936. *J. Am. Chem. Soc.* 58:463–65
25. Haurowitz, F. 1963. *The Chemistry and Function of Proteins.* New York: Academic. 455 pp.
26. Heremans, K. 1982. *Ann. Rev. Biophys. Bioeng.* 11:1–21
27. Heymann, E. 1936. *Biochem. J.* 30:127–31
28. Heymann, E. 1936. *Trans. Faraday Soc.* 32:462–73
29. Holcomb, D. N., Van Holde, K. E. 1962. *J. Phys. Chem.* 66:1999–2006
30. Hotta, K., Terai, F. 1966. *Arch. Biochem. Biophys.* 114:288–98
31. Ikkai, T., Ooi, T., Noguchi, H. 1966. *Science* 152:1756–57
32. Jaenicke, R., Lauffer, M. A. 1969. *Biochemistry* 8:3083–92
33. Jolicoeur, C., Boileau, J. 1978. *Can. J. Chem.* 56:2707–13
34. Katz, S. 1972. *Methods Enzymol.* 26: 395–406
35. Kauzmann, W. 1959. *Adv. Protein Chem.* 14:1–64
36. Kharakoz, D. P., Sarvazyan, A. P. 1980. *Stud. Biophys.* 79:179–80
37. Kirchnerova, J., Farrell, P. G., Edvard, J. T. 1976. *J. Phys. Chem.* 80:1974–80
38. Klotz, I. 1950. *Chemical Thermodynamics,* pp. 183–204. Englewood Cliffs, NJ: Prentice-Hall. 369 pp.
39. Klotz, I. 1958. *Science* 128:815–22
40. Kopp, H. 1839. *Poggendorf's Ann.* 47:133–52
41. Kratky, O., Leopold, H., Stabinger, H. 1969. *Angew. Phys.* 27:273–77
42. Kratky, O., Leopold, H., Stabinger, H. 1973. *Methods Enzymol.* 27:98–110
43. Kupke, D., Beams, J. W. 1972. *Methods Enzymol.* 26:74–107
44. Lamb, A. B., Lee, R. E. 1913. *J. Am. Chem. Soc.* 35:1666–93
45. Lapanje, S. 1978. *Physicochemical Aspects of Protein Denaturation.* New York: Wiley Intersci. 331 pp.
46. Lapanje, S., Skerjanc, 1975. *J. Immunol. Methods* 9:195–200
47. Le Bas, G. 1915. *The Molecular Volumes of Liquid Chemical Compounds.* London: Longmans. 275 pp.
48. Leeman, S. E., Mroz, E. A., Carraway, R. E. 1977. In *Peptides in Neurobiology,* ed. H. Gainer, pp. 99–144. New York: Plenum. 464 pp.
49. Lehninger, A. L. 1975. *Biochemistry.* New York: Worth. 1104 pp.
50. Linderstrøm-Lang, K. 1962. *Selected Papers.* Copenhagen: Danish. 584 pp.
51. Linderstrøm-Lang, K., Jacobsen, C. F. 1941. *C. R. Trav. Lab. Carlsberg Ser. Chim.* 24:1–48
52. Linderstrøm-Lang, K., Lanz, H. 1935/38. *C. R. Trav. Lab. Carlsberg Ser. Chim.* 21:315–38
53. Makino, S., Noguchi, H. 1971. *Biopolymers* 10:1253–60
54. Marks, N. 1977. See Ref. 48, pp. 221–25
55. Marsland, D. A., Brown, D. E. S. 1942. *J. Cell. Comp. Physiol.* 20:295–305
56. Millero, F. J., Surdo, A. L., Shin, C. 1978. *J. Phys. Chem.* 82:784–92
57. Moelwyn-Hughes, E. A. 1961. *Physical Chemistry.* London: Pergamon. 1295 pp.
58. Nagai, K., Murakami, H., Sano, A., Kabutake, H. 1970. *Arzneim. Forsch.* 20:1876–78
59. Nair, R. M. G., Kastin, A. J., Shally, A. V. 1972. *Biochem. Biophys. Res. Commun.* 47:1420–25
60. Noguchi, H. 1966. *Biopolymers* 4: 1105–13
61. Noguchi, H. 1975. *Prog. Polym. Sci. Jpn.* 8:191–232
62. Noguchi, H., Yang, J. T. 1963. *Biopolymers* 1:359–70
63. Ohta, Y., Gill, T. J., Leung, C. S. 1970. *Biochemistry* 9:2708–13
64. Ostanevich, Y. M., Serdyuk, I. N. 1982. *Usp. Fiz. Nauk* 137:85–116
65. Pearse, A. G. E. 1978. In *Centrally Acting Peptides,* ed. J. Hughes, pp. 49–57. London: Macmillan. 259 pp.
66. Pickart, L., Thayer, L., Thaler, M. M. 1973. *Biochem. Biophys. Res. Commun.* 54:562–66
67. Picker, P., Tremblay, E., Jolicoeur, C. 1974. *J. Solution Chem.* 3:377–84
68. Pleau, J.-M., Dardenne, M., Blouquit, Y., Bach, J.-F. 1977. *J. Biol. Chem.* 252:8045–47
69. Richards, F. M. 1974. *J. Mol. Biol.* 82: 1–14
70. Richards, F. M. 1977. *Ann. Rev. Biophys. Bioeng.* 6:151–76
71. Sacura, J. D., Reithel, F. J. 1972. *Methods Enzymol.* 26:107–19
72. Sakurai, A., Sakata, K., Tamura, S.,

Aizawa, K., Yanagishima, N., Shimoda, C. 1976. *Agric. Biol. Chem.* 40:1451–52
73. Sarvazyan, A. P., Hemmes, P. 1979. *Biopolymers* 18:3015–24
74. Schaller, H. C., Bodenmüller, H. 1981. *Proc. Natl. Acad. Sci. USA* 78:7000–4
75. Schalley, A. V. 1978. *Science* 202:18–28
76. Scheraga, H. A. 1961. *Protein Structure*, pp. 1–30. New York: Academic. 305 pp.
77. Scheraga, H. A., Neméthy, G., Steinberg, I. Z. 1962. *J. Biol. Chem.* 237: 2506–8
78. Shahidi, F., Farrell, P. G. 1978. *J. Chem. Soc. Faraday Trans. 1* 72:858–68
79. Shahidi, F., Farrell, P. G. 1981. *J. Chem. Soc. Faraday Trans. 1* 77:963–68
80. Snyder, S. H., Uhl, G. R., Kuhar, M. J. 1978. See Ref. 65, pp. 85–97
81. Sreenivasaya, M., Bhagvhat, K. 1937. *Ergeb. Enzymforsch.* 6:234–43
82. Stauff, J., Rasper, J. 1958. *Kolloid Z.* 159:97–108
83. Stevens, C. L., Lauffer, M. A. 1965. *Biochemistry* 4:31–37
84. Stone, J. V., Mordue, W., Batley, K. E., Morris, H. R. 1976. *Nature* 263:207–11
85. Sturtevant, J. M. 1972. *Methods Enzymol.* 26:227–53
86. Sutoh, K., Noda, H. 1974. *Biopolymers* 13:2391–2404
87. Tanford, C. 1958. *Adv. Protein Chem.* 23:121–282
88. Tanford, C. 1962. *J. Am. Chem. Soc.* 84:4240–47
89. Tanford, C. 1980. *The Hydrophobic Effect: Formation of Micelles and Biological Membranes.* New York: Wiley. 233 pp.
90. Traube, J. 1896. *Liebig's Ann.* 290:43–122
91. Trygstad, O., Foss, I., Edminson, P. D., Johanson, J. H., Reihelt, K. L. 1978. *Acta Endocrinol. Copenhagen* 89:196–208
92. Ulrich, D. V., Kupke, D. W., Beams, J. W. 1964. *Proc. Natl. Acad. Sci. USA* 52:349–56
93. Weber, G., Drickamer, H. G. 1983. *Q. Rev. Biophys.* 16:89–112
94. Weitzel, G., Eisele, K., Guglielmi, H., Stock, W., Renner, R. 1971. *Hoppe-Seyler's Z. Physiol. Chem.* 352:1735–38
95. Wolthuis, O. L., de Wied, D. 1976. *Pharmacol. Biochem. Behav.* 4:273–78
96. Worcester, D. L., Gills, J. M., O'Brien, E. J., Ibel, K. 1976. *Brookhaven Symp. Biol.* 27(3):101–14
97. Yajima, H. 1977. *Gastroenterology, Baltimore* 72:793–96
98. Yamamura, H. I., Enna, S. J., eds. 1980. *Neurotransmitter Receptors*, Pt. 1, p. 73. London: Chapman & Hall. 212 pp.
99. Zamyatnin, A. A. 1968. USSR Patent No. 276484
100. Zamyatnin, A. A. 1968. USSR Patent No. 276487
101. Zamyatnin, A.. 1969. USSR Patent No. 280926
102. Zamyatnin, A. A. 1969. *Stud. Biophys.* 17:165–72
103. Zamyatnin, A. A. 1970. *Stud. Biophys.* 24/25:53–60
104. Zamyatnin, A. A. 1971. *Biofizika* 16: 163–71
105. Zamyatnin, A. A. 1971. *Stud. Biophys.* 28:99–104
106. Zamyatnin, A. A. 1971. *Zh. Fiz. Khim.* 45:1007–9
107. Zamyatnin, A. A. 1972. *Progr. Biophys. Mol. Biol.* 24:107–23
108. Zamyatnin, A. A. 1973. *The Dilatometry of Protein Solutions.* Moscow: Nauka. 101 pp. (In Russian)
109. Zamyatnin, A. A. 1973. *Stud. Biophys.* 40:221–28
110. Zamyatnin, A. A. 1976. In *Biophysics and Biochemistry of Muscle Contraction*, pp. 138–44. Moscow: Nauka. 291 pp. (In Russian)
111. Zamyatnin, A. A. 1977. In *Molecular and Cell Biophysics*, ed. G. Frank, pp. 178–86. Moscow: Nauka. 310 pp. (In Russian)
112. Zamyatnin, A. A. 1978. In *6th Int. Biophys. Congr. Abstr., V-32, Kyoto, Japan*
113. Zamyatnin, A. A. 1982. In *First All-Union Congress on Biophysics, Abstr.*, 1:17–18. Moscow: Inst. Biol. Phys. Acad. Sci. USSR. 372 pp. (In Russian)

Ann. Rev. Biophys. Bioeng. 1984. 13: 167–89

STRUCTURAL IMPLICATIONS OF THE MYOSIN AMINO ACID SEQUENCE

A. D. McLachlan

Medical Research Council Laboratory of Molecular Biology, Hills Road, Cambridge CB2 2QH, England

INTRODUCTION

Many questions about the action of muscle would be answered if we knew the atomic structures of both myosin and actin. The analysis of complete myosin genes and the sequencing of the amino acids are vital steps towards this end. Genetic analysis identifies the different variants of myosin that exist in each animal, and the study of mutants will help to distinguish essential parts of the molecule. New cloned genes with changed functions will soon be constructed. The amino acid sequence places the important active groups and structural units of this very large protein in their correct framework. It also helps to show how the individual molecules form into regular arrays in the thick filaments of muscle.

Under the electron microscope, individual myosin molecules appear to have long, thin rodlike tails (78) with two globular heads emerging in a forked configuration (29) at one end. Each molecule is a doublet containing two, paired myosin chains. The rod part is approximately 1500 Å long and 20 Å in diameter, while the heads are elongated, with a diameter of 70 Å and length of up to 200 Å. The main protein subunit of myosin is called the heavy chain. The unc-54 gene heavy chain (66, 92) from the soil nematode worm *Caenorhabditis elegans* contains 1966 amino acids (Figures 1 and 2). Figure 1 also shows part of M. Elzinga's chemical sequence from the head of rabbit skeletal muscle (13, 30, 31, 134) for comparison. The rod sequence from nematode alone (Figure 2) contains 7-residue and 28-residue repeats, so it is laid out in zones of 28 amino acids with its own local numbering system (1′ to 1117′) indicated in the rest of this article by primes. Other sequences and special features marked in the Figures are described later.

0084–6589/84/0615–0167$02.00

```
    "Head"  1-849
    "25K Segment"       Methyl-Lys↑
----MEHEKDPGWQYLRRTREQVLEDQSKPYDSKKNVWIPDPEEGYLAGEITAT
SSDADMAVFGEAAPYLRKSEKERIEAQNKPFDAKNSVFVADPKESYVKATVQSR
            10        20        30        40        50
                           Reactive Lys↑
    KGDQVTIVTAREMSVIQVTLKKELVQEMNPPKFEKTEDMSNLSFLNDASV
    EGGKVTVKTEAGASVTVKED---QVFPMNPPKYDKIEDMAMMTHLHEPAV
            60        70        80        90       100
                  Trimethyl-Lys↑*Photolabel          [-
    LHNLRSRYAAMLIYTYSGLFCVVINPYKRLPIYTDSCARMFMGKRKTEMP
    LYNLKERYAAWMIYTYSGLFCVTVNPYKWLPVYNAEVVTAYRGKKRQEAP
           110       120       130       140       150
    -----------ATP binding ?-------------]
    PHLFAVSDEAYRNMLQDHENQSMLITGESGAGKTENTKKVICYFAAVGAS
    PHIFSISDNAYQFMLTDRENQSILITGESGAGKTVNTKRVIQYFATIAIT
           160       170       180       190       200
    [---"Loop"---] "50K Segment"
    QQEGGAEVDPNKKKVTLEDQIVQTNPVLEAFGNAKTVRNNNSSRFGKFIR
    GDK...............................................
           210       220       230       240       250
    IHFNKHGRLASCDIEHYLLEKSRVIRQAPGERCYHIFYQIYSDFRPELKK
           260       270       280       290       300
    ELLLDLPIKDYWFVAQAELIIDGIDDVEEFQLTDEAFDILNFSAVEKQDC
           310       320       330       340       350
    YRLMSAHMHMGNMKFKQRPREEQAEPDGTVEAEKASNMYGIGCEEFLKAL
           360       370       380       390       400
    TKPRVKVGTEWVSKGQNCEQVNWAVGAMAKGLYSRVFNWLVKKCNLTLDQ
           410       420       430       440       450
    KGIDRDYFIGVLDIAGFEIFDFNSFEQLWINFVNEKLQQFFNHHMFVLEQ
           460       470       480       490       500
    EEYAREGIQWVFIDFGLDLQACIELIEKPLGIISMLDEECIVPKATDLTL
           510       520       530       540       550
    ASKLVDQHLGKHPNFEKPKPPKGKQGEAHFAMRHYAGTVRYNCLNWLEKN
           560       570       580       590       600
                                        [---"loop"---]
    KDPLNDTVVSAMKQSKGNDLLVEIWQDYTTQEEAAAKAKEGGGGGKKKGK-
          610       620       630       640       650
    "23K Segment"
    SGSFMTVSMLYRESLNNLMTMLNKTHPHFIRCIIPNEKKQSGMIDAALVL
    ............................................EHELVL
           660       670       680       690       700
     SH2↑ Thiols  ↑SH1
    NQLTCNGVLEGIRICRKGFPNRTLHPDFVQRYAILAAKEAKSDDDKKKCA
    HELRCNGVLEGIRICRKGFPSRILYADFKQRYKVLNASAIPEGQFIDSKK
           710       720       730       740       750
    EAIMSKLVNDGSLSEEMFRIGLTKVFFKAGVLAHLEDIRDEKLATILTGF
    ASEKLLGSIDVDHQ--TYKFGHTKVFFKAGLLGLLEEM............
           760       770       780       790       800
         [----"Swivel"----]
    QSQIRWHLGLKDRKRRMEQRAGLLIVQRNVRSWCTLRTWEWFKLYGKVK
           810       820       830       840       849
```

Figure 1 Myosin head amino acid sequences of nematode (*top*) and rabbit aligned, in one-letter code. Incomplete parts (...) and alignment gaps (–) are marked. Special features are labelled above the sequence, and position numbers come below.

```
"Rod"   Zones of 28 Residues
                                              rod        full
   defgabc defgabc defgabc defgabc          numbers    numbers
   "S-2"
 1                 PMLKAGK EAEELEK             14'        863
 2 INDKVKA LEDSLAK EEKLRKE LEESSAK             42'        891
 3 LVEEKTS LFTNLES TKTQLSD AEERLAK             70'        919
 4 LEAQQKD ASKQLSE LNDQLAD NEDRTAD             98'        947
 5 VQRAKKK IEAEVEA LKKQIQD LEMSLRK            126'        975
 6 AESEKQS KDHQIRS LQDEMQQ QDEAIAK            154'       1003
 7 LNKEKKH QEEINRK LMEDLQS EEDKGNH            182'       1031
 8 QNKVKAK LEQTLDD LEDSLER EKRARAD            210'       1059
 9 LDKQKRK VEGELKI AQENIDE SGRQRHD            238'       1087
10 LENNLKK KESELHS VSSRLED EQALVSK            266'       1115
                            .....
11 LQRQIKD GQSRISE LEEELEN ERQSRSK            294'       1143
   ...."Short S-2.....]     [-"Weak"
12 ADRAKSD LQRELEE LGEKLDE QGGATAA            322'       1171
   --]["Long S-2"....
13 QVEVNKK REAELAK LRRDLEE ANMNHENQ        *  351'       1200
14 LGGLRKK HTDAVAE LTDQLDQ LNKAKAK            379'       1228
15 VEKDKAQ AVRDAED LAAQLDQ ETSGKLN            407'       1256
16 NEKLAKQ FELQLTE LQSKADE QSRQLQD            435'       1284
17 FTSLKGR LHSENGD LVRQLED AESQVNQ            463'       1312
18 LTRLKSQ LTSQLEE ARRTADE EARERQT            491'       1340
                 ....."Long S-2"..][.
19 VAAQAKN YQHEAEQ LQESLEE EIEGKNE            519'       1368
   ..."LMM".....
20 ILRQLSK ANADIQQ WKARFEG EGLLKADE        *  548'       1397
21 LEDAKRR QAQKINE LQEALDA ANSKNAS            576'       1425
22 LEKTKSR LVGDLDD AQVDVER ANGVASA            604'       1453
23 LEKKQKG FDKIIDE WRKKTDD LAAELDG            632'       1481
24 AQRDLRN TSTDLFK AKNAQEE LAEVVEG            660'       1509
25 LRRENKS LSQEIKD LTDQLGE GGRSVHE            688'       1537
26 MQKIIRR LEIEKEE LQHALDE AEAALEA            716'       1565
27 EESKVLR AQVEVSQ IRSEIEK RIQEKEEE        *  745'       1594
28 FENTRKN HARALES MQASLET EAKGKAE            773'       1622
29 LLRIKKK LEGDINE LEIALDH ANKANAD            801'       1650
30 AQKNLKR YQEQVRE LQLQVEE EQRNGAD            829'       1678
31 TREQFFN AEKRATL LQSEKEE LLVANEA            857'       1706
32 AERARKQ AEYEAAD ARDQANE ANAQVSS            885'       1734
33 LTSAKRK LEGEIQA IHADLDE TLNEYKA            913'       1762
34 AEERSKK AIADATR LAEELRQ EQEHSQH            941'       1790
35 VDRLRKG LEQQLKE IQVRLDE AEAAALKG        *  970'       1819
36 GKKVIAK LEQRVRE LESELDG EQRRFQD            998'       1847
37 ANKNLGR ADRRVRE LQFQVDE DKKNFER           1026'       1875
38 LQDLIDK LQQKLKT QKKQVEE AEELANL           1054'       1903
39 NLQKYKQ LTHQLED AEERADQ AENSLSK           1082'       1931
40 MRSKSRA SASVAP                            1095'       1944
   "Tailpiece"
   GLQSSASAAVIRSPSRARASDF                    1117'       1966
```

Figure 2 Nematode rod sequence displayed as forty 28-residue zones with both local (primed) and full chain numbering on the right. Skip residues (*) and coil heptad positions *a*, *b*, *c*, *d*, *e*, *f*, *g* are marked.

This review concentrates on structural aspects of muscle, excluding chemical kinetics (39, 83, 130) and the dynamics of contraction (34, 48, 56, 57, 59). We begin with a short account of basic facts. Next we consider the myosin genes and the topography of the active regions in the head. An analysis of regularities in the rod sequence then leads on to questions about thick filament packing and the mechanical flexibility of the rod.

MUSCLE STRUCTURE

STRIATED AND SMOOTH MUSCLE The motive power for contraction in all known muscles is the movement of myosin molecules along actin filaments, energized by the hydrolysis of ATP. The two kinds of molecule are organized in separate filaments (47, 58) that are attached to parts of the cell and transmit the tension, but the detailed filament arrangement varies in different animals and different muscle types (46). The most important structural division is between striated muscle, which has a striped appearance under the microscope, and smooth muscle. Striated muscle is built up of long lines of contractile units, the sarcomeres, placed end to end, which contract in concert. The skeletal and heart muscles of vertebrates, the wing muscles of insects, and the body-wall muscles of nematodes are all striated. They do, however, differ considerably in their control mechanisms. Skeletal muscles are activated by nerve impulses; heart muscles undergo waves of contraction partly under local control; and insect wing muscles are designed to oscillate rapidly. Smooth muscles are much simpler in structure, without sarcomeres, and often consist of a network of long spindle-shaped cells. Their contraction is often activated by chemical stimuli, such as hormones. The muscles in the blood vessels and uterus of vertebrates are smooth. So too are the catch muscles, which are used to close the shells of mollusks such as scallops and oysters. In spite of the anatomical differences between muscle types (123) and their different external activation mechanisms, the interaction between myosin and actin seems to be much the same in all of them. Most of our detailed structural knowledge comes from work on vertebrate skeletal muscles (54, 110), such as the frog's leg sartorius muscle, but many of the findings should apply in other types, including the obliquely striated body-wall muscle myosin from the nematode worm unc-54 gene.

THE SARCOMERE A typical vertebrate sarcomere is a cylindrical structure about 20–30,000 Å long and 20,000 Å in diameter, with circular disks (the Z-disks) at each end and another disk, the M-disk in the middle. From each Z-disk many attached thin filaments of actin project inwards towards the center. Each actin filament is about 10,000 Å long. The thick myosin filaments are double-ended structures which point out from the M-disk symmetrically in both directions up to a total length of about 16,000 Å. The Z-disk and M-disk keep the ends of their respective filaments in register and help to maintain a regular cross-section lattice, so that the thin filaments fit into the gaps between the thick ones and overlap them. The M-disk is a network of protein units which cross-link the midpoints of adjacent thick filaments. Under the microscope a side view of the sarcomere shows a series of bands or zones; the I-bands next to each Z-disk contain the near attached ends of the actin filaments, while the denser A-band contains the overlapped actin and myosin filaments. However, the actin filaments do not usually project far enough to reach all the way to the M-disk and so there is a central region, called the H-zone, where only the inner parts of the myosin shafts appear.

CROSS-BRIDGES The myosin filament is covered with small projecting knobs or cross-bridges (50, 51), which reach across to the neighboring actin filaments in the overlap zone and exert a tension on them when the muscle is active (52). In a highly stretched muscle with a sarcomere length of 36,500 Å, the length of overlap is almost zero and here the tension is extremely weak; but as the sarcomere shortens, the number of cross-bridge connections increases until the tension reaches a maximum at a length of 22,000 Å. At this length the whole cross-bridge array is overlapped by actin filaments. The lateral spacing between frog thick and thin filaments (50) is 200–280 Å.

REGULATION The energy-generating reaction of ATP hydrolysis to ADP takes place at an active site within the myosin molecule. The cross-bridge attaches itself to one or more actins on the thin filament and pulls itself along, so that the whole thick filament climbs along the thin one (52). The contraction event is regulated by calcium ions (68). In vertebrate skeletal muscle a nerve impulse generates a twitch by releasing a burst of calcium into the cell (49). The actin filament carries accessory proteins, tropomyosin (14), and the three subunits I, T, C of troponin (144), which normally inhibit the reaction with myosin, probably by covering up a receptor site on actin (53), but this explanation is controversial (13a, 137b). Calcium binds to the C subunit of troponin and thereby releases the inhibition so that the muscle is active. Activity continues briefly till the calcium is pumped out or the ATP is used up (26). In mollusks, such as scallop (69), calcium switches on the myosin molecule itself. It binds to a pair of small proteins—the regulatory light chains—which are attached to the cross-bridge region of myosin. These myosin light chains are closely related to troponin C (17). When the supply of ATP is exhausted under artificial conditions, the contraction cycle ends abnormally in the tense state of rigor (130). Here the ADP and phosphate have left the myosin active site and the cross-bridge is permanently bound to actin at the end of its working stroke. Addition of further ATP will now release the actin and recharge the head for further reaction cycles.

SUBUNITS OF THE MOLECULE The myosin molecule is a composite dimeric structure made up of two heavy chains (77, 79), each of molecular weight 230,000 daltons, and two pairs of smaller light chains of 20,000 daltons each. The whole molecule therefore has a weight of about 540,000. The heavy chain protein subunit consists of one complete globular head at the amino end and a long fibrous alpha-helical tail at the carboxyl end. The two helices in a complete molecule lock together, probably along their full length, to form a single rigid coiled coil, while the two heads fork apart at the tip. The light chains both belong to the calmodulin family of calcium-binding proteins and are of two kinds. The two tightly-bound essential light chains in each molecule [alkali light chains of rabbit muscle (35) or thiol light chains of scallop] were once thought to be required for the enzyme function of the head, but can in fact be removed (120a, 137a) without destroying the actin ATPase activity. The other two regulatory light chains [DTNB chains of rabbit (145) and EDTA chains of scallop (69)] are much more easily detached. They can bind calcium ions and in some muscles these light chains become phosphorylated (1). The light chains probably bind to the head near the neck of the rod. The junction between the rod and the two heads appears to be a flexible joint, since the heads are seen to splay out at various angles.

PROTEOLYTIC FRAGMENTS Enzymes such as trypsin and papain (80, 146) cut the myosin heavy chain into characteristic structural fragments (77). The rod is easily cut (146) in an extended region about one third of the way along its length to remove the carboxyl end portion (light meromyosin or LMM) and leave the two intact heads attached to a truncated rod (heavy meromyosin or HMM). A further cut of the HMM separates the globular head (subfragment 1, or S-1) from the first part of the tail (subfragment 2 or S-2). It is found that the S-1 fragment carries all the ATPase and actin-binding activity of the molecule. Complete rods can also be prepared from intact heavy chains by severing the S-1 with papain (80). The separation of the heavy chain into fragments is closely related to the mechanical properties of the molecule. The junction between S-1 and S-2 probably represents a pivot or swivel that allows each head to rotate independently of the tail. The junction between S-2 and LMM may contain a molecular hinge that lets the rod bend. Fluorescence depolarization measurements (95) and electron spin resonance spectra (132, 133) from labels attached to the head show that heads can rotate rapidly relative to the rest of the molecule (61) with a correlation time of 10^{-6} to 10^{-7} sec. The globular head also has an internal substructure, since it can be cleaved by enzymes into three principal segments (3), described, according to their molecular weights as the 25K, 50K, and 23K segments. Each segment has its own functions, and so the head may be built from three or more structural domains.

FUNCTIONS OF THE ROD The head of myosin is the active power element of muscle contraction where the energy of ATP is converted into mechanical work, but the rod is also of interest for several reasons.

1. The rods build the shaft of the thick filament and so their amino acid sequences contain much of the information which specifies the shaft structure. The heads of parallel molecules on the shaft lie on a regular helical surface lattice, which is different in different animals, but certain universal molecular spacings occur in all species. These are the 143–145 Å axial spacing between adjacent heads and the 430–440 Å helix repeat of heads which point in the same direction out of the surface.
2. In a middle "bare zone" of the filaments near the M-disk where there are no projecting heads, the rods are assembled back-to-back. The rods therefore have to pack together in both parallel and antiparallel groupings.
3. The rod tethers the head to the thick filament backbone, so that tension can be developed.
4. Near the head-rod junction is a pivot that allows the head to rotate freely, both when it seeks a contact to the actin filament and if it tilts during the working stroke.
5. There is evidence for a molecular hinge or joint where the rod can bend sharply, about one third of the way along its length. Electron micrographs of individual molecules can show kinks (29). A joint could be important during contraction (52). The myosin head would sit at the end of a long level arm or flexible stalk that lets it move out to reach a distant actin on a neighboring thin filament.
6. The rod itself is flexible. It can bend and possibly stretch under tension.

Analysis of the amino acid sequence is helping to answer questions about these functions of the rod:

1. How does the sequence specify the axial spacings of parallel molecules in the filaments and the length of the complete shaft?
2. What are the antiparallel interactions between rods in the bare zone?
3. What is the pitch of the coiled-coil, and how is it related to the three-dimensional packing of rods in the filament?
4. Is the coiled-coil structure of the rod regular or not; are there breaks in it?

5. Is there a hinge point in the alpha helix?
6. How does the rod behave mechanically; is it rigid, or can it behave like an elastic spring?

The rod sequence has unusual characteristics which point to some answers.

MYOSIN GENES

NEMATODE unc-54 GENE The soil nematode *C. elegans* is a small simple worm that is highly suitable for studying the genetics of muscle proteins and cell differentiation, through the molecular properties of mutant animals, especially ones with defective muscles. Since Brenner (9) began the study of nematode genetics, hundreds of paralyzed mutants or "uncoordinated" unc strains have been examined (154). They are the result of changes in over 100 genetic loci, which include at least 22 muscle protein genes (142). These strains include mutants of the myosin heavy chain (32, 86) and of paramyosin (42, 84, 139, 140). The normal unc-54 myosin gene codes for the most abundant heavy chain of the obliquely striated body wall muscle (32, 86). In the paralyzed unc-54 mutants the body muscles are usually highly disorganized, with only 20–25% of the usual thick filament content. Another gene, unc-15, produces paramyosin, which is required in the core of the thick filaments. Worms with unc-54 mutations can survive only because the pharynx muscle myosin is coded by two other genes and so the animal can still feed. The body-wall muscle is affected in two ways. Some mutants are recessive and fail to produce any stable protein. They include the E190 deletion mutant (85), which lacks the last 350 amino acids of the heavy chain and produces a very unstable messenger RNA; also the amber E1300 and nonsense E1092 mutants, which generate truncated heavy chains with unstable myosin molecules. The affected animals use a second minor body-wall myosin gene to produce a few residual thick filaments. Mutants of a second class are genetically dominant. They produce normal amounts of heavy chains which do not assemble correctly (86, 97, 141). Examples of this type are the E675 mutant (85), which has a deletion of 90 amino acids near the end of the rod, and E1157, where a point mutation to proline weakens the helical structure of LMM. Further genetic analysis of other myosin mutants is in progress (66).

CLONING THE NEMATODE GENE The unc-54 gene is an attractive starting point (65, 85) for the analysis of a genetically pure heavy chain amino acid sequence. Nematode genes contain many fewer intervening sequences than mammals such as rat or rabbit and so the gene sequence is relatively short. The number of heavy chain genes is small (two body-wall and two pharynx myosins), with only one copy of each gene, and the genetic map of the unc-

54 mutants is well understood. Mammals, in contrast, may produce twenty or more different variants of myosin. Genes often occur in multiple copies and are hard to distinguish from pseudogenes.

In order to clone the gene it was first necessary to distinguish the unc-54 messenger from the others. RNA was isolated and enriched from worms with normal, E675, and E190 myosin genes and was tested with antibodies to check whether these did indeed synthesize protein in vitro (85). Plasmid cDNA fragments specific for the unc-54 message were synthesized and selected, which could hybridize to both the normal unc-54 and the shorter E675 messengers. The plasmids were then used as probes to identify sections of nematode DNA which had been cloned randomly into bacteriophage lambda in lengths of 15–20,000 base pairs (65, 85). This method of analysis picked out 22 overlapping sections which covered the entire unc-54 gene region of 9,000 base pairs. Finally, short segments of unc-54 DNA were transferred into the single-stranded DNA of bacteriophage M13 and sequenced.

INTERVENING SEQUENCES The unc-54 gene has eight intervening sequences (64), five in the head after residues 21, 64, 114, 266, 528 and three short ones in the last 200 residues of the rod at positions 1750, 1822, 1897. Each intron begins with a typical 5′ sequence (8, 101) of the type GTRAGTTTT (R stands for A or G) and ends with a 3′ sequence TTCAG. Some of these introns also occur at corresponding positions in the other nematode myosin genes, but there are several variations. None of the intervening sequences appears to mark out any known important structural divisions of the molecule such as the S-1/S-2 or S-2/LMM junctions. As we have noted (Figure 2), the rod sequence has a regular 28-residue repeat, but the introns in the rod do not match it at all. The random exon pattern suggests that the main structural divisions of the molecule evolved into their present form without using any introns to splice the parts together. In spite of proposals that large proteins evolve by combining separate exons that code for structural domains with particular functions (23, 38), the gene sequences of several other proteins, such as leghemoglobin (62) lysozyme (63) and proinsulin (4), show that introns do not generally mark out self-sufficient folding domains but merely tend to lie on molecular surfaces (22). Several genes that code for proteins with repeated domains, such as collagen (153), bird ovomucoid protease inhibitor (124), or serum albumin (27), do contain repeated groups of introns, but the DNA that codes for the myosin rod clearly does not belong to this class.

OTHER MYOSINS Higher animals contain large numbers of muscle protein genes (103) which are used for special purposes and switched on in different parts of the body at certain stages of development (150, 151). There are many levels of complexity; multiple myosin heavy chains in each tissue; multiple copies of nearly identical genes; pseudogenes; messenger-RNA and protein processing; chemical modifications to amino acids. It is very difficult to isolate and sequence unique genes. A chemically determined amino acid sequence represents an average over some unspecificed mixture of gene products. In the rat and other vertebrates there are myosins for skeletal muscles (43), heart muscle (120), and smooth muscle, as well as non-muscle myosins. Each type can include several genes (74, 102). The reasons for having so many genes are not clear; functional differences between fast and slow muscles (37, 147), and the use of different types of rod sequence to build distinct zones of the thick filament, are possibilities. Nadal-Ginard et al (87) have identified several heart muscle genes in the rat and have sequenced part of the LMM (rod residues 673′–1100′) in two adult genes. The rat LMM sequence is remarkably similar to the nematode, with 48% of its amino acids the same. *Drosophila* has only one myosin heavy chain gene (7, 119), and those parts of the rod that have been analyzed are also clearly homologous to the other species. Variations in this insect seem to be produced by modifying the C-terminal part of the mRNA, possibly past the end of the protein coding sequence. Elzinga and his colleagues (13, 30, 31, 134) have chemically sequenced more than half of the rabbit skeletal heavy chain, including almost all of the head and most of S-2. They have positioned some interesting modified residues, such as the trimethyllysines in the head.

Comparisons between these sequences show that myosin has evolved very slowly (92). Seventy-five percent of the rat and rabbit amino acids match in residues 787′–1098′ of the LMM and 43% match between rabbit and nematode in the S-2. A more restricted comparison between the nematode unc-54 and gene-1 shows that the head is even more invariant than the rod, with 82% of 828 known residues identical, and that 64% of the nematode residues match the known rabbit head sequence. Variations in the head are concentrated in a few patches close to flexible loops (see below). The rod is significantly more variable, with only 54% of residues matching between unc-54 and gene-1. One exceptional highly conserved part of the rod lies in LMM between residues 581′ and 706′. Sequence variations in the rod are highly conservative when they occur; charged residues and hydrophobic groups in the coil backbone seldom change.

TOPOGRAPHY OF THE HEAD

Structural Segments

The whole heavy chain sequence of nematode falls naturally into three divisions. The head, residues 1–849, includes the active site for ATP hydrolysis and the actin-binding areas. The helical rod, whose ends are marked out by prolines 850 and 1944, is 1095 residues long (92) and this is followed by a short nonhelical tailpiece (residues 1945–1966), similar to one found in *Drosophila.*

The head itself (Figure 1) contains three major proteolytic segments (3) that are probably structural domains. The 25K segment (nominal molecular weight 25,000), residues 1–212, ends in a hydrophilic surface loop of variable length containing many lysines whose position corresponds to known tryptic cleavages of rabbit skeletal myosin after residues 204 and 212. The 50K segment, 213–645, ends in another long surface loop that contains clusters of Ala, Gly, and Lys. The 23K segment 646–849 extends right to the beginning of the rod. The molecular weights of the three segments, as described, are 22,690, 49,840, and 22,898. The pepsin cleavage site (80) just inside the beginning of the rod is after residue 853. The final portion of the 23K segment includes a further surface loop that probably represents the "swivel" where the head turns. The swivel is identified by a variable section of sequence (residues 808–824) and a tryptic cleavage site in rabbit (Arg 821). The remainder of the 23K segment (residues 822–849) may be a short neck that connects the main part of each head to the rod.

THE 25K SEGMENT This region contains important parts of the binding site for ATP which have been identified by photochemical labelling experiments (128) and spectroscopic studies. Trimethyllysine 128 is a particularly interesting position. The methyl groups surround the nitrogen atom and would ensure that it remains positively charged, even when buried inside the protein and placed close to a phosphate group. Okamoto & Yount (106) have designed a special nitroanilinoethyl diphosphate labelling compound with a size and shape similar to ATP, which fits specifically into the active site of rabbit myosin. The compound contains an azide group, so that photolysis makes the nitro group of the label form a covalent bond with any nearby amino acid side chain. Chemical analysis showed that the label binds to the 25K segment in the neighborhood of residues 124–133. Here it links to a tryptophan which corresponds to Arg 129 of nematode myosin. Another lysine in the 25K fragment (33, 44, 70) is noted for its ability to react with 2,4,6-trinitrobenzene sulphonate, with effects on the ATPase activity. This residue is known (71) to lie in a peptide Asn-Pro-Pro-Lys and can thus be identified as Lys 82. Lys 30 corresponds to a site which is 60% singly methylated in the rabbit. There is a short apparent sequence homology (138) between residues 139–206 in the 25K segment and the ATP binding sites from both the mitochondrial membrane ATP synthetase and adenylate kinase (107). The glycine-rich loop 177–182 is typical of many enzyme nucleotide binding sites. This observation raises the hope that myosin will turn out to be related to other ATP-driven enzymes.

THE 50K SEGMENT The main known function of the 50K segment is to bind actin. Kassab and his co-workers (100) have taken actin filaments activated with carbodiimide or bis-imido ester reagents and cross-linked them to S-1 which had been nicked by trypsin. Afterwards they analyzed the linked products. Kassab found that the 50K–23K part of S-1 was linked to two or more actin monomers, depending on the conditions (98, 99). Actin probably bound close to the 50K/23K joint, since this loop was protected from proteolysis. The most striking result, however, was that the cross-linking to actin enormously enhances the ATPase activity of myosin (73). On the other hand the 50K/23K cut abolishes this activation. These experiments suggest that contacts between actin and two movable segments of the head are needed to maintain the active conformation of myosin.

THE 23K SEGMENT This segment is not only concerned in the binding of actin. It also contains two reactive SH groups, called SH1 (Cys 715) and SH2 (Cys 705), that strongly affect ATPase activity (12, 148, 149). The groups are known to lie close in space to the ATP-binding region of the N-terminal segment, near Lys 82 (129) and Lys 128. The polypeptide chain between them (30), which has the sequence

T Ĉ N G V L E G I Ĉ R (residues 704–716),

appears to be flexible, since Yount has been able to cross-link the SH groups together at distances ranging from 2 to 14 Å. In fact this cross-linking (106) is a necessary condition for the ADP photolabel to bind near trimethyl-lysine 128. The picture which emerges from these and other experiments is that Mg^{2+}-ATP binds to S-1 in a pocket or cleft which can open and close in concert with a movement of the connection between Cys 705 and Cys 715. The pocket could lie entirely within the 25K segment, or else between it and the 23K segment. The active sites of yeast hexokinase, where a cleft opens and shuts between two large domains (6), and of phosphoglycerate kinase (143), where two widely separated lobes move together in the active form, suggest interesting analogies.

The other important suggested function of the 23K segment is to bind the light chains. Thus, the light chains may be in a position to influence the conformation of the reactive SH groups and control the ATPase activity. Further cross-linking experiments (40) show that, at least in scallop myosin, the two regulatory (EDTA) light chains lie close together when muscle is in the relaxed state. The light chains are also concerned in a curious interaction between the myosin heads of smooth muscle and the tails of the rods (105, 135). The regulatory light chains can exist in two states, phosphorylated or not. Phosphorylated filaments are quite stable (67) and the individual molecules have a normal straight rod. Nonphosphorylated filaments are unstable in solutions with magnesium-ATP and dissociate into single molecules that have a curious folded-up tail. The rod bends sharply at two hinge regions approximately 445 Å and 950 Å down its length (21). A section of LMM near the second bend then appears to stick between the necks of the two heads. So far there is no evidence that this folding occurs in striated muscle.

THE ACTIN-MYOSIN COMPLEX The three-dimensional structure of the actin-myosin complex has been analyzed by electron microscopy of thin filaments decorated with the S-1 heads of myosin in the rigor state (2, 131). This state, without ATP, may correspond to the end point of the working stroke in muscle, when the head becomes rigidly attached to actin (18). The head appears as a comma-shaped object about 150 Å long and 60 Å thick, while the actin monomers appear the same shape as in actin crystals (125), with a large and small lobe. The broad part of S-1 makes one main contact to an actin on one strand and a second extensive contact to the other lobe of an actin nearer the Z-disk on the opposite strand.

THE ROD

THE THICK FILAMENT In the double-ended structure of the filament (51), the tails of myosin molecules assemble side by side along the length of the shaft, and the heads project out from the surface (19). These are the active cross-bridge units themselves and are not rigidly attached to the shaft. The rods near the M-disk are packed antiparallel, with their heads pointing away from the middle, so that the central part of the thick filament, viewed from the side, contains a "bare zone" free of cross-bridges and approximately 1600 Å wide (19, 121). The outer regions of the filament contain rows of parallel myosin molecules with cross-bridges spaced at regular intervals of 143–145 Å. The diameter of the filament, the number of strands within it, and the precise cross-bridge spacings (55, 123) depend on the animal and the muscle type (96, 123, 152).

A typical vertebrate myosin filament is found in frog skeletal muscle (55). The mean diameter is 160 Å and the cross-bridge spacing is 143 Å. The frog filament has a helical arrangement of cross-bridges on the surface of the shaft with a threefold symmetric arrangement. At each cross-bridge level along the shaft, three cross-bridges project symmetrically outwards at angles of 120° to one another. There is also a helical repeat of 430 Å, which arises because every third set of cross-bridge levels is in the same angular orientation. Thick filaments are not always made entirely of myosin. In mollusks, insects, and nematodes they can be much thicker and they contain a cylindrical core of paramyosin (16, 36). Paramyosin is another alpha-helical coiled coil protein of two strands and itself assembles with axial staggers of 725 Å and 435 Å. Vertebrate muscles contain other proteins on the filament surface. A particularly important one is the C-protein (104), which may help to stabilize the filament structure.

Frog filaments divide into several structural regions as one moves from the central M-disk towards the tip. The cross section is hexagonal at the center and gradually becomes triangular in the bare zone (82). Next comes a succession of 50 cross-bridge levels, beginning with a set of inner levels, probably 12. Then there is a uniform zone 19 bridge-spacings long in which molecules of C-protein bind (20) at seven regular intervals of 430 Å. Lastly comes a tapered outer zone of 19 more levels, with a set of bridges missing close to the tip. These three distinct surface zones probably reflect differences in the internal packing of the shaft, and the distribution of C-protein shows that the surface repeat distance is 430 Å rather than 143 Å.

ASSEMBLY MECHANISMS Little is known in detail about the molecular events in rod assembly (24a, 113, 114). Each filament probably starts to grow from the central antiparallel packed bare zone (51) which is the most stable part of the filament under high pressures (24a), but there is evidence (10, 25) that myosin molecules (two chains each) are paired into parallel dimers staggered by 430 Å. Synthetic myosin filaments prepared without C-protein grow to limiting lengths of up to 16,000 Å under suitable conditions. Thus it appears that the myosin molecules by themselves can specify the approximate length and taper of the thick filament. Abortive double-ended minifilaments can form with only 16–18 complete molecules (118). Paracrystals of aligned rod fragments (5) have the same fundamental 430-Å and 143-Å parallel spacings as natural muscle. The driving force for assembly resides mainly in the LMM fragments, which have a strong tendency to aggregate. The S-2 shows a more complicated behavior. Proteolytic cleavage of rabbit skeletal rod into S-2 and LMM produces fragments of various lengths (79, 146). The biggest, called "long S-2" (127, 146) is about 517 residues long (80) and does aggregate in solution, but its C-terminal section can be cut further in many places to leave a "short S-2" fragment of only about 320 residues. Short S-2 does not normally associate and is only loosely held onto the filament surface (137).

Analysis of the Rod Sequence

The amino acid sequence of a regular alpha-helical protein yields direct structural information in a way which is not possible for globular proteins. At various levels of precision we can treat the rod as a one-dimensional object (for instance, a linear pattern of charged and uncharged groups), a cylindrical surface on the helix net, or a twisted coiled coil with a definite pitch. Simple exact methods can therefore be used to analyze the information in the amino acid sequence: Fourier analysis of amino acid distributions (91, 94, 108), counting of electrostatic interactions between charges (45), and searches for sequence repeat patterns (90).

HELICAL STRUCTURE The rod sequence in nematode (92) appears (Figure 2) as a very long alpha-helical region of 1094 amino acids uninterrupted by proline residues and about 1600 Å long. Secondary structure predictions suggest that almost the whole region is helical. Because the molecule is a dimer, the rod is a pair of helices that lie side by side in register with their axes approximately 9.8 Å apart (36). The two twisted strands are held together by a large number of hydrophobic side chains in contact. The outer surface of the coiled coil is covered with clusters of charged and polar amino acids, which are used in bridging interactions between each rod and its neighbors.

The hydrophobic groups have the characteristic cyclic 7-residue pattern *a*, *b*, *c*, *d*, *e*, *f*, *g* found in many coiled-coil proteins (24, 93, 111), in which the residues at positions *a* and *d* form a zig-zag pattern of knobs and holes down one side of the helix that interlock with the other strand and form a close-packed core. The core is rich in Leu, Val, and Ala and contains very few negatively charged amino acids. Side-chains of Arg and Lys, which have long aliphatic sections, occur fairly often. The outermost positions *b*, *c*, *f* are highly charged.

GENE DUPLICATION The sequence also contains a strong 28-residue repeat pattern which recurs, with variations, 40 times in the entire coil. This feature has probably evolved by gene duplication, like the repeats in collagen (153) and tropomyosin (94). The amino acids therefore have a periodic distribution along the rod. The pattern has two other features which may be important in the thick filament structure. First, the repeats are interrupted at four positions by the insertion of one extra "skip residue," which may distort the coiled coil. The skips occur at intervals of 196, 196, and 224 residues, which are multiples of 28. Second, a detailed analysis of the pattern (91) shows evidence for a longer repeated unit of 197 amino acids (seven 28's and one skip). The rod seems to have evolved from a 28-residue ancestral peptide that has been multiplied and modified to give the present structure. It is interesting that in *Acanthamoeba* myosin II the rod is only 860 Å long and therefore contains fewer of these repeats (117a).

BANDS OF CHARGE The banding of positive and negative charges along the rod is unusually regular. A typical 28-residue zone contains a strong band of negative charges with a band of positive charges 14 residues away from it. There are also subsidiary alternating bands of charge in between so that each zone has six bands in all. Fourier analysis of the charge distribution reveals strong periods of 28 and 28/3 residues, with a weak period of 196/9 in the negative charges only. These strong periodicities are peculiar to myosin, although weaker patterns occur in keratins (111), and we shall see that they are closely related to the filament structure.

WEAK POINTS IN THE HELIX The hydrophobic zigzag of the coil is not strictly regular. Charged groups sometimes invade core positions *a* or *d* in order to maintain the regular banding pattern. At each skip position, the extra residue may be accommodated in a wide turn of π helix with some perturbation of the coiled-coil pitch. The sequences in these regions are highly conserved. In some other parts of the coil (92) the hydrophobic core appears to be weakened locally by the presence of several polar or acid residues, and one such region, 316′–327′, which lies very close to the place where rods of both skeletal and smooth muscle appear to kink (21, 29), may be a "weak spot" or molecular hinge in the rod. The coiled coil shows breaks at this position in three nematode genes which each have a similar pattern of hydrophobic groups and charges. Thus the sequence may have adapted itself locally to fold in alternative helical and nonhelical forms. The short S-2 fragment ends between residues 289′ and 325′ (80, 81), while long S-2 probably ends near residue 519′. So the weak spot correlates well with the end of short S-2. The mechanical implications of a hinge in myosin are discussed later. So far, analysis of the nematode rod fails to show why enzymes so easily cut the long S-2 into pieces.

HALF-STAGGERED INTERACTING UNITS The "magic numbers" of 7, 28, and 197 residues in the sequence repeats, together with the regular bands of charge, suggest (109) that there will be strong electrostatic attractions if two 28-residue zones are displaced axially by 14 residues or if two complete rods are shifted by any odd multiple of 14. Assuming a helical rise (36) of 1.485 Å, each zone is 41.6 Å long. So the 143-Å spacing in muscle corresponds to a stagger of $3\frac{1}{2}$ zones or 98 residues (98.5 if we include the skips), while the 430 Å repeat corresponds to 294 or 295 residues. The 143-Å stagger may then arise from charge pairings between structural units of 197 amino acids staggered by half their length. The electrostatic force between two rods side by side has been calculated very simply (92) by treating the helices as linear arrays of charged and uncharged amino acids with short-range attractions and repulsions. The net attraction oscillates strongly with the stagger and has peaks at all odd multiples of 14 residues, but the strongest peaks of all are at 98 and 294 as expected. The attractions are rather weaker near the beginning of the S-2 than elsewhere, in keeping with the fact that S-2 does not aggregate with itself in solution whereas LMM does. A similar analysis (89) of the strongest attractions between two antiparallel rods suggests that in the bare zone the head ends of the rods are 1626 Å apart.

These periodic electrostatic forces lead to a simple linear view of the thick filament in which arrays of parallel one-dimensional rods are staggered by 143 Å in such a way as to match all the bands of positive and negative charge. For example, rabbit skeletal filaments are believed to contain three trigonally related smaller arrays, or subfilaments (82, 88), so that at every 143-Å cross-bridge level the three subfilaments each contribute one myosin molecule. As the rod is 1600 Å long, one subfilament would be at most twelve rods thick, but since the S-2 end of the rod is only loosely attached it is more realistic to picture a subfilament as an array of rods with loose ends that has a firm core of staggered LMM units 6–8 rods thick.

PACKING OF THE RODS The conclusion that the thick filament is held together largely by electrostatic forces suggests that the three-dimensional cross section must be such as to avoid the strong repulsions between rods which are staggered by even multiples of 143 Å. The closest-packed suitable lattice would be a square one (76), as has been proposed for paramyosin (28), but electron micrographs of muscles (115, 116) suggest that the packing is of a hexagonal type with a cross section that contains straight microfilaments (smaller than subfilaments) spaced 35–40 Å apart (152). The proposed microfilament could be a two-stranded rope of four alpha-helices made of two staggered myosin molecules with complementary charge bands. Rope structures pose problems about the continuity of the strands and the symmetrical equivalence of molecules, because a single pair of parallel rods staggered by, say 430 Å, has a leading and a lagging set of heads, which are not equivalent. An infinite two-stranded rope with this stagger could be built by allowing an 860-Å length of LMM to form two short successive 430-Å sections, each paired to the next staggered section of a near-neighbor LMM. The S-2 section of each rod would be more loosely bound. Every level of such a two-stranded rope includes both positive and negative charges, so two ropes would need to pack with their "plus" and "minus" strands correctly phased. One possibility (91) is a hexagonal lattice of twisted ropes in which the pattern of side-to-side interactions changes at every 60° of twist.

There is no agreed model for the internal structure of any thick filament. Although the arrangement of cross-bridges on the surface is known in scallop, horse-shoe crab, and tarantula, the shafts do not show up clearly in electron micrographs. Many models have been proposed (112, 122, 152), notably by Pepe (112), who favors two-stranded microfilaments, and Squire (122), who starts from simple arrangements of parallel myosin molecules on a flat foundation surface. Each proposal has difficulties with either filament symmetry or charge pairing. A plausible structural unit for the rabbit muscle subfilament (one third of a filament section) might be based on three long parallel two-stranded ropes (or microfilaments), each staggered by 143

Å past the next. Every rope could be a 430-Å-staggered series of LMM units with all its double heads facing to the same edge.

COILED-COIL PITCH The pitch of a helical coiled coil depends on the small difference between the 3.5-residue period of its hydrophobic seam and the 3.6-residue period of the alpha helix itself. X-ray data give 137 Å for the pitch of tropomyosin (117) and 140–190 Å for paramyosin (15, 28). In myosin itself, a complication is that the skip residues may modulate the pitch. The 28-residue sequence repeat suggested that the pitch might be a simple multiple of 28; for instance, a pitch of 112 residues would give each zone a 90° twist. The actual amount of twist might reveal itself through an optimal distribution of paired charges along edges of the helix surfaces when two interacting rods are placed side by side with a 143-Å or 430-Å stagger and the correct pitch. Calculations based on this idea (89) show evidence for preferred interactions along edges with pitches of 100.8 and 131.1 amino acids. The first value suggests a coil pitch close to 143 Å, whereas the second, which is 4/3 times 143 Å, is considerably larger (191 Å). A possible interpretation is that this long pitch comes from interactions between two rods in a twisted rope. A rope pitch of 4 times 143 Å, combined with a coil pitch of 143 Å, would give an apparent pitch of 191 Å, but another model that agrees better with the 430 Å helix repeat in muscle is a rope pitch of twice 430 Å combined with a coil pitch of 12/11 times 143 Å or 156 Å. This last type of rope would allow all cross-bridges staggered by 430 Å to emerge in the same direction.

Mechanical Flexibility of the Rod

Let us consider how the stiffness of an alpha-helical coiled coil compares with the force exerted on one double-headed myosin molecule when a muscle contracts. One way to estimate the force is through the energy of ATP hydrolysis (72). Assume that one ATP molecule produces 50 kJ/mol of free energy, which is used with 50% efficiency, and that 50% of the heads are attached to actin during a working stroke of 120 Å. The work done by each molecule is then 4.2×10^{-13} ergs, and the mean force on it is 3.5×10^{-7} dyn. Alternatively, we could take a maximum force of 3×10^{-6} dyn exerted by a square centimeter of frog muscle with approximately 6×10^{-10} thick filaments in its cross section. If each filament has 3 rods per cross-bridge level at 49 levels along the half-sarcomere, the tension in each rod would be 3.3×10^{-7} dyn.

An alpha-helix can be thought of either as a framework of peptide units held together by valence forces and hydrogen bonds or as a solid elastic cylinder with radius a, cross-sectional area A, and Young's modulus Y. The force F needed to stretch an elastic cylinder of length L by ΔL is then

$F = S(\Delta L/L)$, where S is the stretching stiffness constant. The couple G needed to bend the same cylinder into a circle of radius R is $G = B/R$, where B is the bending stiffness constant:

$$S = YA, \qquad B = \tfrac{1}{4}YAa^2. \qquad 1.$$

The stiffness of an alpha-helix of poly-Ala or poly-Gly has been calculated directly from the force constants of the chemical bonds (75, 126), with the result for a single helix of poly-Ala that

$$S_h = 1.96 \times 10^{-3}\ \text{dyn}, \qquad B_h = 4.05 \times 10^{-19}\ \text{dyn} \cdot \text{cm}^2\ \text{(helix)}.$$

The equivalent hypothetical cylinder with these constants would have a radius of 4.55 Å and a Young's modulus of 3.0×10^{11} dyn · cm^{-2}.

To extend this analysis to a coiled coil we now take two parallel cylinders of the same radius a in contact and twisted round one another. There are two local stiffness constants B_x and B_y for bending across the narrow and wide cross-sections respectively, which are calculated to be $B_x = \frac{1}{2}YAa^2$ and $B_y = 5YAa^2/2$. The mean effective bending constant B_c for a twisted coil with a long pitch is given by $2/B_c = 1/B_x + 1/B_y$, and the stretching stiffness is $2S_h$. So $S_c = 2YA$ and $B_c = 5YAa^2/6$. The theoretical estimated constants are thus:

$$S_c = 3.92 \times 10^{-3}\ \text{dyn}, \qquad B_c = 1.35 \times 10^{-18}\ \text{dyn} \cdot \text{cm}^{-2}\ \text{(coil)}.$$

The flexing of rodlike molecules in solution contributes a term to the frequency-dependent shear modulus of the fluid, which can be measured and depends on a temperature-dependent quantity called the persistence length $p = B/kT$. A rod of length p bent round a circle of radius p would have an elastic deformation energy equal to $\frac{1}{2}kT$. Hvidt (60, 61) has measured the flexibility of the rods of myosin, paramyosin, and tropomyosin at 5–10°C and finds persistance lengths of approximately 1300 Å, which correspond to $B_c = 10 \times 10^{-19}$ dyn · cm^2. This figure fits the theoretical estimates from bond energies within a factor of two. The experiments give no support to the idea that the myosin rod contains two hinged rigid segments.

An ideal coiled coil is virtually unstretched under the forces of muscle contraction. If we take the S-2 region of myosin to be 500 Å long and apply the maximum tension of 3.3×10^{-7} dyn with $S_c = 3.92 \times 10^{-3}$ dyn, the extension is only 0.042 Å. The coil does, however, bend very easily, and even the random thermal displacements of a flexible rod are large. They can be estimated (60) by calculating the behavior of a bent beam of length L, which is clamped at one end. The energy needed to move the far end sideways by a distance y is $3By^2/L^3$, and if this is set equal to $\frac{1}{2}kT$ the root mean square displacement comes out as $y = (L^3/3p)^{1/2}$. The experimental persistence length of 1300 Å yields $y = 180$ Å, which is comparable with the lateral actin-myosin filament spacings. The force needed to produce a static displacement of this magnitude would be very small, only 2.3×10^{-8} dyn.

The idea of a hinged joint between rigid S-2 and LMM segments was originally proposed to allow the myosin heads to move off the thick filament surface and reach actin (52). It was also used to account for the mechanical elasticity of the linkage between actin and myosin filaments when the length of a muscle changes suddenly (48). A more controversial idea (11, 41, 136) was that a long section of rod, which alternately melts and refolds during muscle contraction, acts as a spring. Our discussion of stiffness shows that there is no need for a hinge, since a normal alpha-helical coiled coil is so flexible. A clamped LMM with a flexing S-2 can provide more than sufficient thermal movement, unless, that is, the weak electrostatic forces between S-2 and the shaft provide a restraint. It may be better to think of the S-2 as a flexible rod weakly bound to the filament surface by an electrostatic spring, which can "peel off" for a certain length at the head end under the influence of outward tensions. The mechanical elastic stretch in the head and S-2 linkage (34) is now thought to be 39 Å, which is enormous compared with the stretching of a regular helix, and it probably occurs between the segments of the head or in the pivot. The weak points and kinks in S-2 may perhaps have other uses, for example in the assembly of the filament shaft (112).

ACKNOWLEDGMENTS

I thank Jon Karn, Murray Stewart, Marshall Elzinga, and Hugh Huxley for their comments and explanations.

Literature Cited

1. Adelstein, R. S., Eisenberg, E. 1980. *Ann. Rev. Biochem.* 49:921–56
2. Amos, L. A., Huxley, H. E., Holmes, K. C., Goody, R. S., Taylor, K. A. 1982. *Nature* 299:467–69
3. Balint, M., Wolf, L., Tarcsafalvi, A., Gergely, J., Sreter, A. 1978. *Arch. Biochem. Biophys.* 190:793–99
4. Bell, G. I., Pictet, R. L., Rutter, W. J., Cordell, B., Tischer, E., Goodman, H. M. 1980. *Nature* 284:26–32
5. Bennett, P. M. 1981. *J. Mol. Biol.* 146:201–21
6. Bennett, W. S., Steitz, T. A. 1978. *Proc. Natl. Acad. Sci. USA* 75:4848–52
7. Bernstein, S. I., Mogami, K., Donady, J. J., Emerson, C. P. 1983. *Nature* 302:393–97
8. Breathnach, R., Benoist, C., O'Hare, K., Gannon, F., Chambon, P. 1978. *Proc. Natl. Acad. Sci. USA* 75:4853–57
9. Brenner, S. 1974. *Genetics* 77:71–94
10. Burke, M., Harrington, W. F. 1972. *Biochemistry* 11:1456–62
11. Burke, M., Himmelfarb, S., Harrington, W. F. 1973. *Biochemistry* 12:701–10
12. Burke, M., Reisler, E. 1977. *Biochemistry* 16:5559–63
13. Capony, J. P., Elzinga, M. 1981. *Biophys. J.* 33:148a

13a. Chalovich, J. M., Eisenberg, E. 1982. *J. Biol. Chem.* 257:2432–37

14. Cohen, C., Caspar, D. L. D., Johnson, J. P., Nauss, K., Margossian, S. S., Parry, D. A. D. 1972. *Cold Spring Harbor Symp. Quant. Biol.* 37:287–97

15. Cohen, C., Holmes, K. C. 1963. *J. Mol. Biol.* 6:423–32
16. Cohen, C., Szent-Gyorgyi, A. G., Kendrick-Jones, J. 1971. *J. Mol. Biol.* 56:223–37
17. Collins, J. H. 1974. *Biochem. Biophys. Res. Commun.* 58:301–8
18. Cooke, R. 1981. *Nature* 294:570–71
19. Craig, R. 1977. *J. Mol. Biol.* 109:69–81
20. Craig, R., Offer, G. 1976. *Proc. R. Soc. London Ser. B* 192:451–61
21. Craig, R., Smith, R., Kendrick-Jones, J. 1983. *Nature* 302:436–39
22. Craik, C. S., Sprang, S., Fletterick, R., Rutter, W. J. 1982. *Nature* 299:180–82
23. Crick, F. 1979. *Science* 204:264–71
24. Crick, F. H. C. 1953. *Acta Crystallogr.* 6:689–97
24a. Davis, J. S. 1981. *Biochem. J.* 197:301–14
25. Davis, J. S., Buck, J., Greene, E. P. 1982. *FEBS Lett.* 140:293–97
26. Ebashi, S. 1976. *Ann. Rev. Physiol.* 38:293–313
27. Eiferman, F. A., Young, P. R., Scott, R. W., Tilghman, S. M. 1981. *Nature* 294:713–18
28. Elliott, A., Lowy, J., Parry, D. A. D., Vibert, P. J. 1968. *Nature* 218:656–59
29. Elliott, A., Offer, G. 1978. *J. Mol. Biol.* 123:505–19
30. Elzinga, M., Collins, J. H. 1977. *Proc. Natl. Acad. Sci. USA* 74:4281–84
31. Elzinga, M., Trus, B. L. 1980. In *Methods in Peptide and Protein Sequence Analysis*, ed. C. Birr, pp. 213–24. Amsterdam: Elsevier/North-Holland
32. Epstein, H. F., Waterston, R. H., Brenner, S. 1974. *J. Mol. Biol.* 90:291–300
33. Fabian, F., Muhlrad, A. 1968. *Biochim. Biophys. Acta* 162:596–603
34. Ford, L. E., Huxley, A. F., Simmons, R. M. 1981. *J. Physiol.* 311:219–49
35. Frank, G., Weeds, A. 1974. *Eur. J. Biochem.* 44:317–34
36. Fraser, R. D. B., Macrae, T. P. 1973. *Conformation in Fibrous Proteins*. New York: Academic
37. Gauthier, G. F., Lowey, S. 1979. *J. Cell Biol.* 81:10–25
38. Gilbert, W. 1978. *Nature* 271:501
39. Goldman, Y. E., Hibberd, M. G., McCray, J. A., Trentham, D. R. 1982. *Nature* 300:701–5
40. Hardwicke, P. M. D., Wallimann, T., Szent-Gyorgyi, A. G. 1983. *Nature* 301:478–82
41. Harrington, W. F. 1979. In *The Proteins IV*, ed. H. Neurath, R. L. Hill, pp. 245–409. New York: Academic. 3rd ed.
42. Harris, H. E., Epstein, H. F. 1977. *Cell* 10:709–19
43. Hoh, J. F. Y., Yeoh, G. P. S. 1980. *Nature* 280:321–23
44. Hozumi, T., Muhlrad, A. 1981. *Biochemistry* 20:2945–50
45. Hulmes, D. J. S., Miller, A., Parry, D. A. D., Piez, K. A., Woodhead-Galloway, J. 1973. *J. Mol. Biol.* 79:137–48
46. Huxley, A. F. 1974. *J. Physiol.* 243:1–43
47. Huxley, A. F., Niedergerke, R. 1954. *Nature* 173:971–72
48. Huxley, A. F., Simmons, R. M. 1971. *Nature* 233:533–38
49. Huxley, A. F., Taylor, R. E. 1955. *Nature* 176:1068
50. Huxley, H. E. 1957. *J. Biophys. Biochem. Cytol.* 3:631–48
51. Huxley, H. E. 1963. *J. Mol. Biol.* 7:281–308
52. Huxley, H. E. 1969. *Science* 164:1356–66
53. Huxley, H. E. 1972. *Cold Spring Harbor Symp. Quant. Biol.* 37:361–76
54. Huxley, H. E. 1979. See Ref. 110, 1:71–95
55. Huxley, H. E., Brown, W. 1967. *J. Mol. Biol.* 30:383–434
56. Huxley, H. E., Faruqi, A. R. 1983. *Ann. Rev. Biophys. Bioeng.* 12:381–417
57. Huxley, H. E., Faruqi, A. R., Kress, M., Bordas, J., Koch, M. H. J. 1982. *J. Mol. Biol.* 158:637–84
58. Huxley, H. E., Hanson, J. 1954. *Nature* 173:973–76
59. Huxley, H. E., Simmons, R. M., Faruqi, A. R., Kress, M., Bordas, J., Koch, M. H. J. 1981. *Proc. Natl. Acad. Sci. USA* 78:2297–2301
60. Hvidt, S., Ferry, J. D., Roelke, D. L., Greaser, M. L. 1983. *Macromolecules* 16:740–45
61. Hvidt, S., Nestler, F. H. M., Greaser, M. L., Ferry, J. D. 1982. *Biochemistry* 21:4064–72
62. Jensen, E. O., Paludan, K., Hyldig-Nielsen, J. J., Jorgensen, P., Marcker, K. A. 1981. *Nature* 291:677–79
63. Jung, A., Sippel, A., Grez, M., Schutz, G. 1980. *Proc. Natl. Acad. Sci. USA* 77:5759–63
64. Karn, J., Brenner, S., Barnett, L. 1983. *Proc. Natl. Acad Sci. USA* 80:4253–57
65. Karn, J., Brenner, S., Barnett, L., Cesareni, G. 1980. *Proc. Natl. Acad. Sci. USA* 77:5172–76
66. Karn, J., McLachlan, A. D., Barnett, L. 1982. In *Muscle Development: Molecular and Cellular Control*, ed. M. L. Pearson, H. F. Epstein, pp. 129–42. Cold Spring Harbor, NY: Cold Spring Harbor Lab.
67. Kendrick-Jones, J., Cande, W. Z.,

Tooth, P. J., Smith, R. C., Scholey, J. M. 1983. *J. Mol. Biol.* 165:139–62
68. Kendrick-Jones, J., Scholey, J. M. 1981. *J. Muscle Res. Cell Motility* 2:347–72
69. Kendrick-Jones, J., Szent-Kiralyi, E. M., Szent-Gyorgyi, A. G. 1976. *J. Mol. Biol.* 104:747–79
70. Kubo, S., Tokura, S., Tonomura, Y. 1960. *J. Biol. Chem.* 235:2835–39
71. Kubo, S., Tokuyama, H., Tonomura, Y. 1965. *Biochim. Biophys. Acta* 100:459–70
72. Kushmerick, M. J., Davies, R. E. 1969. *Proc. R. Soc. London Ser. B* 174:315–53
73. Labbe, J. P., Mornet, D., Roseau, G., Kassab, R. 1982. *Biochemistry* 21: 6897–6902
74. Leinwand, L. A., Saez, L., McNally, E., Nadal-Ginard, B. 1983. *Proc. Natl. Acad. Sci. USA* 80:3716–20
75. Levy, R. M., Karplus, M. 1979. *Biopolymers* 18:2465–95
76. Longley, W. 1975. *J. Mol. Biol.* 93:111–15
77. Lowey, S. 1979. See Ref. 110, pp. 1–26
78. Lowey, S., Cohen, C. 1962. *J. Mol. Biol.* 4:293–308
79. Lowey, S., Slayter, H. S., Weeds, A., Baker, H. 1969. *J. Mol. Biol.* 42:1–29
80. Lu, R. C. 1980. *Proc. Natl. Acad. Sci. USA* 77:2010–13
81. Lu, R. C., Wong, A. 1982. *Biophys. J.* 37:52a
82. Luther, P. K., Munro, P. M. G., Squire, J. M. 1981. *J. Mol. Biol.* 151:703–30
83. Lymn, R. W., Taylor, E. W. 1971. *Biochemistry* 10:4617–24
84. Mackenzie, J. M., Epstein, H. F. 1980. *Cell* 22:747–55
85. MacLeod, A. R., Karn, J., Brenner, S. 1981. *Nature* 291:386–90
86. MacLeod, A. R., Waterston, R. H., Fishpool, R. M., Brenner, S. 1977. *J. Mol. Biol.* 114:133–40
87. Mahdavi, V., Periasamy, M., Nadal-Ginard, B. 1982. *Nature* 297:659–64
88. Maw, M. C., Rowe, A. J. 1980. *Nature* 286:412–14
89. McLachlan, A. D. 1984. *J. Mol. Biol.* In press
90. McLachlan, A. D. 1983. *J. Mol. Biol.* 169:15–30
91. McLachlan, A. D., Karn, J. 1983. *J. Mol. Biol.* 164:605–26
92. McLachlan, A. D., Karn, J. 1982. *Nature* 299:226–31
93. McLachlan, A. D., Stewart, M. 1975. *J. Mol. Biol.* 98:293–304
94. McLachlan, A. D., Stewart, M. 1976. *J. Mol. Biol.* 103:271–98
95. Mendelson, R. A., Morales, M. F., Botts, J. 1973. *Biochemistry* 12:2250–55
96. Miller, A., Tregear, R. T. 1972. *J. Mol. Biol.* 70:85–104
97. Moerman, D. G., Plurad, S., Waterston, R. H., Baillie, D. L. 1982. *Cell* 29:773–81
98. Mornet, D., Bertrand, R., Pantel, P., Audemard, E., Kassab, R. 1981. *Nature* 292:301–6
99. Mornet, D., Bertrand, R., Pantel, P., Audemard, E., Kassab, R. 1981. *Biochemistry* 20:2110–20
100. Mornet, D., Pantel, P., Audemard, E., Kassab, R. 1979. *Biochem. Biophys. Res. Commun.* 89:925–32
101. Mount, S. M. 1982. *Nucleic Acids Res.* 10:459–72
102. Nguyen, H. T., Gubits, R. M., Wydro, R. M., Nadal-Ginard, B. 1982. *Proc. Natl. Acad. Sci. USA* 79:5230–34
103. Nudel, U., Katcoff, D., Carmon, Y., Zevin-Sonkin, D., Levi, Z., Shaul, Y., et al. 1980. *Nucleic Acids Res.* 8:2133–46
104. Offer, G., Moos, C., Starr, R. 1973. *J. Mol. Biol.* 74:653–76
105. Ohnishi, H., Wakabayashi, T. 1982. *J. Biochem. Tokyo* 92:871–79
106. Okamoto, Y., Yount, R. G. 1983. *Biophys. J.* 41:298a
107. Pai, E. G., Sachsenheimer, W., Schirmer, R. H., Schulz, G. E. 1977. *J. Mol. Biol.* 114:37–45
108. Parry, D. A. D. 1979. See Ref. 110, pp. 393–427
109. Parry, D. A. D. 1981. *J. Mol. Biol.* 153: 459–64
110. Parry, D. A. D., Creamer, L. K., eds. 1979. *Fibrous Proteins: Scientific Industrial and Medical Aspects*, Vol. 1. New York: Academic
111. Parry, D. A. D., Crewther, W. G., Fraser, R. D. B., MacRae, T. P. 1977. *J. Mol. Biol.* 113:449–54
112. Pepe, F. A. 1967. *J. Mol. Biol.* 27:203–25
113. Pepe, F. A. 1982. In *Cell and Muscle Motility*, ed. R. M. Dowben, J. W. Shay, 2:141–71. New York: Plenum
114. Pepe, F. A. 1983. In *Muscle and Nonmuscle Motility*, ed. A. Stracher, 1: 105–49. New York: Academic
115. Pepe, F. A., Ashton, F. T., Dowben, P., Stewart, M. 1981. *J. Mol. Biol.* 145: 412–40
116. Pepe, F. A., Dowben, P. 1977. *J. Mol. Biol.* 113:199–218
117. Phillips, G. N., Fillers, J. P., Cohen, C. 1980. *Biophys. J.* 32:485–502
117a. Pollard, T. D. 1982. *J. Cell Biol.* 95: 816–25
118. Reisler, E., Smith, C., Seegan, G. 1980. *J. Mol. Biol.* 143:129–45
119. Rozek, C. E., Davidson, N. 1983. *Cell* 32:23–34

120. Sartore, S., Pierobon-Bormioli, S., Schiaffino, S. 1978. *Nature* 274:82–83
120a. Sivaramakrishnan, M., Burke, M. 1982. *J. Biol. Chem.* 257:1102–5
121. Sjostrom, M., Squire, J. M. 1977. *J. Mol. Biol.* 109:49–66
122. Squire, J. M. 1973. *J. Mol. Biol.* 77:291–323
123. Squire, J. M. 1981. *The Structural Basis of Muscle Contraction*. New York: Plenum
124. Stein, J., Catterall, J., Kristo, P., Means, A., O'Malley, B. 1980. *Cell* 21:681–87
125. Suck, D., Kabsch, W., Mannherz, H. G. 1981. *Proc. Natl. Acad. Sci. USA* 78:4319–23
126. Suezaki, Y., Go, N. 1976. *Biopolymers* 15:2137–53
127. Sutoh, K., Karr, T., Harrington, W. F. 1978. *J. Mol. Biol.* 126:1–22
128. Szilagyi, L., Balint, M., Sreter, F. A., Gergely, J. 1979. *Biochem. Biophys. Res. Commun.* 87:936–45
129. Takashi, R., Muhlrad, A., Botts, J. 1982. *Biochemistry* 21:5661–68
130. Taylor, E. W. 1979. *CRC Crit. Rev. Biochem.* 6:103–64
131. Taylor, K. A., Amos, L. A. 1981. *J. Mol. Biol.* 147:297–324
132. Thomas, D. D., Ishiwata, S., Seidel, J. C., Gergely, J. 1980. *Biophys. J.* 32:873–89
133. Thomas, D. D., Seidel, J. C., Hyde, J. S., Gergely, J. 1975. *Proc. Natl. Acad. Sci. USA* 72:1729–33
134. Trus, B. L., Elzinga, M. 1981. In *Structural Aspects of Recognition and Assembly in Biological Macromolecules*, ed. M. Balaban, J. L. Sussman, W. Traub, A. Yonath, 1:361. Boston: Int. Sci. Serv.
135. Trybus, K. M., Huiatt, T. W., Lowey, S. 1982. *Proc. Natl. Acad. Sci. USA* 79:6151–55
136. Ueno, H., Harrington, W. F. 1981. *Proc. Natl. Acad. Sci. USA* 78:6101–5
137. Ueno, H., Harrington, W. F. 1981. *J. Mol. Biol.* 149:619–40
137a. Wagner, P. D., Giniger, E. 1982. *Nature* 292:560–62
137b. Wagner, P. D., Stone, D. B. 1982. *Biochemistry* 22:1334–42
138. Walker, J. E., Saraste, M., Runswick, M. J., Gay, N. J. 1982. *EMBO J.* 1:945–51
139. Waterson, R. H., Epstein, H. F., Brenner, S. 1974. *J. Mol. Biol.* 90:285–90
140. Waterston, R. H., Fishpool, R. M., Brenner, S. 1977. *J. Mol. Biol.* 117:679–97
141. Waterston, R. H., Smith, K. C., Moerman, D. G. 1982. *J. Mol. Biol.* 158:1–15
142. Waterston, R. H., Thomson, J. N., Brenner, S. 1980. *Dev. Biol.* 77:271–302
143. Watson, H. C., Walker, N. P. C., Shaw, P. J., Bryant, T. N., Wendell, P. L., Fothergill, L. A., et al. 1982. *EMBO J.* 1:1635–40
144. Weber, A., Murray, J. 1973. *Physiol. Rev.* 53:612–73
145. Weeds, A. G., Lowey, S. 1971. *J. Mol. Biol.* 61:701–25
146. Weeds, A. G., Pope, B. 1977. *J. Mol. Biol.* 111:129–57
147. Weeds, A. G., Trentham, D. R., Kean, C. J. C., Buller, A. J. 1974. *Nature* 247:135–39
148. Wells, J. A., Sheldon, M., Yount, R. G. 1980. *J. Biol. Chem.* 255:1598–1602
149. Wells, J. A., Yount, R. G. 1979. *Proc. Natl. Acad. Sci. USA* 76:4966–70
150. Whalen, R. G., Schwartz, K., Bouveret, P., Sell, S. M., Gros, F. 1979. *Proc. Natl. Acad. Sci. USA* 76:5197–5201
151. Whalen, R. G., Sell, S. M., Butler-Browne, G. S., Schwartz, K., Bouveret, P., Pinset-Harstrom, I. 1981. *Nature* 292:805–9
152. Wray, J. S. 1979. *Nature* 277:37–40
153. Yamada, Y., Avvedimento, V. E., Mudryj, M., Ohkubo, H., Vogeli, G., Irani, M., et al. 1980. *Cell* 22:887–92
154. Zengel, J. M., Epstein, H. F. 1980. *Cell Motility* 1:73–97

Ann. Rev. Biophys. Bioeng. 1984. 13: 191–219

OPTICAL SECTIONING MICROSCOPY: Cellular Architecture in Three Dimensions

David A. Agard

Department of Biochemistry and Biophysics, University of California, San Francisco, California 94143

PERSPECTIVE AND OVERVIEW

The ability to analyze biological specimens in three dimensions represents one of the major achievements of modern structural biology. For all but the simplest repeating structures, three-dimensional analysis is a crucial prerequisite for understanding complex biological assemblies. Macromolecular X-ray crystallography has provided three-dimensional structures at atomic resolution for proteins, nucleic acids, and viruses. Indeed, most of our fundamental knowledge on the relation of structure to function of individual biological molecules has come from these studies. Electron microscopy has been a major tool for the study of cell components. With the development of image reconstruction methods by Klug and coworkers (9, 10, 17) it is now routine to study two-dimensional crystalline or high symmetry objects in three dimensions at moderate resolutions (7–20 Å). Recently, several workers (22) have even begun to reconstruct relatively large and noncrystalline objects ($\sim$1000 Å) from cut sections, although at somewhat lower resolutions ($\sim$50–100 Å). The advent of relatively inexpensive computers and digital image acquisition systems has now made possible the three-dimensional reconstruction of images taken from the optical microscope. Three-dimensional data is collected by recording a series of images taken at different focal planes throughout the specimen (optical sectioning). Most of the common imaging methods available with light microscopy may be used. Each recorded image represents the sum of

0084–6589/84/0615–0191$02.00

in-focus information from the focal plane and out-of-focus information from the remainder of the specimen. Much of the out-of-focus information can be removed computationally. Thus, cellular architecture in intact specimens may now be subject to detailed analysis.

Optical microscopy uniquely allows complex, nonsymmetric, noncrystalline structures to be studied in a native, hydrated environment. This can be quite important, as many biological structures (such as chromatin) are exquisitely sensitive to subtle variations in ionic composition and concentration. By contrast, electron microscopy is routinely carried out in a vacuum. Furthermore, the preparation of thick specimens for electron microscopy generally involves a dehydration step followed by embedding and sectioning. All of these steps can potentially distort the observed structure. Difficulties in aligning the sections can lead to further errors. In some cases, these problems may be avoided by using optical microscopy, although at lower resolution.

Radiation at visible wavelengths does not damage biological specimens, thus making optical microscopy ideal for examining large (>1000 Å) structures. In addition, the analysis of heterogeneous complexes is greatly simplified in the light microscope by the availability of a wide range of specific fluorescent dyes and stains. Fluorescently labeled polyclonal or monoclonal antibodies further extend the possible versatility of optical microscopy.

Recording three-dimensional information with the optical microscope is difficult because of the inherent contradictions between the requirements for high resolution and large depth-of-field. For very thin specimens (less than or equal to the depth of field) it is possible to reconstruct the object by recording images at many different specimen tilts. This approach is commonplace in electron microscopic imaging (9, 17). To date, in optical microscopy this method has only been used to analyze DNA packing in *Drosophila melanogaster* sperm heads (1). The *Diosophila* sperm heads, which are ~0.5 μm × 13 μm, were mounted in quartz capillaries and tilted about their long axis, recording pictures every 10°. Unfortunately, this method will not work with thicker samples where the recorded images no longer represent true projections of the specimen as is required by the tilted-view reconstruction method. One approach that can be used is to physically section the specimen, image each section, and reconstruct the entire object by stacking the sections. This approach suffers from many of the problems already mentioned for electron microscopy. An alternative to physical sectioning is to optically section the object and take pictures at many different focal planes throughout the specimen. Each image will then represent the sum of in-focus information from the portion of the specimen located at the focal plane and out-of-focus information from the remainder of the specimen. For any given image, only 1–2% of the information recorded may be derived from the in-focus component. There are several different imaging schemes that can be used with optical sectioning to extract three-dimensional information from intact thick specimens: fluorescence or bright-field optical section microscopy (OSM), Normarski or Differential Interference Contrast (DIC), and Scanning Con-focal Microscopy (SCM).

The purpose of this review is to discuss what is likely to become the most broadly applicable three-dimensional reconstruction method now available for the optical microscope (OSM). Unlike other techniques, OSM can be used with a wide variety of optical imaging methods: bright field, polarization, phase contrast, and fluorescence. And it can be used with a conventional microscope. In addition, very thick specimens (0.1–1 mm) can be imaged. As this is a very new area, this article is concerned primarily with introducing the technique and discussing the capabilities of various reconstruction methods. Finally, a comparison is made with other methods, and the biological results obtained thus far are briefly discussed.

Data Collection

Collecting optical section data requires that images be taken at many (20–100) different focal planes throughout the specimen. It is very important that the exact focal spacing between adjacent images be accurately controlled. For this reason, it is desirable to attach either a high-resolution stepping motor or a high-resolution encoded DC-servo motor to the microscope's focus control. Several groups have developed either stand-alone or computer-controlled hardware for this purpose (2, 21, 23, 29).

The actual image data can be recorded on film and subsequently digitized on a computer-based microdensitometer. Following digitization, it is necessary to accurately align the images before they can be processed further. Although this is a time-consuming process, a reliable algorithm has been developed (1) and used for this purpose (2).

A much better method is to directly digitize the images from the microscope using a high quality video camera (equipped with either a chalnacon or silicon-integrated-target tube) attached to a video digitizer and frame buffer store (20, 21). Several scan frames (each 1/30 sec) can be averaged to improve the signal-to-noise ratios and then transferred to a host computer for further processing.

THEORY OF IMAGE FORMATION

Thin Specimens

Much of the following treatment parallels Castleman (8). A highly simplified microscope consisting of a single element, the objective lens, is depicted in Figure 1. Naturally, any real microscope will be more complex and will incorporate at least a condenser lens and an eyepiece in addition to the objective lens shown. As almost all of the critical image-forming properties of the microscope arise from the objective lens, for our purposes the effects of the remaining lenses can be neglected. The distance between the objective lens and the image plane, d_i, is fixed in a microscope and usually set to 160 mm. Together with the focal length of lens, this specifies the distance of the lens to the focal plane within the object according to the lens law:

$$1/d_i + 1/d_f = 1/f.$$

The magnification M or power of the objective lens is then determined by the ratio of the image distance d_i to the focal distance of the lens d_f:

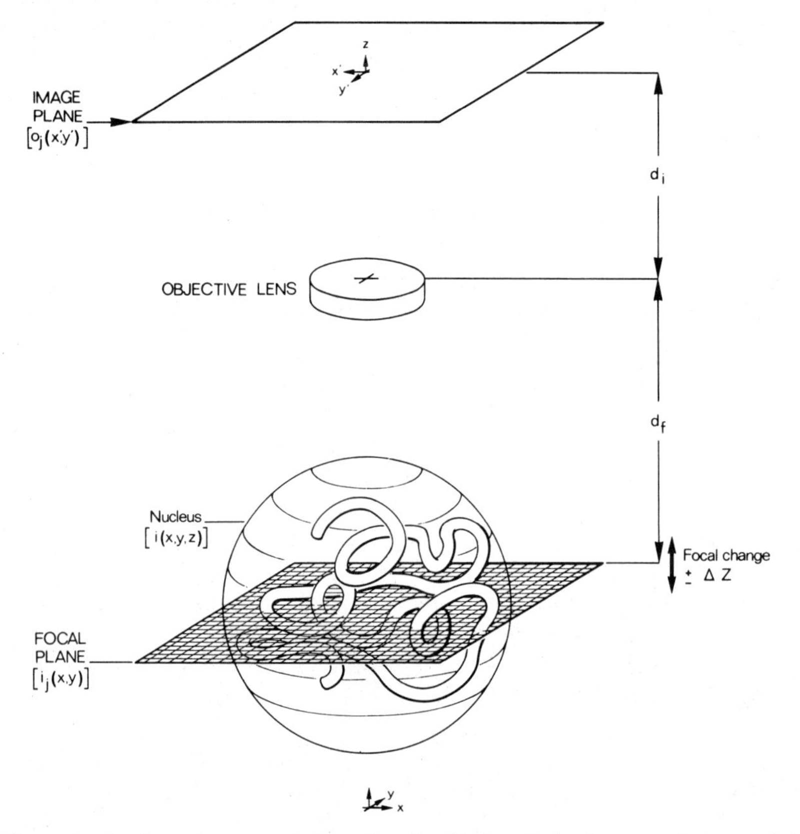

Figure 1 A schematic representation of a simplified optical microscope being used to examine a thick specimen [from Reference (2)]. The recorded image o_j is a composite of in-focus information from the focal plane i_j and out-of-focus detail from the rest of the specimen $i(x, y, z)$.

$M = d_i/d_f$. For convenience, we can calculate d_f directly from the easily obtained objective lens values (M, f):

$$d_f = \frac{M+1}{M} f. \qquad 1.$$

We shall now begin to consider the actual imaging properties of the microscope for the case of an infinitely thin (two-dimensional) specimen. Let $i(x, y)$ represent the two-dimensional distribution of optical density within the specimen. The corresponding optical density observed at the image plane will be $o(x', y')$. The coordinate system at the image plane is related to that at the object plane by both an inversion and the magnification factor M:

$$x' = -Mx,$$

$$y' = -My.$$

As this transformation is trivial and in practice is modified by the concatination of similar effects arising from the eyepiece and any other lenses in the system following the objective lens, we can simplify the ensuing discussions by projecting the image back into the object plane as $o(x, y)$. This simplifies the notation by eliminating the magnification and 180° rotation effects, thereby establishing a consistent coordinate system for both the object and image. All spacings and dimensions can then be kept in the object's frame of reference.

In general, the optical behavior of the microscope can be described in terms of its point-spread function or PSF. The PSF describes the appearance in the image plane of a purely impulsive object (idealized point or delta function). Any deviation from ideality of the microscope will result in an image that deviates from a delta function (point). The effect on an arbitrary image is then given in terms of the convolution of the PSF, here called the smearing function $s(x, y)$, with the object:

$$o(x, y) = \int_{-\infty}^{\infty} \int_{-\infty}^{\infty} i(x', y') s(x-x', y-y')\, dx'\, dy'; \qquad 2.$$

or more compactly:

$$o(x, y) = i(x, y) * s(x, y), \qquad 3.$$

where the $*$ represents the convolution operation of Equation 2.

Another very useful way to describe the behavior of the microscope is by using the contrast-transfer function (CTF). The CTF, $S(u, v)$, is simply the Fourier transform of the PSF and defines the manner in which the various spatial frequency components of the object are altered by the optical system. The convolution theorem (e.g. see 6) indicates that taking the Fourier transform of both sides of Equation 3 converts the convolution into a multiplication:

$$O(u, v) = I(u, v) \cdot S(u, v), \qquad 4.$$

where $O(u, v)$, $I(u, v)$, and $S(u, v)$ are the two-dimensional Fourier transforms of $o(x, y)$, $i(x, y)$, and $s(x, y)$ respectively. For example,

$$O(u, v) = \int_{-\infty}^{\infty} \int_{-\infty}^{\infty} o(x, y) e^{2\pi i(xu + yv)}\, dx\, dy,$$

and conversely, by inverse Fourier transform,

$$o(x, y) = \int_{-\infty}^{\infty} \int_{-\infty}^{\infty} O(u, v) e^{-2\pi i(xu + yu)}\, du\, dv. \qquad 5.$$

We shall now proceed to discuss the relationship between $o(x, y)$ and $i(x, y)$ for the special case of an incoherently illuminated, correctly focussed, aberration-free optical system [for a more general treatment, see (14)]. This actually corresponds reasonably well to the situation for a modern, high-quality microscope that records bright-field or fluorescent images at low or moderate power.

Such an aberration-free system is said to be diffraction limited. That is, all of the rays from the object captured by the lens will reach the image plane undistorted. The only distortions arise from the fact that not all of the rays are captured by the objective lens; in other words, the system has a finite resolving power. In most cases, no physical receiving aperture is provided, and the diameter of the objective lens itself acts as the aperture. The acceptance angle α (see Figure 2) is related to the numerical aperture of the lens by

$$\text{N.A.} = \eta \sin \alpha,$$

where η is the refractive index of the medium between the lens and the specimen (1.0 for air, 1.515 for an oil immersion lens). The highest spatial frequency that can be passed by the system f_c is

$$f_c = (2\eta \sin \alpha)/\lambda \qquad 6.$$

with λ being the wavelength of the light used. Using Raleigh's criterion, the smallest separation between two points that can be resolved is $1/(1.22 \cdot f_c)$ or

$$d_{min} = \frac{\lambda}{2.44\ \text{N.A.}},$$

$$d_{min} = 0.146\ \mu\text{m for } \lambda = 0.5\ \mu\text{m and using a 1.4 N.A. oil immersion lens.} \qquad 7.$$

For example, the way in which a circular aperture function affects the image is most simply described in terms of its CTF. Because the pupil is circular, the CTF also has circular symmetry and can be shown to be (14):

$$S(q) = \frac{(2\beta - \sin 2\beta)}{\pi}, \qquad 8.$$

where $\beta = \cos^{-1}(q/f_c)$, $q = (u^2 + v^2)^{1/2}$, and f_c as in Equation 6. This equation is only strictly correct for lenses where α is $< 30°$. For a derivation of the considerably more complex relationship for lenses with large α and for inclusion of effects arising from the condenser lens, see Sheppard & Wilson (25). Fortunately, the deviations from the behavior predicted by Equation 8 are sufficiently small, even for a N.A. 1.4 lens ($\alpha = 56°$), that they may be safely neglected in our treatment. The contrast-transfer functions for several different lenses are plotted in Figure 3. The corresponding point-spread functions can be readily determined by numerically taking the Fourier transform of Equation 8; two examples are shown in Figure 4.

Thick Specimens

Before analyzing the nature of image formation from thick specimens, it is first necessary to consider what happens to the recorded image of an infinitely thin specimen when recorded under out-of-focus conditions. Just as the ideal image $i(x, y)$ was distorted by the in-focus point-spread function

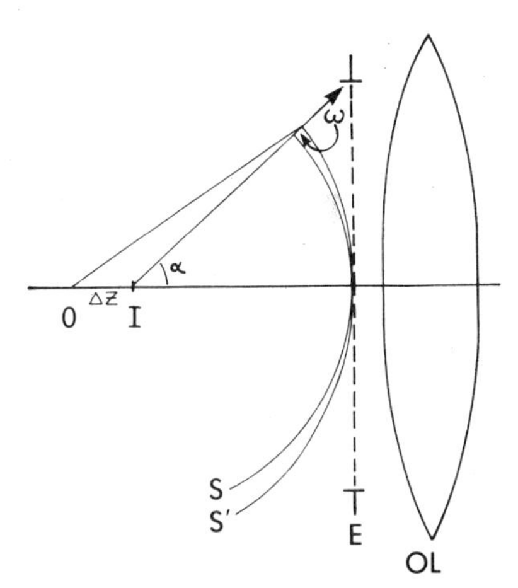

Figure 2 Geometrical analysis for a defocussed optical system. OL, the objective lens, has an entrance pupil of diameter E. The extreme ray that can pass through E makes an angle α to the optical axis. The spherical wavefront S is derived from the in-focus point I, whereas S' comes from a point ΔZ out of focus located at O. The maximum optical path length error is given by w. After Stokeseth (27).

$s_i(x, y)$ to yield $o_i(x, y)$ (Equation 3; where the subscript i connotes in-focus), so we must further distort $o_i(x, y)$ with the out-of-focus PSF $s_o(x, y, \Delta z)$:

$$o(x, y) = o_i(x, y) * s_o(x, y, z), \qquad 9.$$

where Δz represents the displacement of the object plane from the focal plane. Combining Equations 3 and 9 yields the more complete expression:

$$o(x, y) = i(x, y) * s_i(x, y) * s_o(x, y, \Delta z).$$

For convenience, the two different point-spread functions can be merged into a total PSF, thereby simplifying this equation to

$$o(x, y) = i(x, y) * s(x, y, \Delta z). \qquad 10.$$

We can consider a thick specimen $i(x, y, z)$ to be the sum of a stack of N thin specimens $i(x, y, z_j)$ located at different positions (z_j) along the z-axis separated by a spacing Δz. The image recorded from the entire stack with the focal plane set to z_o is thus the sum of each of the individual planes distorted by the appropriate blurring function:

$$o(x, y)|_{z_o} = \sum_{j=1}^{N} i(x, y, z_j) * s(x, y, z_o - z_j)\Delta z. \qquad 11.$$

In the limit as $\Delta z \to 0$ and $N \to \infty$, the finite sum of Equation 11 becomes the continuous definite integral over the thickness t:

$$o(x, y)|_{z_o} = \int_0^t i(x, y, z') * s(x, y, z_o - z')\, dz'.$$

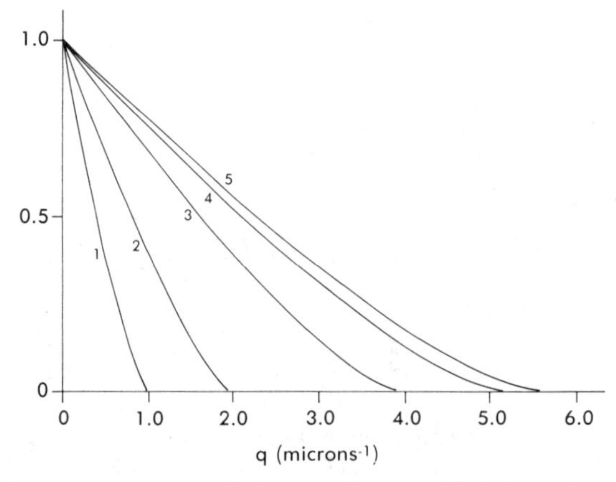

Figure 3 The in-focus contrast-transfer functions for several Zeiss lenses are compared. They are (*1*) 10 × 0.25 N.A., (*2*) 25 × 0.5 N.A., (*3*) 50 × 1.0 N.A. oil, (*4*) 100 × 1.25 N.A. oil, (*5*) 63 × 1.4 N.A. oil.

If $i(x, y, z)$ is 0 for $z < 0$ and $z > t$, as we can assume for most real specimens, then the finite integral can be rewritten as an infinite integral:

$$o(x, y)|_{z_o} = \int_{-\infty}^{\infty} i(x, y, z') * s(x, y, z_o - z')\, dz', \qquad 12.$$

which (by reference to Equation 2) is seen to be just another convolution operation. Therefore, the imaging of a thick specimen in the optical microscope reduces to the three-dimensional convolution:

$$o(x, y, z) = i(x, y, z) * s(x, y, z), \qquad 13.$$

where $s(x, y, z)$ represents the microscope's total three-dimensional point-spread function.

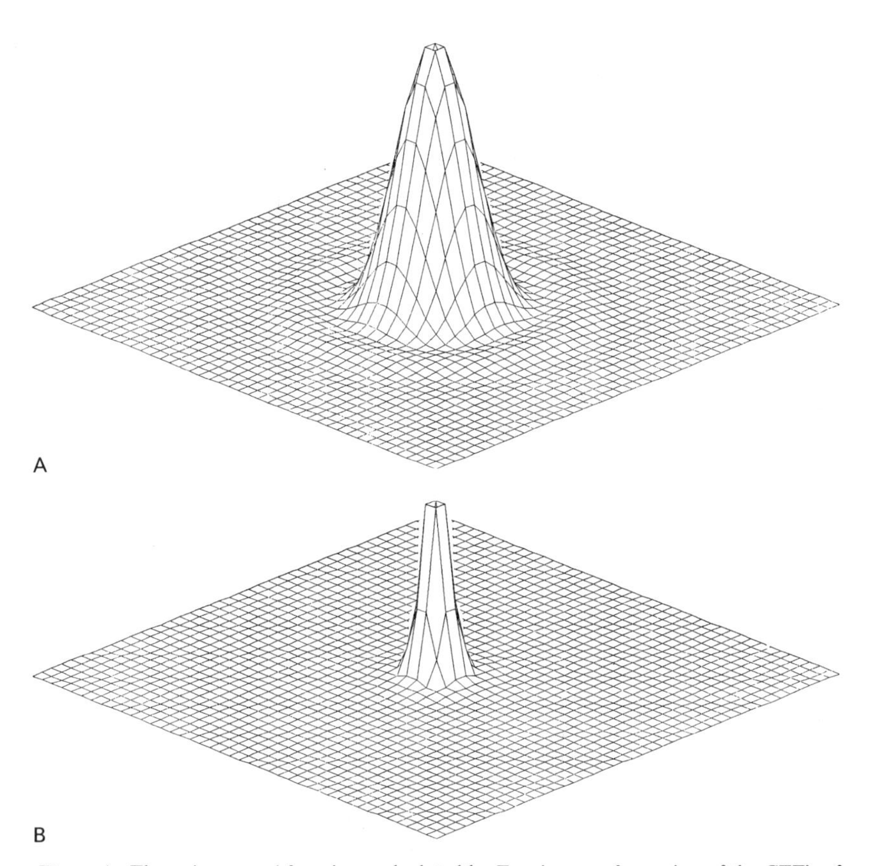

Figure 4 The point-spread functions calculated by Fourier transformation of the CTF's of Figure 3. (*A*) 25 × 0.5 N.A. lens (#2); (*B*) 63 × 1.4 N.A. lens (#5). Each grid square is $\lambda/6$ or 833 Å. Note how much sharper the 1.4 N.A. lens is than the 0.5 N.A. lens.

THE OUT-OF-FOCUS CONTRAST TRANSFER FUNCTION

Unfortunately, the derivation of the contrast transfer function for a defocussed optical system is beyond the scope of this review. I refer the interested reader again to Goodman (14) for general information and to Stokseth (27) and Hopkins (18) for a more detailed treatment of this topic. Only the results of their calculations shall be reported here.

For an optical system with a circular aperture, Hopkins (18) has shown the contrast transfer function to be

$$S(b) = \frac{4}{\pi a}\cos\left(\frac{ab}{2}\right) \times \left\{\beta J_1(a) + \sum_{n=1}^{\infty}(-1)^{n+1}\frac{\sin(2n\beta)}{2n}[J_{2n-1}(a) - J_{2n+1}(a)]\right\} - \frac{4}{\pi a}\sin\frac{(ab)}{2}\sum_{n=0}^{\infty}(-1)^n\frac{[\sin(2n+1)\beta}{2n+1}[J_{2n}(a) - J_{2n+2}(a)] \qquad 14.$$

$$a = 4\pi wb \qquad b = \frac{2q}{f_c},$$

where $a = 8\pi wq/f_c$ with q, f_c, β as defined previously. The maximum optical path length error due to defocussing, w (see Figure 2), for a given lens $(d_f \cdot \alpha)$ and defocus (Δz) is given by Stokseth (27):

$$w = -d_f - \Delta z\cos\alpha + (d_f^2 + 2d_f\Delta z + \Delta z^2\cos^2\alpha)^{1/2}. \qquad 15.$$

Stokseth (27) has derived an approximate expression to Equation 14, which is especially good for large amounts of defocus:

$$S(q) = (1\text{–}0.696b + 0.000766b^2 + 0.0436b^3)\text{jinc}\left[\frac{8\pi w}{\lambda}\left(1 - \frac{q}{f_c}\right)\frac{q}{f_c}\right] \qquad 16.$$

$$\text{junc}(x) = 2\frac{J_1(x)}{x}.$$

Castleman (8) has suggested replacing the polynomial expression in Equation 16 by the correct in-focus CTF to provide a more accurate function at small amounts of defocus:

$$S(q) = \frac{1}{\pi}(2\beta - \sin 2\beta)\text{jinc}\left[\frac{8\pi w}{\lambda}\left(1 - \frac{q}{f_c}\right)\frac{q}{f_c}\right]. \qquad 17.$$

Because of the circular symmetry of the aperture, the contrast transfer function is also circularly symmetric. Equation 17 is the expression that we have used in our own work (2) and that will be used as a basis for all of the following calculations. Figure 5 shows the CTF's for two different lenses as a function of defocus. The contrast transfer functions calculated using Equation 17 can be inverse Fourier transformed to generate the corresponding real-space point-spread functions (see Figure 6). Alternatively, it is possible to take a set of CTF's, for different defocus values, and Fourier transform them in the z direction to obtain the complete, cylindrically symmetric CTF describing a stack of images [$S(q, w)$] (Figure 7). Note how the effect is to lose information (resolution) in a wedged-shaped region (a cone in three dimensions) along the axis perpendicular to the image plane. The overall effect of optically sectioning the specimen is to substantially elongate and distort the image in the direction of sectioning. For a general discussion of the missing cone problem, see Reference 4.

METHODS FOR RECOVERING IN-FOCUS INFORMATION

We have seen how the image of a thick specimen is distorted by an optical microscope. The severity of this distortion depends both upon the numerical aperture of the objective lens as well as upon the required spacing between sections. We now consider several approaches for recovering the in-focus information from a single observed data set. In all cases the observations are a stack of digital images recorded at different focal planes throughout the thick specimen. The choice of which technique to use depends very much on the problem being studied. In general, there is a direct relationship between accuracy (and z-axis resolution) and amount of computing time. Although very simple schemes will suffice for sections 5 μm apart, considerably more sophisticated approaches are required when dealing with images recorded every 0.75 μm or less.

Where possible, the various methods are compared, and representative reconstructions shown. For the purposes of viewing the effects of defocus, and the capabilities of differing reconstruction algorithms, all pictures correspond to two-dimensional reconstructions from a "one-dimensional" microscope. That is, our test object is a two-dimensional specimen in the x–z plane. Each "picture" recorded through the hypothetical microscope is a horizontal line; the stack of images can then be displayed stimultaneously as a two-dimensional image. All images and reconstructions are calculated using the CTF of Equation 17. The lens parameters correspond to actual Zeiss lenses (either a 25 × 0.5 N.A. lens or a 63 × 1.4 N.A. lens). Figure 8 shows a set of "observed" images at two different section spacings with these lenses.

In order to recover the in-focus information from a stack of images, it is

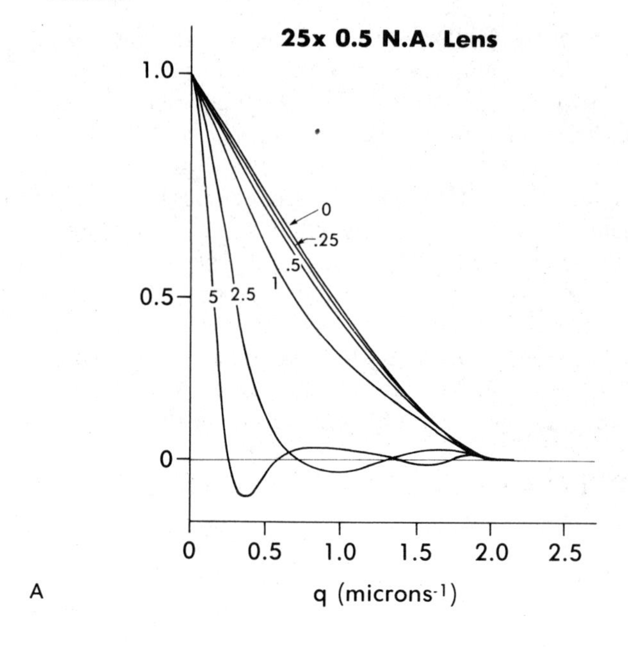

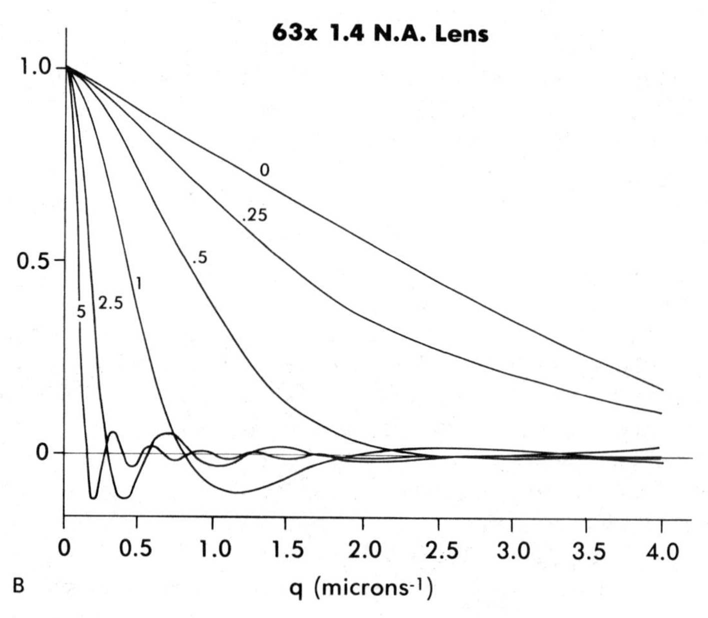

Figure 5 Panels *A* and *B* show the contrast transfer functions for two different lenses at varying degrees of defocus. The amount of defocus in microns is shown for each curve ($\lambda = 0.5\ \mu m$).

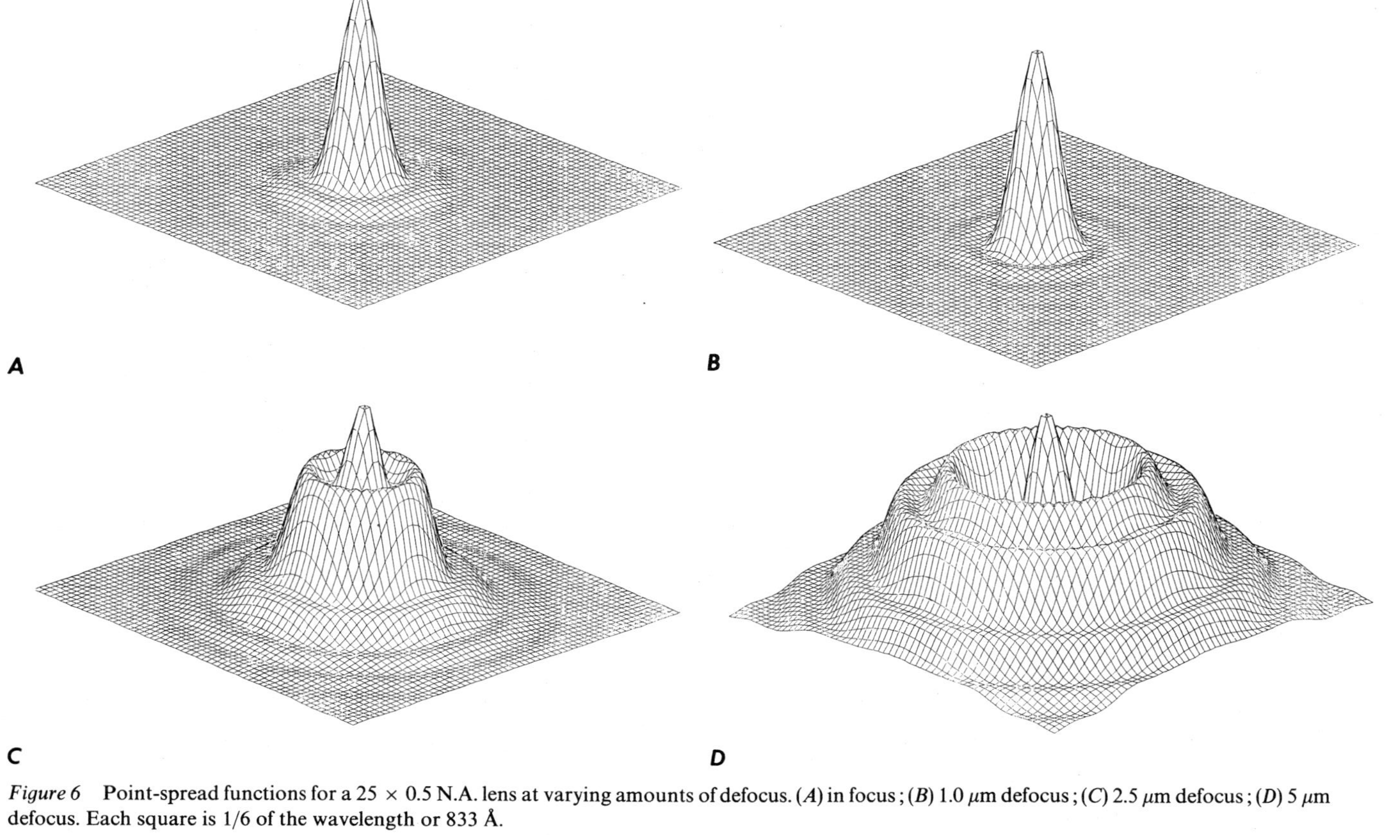

Figure 6 Point-spread functions for a 25 × 0.5 N.A. lens at varying amounts of defocus. (*A*) in focus; (*B*) 1.0 μm defocus; (*C*) 2.5 μm defocus; (*D*) 5 μm defocus. Each square is 1/6 of the wavelength or 833 Å.

necessary to reverse the convolution operation described by Equations 12 and 13; extracting $i(x, y, z)$ from $o(x, y, z)$ and a knowledge of $s(x, y, z)$. Accomplishing this task can be extremely difficult and very expensive, computationally. This is especially true if the stack is very large and the spacing between sections is small. Before discussing "exact" solutions to the problem, we begin with approximate methods. Although the recovery of $i(x, y, z)$ is not as good as with more exact methods, the ease of computation may justify such an approach for many problems.

Simple Real and Reciprocal Space Methods

APPROXIMATE METHODS To simplify the notation, we will drop the (x, y, z) and use subscripts to refer to a particular plane. Thus Equation 11 can be rewritten as

$$o_j = \sum_{k=1}^{N} i_k * s_{k-j}, \tag{18.}$$

where the plane spacing Δz is implicitly assumed in the out-of-focus PSF. Equation 18 can be rearranged to explicitly separate the contributions of the focal plane section from those of the m planes above and the m' planes below the focal plane:

$$o_j = i_j * s_0 + \sum_{\substack{k=j-m' \\ k \neq j}}^{j+m} i_k * s_{k-j}.$$

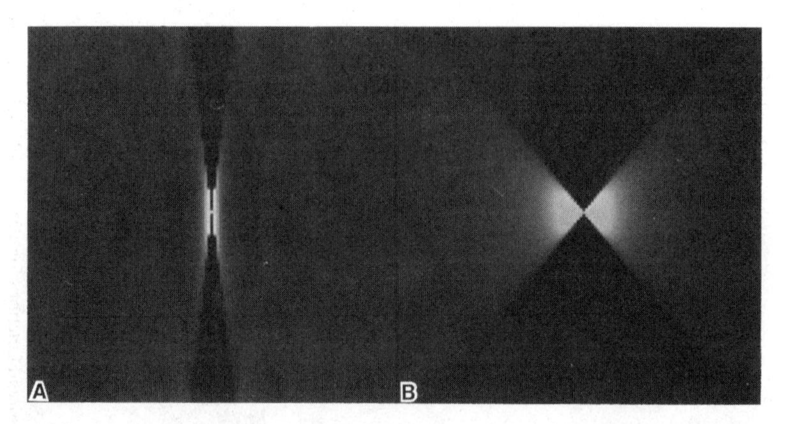

Figure 7 The total optical sectioning CTF for the 63 × 1.4 N.A. lens. The spacing between adjacent images is 5 μm in A and 0.75 μm in B. The in-plane spacing is 0.5 μm and $\lambda = 0.5$ μm. Note how the effect of optical sectioning is to remove a wedge of information (a cone in three dimensions). The in-focus falloff can be seen along the central horizontal line. Decreasing the numerical aperture for the lens produces an effect similar to decreasing the section spacing.

To reduce the computation, we assume that only the neighboring sections will contribute significantly to the observed images; hence we can keep m and m' rather small. This is reasonably correct when using high numerical aperture lenses with large inter-planar spacing (see Figure 3). Further, we can consider only the effects due to out-of-focus information and ignore the deterioration arising from the in-focus PSF (i.e. including that term in the "true" image). Thus the observed image o_j for any plane j is a composite of the "true" density function i_j (now corresponding to the in-focus image) for that plane with the blurred contributions from the m planes above and the m' planes below:

$$o_j = i_j + \sum_{\substack{k=j-m' \\ k \neq j}}^{j+m} i_k * s_{k-j}. \quad 19.$$

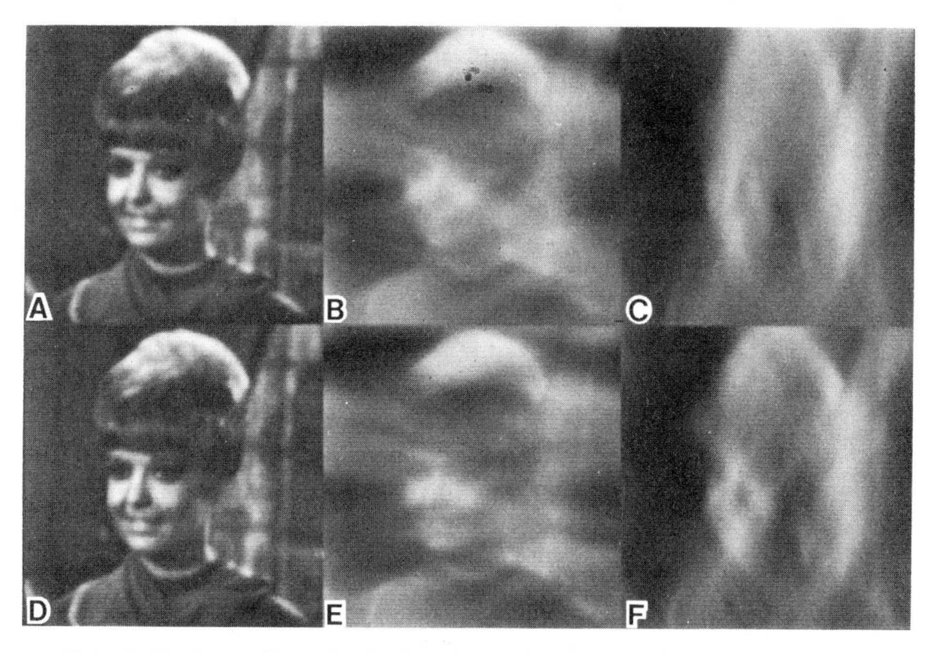

Figure 8 Test images illustrating the distortion produced during optical section data collection. Undistorted images (see text) of the standard image processing GIRL are shown in *A*, *D*. (*B*) 25 × 0.5 N.A., 5 μm interplanar spacing (Δz); (*C*) 25 × 0.5 N.A., 0.75 μm Δz; (*E*) 63 × 1.4 N.A., 5 μm Δz; (*F*) 63 × 1.4 N.A., 0.75 μm Δz. Although the interplanar spacing is varied, the images are shown as square to facilitate direct comparison. The degree of distortion is very dependent on the desired resolution in the focal direction as well as on the numerical aperture of the lens. All subsequent reconstructions are shown using only the best lens (63 × 1.4 N.A.). Important regions to examine when comparing the different reconstruction methods are the hair and around the mouth and eyes.

Rearranging Equation 19 suggests a means for generating a restored image:

$$i_j = c_2 \left[o_j - c_1 \sum_{\substack{k=j-m' \\ k \neq j}}^{j+m} i_k * s_{k-j} \right]. \qquad 20.$$

That is, the correct value for the density on a given plane j can be derived by subtracting the blurred contributions of all the other planes from the observed values for the selected plane. The constants c_1, c_2 are used to balance out the relative contributions of focal plane and the adjacent planes and need to be empirically determined. Unfortunately because the i_k are unknown, it is not possible to directly compute the necessary $i_k * s_{k-j}$. Weinstein & Castleman (30) suggested that as a very first approximation this problem could be ignored and also that only the immediately adjacent planes ($m = m' = 1$) needed to be considered:

$$i_j = c_2[o_j - c_1(s_1 * o_{j-1} + s_1 * o_{j+1})]. \qquad 21.$$

More recently, Shantz (23) proposed that the maximum value of the blurred contribution from the neighboring planes be used instead of the linear combination of Equation 21:

$$i_j = c_2[o_j - c_1 \max(s_1 * o_{j-1}, s_1 * o_{j+1})], \qquad 22.$$

where max (x, y) is the point-by-point maximum of the two values. It should be recognized that most of this work was done with a rather large spacing between sections ($> = 5$ μm). I found the results to be best with $c_1 = 0.45$ and 0.9, respectively. Reconstructions of Figure 8 using these methods are shown in Figure 9.

Somewhat more exact versions of this method treat larger numbers of adjacent sections simultaneously. Again, an empirical scale constant needs to be determined. Examples of a few different reconstructions are shown in Figure 10.

In his discussion of various reconstruction methods, Castleman (8) suggested that a better approximation to the i_k required by Equation 20 could be obtained by applying a high-pass filter function f to the i_k:

$$i_j = c_2 \left[o_j - c_1 \sum_{\substack{k=j-m' \\ k \neq j}}^{j+m} (i_k * f) * s_{k-j} \right]. \qquad 23.$$

The high-pass filter acts to remove the excess low-frequency information present in blurred sections. A reasonable choice for the filter might have a CTF that was given by $1 - e^{-\alpha q^2}$ where the value of α would be empirically determined.

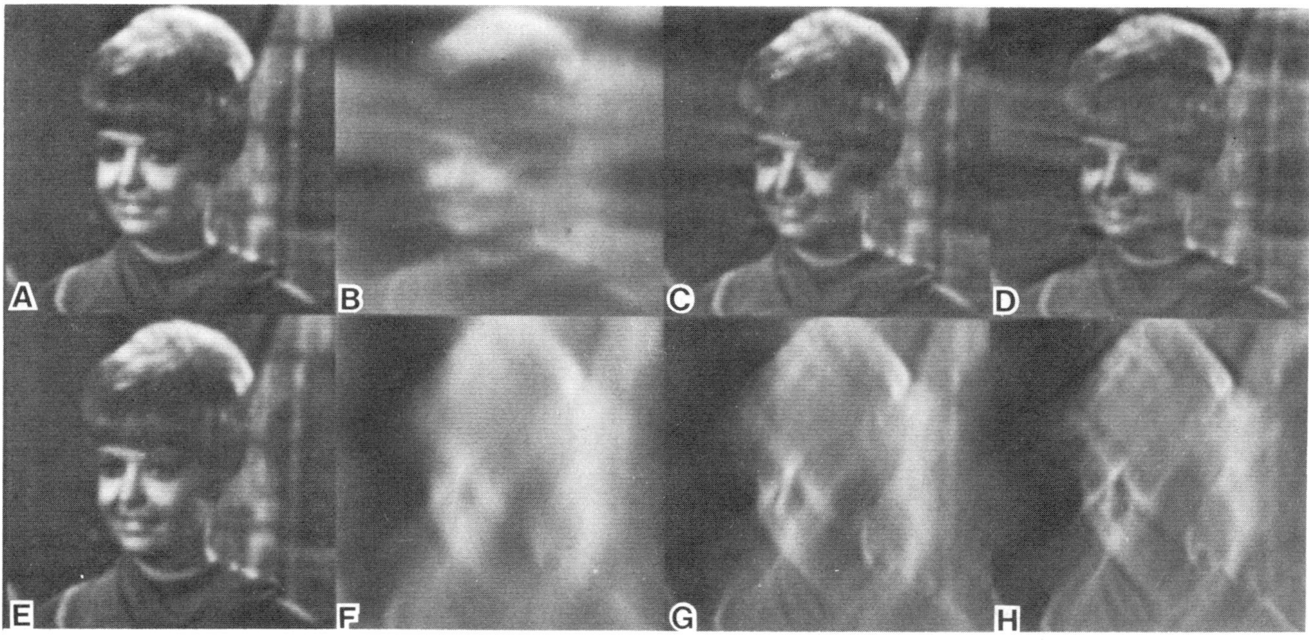

Figure 9 Reconstructions made from the observed data of Figure 8 (*E*, *F*) using the methods suggested by Castleman and co-workers (23, 30). As before, *A*, *E* are undistorted images, *B*. *F* are the "observed" images calculated for inter-planar spacings of 5 μm and 0.75 μm, respectively. The corresponding reconstructions made with either the method of Equation 21 (panels *C* and *G*) or with Equation 22 (*D* and *H*) are shown. These methods (especially Equation 21) work very well for the large interplanar spacing.

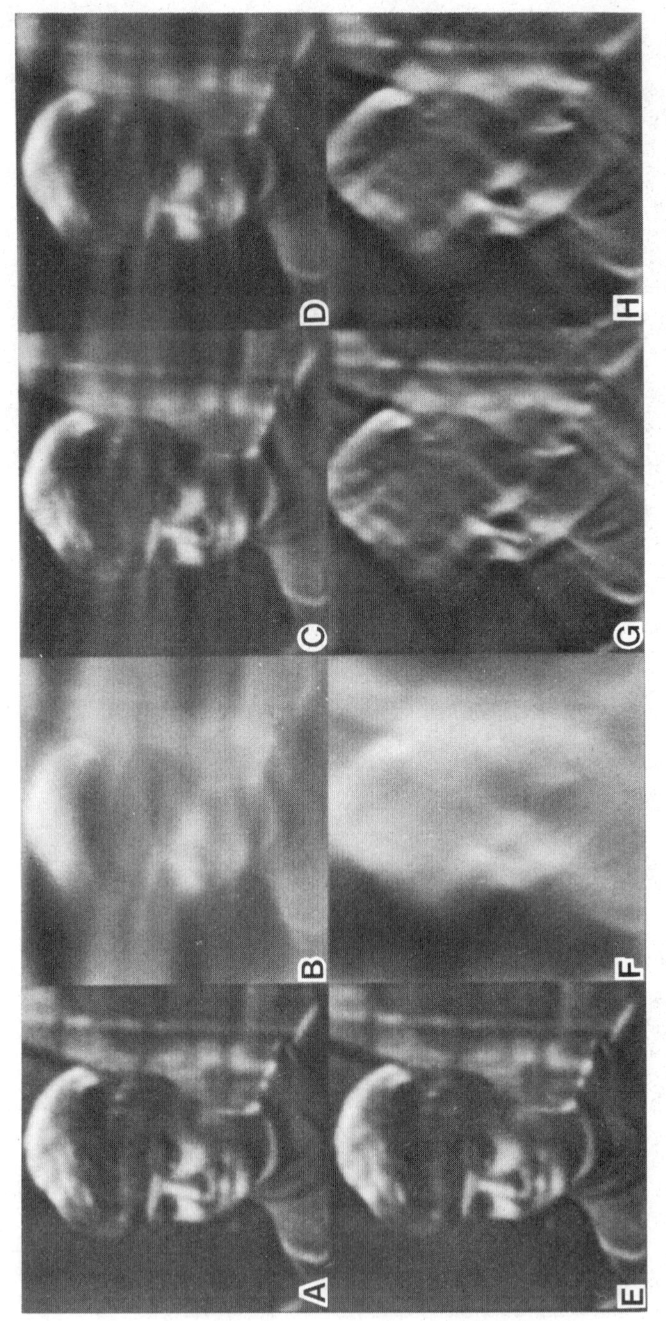

Figure 10 Reconstructions similar to those in Figure 9 but using more than a single adjacent section (Equation 20). Panels *A,B,E,F* as before. Reconstructions with $\Delta z = 5$ μm and $m = m' = 3$ (*C*) or $m = m' = 5$ (*D*) as shown. For $\Delta z = 0.75$ μm reconstructions using $m = m' = 5$ (*G*) and $m = m' = 10$ (*H*) are shown. Best results were obtained with $c_1 = 0.95$. At high resolution, including more sections produces a more accurate result.

A LESS APPROXIMATE METHOD Previously, we (2) developed an iterative scheme based on several modifications to Equation 20. Equation 20 can be recast by taking the Fourier transform of both sides of the equation, thus converting the convolution into a multiplication:

$$I_j = c_2 \left[O_j - c_1 \sum_{\substack{k=j-m' \\ k \neq j}}^{j+m} I_k \cdot S_{k-j} \right], \tag{24.}$$

where O, I, and S, respectively, are the two-dimensional Fourier transforms of o, i, and s. Given some initial guess ($n = 0$) for the I (I^0), it is possible to generate an improved guess ($n+1 = 1$)(I^1) as follows:

$$I_j^{n+1} = c_2 \left[O_j - c_1 \sum_{\substack{k=j-m' \\ k \neq j}}^{j+m} I_k^n \cdot S_{k-j} \right]. \tag{25.}$$

Thus in an iterative fashion it is possible to develop a self-consistent solution to the three-dimensional convolution implied by Equation 20. Values for the deblurred images (i_j) are determined by inverse two-dimensional Fourier transformation of the final I_j. The result is the determination of a set of images that when appropriately blurred and summed will equal, within some predetermined error, the observed images (o_j).

In practice, the initial guess (I^0) was set equal to the observed data (O). The number of sections that need to be considered simultaneously ($m+m'$) depends on the choice of objective lens and the spacing Δz between adjacent sections. For those studies (using a 63 × 1.25 N.A. oil immersion lens and $\Delta z = 1.2$ μm, it was only necessary to treat simultaneously a stack of 7 planes ($m = m' = 3$). For an example of this method, see Figure 14. For most applications this approach is not recommended because of the extreme sensitivity to scaling factors. The method of Equation 20 is preferred.

A Full Matrix Linear Filtering Approach

Again, by reference to Equation 13, and Fourier transforming both sides:

$$O_j = \sum_{k=1}^{N} I_k \cdot S_{k-j}. \tag{26.}$$

If we consider the set of $N\{O_j, I_k\}$ and $N^2\{S_{k-j}\}$ for each in-plane coordinate (u, v) (where N is the number of sections), Equation 26 can be expressed in matrix notation:

$$[O] = [S][I], \tag{27.}$$

where $[O]$, $[I]$ are vectors N long and $[S]$ is an $N \times N$ matrix. Equation 27 states that in Fourier space there is a set of simultaneous linear equations that describes the relationship between the observed data and the "true" image. The in-focus component $[I]$ can be found by solving this set of simultaneous equations at each point (u, v). In principle, this is done by matrix inversion:

$$[I] = [S]^{-1}[O]. \qquad 28.$$

Unfortunately, for those values (u, v) that are at low spatial frequency, the matrix $[S]$ is singular and cannot be simply inverted. The optimal way of handling this problem is to use linear filtering methods (11, 12). That is, first we decompose matrix $[S]$ into its component eigenfunctions, each of which has a given eigenvalue. All those components whose eigenvalues are less than a given percentage (generally 0.1%) of the maximum eigenvalue are discarded. The remaining eigenfunctions are used to develop the inverse matrix $[S]^{-1}$. This is a very general approach that allows the maximum amount of information to be extracted while avoiding serious problems due to noise and ill-conditioning.

It should be pointed out that since the $S(u, v)_j$ are radially symmetric $(= S(q)_j)$, only a limited number of $[S]^{-1}$ need to be calculated. Furthermore, as these matrices only depend on the experimental setup (lens parameters and Δz) and not on the actual image data, they can be precomputed and stored on disk.

This method should provide the best reconstruction that can be obtained using a linear technique. It has the further advantage that there are no parameters to adjust or optimize. At low values of q there are very few non-zero eigenvalues. As a consequence, there tend to be low-frequency ripples running through the reconstruction along the z-axis. It can be shown that the sum of values along z is correct, yet the absence of sufficient terms causes them to be incorrectly distributed, with both positive and negative values appearing. The results obtained with this method are shown in Figure 11.

A Constrained-Iterative Method

To obtain a more accurate reconstruction than that provided by the full-matrix, linear-filtering method described above requires the use of non-linear methods. All these methods seek to find a solution to the deconvolution problem that satisfies additional constraints.

One very powerful and general class of constraints that is useful here is to demand that the solution $[i(x, y, z)]$ be positive everywhere. If a proper background has been subtracted from the observed data, then the recorded optical density as well as the "true" optical density will always be a positive value. Under these conditions, negative optical densities are physically

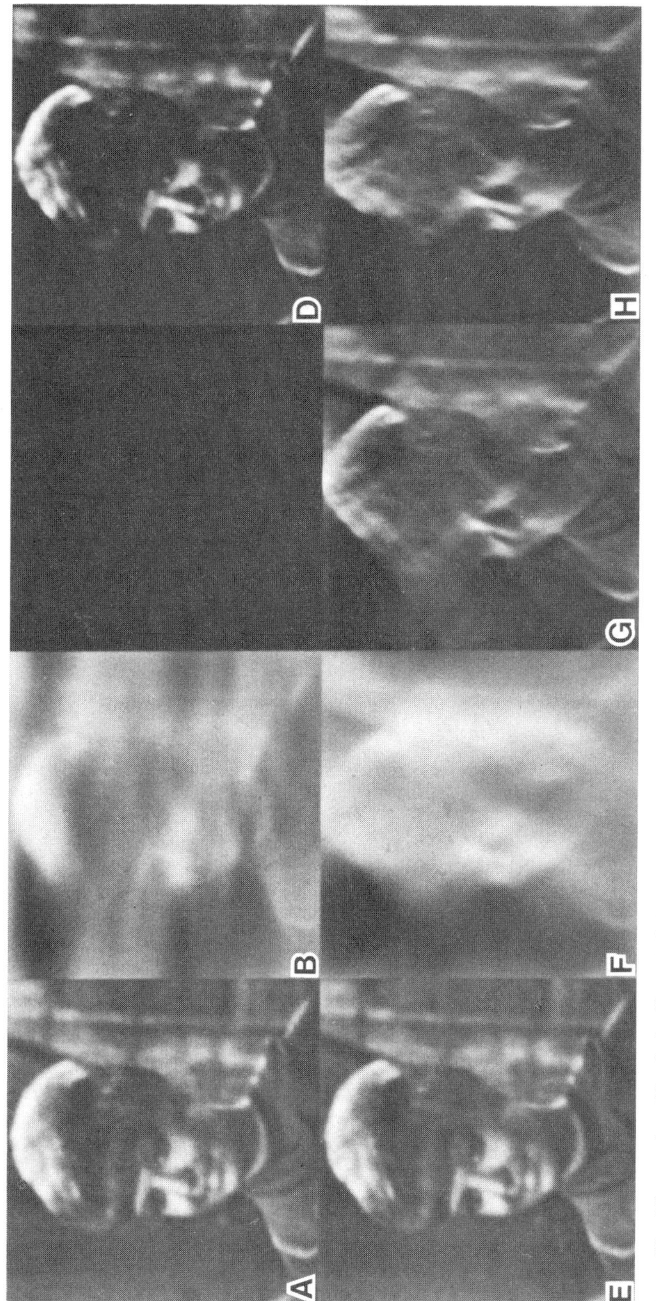

Figure 11 The results of both the filtered, full-matrix (Equation 28) D and constrained, iterative (Equation 29) G and H methods are contrasted. For technical reasons, the matrix reconstruction for $\Delta z = 0.5\ \mu m$ is not shown. The deconvolution scheme produces a higher contrast but probably better reconstruction than does the matrix method. These methods should only be used when high resolution (fine interplanar spacing) is desired.

meaningless, although they are mathematically allowed. The presence of noise in the original image serves to further compound this problem. If the positivity constraint is utilized, much of the aberrations due to ripples and noise can be effectively eliminated.

Unfortunately, nearly all methods that employ positivity constraints are iterative, and those that are not are even more costly to compute. With the iterative scheme, the general approach is to seek a set of i_j consistent with all additional constraints that when distorted by the optical system (Equation 18) will approximate the observed data to an arbitrary degree of accuracy. We have used this approach to solve many different problems in resolution enhancement, structural analysis, and imaging (1, 3, 4). The latter reference, dealing with limited-angle tilted-view reconstructions from electron microscope data, poses a reconstruction problem surprisingly similar to that of optical section microscopy.

The basic algorithm, after Frieden (13), is described here as a one-dimensional problem for simplicity:

$$(a)\ o^k(x) = i^k(x) * s(x),$$

$$(b)\ i^{k=1}(x) = i^k(x) + \gamma(x)\,[o(x) - o^k(x)],$$

$$(c)\ \text{if} \quad i^{k+1}(x) < 0, \qquad \text{then} \quad i^{k+1}(x) = 0,$$

$$(d)\ k = k+1,$$

$$\text{and} \quad \gamma(x) = 1 - [o^k(x) - A]^2/A^2, \tag{29.}$$

where A is a constant set to the maximum value of $o(x)/2$. The starting point for the refinement is $k = 0$ and $i^0(x) = o(x)$. The current guess, $i^k(x)$, is smeared by convolution with the PSF [step (a)]. A new guess, $i^{k=1}(x)$, to the correct $i(x)$ is generated by comparison of the blurred $i^k(x)\,[= o^k(x)]$ with the observed data $o(x)$ in step (b). If $o^k(x)$ is greater than $o(x)$, as occurs when $i^k(x)$ is too blurred, $i^{k+1}(x)$ will be sharpened by subtracting an amount proportional to $o^k(x) - o(x)$. The applied correction is modulated by $\gamma(x)$; when $o^k(x)$ approaches either the lower or upper allowable density limits $(0, 2A)$, the magnitude of the correction term approaches zero. The entire procedure is iterated until there is no significant difference between $o_k(x)$ and $o(x)$. At this point, $i_k(x)$ becomes equivalent to the desired $i(x)$. Both the $\gamma(x)$ function and step (c) are used to apply the non-negativity constraint. As before, the convolution operation of step (a) is performed as a multiplication operation in Fourier space for efficiency. Using Fast Fourier Transforms for this operation provides a 300–1000 fold increase in speed over the corresponding real-space convolution calculation. Figure 11 depicts the results obtained using the constrained-iterative technique. This method tends to overemphasize the contrast in the image. Although this actually yields the best reconstruction for high-contrast images, it appears to do somewhat less well for the continuous contrast example shown here.

The Use of Two or More Views

All of the reconstruction methods described so far assume that the observed data set consists of only a single stack of images. Inspection of the total CTF shown in Figure 7 indicates that most of the distortion results from the loss of data in a conical region about the section axis. If the sample were to be rotated by 90° and a new stack of images recorded, it should be possible to fill in much of the missing data. A very accurate reconstruction should result!

Several different schemes can be envisioned for the merging of the data sets. This operation is best performed in Fourier space where the relative contributions to the final transform can be easily sorted out from a knowledge of the CTF. One approach is to combine the individual three-dimensional Fourier transforms in a manner that uses knowledge of the relative values of the CTF for the different data sets at each point in Fourier space. A reasonable weighting scheme is

$$I(u, v, w) = \Sigma w_j O'_j(u, v, w),$$

$$w_j = S_j(u, v, w)/\Sigma|S_j(u, v, w)|, \qquad 30.$$

where O'_j is the correctly aligned Fourier transform of data set j. This should help to maximize the signal-to-noise ratio in the reconstructed image (see Figure 12). For example, if the ratio of one $S(u, v, w)$ to another were large, then the corresponding data value would contribute more strongly.

There are two practical problems with this approach. First, it is necessary to be able to rotate the specimen by an angle close to 90°. Obviously, the sample cannot be mounted on a slide (maximum tilt with a high numerical aperture lens is ± 10 to $\pm 15°$). It is possible to mount the specimen in a thin-walled glass or quartz X-ray capillary. The optical distortions arising from the curved surface can be greatly reduced by the appropriate choice of the index of refraction of the immersion oil. Further improvement results if the space between the specimen and the inside of the capillary is also filled with a matching fluid.

A very significant second problem is the need for either a very accurate eucentric tilt stage or the ability to align the numerical data sets after they have been collected. It is probably necessary to align the data sets to 1/5–1/10 the desired resolution (~500 Å)! Computational difficulties could be greatly simplified if several fiducial points can be accurately located within the specimen, or by the use of refractile or fluorescent beads. The exact

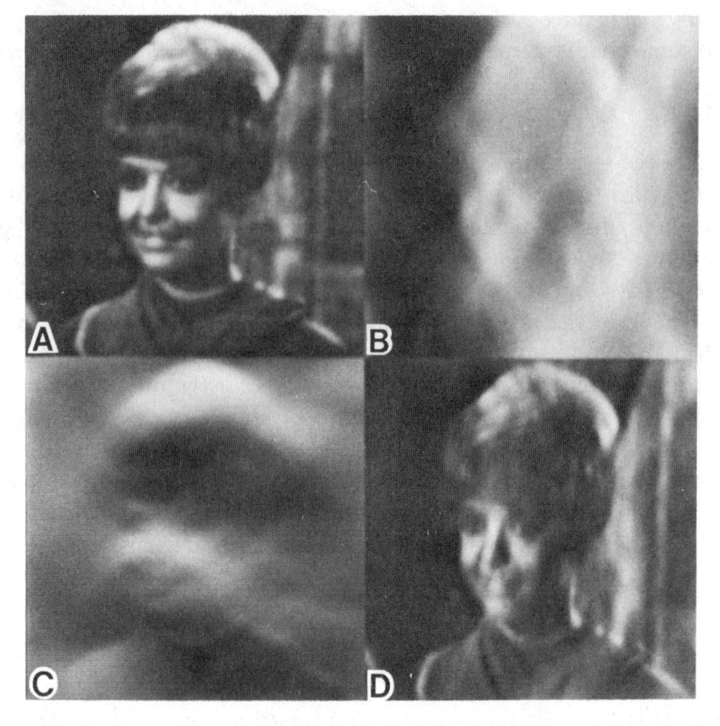

Figure 12 A method that uses two data sets, rotated by 90°, is shown. *A* is the undistorted image, *B* and *C* correspond to the "observed" data taken with either vertical or horizontal sections, and *D* is the reconstructed image. Notice that the nose shows up clearly in *B* and not the mouth while in *C* the converse is true. This method produces by far the best reconstructions possible and promises to be useful even at the resolution limits of optical microscopy.

transformation could then be calculated from the (x, y, z) coordinates of these points. We are currently pursuing this problem further.

OTHER METHODS

The problem of reconstructing three-dimensional images from optical sections arises because there is not a sufficient difference between the contrast-transfer function for in-focus and out-of-focus images. Two methods exist for enhancing this difference by purely optical means: differential interference contrast (DIC) and, with a radically new microscope design, the scanning con-focal microscope (SCM).

DIC or Normarski (5) imaging is a very powerful method whereby the gradient of phase contrast (gradient of the index of refraction) is recorded.

This is accomplished by the insertion of two modified Wollaston prisms into the optical path (one before the condenser and the other after the objective lens). These prisms act to first split the incoming beam while producing a small shear in index of refraction between adjacent points and to then recombine the split beams, interfering them so as to produce an image proportional to the phase gradient in the specimen. As this is a gradient method, the amount of interference arising from overlapping planes in a thick specimen is greatly reduced. If the specimen is rather thick (>10 μm) then phase errors will build up to a point where detail can be obscured.

Although this method is generally applicable and does not require any computation, it does suffer from a few drawbacks: (*a*) the thickness of the specimen must be less than 10–20 μm; (*b*) only phase images can be recorded; and (*c*) the shear produced by the Wollaston prisms is in one direction only, and thus high resolution can only be obtained in that direction. However, for many applications DIC is ideal.

The scanning con-focal microscope (7, 24) uses a doubly focussed objective lens system and a pin-hole aperture to effectively image only a single point within the specimen. The PSF of the system determines the volume of the specimen that is imaged. The use of two high-numerical lenses leads to PSF that is considerably sharper (the product of the PSF's for each lens) in all directions. Three-dimensional data is collected by scanning the sample in three dimensions while measuring the light transmitted through (or emitted by) the specimen with a photomultiplier (16). The main difficulty is that the instrumental tolerances required to obtain high resolution imaging are very difficult to achieve. Once this significant problem is solved, the SCM will undoubtedly prove to be a very valuable research tool.

BIOLOGICAL RESULTS

Noncomputational Methods

As stated earlier, the entire field of three-dimensional optical microscopy is quite new. To date, there are relatively few examples of biological problems that have been solved using these methods. By far the most extensively used approach employs DIC microscopy. Perhaps the most impressive work involves the mapping of the entire pattern of cell lineage for all of the cells in the worm *Caenorhabditis elegans* (19, 28). As an adult, this transparent worm is 20 μm in diameter and contains 1000 somatic cells. This *tour de force* required the painstaking observation of hundreds of living worms and the examination of each at many different focal planes using Nomarski optics. Other questions in embryology have been approached using through-focal series, followed by compilation of a photographic mosaic (15). Skaer & Whytock (6) used Nomarski optics and a limited tilt stage ($\pm 15°$) to follow the path of the giant polytene chromosomes in the salivary gland nuclei of the insect *Chironomus dorsalis*. A three-dimensional wire model was built to the data collected.

Computational Methods

Castleman & collaborators (8, 23, 30) have used optical section microscopy and the reconstruction methods of Equations 21 and 22 to examine the path of stained neurons. A stereo image of one of their reconstructions (8) is shown in Figure 13. The use of a very high-contrast stain simplifies the reconstruction problem and allows much of the very fine detail to be recovered. Recently, we (2) have been examining the three-dimensional topology of polytene chromosomes within nuclei of *Drosophila melanogaster* salivary glands. To provide contrast, the DNA within the chromosomes was stained using a DNA-specific non-intercalative fluorescent dye. Originally, section data were collected on film and digitized prior to processing (using the algorithm of Equation 25). An example of the results obtained in this manner is shown in Figure 14. More recently, a SIT video camera and a video digitizer have been added to greatly facilitate data acquisition (20, 21). We are now in the process of applying the constrained deconvolution and the linear filtering algorithms to real data sets.

CONCLUSIONS

Optical section microscopy uniquely allows the three-dimensional analysis of intact biological specimens using a wide variety of imaging methods. The price paid for this versatility is the need to digitally process the images in an effort to remove the out-of-focus information that contaminates each image. Several different methods for solving this problem, ranging from the simple to the complex, have been discussed. Although the best results can be obtained by taking at least two data sets at orientations 90° to each other, significant improvements can be obtained for a single stack of images. Especially reasonable are the simple approaches that treat several sections simultaneously (Fig. 10). For the simple methods, a stack of 64 256×256 pixel images can be processed in about 30 min on a VAX 11/780. More sophisticated algorithms or larger data sets can lead to computation times of up to 24 hr. Array processors are especially suited to such problems and can routinely provide a 5- to 10-fold increase in speed. Although such times may seem excessive, they are comparable to that required to process three-dimensional e.m. data, and negligent compared to the time involved in either solving or refining a simple protein structure.

As of now, the cost of setting up a three-dimensional data collection and

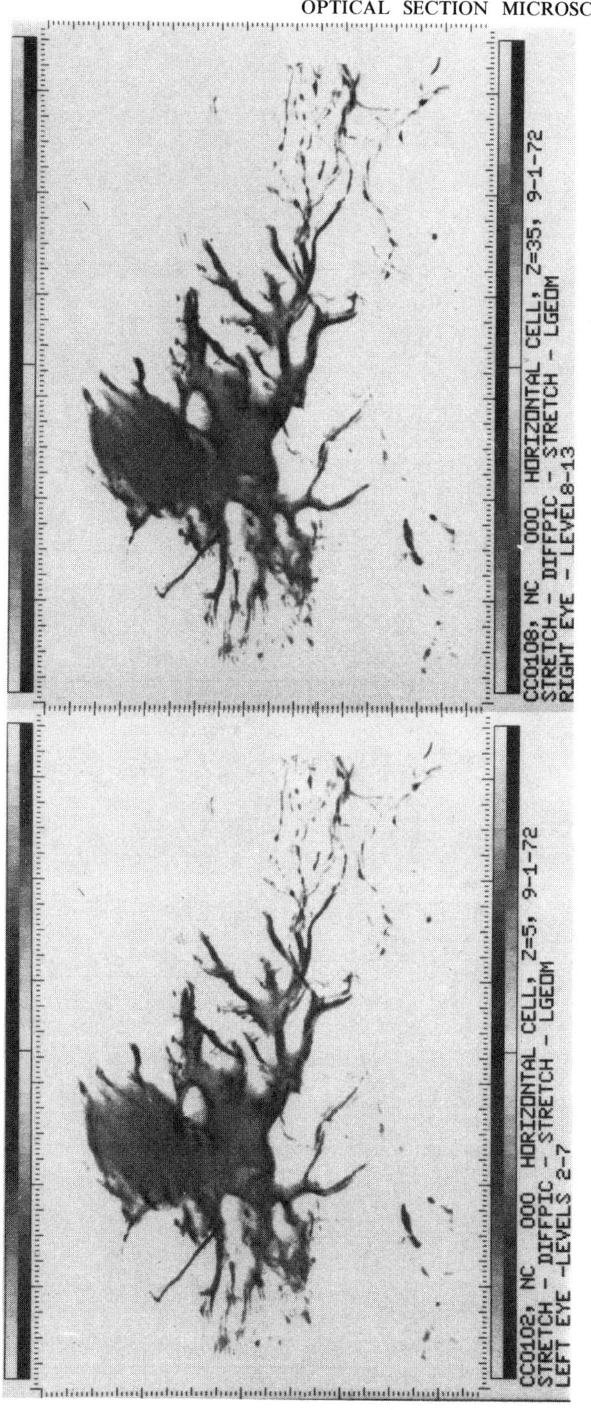

Figure 13 A stereo pair of a reconstructed (Equation 21) stack of images derived from stained horizontal cells [from Castleman (8)].

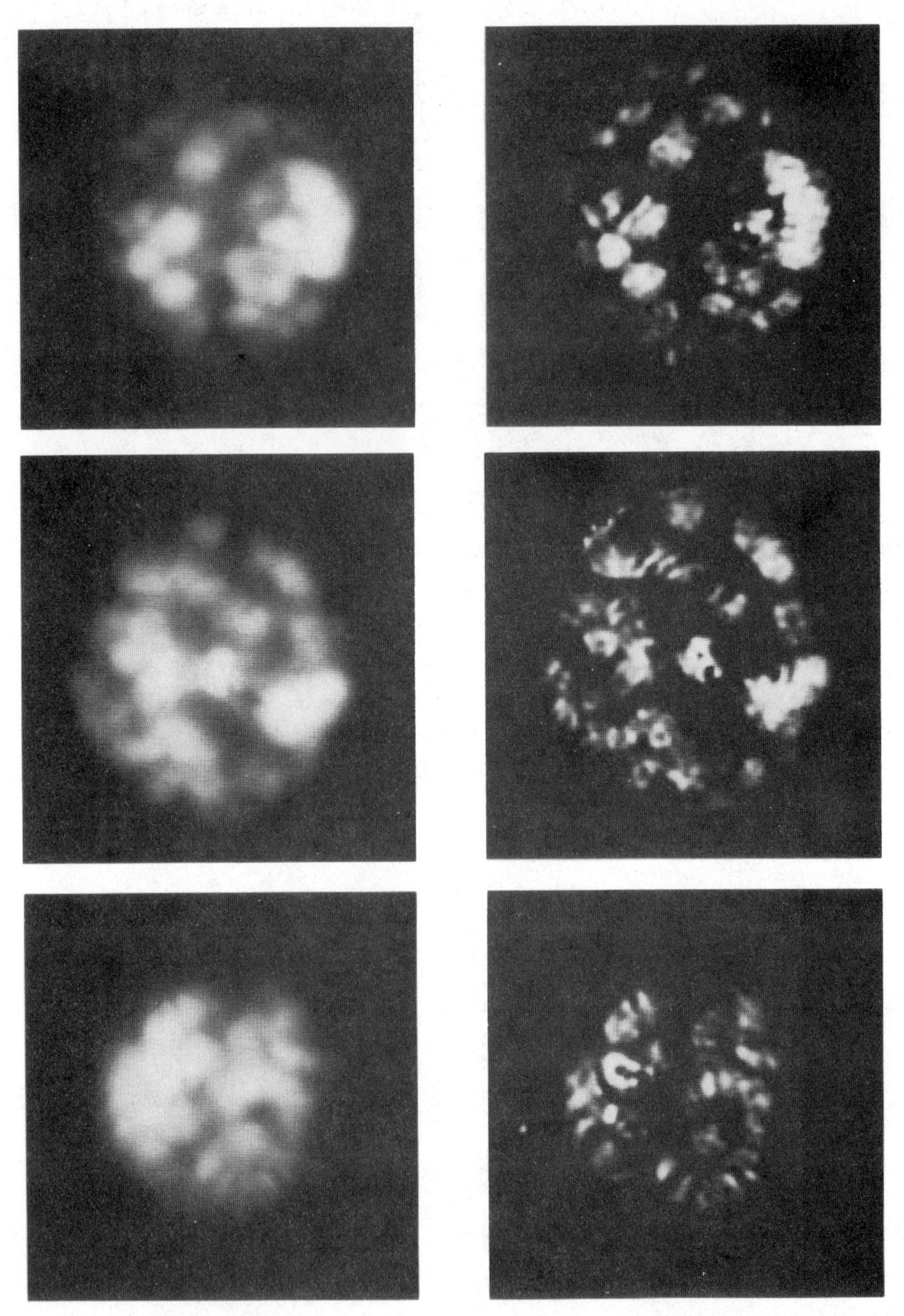

Figure 14 Before and after comparisons for three noncontiguous sections from a stack of 30 planes showing the arrangement of polytene chromosomes within a salivary gland nucleus. The data have been processed to remove out-of-focus information (Equation 25) as well as flare due to fluorescence imaging. From Reference (2).

processing facility is rather high. Yet with the rapid advances being made in microprocessor and display technology, the costs should fall rapidly. The power of this method to help understand the complex structures and interactions taking place within the living cell has only begun to be exploited.

ACKNOWLEDGMENTS

I thank Dr. John W. Sedat for many helpful discussions. This work was supported in the early stages by a Helen Hay Whitney Postdoctoral Fellowship and more recently by the National Institutes of Health, grant no. GM 31627-01. D.A.A. is currently a Searle Scholar.

Literature Cited

1. Agard, D. A., Sedat, J. W. 1980. *Proc. Soc. Photo-Opt. Instrum. Eng.* 264:110–17
2. Agard, D. A., Sedat, J. W. 1983. *Nature* 302:676–81
3. Agard, D. A., Steinberg, R. A., Stroud, R. M. 1980. *Anal. Biochem.* 111:257–68
4. Agard, D. A., Stroud, R. M. 1982. *Biophys. J.* 37:589–602
5. Allen, R. D., David, G. B., Nomarski, G. 1969. *Z. Mikorsk. Tech.* 69:193–221
6. Bracewell, R. 1965. *The Fourier Transform and Its Applications*, pp. 108–12. New York: McGraw-Hill. 381 pp.
7. Brakenhoff, G. J., Bloom, P., Barends, P. 1979. *J. Microsc.* 117:219–32
8. Castleman, K. R. 1979. In *Digital Image Processing*, pp. 35–360. Englewood Cliffs, NJ: Prentice-Hall. 429 pp.
9. Crowther, R. A., Klug, A. 1975. *Ann. Rev. Biochem.* 44:161–82
10. DeRosier, D. J., Klug, A. 1968. *Nature* 217:130–34
11. Diamond, R. 1958. *Acta Crystallogr.* 11:129–38
12. Diamond, R. 1966. *Acta Crystallogr.* 21:253–66
13. Frieden, B. R. 1975. *Top. Appl. Phys.* 6:177–248
14. Goodman, J. W. 1968. *Introduction to Fourier Optics*, pp. 101–25. San Francisco: McGraw-Hill. 287 pp.
15. Gordon, R. 1983. In *Computing in Biological Science*, ed. M. J. Geisow, A. N. Baredt, pp. 23–70. New York: Elsevier Biomed.
16. Hamilton, D. K., Wilson, T., Sheppard, C. J. R. 1981. *Opt. Lett.* 6:625–26
17. Henderson, R., Unwin, P. N. T. 1975. *Nature* 257:28–32
18. Hopkins, H. H. 1955. *Proc. R. Soc. London* A231:91–103
19. Krieg, C., Cole, T., Deppe, U., Schierenberg, E., Schmitt, D., Yoder, B., Ehrenstein, G. 1978. *Dev. Biol.* 65:193–215
20. Mathog, D., Hochstrasser, M., Gruenbaum, Y., Saumweber, H., Sedat, J. W. 1983. *Nature*. In press
21. Mathog, D., Sedat, J. W. 1984. Manuscript in preparation
22. Olins, D. E., Olins, A. L., Levy, H. A., Dunfee, R. C., Margel, S. M., Tinnel, E. P., Dover, S. D. 1983. *Science* 220:498–500
23. Shantz, M. J. 1976. In *Computer Technology in Neuroscience*, ed. P. B. Brown, pp. 113–29. New York: Wiley
24. Sheppard, C. J. R., Wilson, T. 1981. *J. Microsc.* 124:107–17
25. Sheppard, C. J. R., Wilson, T. 1982. *Proc. R. Soc. London* A379:145–58
26. Skaer, R. J., Whytock, S. 1975. *J. Cell Sci.* 19:1–10
27. Stokseth, P. A. 1969. *J. Opt. Soc. Am.* 59:1314–21
28. Sulston, J. E., Horvitz, H. R. 1977. *Devel. Biol.* 56:110–56
29. Van Haarlem, R., Lagerweij, C., Ten Horn, L. C. J. E. M. 1982. *J. Microsc.* 127:265–69.
30. Weinstein, M., Castleman, K. R. 1971. *Proc. Soc. Photo-Opt. Instrum. Eng.* 26:131–38

Ann. Rev. Biophys. Bioeng. 1984. 13: 221–46

NMR STUDIES OF INTRACELLULAR METAL IONS IN INTACT CELLS AND TISSUES

Raj K. Gupta and Pratima Gupta

Department of Physiology and Biophysics, Albert Einstein College of Medicine, New York, New York 10461

Richard D. Moore

Biophysics Laboratory, State University of New York, Plattsburgh, New York 12901

INTRODUCTION

Nuclear magnetic resonance (NMR), a spectroscopic technique that measures magnetic nuclei and their environment, has gained considerable success and popularity in studying the dynamic state of inorganic ions and organic metabolites in intact organisms as well as in isolated living cells, tissues, and organs (4, 8, 10a, 11, 14–20, 23–28, 30–32, 34, 39, 40, 46, 48, 52, 53, 57, 58, 63–66). The success of such studies was made possible to a large extent by the incorporation of Fourier transform methods into NMR spectroscopy (13, 29, 49, 50). The main advantage of NMR is that it is noninvasive, so that ionic and metabolic changes in the cellular environment can be observed as they take place within an essentially unperturbed living system. It is also possible under suitable conditions to study the steady state dynamics of the living cell using time-resolved NMR experiments (8, 19, 46). With the introduction of sophisticated and continuously improving high field superconducting spectrometers, the NMR technique is approaching the status of a common, noninvasive analytical tool for cellular studies and research.

0084–6589/84/0615–0221$02.00

The work of Moon & Richards (39) on human red blood cells appears to be the first published account of a high resolution NMR spectrum of a cell system, although earlier reports of low resolution NMR observations on water and electrolytes in blood cells are available in the literature (35, 47). Several reviews of the applications of NMR spectroscopy to intact cells, tissues, and organs have already appeared (11, 14, 34, 52, 57). Because of the rapid expansion and intensive research being carried out in the cellular applications of NMR, progress in this area needs periodic reviewing. There have been notable advances during the last few years; the most outstanding have been the applications of NMR to the study of intracellular pH, free Mg^{2+}, free Ca^{2+}, free Na^+ and K^+ ions, and the application of NMR to the study of ^{31}P metabolites in humans (4, 14, 18–32, 34, 40, 43, 48, 52, 53, 58, 66).

Intracellular metal ions (Na^+, K^+, Mg^{2+}, Ca^{2+}) and pH appear to modulate a variety of cellular and tissue functions (7, 45). It is important to explore possible regulatory roles of these ions in cellular proliferation, differentiation, volume regulation, and hormonal control of various cellular processes as well as to understand how the concentrations of intracellular ions are managed by various membrane transport processes. How such regulation goes astray in disorders such as cancer (9, 26), hypertension (6, 33), diabetes (41), and sickle cell disease (60) is of clinical significance. NMR spectroscopy provides a unique noninvasive tool for such studies.

Nuclei used for NMR studies of intact cells, tissues, and organisms have included ^1H, ^{19}F, ^{31}P, ^{13}C, ^{15}N, ^{23}Na, and ^{39}K. Recent research from our laboratory resulted in the development of new NMR approaches for measuring free Mg^{2+} (by analyzing ^{31}P NMR of cellular ATP) (18, 20, 23, 26a, 26b, 30, 32) and Na^+ ions (using an anionic hyperfine shift reagent $Dy(PPP_i)_2^{7-}$ to distinguish between intra- and extracellular ions) (24–26, 27). An NMR method for estimating intracellular free Ca^{2+} level has been suggested by Smith et al (58). The cellular usage of NMR, however, is not limited to measurements of intracellular metal ions. NMR studies of various intracellular metabolites, intracellular pH, activities of membrane transport processes, and steady state rates of intracellular reactions by saturation transfer, as well as detection and identification of new metabolites by NMR and mapping of metabolic pathways for the processing of organic molecules by ^{13}C NMR have been described (4, 8, 14–20, 23–28, 30–32, 34, 39, 40, 46, 48, 52, 53, 57, 66). The feasibility of using surface coils and topical NMR techniques for in vivo studies of the ionic and metabolic state of tissue has been demonstrated (16, 17, 53).

This review describes the applications of the multinuclear NMR techniques to the study of metal ions (Na^+, K^+, Ca^{2+}, Mg^{2+}) in living cells and discusses the physiological significance of such studies. Since new developments in the area of ^{23}Na NMR have been rapid and are very recent, a large part of the review is devoted to the study of this intracellular cation. Representative examples from the work of our own laboratory are used for illustrative purposes where appropriate.

The information available from NMR often complements that available from other techniques. Unlike the atomic absorption and flame emission techniques that yield the total amounts of various ions in a cell, NMR may in suitable cases provide information on free ions. It should be mentioned that while Mg^{2+}, H^+, and Ca^{2+} ions in intact cells are at present observed only indirectly through their effects on the NMR spectra of either ^{31}P (Mg^{2+}, H^+) or ^{19}F (Ca^{2+}) nuclei, Na^+ and K^+ ions can be observed directly via their own resonance absorptions.

INTRACELLULAR Na^+ IONS IN LIVING CELLS

Significance of Na^+ Ion Measurements

Na^+ ions in animal cells play an important role in a variety of vital cell functions such as nerve transmission and generation of action potentials. Abnormal transport of Na^+ ions and the resulting intracellular ionic imbalance appear to be associated with cancer, hypertension, and diabetic states. Intracellular Na^+ ion concentrations appear related to the cell proliferation state and have been implicated in the mechanisms of mitogenic as well as oncogenic phenomena (7, 9, 26). The association between Na^+ ions and hypertension has long been recognized in the biomedical literature (6, 33). In fact it has been hypothesized that a primary defect in hypertension is an increase in intracellular Na^+ of arteriolar smooth muscle due to the presence of an abnormality in membrane sodium transport (6, 33). A recent study also suggests the association of an alteration in intracellular Na^+ ions with the diabetic state (41).

Changes in Na^+ ion concentration and electrochemical gradient could affect cell function in a variety of ways. Intracellular Na^+ concentration is a direct regulator of the plasma membrane Na^+-pump, which in turn influences the energetics of the cell. The Na^+ electrochemical gradient appears to be coupled to the transport of type A amino acids, bases, and other nutrients into certain types of cells and may thereby exert an influence on cell growth. Further, changes in Na^+ ion concentration could profoundly alter other intracellular ions. For example, discharge of H^+ ions from the cell, with a concomitant rise in intracellular pH, could accompany Na^+ influx via the Na : H exchange mechanism; or a change in intracellular Ca^{2+} or Mg^{2+} may occur during Na^+ influx via a Na:Ca or Na:Mg exchange mechanism. Intracellular pH, Ca^{2+}, or Mg^{2+} ions may in turn be intimately involved in fine control of cellular functions (7, 45). For example, an increase in intracellular Na^+ ion in the hypertensive state would be expected to lead to a significant rise in cellular free Ca^{2+} through a $Na^+:Ca^{2+}$ exchange mechanism in arteriolar smooth muscle, with a consequent increase in arterial tone and vasoconstriction (6). It is therefore of considerable physiological interest to study in a noninvasive manner Na^+ ions and their transport across surface membrane in intact cells and tissues.

NMR Observation of Intracellular Na^+ Ions

Although NMR spectroscopy offers a useful noninvasive technique for studies of intact cells and tissues, the interesting small ^{23}Na resonance of intracellular ions and the uninteresting, but large, resonance of extracellular Na^+ ions in physiological cell suspensions occur at the same frequency in the spectrum. This lack of spectral resolution has in the past limited the applicability of the NMR technique to the study of the state of intracellular Na^+ ions. Our recent discovery (25) of the highly anionic hyperfine shift reagent bis(tripolyphosphate)dysprosium(III) $Dy(PPP_i)_2^{7-}$ circumvents this problem and allows the detection of separate NMR resonances from intra- and extracellular Na^+ ions in living cells that would otherwise be indistinguishable from each other (24–26) (Figure 1). This has paved the way for future NMR studies of monovalent cations in cells, tissues, and organisms. The detection of separate resonances from intra- and extracellular Na^+ ions depends on the fact that the highly anionic paramagnetic reagent causes a hyperfine shift in the frequency of the resonance of Na^+ ions accessible to it. Since this reagent does not penetrate the cell membrane over the time scale of NMR measurements and remains on the outside of the cells, only extracellular Na^+ ions experience the resonance shift. The ability to separate intra- from extracellular Na^+ ions enables us to study and quantitate the intracellular ions without contamination from extracellular ions (25).

It should be mentioned that Springer and associates from SUNY at Stony Brook have independently described hyperfine shift reagents for cations (4). The first reports of the discovery of the hyperfine shift reagents for Na^+ from the two laboratories appeared simultaneously (24, 59). $Dy(PPP_i)_2$, however, appears to be more effective by an order of magnitude, because it shifts the Na^+ resonance more and is less toxic [judged by the maintenance of cellular ATP and P-creatine levels and by the maintenance of tissue electrophysiological characteristics (10a)] than the reagents reported by Springer's group (R. K. Gupta, P. Gupta, unpublished results).

$Dy(PPP_i)_2$ causes an upfield hyperfine shift in the frequency of the extracellular Na^+ resonance. A complex of the same ligand with thulium

$Tm(PPP_i)_2$ causes a downfield hyperfine shift presumably due to different paramagnetic properties of the metal ion; but it is 2-fold less effective as judged by the magnitude of the resulting hyperfine shifts (R. K. Gupta, P. Gupta, unpublished observations). Shulman and associates (48) have independently tested the ability of $Dy(PPP_i)_2$ to distinguish between intra- and extracellular ions and have found it useful in studies of intracellular K^+ ions as well. We (R. K. Gupta, P. Gupta, unpublished) have found $Dy(PPP_i)_2$ to be equally effective in separating the resonances of intra- and extracellular Li^+ and Cs^+ ions, which may be useful in some transport studies.

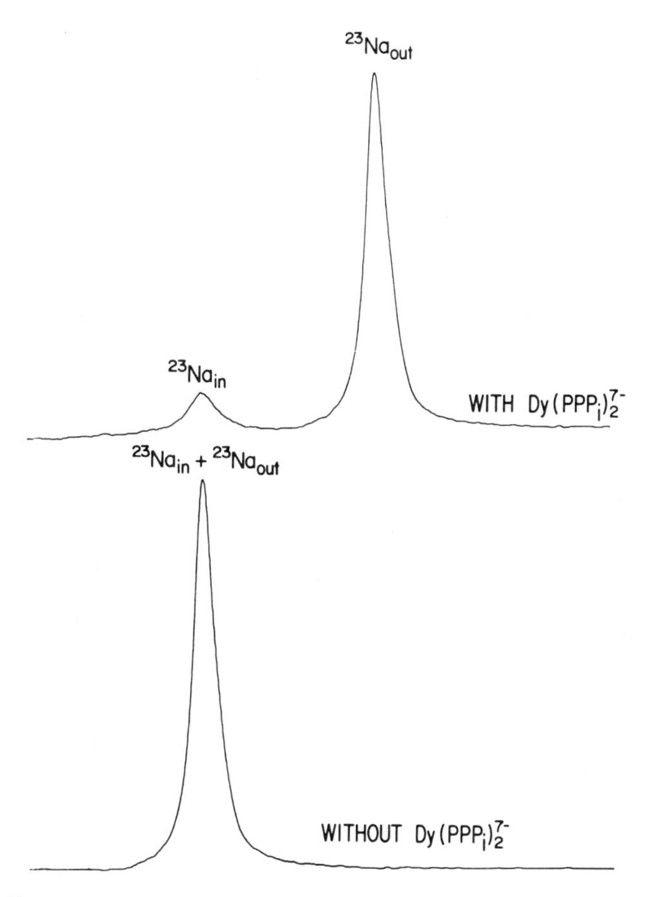

Figure 1 ^{23}Na NMR spectra of red cells in heparinized whole human blood with (*top*) and without (*bottom*) 3 mM $Dy(PPP_i)_2^{7-}$ showing resolution of extra- and intracellular Na^+ ions by the hyperfine shift reagent [adapted from (25)].

Quantitation of Intracellular $^{23}Na^{+}$ *Ions by NMR*

Direct observation of separate resonances from intra- and extracellular $^{23}Na^{+}$ provides a convenient means of studying the state of intracellular Na^{+} and the net influx or efflux of NMR-visible Na^{+} ions. Since $Dy(PPP_i)_2^{7-}$ causes effective resonance separation at sufficiently low concentrations (approximately 3 ppm with 1 mM of reagent) at a physiological level of extracellular Na^{+} and pH, and since essentially all of the Dy^{3+} is complexed to PPP_i (a biometabolite), no significant perturbation of the cellular system is expected. A comparison of the intensity of the resonance from extracellular ions (A_{out}) with that of a noncellular control (A_0) containing the same concentration of Na^{+} ions as is present in the extracellular medium [Na_{out}] directly yields the fractional space in the NMR window that is extracellular (S_{out}). The intensities of the $^{23}Na^{+}$ resonances of intracellular (A_{in}) and extracellular ions and a knowledge of the fractional space that is extracellular then directly yield the concentration of intracellular $^{23}Na^{+}$ ions [Na_{in}] that contributes to the observed resonance signal. The following straightforward equations provide the relationship between the observed resonance intensities and the NMR-visible intracellular Na^{+} ion concentration (25, 26):

$$S_{out} = \frac{A_{out}}{A_0} \qquad 1.$$

$$[Na_{in}] = \left\{\frac{A_{in}}{A_{out}} \cdot \frac{S_{out}}{(1 - S_{out})}\right\} [Na_{out}]. \qquad 2.$$

$^{23}Na^{+}$ *Ions in Human Erythrocytes*

Exposure to $Dy(PPP_i)_2$ does not affect the ATP, 2,3-DPG, or free Mg^{2+} levels in the human red blood cell even over a period of several hours (13a), and the reagent would thus appear to be nontoxic to cellular energy metabolism. ^{23}Na NMR spectrum of well-packed human erythrocytes in a physiological medium (145 mM Na^{+}, 5 mM K^{+}, 1.3 mM Ca^{2+}, 0.9 mM Mg^{2+}, 136 mM Cl^{-}, 14.3 mM bicarbonate, 5.6 mM glucose, and 1.6 mM phosphate at pH 7.4) containing 4 mM $Dy(PPP_i)_2$ shows two well-resolved resonances owing to the spectral separation of intra- and extracellular $^{23}Na^{+}$ ions by the reagent (Figure 2). A similar resonance separation can be obtained by adding Dy $(PPP_i)_2$ directly to the whole human blood. The resonance at right (upfield) corresponds to extracellular ^{23}Na ions that interact with the reagent, whereas the resonance at left (downfield) arises from intracellular ^{23}Na ions. The chemical shift and the intensity of the latter resonance are unaffected by varying the concentration of the paramagnetic reagent (25).

From the spectrum in Figure 2, an intracellular $^{23}Na^+$ ion concentration of 4.1 μmol/ml red cells has been estimated (26). This NMR-visible Na^+ concentration in human erythrocytes is lower than the total sodium concentration estimated by atomic absorption (26). These results suggest that part of the intracellular Na^+ is somehow invisible to the NMR technique. Alternatively, the observed resonance may not represent full absorption associated with all of the ^{23}Na nuclear transitions because of broadening of some of the transitions beyond experimental recognition by quadrupolar interactions of the ^{23}Na nucleus with electric field gradients within the cell. Earlier analyses of ^{23}Na NMR spectra indicated that when a population of ^{23}Na ions experiences an asymmetrical electrical field, 40% of the signal is contained in the central peak, and the remainder (60%) is concealed in two satellite peaks broadened beyond experimental recognition (5, 11). Since the ratio between central and satellite peak intensity is fixed by quantum mechanical considerations, the loss of signal intensity owing to uniform first order nuclear quadrupolar interactions of the entire Na^+ pool can be only 60% (5, 11). The observation of an intensity loss of only 30% in comparison to the expected full signal must be interpreted in terms of either compartmentation (only a part of the Na^+ ion pool is affected by quadrupolar interaction) or some sort of binding interaction that results in immobilization of part of the $^{23}Na^+$ ion pool, or combination of both. Immobilization could cause sufficient qudrupolar broadening that would lead to total disappearance of the resonance of bound or

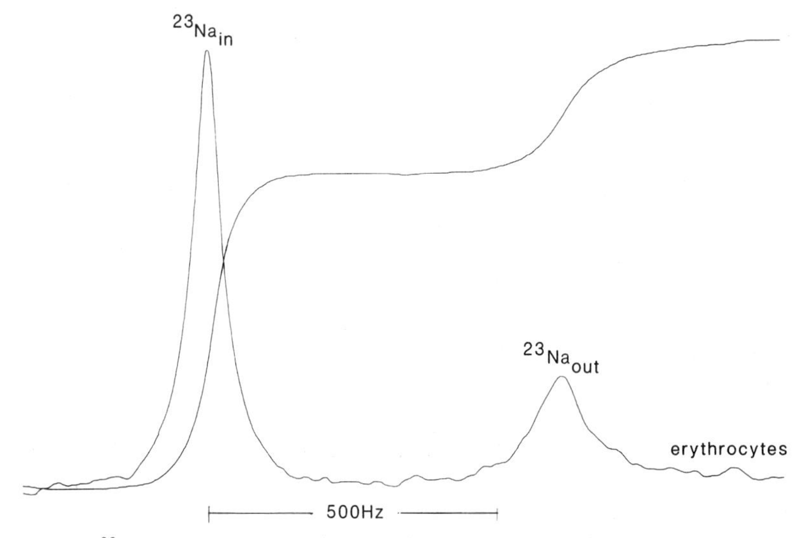

Figure 2 ^{23}Na NMR spectrum of well-packed human erythrocytes in a physiological medium containing 4mM $Dy(PPP)_i)_2^{7-}$ [from (26)].

compartmentalized ^{23}Na, making it NMR-invisible without affecting the signal from free Na^+ ion pool. Such binding or compartmentation of Na^+ ions may be essential for the maintenance of the structural integrity of the cell. In order to explain these observations (26), however, the immobilized or compartmentalized Na^+ must exchange slowly with the remaining NMR-visible Na^+ pool on the T_2 time scale; otherwise, the observed ^{23}Na resonance would be expected to reflect the entire pool of the intracellular Na^+ ions.

In the absence of the paramagnetic reagent, red blood cells exhibit only a single ^{23}Na resonance because of overlap of the signals from intra- and extracellular Na^+ ions. The contribution of the extracellular Na^+ to the observed signal from well-packed erythrocytes (Figure 2) is approximately 30%. Hemolysis of packed human erythrocytes by repeated freeze-thawing caused a $\sim 30\%$ increase in the overall Na^+ signal (Figure 3). This indicates either release of the bound/compartmentalized Na^+ or its rapid exchange with the "free" Na^+ pool upon hemolysis so that most of the cell Na^+ becomes observable by NMR in the lysed state. A similar ($\sim 30\%$) increase in the ^{23}Na NMR signal is observed when packed red cells are lysed by membrane solubilization in a detergent (Figure 3). These observations clearly demonstrate that even in human erythrocytes, which are non-nucleated and lack subcellular organelles, a significant part of the intracellular $^{23}Na^+$ in the intact cell appears immobilized by being bound or sequestered in the cell membrane and/or cytoskeleton. Somewhat surprisingly, this Na^+ is unable to exchange rapidly with the free pool and therefore becomes NMR-invisible. It appears likely from preliminary observations of its release or rapid exchange upon cell lysis that this NMR-invisible Na^+ may, at least in part, be located in the membranes. Binding or sequestering of Na^+ ions in the cell membrane and cytoskeleton could play an important role in the maintenance of cell viability. In any case, the ^{23}Na resonance of the intact cell appears to be quantitatively different from that of the lysate (26).

The hypothesis for the existence of sequestered Na^+ is supported by radioactive tracer exchange kinetic studies in the literature, which indicate that some of the Na^+ in the human red blood cells does not exchange, or exchanges very slowly, with extracellular Na^+ (60). The part of the intracellular Na^+ that is bound, sequestered, or compartmentalized would be unavailable to membrane ion-transport systems.

The availability of intracellular Na^+ ion may also vary with changes in the physiological state of a cell, because of variations in the binding and/or compartmentation of the ion, without any change in the total intracellular Na^+ content (10, 27). The above findings of NMR-invisible Na^+ in red

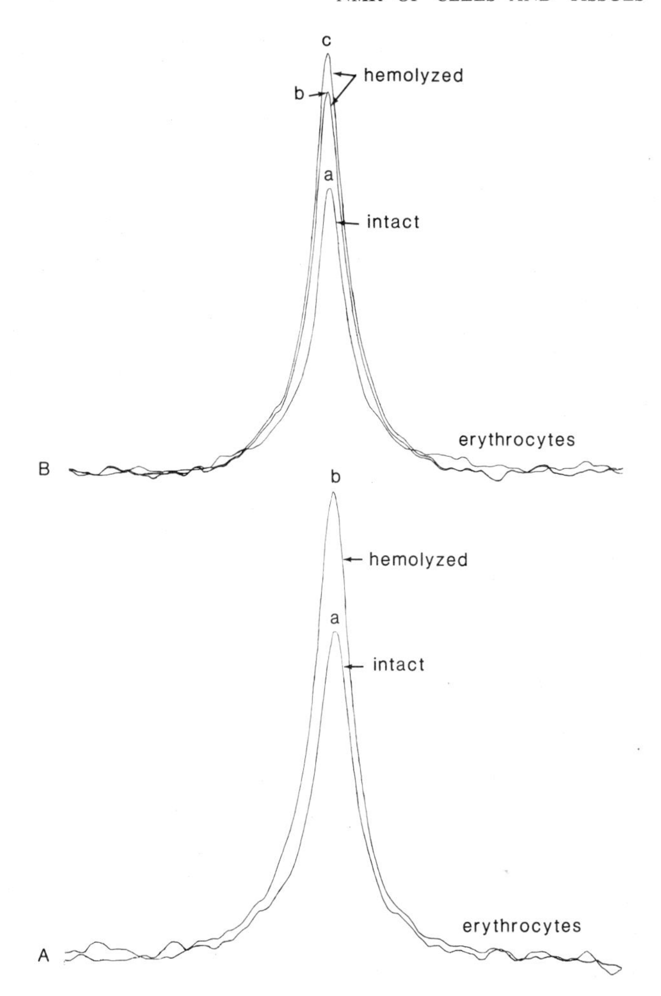

Figure 3 ^{23}Na NMR spectra of well-packed human erythrocytes: (*A*) a, intact cell; b, lysate form repeated freeze thawing; (*B*) a, intact cells; b, detergent-solubilized cells without amplitude correction for dilution of ^{23}Na ions by the addition of detergent; and c, with signal amplitude correction for dilution of ^{23}Na ions by the addition of detergent [from (26)].

blood cells should be pursued in greater depth before definite conclusions about its nature and origin can be drawn.

$^{23}Na^{+}$ Ions in Human Normal and Leukemic Lymphocytes

Lymphocytes are like most cells in that they maintain a high level of K^+ and a low level of Na^+ relative to the concentrations of these ions in the environment. Because human peripheral blood lymphocytes are easily available, can be obtained in the resting state, and are readily stimulated to undergo mitogenic transformation and nearly synchronous cell division, they are one of the most widely used model systems for studies of cellular mitogenesis. Several observations suggest that K^+ and Na^+ are involved in the control of cellular division (36). These include the inhibition of mitogenesis of human lymphocytes when they lose K^+ and gain Na^+ by treatment with ouabain; the increases in the rates of exchange of K^+ and Na^+ induced by mitogens; the inhibition of mitogenesis by K^+ ionophores; the apparent changes in the cellular potential during mitogenic stimulation; and the requirement for Na^+ of mitogen-induced proliferation. Although changes in ion flux and cytoplasmic ion composition have been proposed as mediators of the effects of the mitogenic surface ligands in lymphocytes, and there has been considerable biochemical investigation into the processes of stimulation, not much attention has been directed to membrane transport as a determinant or as a reflection of disordered maturation in leukemic cells. The above observations, however, suggest that the distribution of intracellular monovalent cations may be important in supporting vital cell functions that facilitate cell maturation and mitosis; and that dynamic alterations in uptake and efflux of monovalent cations may be essential concomitants of mitosis and its related changes in cell volume. These considerations have lead to the hypothesis that an abnormality in the control of intracellular monovalent cations could play a role in the disordered mitotic rate and maturation sequence of some leukemic cells (26).

The ^{23}Na NMR technique is readily applicable to living lymphocytes, and acceptable spectral signal-to-noise ratio is easily attainable (26). NMR-visible Na^+ contents of human normal and leukemic lymphocytes have been characterized using the frequency shift reagent. Lymphocytes from patients with chronic lymphocytic leukemia (CLL) have been compared with normal controls (26; Figure 4). In this study lymphocytes were suspended in their own serum in order to reflect the in vivo situation as closely as possible. ^{23}Na NMR spectra from a few normal and abnormal (CLL) samples of lymphocytes yielded significantly different NMR-visible intracellular Na^+ ion concentrations (17.5 ± 1.2 μmol/ml for normal and 8.7 ± 0.8 μmol/ml for abnormal cells) (26).

Although the statistical significance of the observation of a lower level of intracellular Na^+ in CLL lymphocytes in comparison to normal cells should be tested further by acquisition of data on many more samples, the results appear to indicate that NMR-visible Na^+, which is presumed to be predominantly free, in leukemic lymphocytes is $\sim$2-fold lower in comparison to normal lymphocytes. However, whether the observed difference in the state of intracellular Na^+ ion in the leukemic and normal lymphocytes arises from the action of an abnormal extracellular agent in serum of CLL patients or from an intrinsic alteration in the behavior of plasma membrane and subcellular constituents remains to be investigated. Furthermore, whether the decreased NMR-visible intracellular Na^+ ion content of CLL lymphocytes is caused by a more efficient exclusion of Na^+ via Na^+-pump or by a reduced permeability of the plasma membrane for Na^+, or by a combination of both factors, also remains unknown. It has been suggested that an increased Na^+ electrochemical gradient, which would cause abnormal transport of many nutrients into the cell, may be associated with the neoplastic process in vivo (26). Lower free Na^+ in CLL lymphocytes may also be related to the abnormally long life span of these

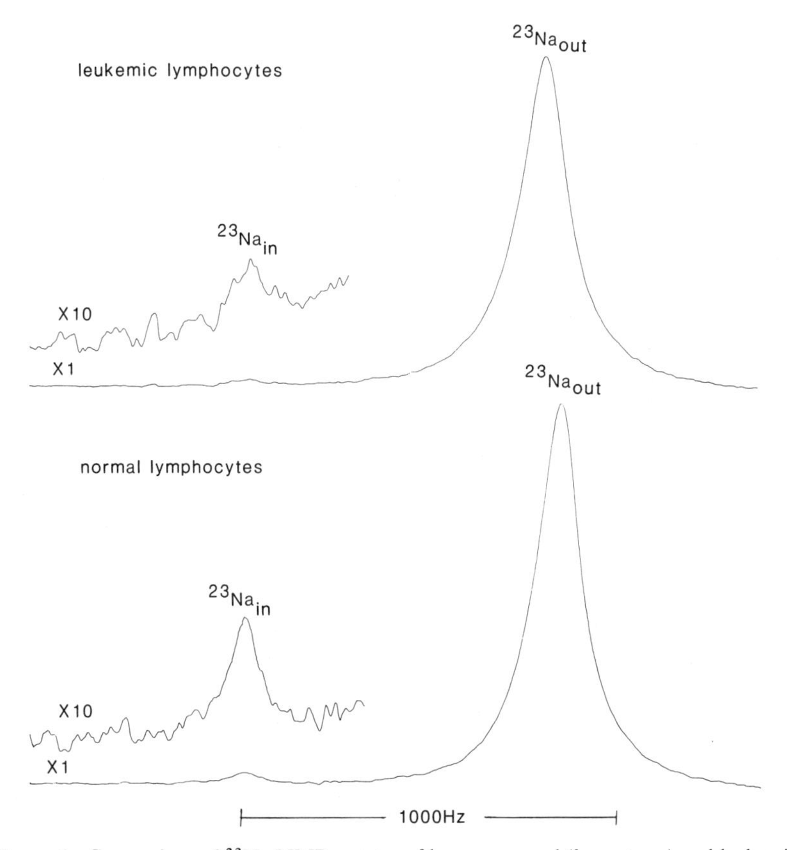

Figure 4 Comparison of ^{23}Na NMR spectra of human normal (*lower trace*) and leukemic (CLL) lymphocytes (*upper trace*) suspended in their own serum containing 4 mM $Dy(PPP_i)_2^{7-}$. Extracellular volume was approximately 85%. [From (26).]

leukemic cells. Adequate characterization of ionic differences between the normal and CLL lymphocytes is of considerable interest since unique structural, biochemical, or biologic features by which CLL lymphocytes can be readily differentiated from normal lymphocytes have not been adequately identified. The ability of the ^{23}Na NMR technique to monitor noninvasively the state of intracellular ions in abnormal lymphocytes may be useful as a simple in vitro assay for determining the effectiveness of action of cancer chemotherapeutic agents (26).

Intracellular $^{23}Na^+$ Ions in Isolated Cardiac Myocytes

The maintenance of a low intracellular Na^+ level is of vital importance to the development of action potentials necessary for cardiac contraction. Since the measurement of cell Na^+ by atomic absorption is complicated by contamination from relatively high levels of extracellular Na^+, Gupta & Wittenberg (31) used the NMR technique and $Dy(PPP_i)_2^{7-}$ to distinguish intra- and extracellular Na^+ ions. From the NMR data on oxygenated myocytes in a Ca^{2+}-containing physiological medium, an NMR-visible Na^+ concentration of 7.8 ± 0.6 μmol/ml cells was obtained, in comparison with the reported atomic absorption total Na^+ of 9.9 ± 2.1 μmol/ml cells (31). These results suggest that most of the intracellular Na^+ in myocytes is NMR-visible and that no appreciable diminution in signal occurs as a result of nuclear quadrupolar interactions. Incubation of myocytes in a Ca^{2+}-free medium resulted in a 3.5-fold increase in NMR-visible Na^+. Upon transfer to a Ca^{2+}-containing medium, the Ca-tolerant myocytes were found to again pump out their intracellular Na^+ to a low value of 9.5 ± 2.4 μmol/ml cells. Ca-intolerant cells, however, became spherical and eventually became permeable to $Dy(PPP_i)_2$. These results were interpreted to indicate that $Dy(PPP_i)_2$ does not inhibit the Na^+-pumping ability of myocytes and that ^{23}Na NMR can be a unique noninvasive technique for investigating $Na^+:Ca^{2+}$ and $Na^+:K^+$ exchanges in heart cells (31).

Intracellular ^{23}Na in Amphibian Oocytes

The amphibian oocyte, which is a kind of uniquely specialized model of the excitable cell, responds to hormones with a series of changes in the plasma membrane permeability and transport of ions such as Na^+, K^+, H^+, and Ca^{2+}. The resulting changes in cytoplasmic ion concentrations appear to act as signals or regulators of cellular events associated with oocyte maturation and development. Because of their large size, easy availability, and physiological stability, frog oocytes constitute a particularly favorable system for NMR studies. Unlike other cells, they can be readily superfused to maintain their physiological state for several hours in the NMR sample tube, and superfusion also allows rapid introduction of various hormones and/or drugs. Adequate signals can be accumulated rapidly on highly synchronous cell populations, and data collection can be continued through different stages of the meiotic and/or mitotic divisions. Using atomic absorption techniques, changes in concentrations of Na^+ and K^+ ions have been reported during the course of amphibian oocyte development. The NMR technique offers the possibility of monitoring "free" (NMR-visible) and "bound" Na^+ during oocyte development (26, 28, 42, 43). It should be mentioned that while there is no problem in getting enough oocytes in the NMR tube for rapid data accumulation, the sensitivity of the current generation of NMR spectrometers is such that we can detect intracellular Na^+ from a single oocyte with acceptable signal-to-noise ratio in about 20 min (R. K. Gupta, unpublished observations).

Gupta et al (27) have used the hyperfine shift reagent $Dy(PPP_i)_2$ for separating extracellular and measuring intracellular Na^+ ions in *Rana* eggs and embryos. $Dy(PPP_i)_2$ did not affect intracellular pH, phosphocreatine level, or membrane potential of oocytes, over a period of an hour, and did not inhibit meiosis or mitosis in this system. Using the NMR technique for simultaneous determination of both extracellular space and intracellular Na^+, only a small part (14–17%) of the Na^+ in the prophase oocyte was detectable. A large fraction of the oocyte Na^+ ($\sim$85%) was NMR-invisible, presumably because of binding, sequestering, and/or compartmentation of the ion. However, NMR-visible Na^+ changed in response to hormonal stimulation and during ovulation and early development. Upon release of the prophase block by progesterone, there was an initial decrease in NMR-visible Na^+ prior to nuclear breakdown, followed by a fourfold increase in NMR-visible Na^+ during nuclear breakdown. These changes in NMR-visible Na^+ occurred without a change in total Na^+ as measured by atomic absorption spectroscopy. By the time of ovulation (second metaphase arrest), total Na^+ increased and NMR-visible Na^+ accounted for 30% of the total Na^+. This increase in NMR-visible Na^+ was largely accounted for by a net increase in Na^+ uptake by the egg during ovulation. During early cleavage there was a small decrease in total Na^+, but NMR-visible Na^+ rose to account for about 70% of the total Na^+. In all cases, however, NMR-visible Na^+ concentration was significantly lower than the total concentration estimated by the atomic absorption technique. Binding and/or compartmentation and immobilization of the NMR-invisible Na^+ could cause line broadening via quadrupolar interactions, leading to disappearance of the resonance of this Na^+ ion pool. The identity of the intracellular organelle(s) responsible for compartmentation of ions within mature oocytes is unknown. The yolk platelets occupying a large part of the oocyte seem to be a particularly likely site of compartmentalization, although the nucleus may also play a role (10). NMR studies, however, provide evidence for the existence of some NMR-invisible Na^+ in the amphibian egg and for its release during early development (27).

Na^+ Ions in Insulin Action

It is now generally recognized that insulin affects the transport of glucose, amino acids, and ions across the plasma membrane (39a). It stimulates a rapid change in the ion transport and electrical properties of the plasma membrane in a number of cell types. This insulin response is seen as membrane hyperpolarization, increased Na^+ and K^+ exchange, and stimulation of $Na^+:K^+$ and $Na^+:H^+$ exchange systems in the plasma membrane (40–43).

In the amphibian oocyte, insulin acts to release the block at prophase arrest and reinitiate the meiotic divisions. It has been found using NMR that insulin produces small but significant increases in intracellular Na^+ ($\sim$20%) and pH_i ($\sim$0.3 unit) in *Rana* oocytes (42, 43), consistent with stimulation of $Na^+:H^+$ exchange in the oocyte plasma membrane (40, 42, 43). However, the possibility that the changes in intracellular Na^+ and pH occur via independent mechanisms has not been ruled out.

Isolated cardiac myocytes that have functionally intact insulin receptors also provide a promising model for the investigation of cardiac cellular responses to insulin. In both *Rana* oocytes and cardiac myocytes, sizable ($\sim$2-fold) insulin-induced increases in intracellular Na^+ were observed in Ca^{2+}-free medium (42, 43; R. K. Gupta, B. A. Wittenberg, unpublished observations). This has generated the hypothesis that insulin increases Na^+ permeability of the plasma membrane but that in a normal Ca^{2+}-medium the Na^+-pump is able to maintain a low level of intracellular Na^+ even in the presence of insulin. In Ca^{2+}-free medium, however, the pump is saturated with Na^+ and therefore a further increase in Na^+ permeability produces a more pronounced effect on intracellular Na^+. A prediction of this hypothesis is that if the Na^+ pump were inhibited with ouabain, the increase in cell Na^+ would be considerably more rapid in the presence of insulin than in its absence.

INTRACELLULAR FREE Ca^{2+} IONS

There is ample evidence that intracellular Ca^{2+} ions play an important regulatory role in directing cell metabolism. Ca^{2+}, at certain optimal concentrations, triggers cascades of enzyme activations and structural changes which, depending on the makeup of the cell, result in hormone production and secretion, muscle contraction, neurotransmitter release, egg activation, or DNA synthesis, and cell division (7). The credibility of Ca^{2+} ions as specific regulators of cell proliferation has been greatly enhanced by the discovery of calmodulin, a small calcium-binding enzyme activator protein that modulates a variety of intracellular enzyme activities. It appears likely that a (Ca^{2+} + Mg^{2+})-dependent regulatory mechanism(s) is altered or changed during carcinogenesis (7).

Measurement of Intracellular Free Ca^{2+} Using ^{19}F NMR Spectroscopy

Following the design of Tsien (61), ^{19}F-labelled intracellular chelators have recently been synthesized by Smith et al to measure the free cytoplasmic Ca^{2+} concentration in intact cells using ^{19}F NMR (58). Although intracellular free Ca^{2+} concentrations can be measured optically with the fluorescent chelator quin 2 (61), ^{19}F chelators have the advantage that the NMR chemical shift may provide identification of the bound ion. They may also be useful for opaque cell suspensions and tissues inappropriate for fluorescence studies. ^{19}F-labelled analogs of 1,2-bis(*o*-aminophenoxy)-ethane-N,N,N′,N′-tetraacetic acid (BAPTA) substituted symmetrically with a single fluorine in each ring are highly sensitive in their NMR chemical shifts to chelation by divalent cations. The chemical shifts of the ^{19}F resonances depend on the identity of the chelated ions. Ca^{2+}, Zn^{2+}, and Fe^{2+} have high affinities for the nFBAPTA (where n denotes the labelled ring positions) chelators while Mg^{2+} has a relatively low affinity.

NMR spectra indicated that the free and bound forms of 3FBAPTA and 5FBAPTA complexes with Ca^{2+} are in slow exchange, whereas for 4FBAPTA the two forms are in fast exchange on the NMR time scale at least at 94 MHz (58). In the slow exchange situation, two separate resonances are observed, weighted in area according to the proportions of the free and bound forms of the chelator. In the fast exchange situation, a single resonance is observed at a chemical shift intermediate between the positions of the free and bound forms weighted according to their proportions. From the NMR chemical shift of the exchange-averaged resonance in the fast exchange situation, and from the relative areas under the resonances of the free and Ca^{2+}-bound forms of the ^{19}F-labelled chelator in the slow exchange situation, one directly obtains the proportion of the molecule that is complexed to Ca^{2+}. A knowledge of the dissociation constant of the Ca^{2+}-chelator complex then yields the concentration of free Ca^{2+} directly from the NMR spectral data.

Two ions that might affect the chemical shifts of the intracellular nFBAPTA analogs are Mg^{2+} and H^{+}. However, all complexes with Mg^{2+} gave single resonances consistent with fast NMR exchange and the low affinity of these chelators for Mg^{2+}. The chemical shift of 4FBAPTA analog is reported to be almost totally insensitive to titration with Mg^{2+} (58). The 5FBAPTA analog showed chemical shifts with added Mg^{2+} that were much smaller in magnitude than those obtained with Ca^{2+}. It is estimated that changes in intracellular free Mg^{2+} concentrations of up to 1 mM will produce chemical shifts of only 0.2 ppm, which is less than 5% of the shift observed when Ca^{2+} binds. Changes in free Mg^{2+} concentration in the cell will therefore not have any significant effect on the free Ca^{2+} measurements.

The 5FBAPTA resonance is reported to shift by about 5.7 ppm between pH 7.5 and pH 4.5; the chemical shift change is about 0.1 ppm between pH 6.9 and 7.3, which is less than 2% of the shift resulting from Ca^{2+} complexation. The 4FBAPTA resonance is reported to be insensitive to pH changes (<0.25 ppm from pH 5 to 8 and 0.03 ppm from pH 6.9 to 7.3) (58). Hence, intracellular pH changes should also not significantly affect the estimation of intracellular free Ca^{2+} with these chelators.

The existence of fast and slow NMR exchange conditions for the different Ca-nFBAPTA complexes may provide two kinds of independent NMR assays of intracellular free Ca^{2+} over a wide range of concentrations (58). With the fast exchange 4FBAPTA chelator, the intracellular free Ca^{2+} is indicated by the measurement of a chemical shift, which is inherently more sensitive than the measurement of the areas of resonances from the free and bound forms for slow exchange complexes. The accessible range of free Ca^{2+} is estimated to be from about 0.3 μM to 30 μM, assuming shift measurements are accurate to 0.05 ppm for the exchange-broadened peaks. Furthermore, the chemical shift of the 4FBAPTA resonance is very insensitive to changes in $[Mg^{2+}]$ and $[H^+]$ and the shift therefore provides a direct measure of the proportion of the chelator complexed with Ca^{2+}, and hence intracellular free Ca^{2+}, without correction for $[Mg^{2+}]$ and $[H^+]$. Other intracellular ions (e.g. Zn^{2+}, Fe^{2+}, Mn^{2+}) were reported to have higher affinities than Ca^{2+} for 4FBAPTA but were in slow exchange. These ions therefore should not affect the Ca-4FBAPTA resonance that indicates free Ca^{2+}.

Free Ca^{2+} in Thymocytes Using ^{19}F NMR

^{19}F NMR spectra at 188 MHz from mouse thymocytes loaded with approximately 1 mM of 5FBAPTA introduced by intracellular hydrolysis of the acetoxymethyl ester derivative consist of two resonances at the positions corresponding to free 5FBAPTA and the Ca-5FBAPTA complex (58). The relative areas of the two resonances corresponded to complexation of 20% of 5FBAPTA with Ca^{2+}, equivalent to an intracellular free Ca^{2+} of 250 nM. Upon addition of succinyl Con A, there was a consistent and significant increase in the proportion of Ca-5FBAPTA, corresponding to an increase in intracellular free Ca^{2+} to about 350 nM. In this study (58), the identity of the Ca-5FBAPTA resonance was confirmed by the addition of 50 μM A23187 to equilibrate the cells with Ca^{2+} in the external medium. A single resonance at the chemical shift of Ca-5FBAPTA was observed in the presence of the ionophore because of saturation of the chelator by intracellular Ca^{2+}.

In thymocytes the chemical shift of 4FBAPTA is reported to be very close to the shift of the free form, indicating a free Ca^{2+} level of less than 300 nM, which is consistent with the value from the slow exchange 5FBAPTA assay of 250 nM. The 4FBAPTA chelator should be useful as an indicator of free Ca^{2+} at concentrations around 1 μM. Free Ca^{2+} of 250 nM indicated by the slow exchange Ca-5FBAPTA complex in thymocytes is higher than the estimate of 120 nM using quin 2 in the same cells (58). The measurement of free Ca^{2+} from the relative areas of the resonances of the free and bound forms is most accurate and sensitive when the ratio is 1 : 1. The chemical shifts of the two resonances, however, provide an identification of the chelated cationic species. Interestingly, the position of the free 5FBAPTA resonance in thymocytes indicated that the free intracellular Mg^{2+} concentration is less than 0.3 mM (58), an observation consistent with published ^{31}P NMR studies of Ehrlich ascites tumor cells (32).

Further development of the fluorinated Ca^{2+} specific indicators should be carried out so that measurements of free Ca^{2+} can be made at intracellular chelator concentrations below 0.1 mM, where they do not stimulate the cells metabolically, which is a major limitation of the method as applied currently (58). A 3-fold improvement in intracellular chelator concentration should be possible with the synthesis of ^{19}F-labelled analogs of BAPTA substituted symmetrically with CF_3 group in each ring.

INTRACELLULAR FREE Mg^{2+} IN LIVING CELLS

Since Mg^{2+} ion plays an important role in the diverse biochemical reactions occurring within a cell, a knowledge of cellular free Mg^{2+} level is essential for an accurate understanding of a functioning intact cell. Although it is easy to measure total Mg^{2+} content of a cell by atomic absorption techniques, such is not the case for the measurement of free Mg^{2+}. A noninvasive ^{31}P NMR technique for determining free Mg^{2+} in intact cells has been suggested (23) and is reviewed here.

^{31}P NMR Measurement of Intracellular Free Mg^{2+}

The ^{31}P NMR technique for measuring free Mg^{2+} is based upon an accurate measurement of the frequency difference between the αP and βP resonances in the ^{31}P NMR spectrum of intracellular ATP (Figure 5). Because the positions of the ^{31}P resonances of ATP depend on its state of complexation with Mg^{2+} (the predominant divalent cation component of the cell), the NMR spectrum allows a direct determination of the fraction of total ATP that exists as the Mg^{2+} complex. An accurate knowledge of the dissociation constant of MgATP (K_D^{MgATP}) under simulated intracellular ionic conditions then yields free Mg^{2+} directly from the NMR spectral data (20, 23, 30, 32).

From a comparison of the measured Mg-dependent separation between the αP and βP resonances of intracellular ATP ($\delta_{\alpha\beta}^{cell}$) with that in extracellular ATP ($\delta_{\alpha\beta}^{ATP}$) and MgATP ($\delta_{\alpha\beta}^{MgATP}$) controls under simulated

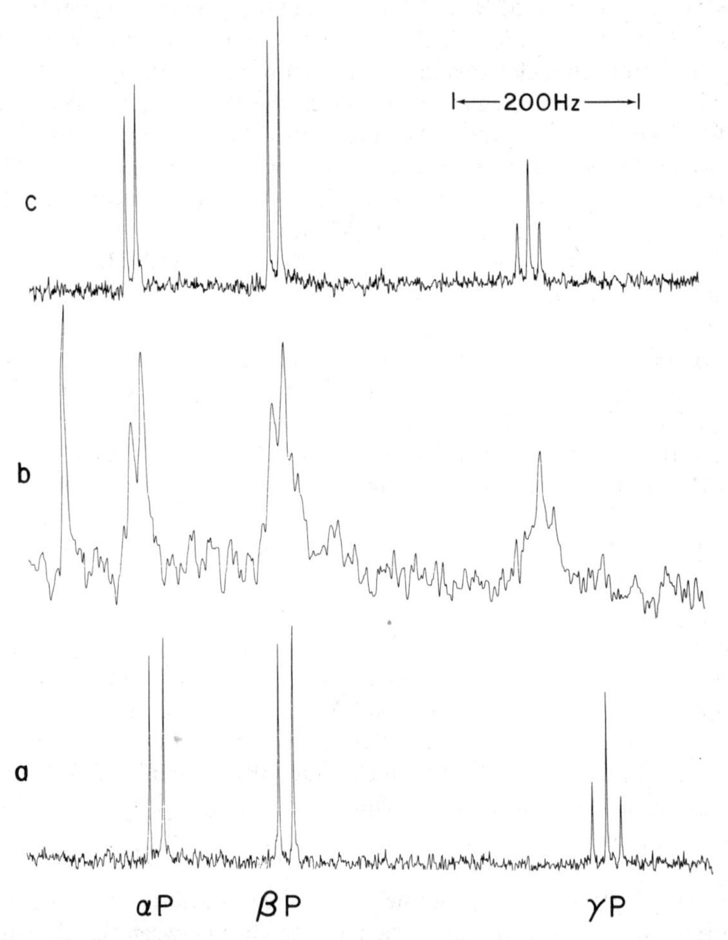

Figure 5 ^{31}P NMR spectra showing the αP, βP, and γP resonances of intracellular ATP in fresh Ehrlich ascites tumor cells with appropriate controls (32, 66). The quantity $\delta_{\alpha\beta}$ is measured as the separation between the central component of the βP triplet and the center of the αP doublet. The spectra were obtained at 10°C. Samples contained (*a*) 4 mM ATP at pH 7.2 and $\mu = 0.15$ M; (*b*) Ehrlich ascites tumor cells; (*c*) same as *a* plus 10 mM Mg^{2+}. [From (32).]

intracellular ionic conditions and pH, one directly calculates a value of ϕ, which is defined as the fraction of total ATP not complexed to Mg^{2+}, according to the following equation:

$$\phi = \frac{[\mathrm{ATP}]_f}{[\mathrm{ATP}]_T} = \frac{\delta_{\alpha\beta}^{\mathrm{cell}} - \delta_{\alpha\beta}^{\mathrm{MgATP}}}{\delta_{\alpha\beta}^{\mathrm{ATP}} - \delta_{\alpha\beta}^{\mathrm{MgATP}}}. \qquad 3.$$

As long as only a single set of resonances is observed, consistent with fast exchange averaging of the resonances of ATP and MgATP, the frequencies of phosphorus resonances reflect the state of Mg^{2+} complexation of ATP. [It should be noted that at higher ^{31}P NMR frequencies (≥ 145 MHz) and lower temperatures ($\leq 10°C$), the assumption of rapid exchange of ATP and MgATP may not be entirely valid (38)]. Equation 4 is then used to derive a value for free Mg^{2+} in the intact cell directly from the spectral data:

$$[\mathrm{Mg}]_f = K_D^{\mathrm{MgATP}}\left\{\frac{1}{\phi} - 1\right\}. \qquad 4.$$

One assumption implicit in the applications of this technique to intact cellular systems is that the chemical shifts of the ^{31}P resonances of intracellular ATP respond to the presence of Mg^{2+} in a manner similar to that observed for isolated ATP. Since, in the cell systems examined so far, much of the total cellular ATP is complexed to Mg^{2+}, a critical part of this assumption is that, upon saturating cells with Mg^{2+}, the separation $\delta_{\alpha\beta}^{\mathrm{cell}}$ will decrease to that observed in isolated MgATP. This assumption was tested (20) by saturating oxygenated erythrocytes with a high level of internal Mg^{2+} (~ 10 μmol/ml cells). The observed separation between the αP and βP resonances changed from its value of (353 ± 1) Hz at 40.5 MHz and 37°C in unperturbed oxygenated erythrocytes to a value of (338 ± 1) Hz, indistinguishable from that observed in extracellular MgATP control (337 ± 1) Hz under simulated intracellular ionic conditions (20). The observation of similar shifts for the intracellular MgATP in human erythrocytes and the extracellular MgATP control indicates that the shifts of the ^{31}P resonances of the red cell ATP are determined predominantly by the normal interactions of Mg^{2+} and ATP. The validity of the ^{31}P NMR method for determining free Mg^{2+} in the red cells suggests that it is likely to be valid in other cell types as well.

Dissociation Constant of MgATP

Accurate knowledge of the affinity of ATP for Mg^{2+} is crucial for calculation of free Mg^{2+} from NMR spectral data, and the value we used (23, 30, 32) has been criticized. We have therefore reinvestigated the issue of the exact value of MgATP dissociation constant under physiologic ionic conditions and over the cellular range of ATP concentrations by a newly developed combination approach that appears uniquely suited for such determinations (18, 26a, 26b). The advantage of this method is that it can be used to measure the apparent dissociation constant of MgATP at physiological levels of the nucleotide. This may be important, since it is possible that additional complexes form at higher concentrations of ATP. The new combination method (18, 26a, 26b) utilizes ^{31}P NMR chemical shifts to determine the degree of Mg^{2+}-chelation of ATP (ϕ) in a solution containing free ATP and MgATP, and it uses a properly calibrated indicator dye antipyrylazo III (56) for optical measurement of free Mg^{2+} in the same solution. NMR-measured ϕ values of 0.07, 0.10, 0.14, 0.17, 0.19, 0.27, and 0.49 in various solutions of ATP and Mg^{2+} at 25°C and pH 7.2 corresponded to optically determined free Mg^{2+} values of 512, 371, 327, 288, 216, 147, and 51 μM, respectively. The data yielded an average value of (50 ± 10) μM (18, 26a, 26b) for the apparent dissociation constant of MgATP, a value similar to those obtained previously by other magnetic resonance methods (21, 22) but quite different from the value determined by Wu et al (140 μM) using the hydroxyquinoline-5-sulfonate dye (63).

It should be reiterated that, in principle, the NMR technique does not require a knowledge of the stability constant for the Mg^{2+}-ATP interaction, since an empirical correlation of the cellular spectra with extracellular controls containing independently known levels of free Mg^{2+} can directly yield the intracellular free Mg^{2+} concentration. However, it is convenient to establish an apparent dissociation constant for this complex under conditions as close as possible to the intracellular conditions, a value that can then be used to calculate the free Mg^{2+} from Equations 3 and 4. An advantage of using an apparent dissociation constant, instead of "intrinsic" constants measured under different conditions but corrected considering all relevant equilibria, is that no assumptions need to be made about the nature of the equilibria involved as long as intracellular ionic conditions are well simulated.

Adolfsen & Moudrianakis (3) have previously measured the MgATP dissociation constants using a divalent cation electrode. They compared these results with those obtained from an indirect spectrophotometric method using hydroxyquinoline as an indicator of free Mg^{2+} concentration. While the electrode measurements conformed to theoretical expectations, the MgATP dissociation constants obtained from the spectrophotometric method were markedly different. The latter continuously increased as the pH was raised, with >30-fold variation over the pH range 7.5 to 9.0. This study by Adolfsen & Moudrianakis (3) clearly indicated that the dye method is afflicted with serious deficiencies. Possible sources of trouble may be the formation of a $Mg(dye)_2$ aggregate, which precipitates, as well as the formation of a Mg-dye-ATP ternary complex. A precipitate was indeed visually detected in some of the solutions in the study by Adolfsen & Moudrianakis (3). Even small amounts of precipitate, which may not be detectable visually, may introduce substantial errors into the data. It should also be mentioned that Na^+ and K^+ ions bind to the metal-ATP complex almost as well as to the metal-free ATP (R. K. Gupta, P. Gupta, unpublished results). Thus the assumption of purely competitive binding of Mg^{2+} and K^+ or Na^+ ions generally made in the literature is not correct.

Because of these problems with hydroxyquinoline dyes, we believe that the values obtained by magnetic resonance methods (21, 22, 26a, 26b) are likely to be more reliable. The combination of NMR with the spectrophotometric titration (26a) is so far the only method capable of yielding an accurate dissociation constant at the relatively high concentrations of ATP found in cells. Measurement of K_D by this comparative method should also minimize any systematic errors and thus permit a more reliable estimation of the free Mg^{2+} level in the intact cell by NMR.

Applications of the ^{31}P NMR Method for Free Mg^{2+} Determination

Intracellular free Mg^{2+} has been measured by the NMR procedure in human normal and sickle red blood cells (23), frog skeletal muscle (30), Ehrlich ascites tumor cells (32), murine lymphoma cells (12a), human lymphocytes (51), dog erythrocytes (64), perfused and ischemic heart muscle (26a, 26b, 63), rabbit urinary bladder and uterus smooth muscles (12), and in amphibian oocytes (42). Our studies consistently indicated that the intracellular concentration of free Mg^{2+} is in the range 0.5 ± 0.3 mM and that a very small fraction of the total cell Mg^{2+} is uncomplexed in each tissue. We concluded that low availability of Mg^{2+} would limit the rates of cellular reactions in which the substrate is the Mg^{2+} complex of ADP or another compound that binds Mg^{2+} weakly. Low cellular free Mg^{2+} levels support an important role for Mg^{2+} in the regulation of metabolic processes in agreement with the Mg^{2+}-coordinated control of metabolism and growth proposed by Rubin and co-workers (54, 55, 62).

In contrast to our estimate of free Mg^{2+} in perfused guinea pig heart muscle of 0.8 mM obtained using the NMR-derived value for the dissociation constant of MgATP (26a, 26b), Wu et al (63) estimated a 4-fold higher value using an inappropriate dissociation constant. A high free Mg^{2+} in heart muscle, such as that suggested by Wu et al, is also inconsistent with the computer modeling of metabolic processes and fluxes (1, 2).

Wu et al (63) also expressed concern about our inability to account for the complexation of the total Mg^{2+} in each tissue by considering its binding only to known soluble phosphorylated metabolites. The values of free Mg^{2+} in frog muscle obtained using the NMR method are significantly lower than those obtained from theoretical calculations based on the simplified assumption that intracellular Mg^{2+} interacts with only ATP, P-creatine, and myosin (44). Such a theoretical approach neglects the binding of Mg^{2+} to phospholipids, phosphoproteins, muscle actin, other soluble proteins, and various cell structures such as surface membranes, sarcoplasmic reticulum, mitochondria, nucleus, and ribosomes. In support of the low NMR value for free Mg^{2+} in skeletal muscle, recent measurements of total diffusable magnesium concentration in muscle cytoplasm using skinned muscle fibers indicate that, of the total Mg^{2+} of 6.2 mmol per Kg present in whole muscle, only 60% is in the diffusable form (37). Thus the remaining 40% of the Mg^{2+} in the living muscle must be bound to various subcellular structures. In the relaxed muscle, the primary moieties of diffusable Mg^{2+}, in addition to free Mg^{2+}, would be MgATP, other Mg-nucleotides, and Mg-creatine phosphate. These results clearly demonstrate that attempting to account for the total Mg^{2+} in terms of its interactions with only a few known cellular components could be very misleading. Maughan (37) concluded from his data that free Mg^{2+} concentration in frog muscle is less than or equal to 0.4 mM, which agrees fairly well with the NMR value of 0.6 ± 0.2 mM in this tissue and argues strongly against the higher value of ~ 3 mM suggested by Wu et al (63).

Studies of Mg^{2+} equilibrium in frog muscle and egg by Ling et al (36a) also indicated that at physiological levels of extracellular Mg^{2+}, only a small fraction (~ 0.4 mM) is in the exchangeable or free form, the rest being tightly bound to intracellular components. Using an ionophore A23187, Flatman & Lew (13b) found free Mg^{2+} in erythrocytes to be low (~ 0.4 mM), in agreement with NMR results (23). Erdos & Maguire (12a) reported that free Mg^{2+} in murine lymphoma cells determined by ^{31}P NMR was ~ 0.2 mM, accounting for only 1–1.5% of the total Mg^{2+}. Cohen & Burt (11a), however, reported a value of ~ 3 mM for free Mg^{2+} in intact frog muscle; they used an NMR method in which the transverse relaxation time (T_2) of the ^{31}P resonance of intracellular P-creatine was used to obtain the free Mg^{2+} value. It has been pointed out (30) that, among other factors, sizable effects of steady state phosphoryl transfer via creatine kinase on the T_2 of P-creatine were neglected in this study. The value of 3 mM obtained by Cohen & Burt must therefore be considered only as an upper limit on possible values of free Mg^{2+}. Detection of sizable saturation transfer effects between P-creatine and γP (ATP) ^{31}P resonances (14) would indicate the existence of sizable contributions of intracellular dynamics on the T_2 of P-creatine. Hess & Weingart (33a), using a Mg^{2+} selective electrode, also produced an estimate of free Mg^{2+} in the millimolar range (3.3 mM). Baylor et al (4a) recently estimated free Mg^{2+} using three metallochromic indicator dyes, but they found that widely different estimates ranging from 0.2 to 6 mM were produced by various dyes used. Baylor et al proposed that the most likely explanation for the variability of the dye methods is that the indicator dyes behave differently inside muscle fibers than in calibrating solutions. The same may well be true for the invasive microelectrodes. Our ability to verify the applicability of MgATP ^{31}P chemical shifts to the intracellular environment in the red cell (20), however, indicates that the noninvasive NMR method may indeed be more reliable than the various invasive procedures in the study of living cells.

Independent evidence to support the conclusion that living cells have a large capacity for complexing divalent cations was obtained by incorporating paramagnetic Mn^{2+} in human erythrocytes and then observing free Mn^{2+} directly by EPR spectroscopy (20). Mn^{2+} replaces Mg^{2+} with varying affinity in many known enzymatic reactions, and it was expected that most of the Mn^{2+} would also be complexed to various cellular constituents. Of the total Mn^{2+} incorporated in ATP-rich oxygenated erythrocytes as determined by the atomic absorption technique, only 3% was free judging from EPR. This was true even though the cells contained 10 times more Mg^{2+} than Mn^{2+}. This Mg^{2+} would weaken the apparent interactions of Mn^{2+} with cellular constituents. Unfortunately, Mn^{2+} can interact not only with Mg^{2+} sites in the cells but also at Ca^{2+} sites, and therefore its interactions with cellular constituents may be somewhat different from those of Mg^{2+}. Nevertheless, these experiments did provide evidence for the large divalent cation complexing ability of the cellular system (20).

It therefore follows that incorrect conclusions would result from attempting to account for all of the Mg^{2+} in a cell without a knowledge of its total Mg^{2+}-complexing capacity. The low intracellular concentrations of free Mg^{2+} indicated by NMR studies support a regulatory role for free Mg^{2+} in some of the cellular reactions that require Mg^{2+} and are consistent with the activation and regulation of the hormone receptor-adenylate cyclase complex by intracellular free Mg^{2+} (9a). Sizable changes in the level of free Mg^{2+} occur during deoxygenation of erythrocytes (23) and upon adenosine-incubation of Ehrlich cells (32), without any measurable change in total Mg^{2+} content, and may also occur in vivo in different physiological states.

CONCLUSIONS

The noninvasive NMR technique can be exploited to measure free Mg^{2+}, Na^{+}, K^{+}, and free Ca^{2+} ions in living cells and tissues. Such studies are leading to a better understanding of the intracellular ions and their environment within intact tissues. ^{23}Na NMR using the hyperfine shift reagent $Dy(PPP_i)_2^{7-}$ provides a rapid method for the measurement of intracellular Na^+ without contamination from extracellular ions. The NMR-visible Na^+ appears to reflect the predominantly free pool of this ion, although bound ions capable of rapid exchange with the free pool will also contribute to the observed resonance intensity. Immobilized and non-exchanging Na^+ ions may become NMR-invisible because of broadening from nuclear quadrupolar interactions. The use of ^{19}F NMR for measuring intracellular free Ca^{2+} appears very promising, although further improvements in ^{19}F-labelled reagents are necessary to enable their use at intracellular concentrations that would be nonperturbing. The ^{31}P NMR technique for measuring free Mg^{2+} has a major advantage of being noninvasive. It, however, appears limited at present by the complexities of the cellular system, such as possible compartmentation of free Mg^{2+} in the cells. Only one set of ^{31}P ATP signals is, however, seen, which must represent the average of nuclear and cytosolic pools. It is likely that ATP in these areas and the mitochondria exchange rapidly. Therefore, the measurements would appear to provide a good approximation to the true level of free Mg^{2+} in intact cells. The NMR observable levels of free ions can vary with changes in physiological state, sometimes without a change in the total ion content. NMR can therefore be a unique technique for studying the role of metal ions in cellular functions.

ACKNOWLEDGMENTS

The preparation of this review and much of the research described herein were supported in part by NIH Grant AM32030 and by NCI Core Grant CA13330.

Literature Cited

1. Achs, M. J., Garfinkel, D. 1979. *Am. J. Physiol.* 236:R21–30
2. Achs, M. J., Garfinkel, D. 1979. *Am. J. Physiol.* 236:R318–26
3. Adolfsen, R., Moudrianakis, E. N. 1978. *J. Biol. Chem.* 253:4378–79
4. Balschi, J. A., Cirillo, V. P., Springer, C. S. Jr. 1982. *Biophys. J.* 38:323–26

4a. Baylor, S. M., Chandler, W. K., Marshall, M. W. 1982. *J. Physiol. London* 331:105–37

5. Berendsen, H. J. C., Edzes, H. T. 1973. *Ann. NY Acad. Sci.* 204:459–85
6. Blaustein, M. P. 1977. *Am. J. Physiol.* 232(3):C165–75
7. Boynton, A. L., McKeehan, W. L., Whitfield, J. F., eds. 1982. *Ions, Cell Proliferation and Cancer.* New York: Academic. 551 pp.
8. Brown, T. R., Ugurbil, K., Shulman, R. G. 1977. *Proc. Natl. Acad. Sci. USA* 74:5551–53
9. Cameron, I. L., Smith, N. K. R., Pool, T. B., Sparks, R. L. 1980. *Cancer Res.* 40:1493–1503

9a. Cech, S. Y., Broadus, W. C., Maguire, M. E. 1980. *Mol. Cell. Biochem.* 33:67–92

10. Civan, M. M. 1978. *Am. J. Physiol.* 234:F261–69

10a. Civan, M. M., Degani, H., Margalit, Y., Shporer, M. 1983. *Am. J. Physiol.* 245:C213–19
11. Civan, M. M., Shporer, M. 1978. *Biol. Magn. Reson.* 1:1–32
11a. Cohen, S. M., Burt, C. T. 1977. *Proc. Natl. Acad. Sci. USA* 74:4271–75
12. Dillon, P. F., Meyer, R. A., Kushmerick, M. J. 1983. *Biophys. J.* 41:252a (Abstr.)
12a. Erdos, J. J., Maguire, M. E. 1983. *J. Physiol. London* 337:351–71
13. Ernst, R. R., Anderson, W. A. 1966. *Rev. Sci. Instrum.* 37:93–102
13a. Fabry, M. E., San Geroge, R. C. 1983. *Biochemistry* 22:4119–25
13b. Flatman, P., Lew, V. L. 1977. *Nature* 267:360–62
14. Gadian, D. G. 1983. *Ann. Rev. Biophys. Bioeng.* 12:69—89
15. Delayre, J. L., Ingwall, J. S., Malloy, C., Fossel, E. T. 1981. *Science* 212:935–36
16. Gordon, R. E., Hanley, P. E., Shaw, D., Gadian, D. G., Radda, G. K., et al. 1980. *Nature* 287:736–38
17. Grove, T. H., Ackerman, J. J. H., Radda, G. K., Bore, P. J. 1980. *Proc. Natl. Acad. Sci. USA* 77:299–302
18. Gupta, P., Gupta, R. K., Yushok, W. D., Rose, Z. B. 1983. *Fed. Proc.* 42:2215 (Abstr.)
19. Gupta, R. K. 1979. *Biochim. Biophys. Acta* 586:189–95
20. Gupta, R. K. 1980. *Int. J. Quant. Chem. Quant. Biol. Symp.* 7:67–73
21. Gupta, R. K., Benovic, J. L. 1978. *Biochem. Biophys. Res. Commun.* 84:130–37
22. Gupta, R. K., Benovic, J. L., Rose, Z. B. 1978. *J. Biol. Chem.* 253:6165–71
23. Gupta, R. K., Benovic, J. L., Rose, Z. B. 1978. *J. Biol. Chem.* 253:6172–76
24. Gupta, R. K., Gupta, P. 1982. *Biophys. J.* 37:76a (Abstr.)
25. Gupta, R. K., Gupta, P. 1982. *J. Magn. Reson.* 47:344–50
26. Gupta, R. K., Gupta, P., Negendank, W. 1982. See Ref. 7, pp. 1–12
26a. Gupta, R. K., Gupta, P., Yushok, W. D., Rose, Z. B. 1983. *Biochem. Biophys. Res. Commun.* 117:210–16
26b. Gupta, R. K., Gupta, P., Yushok, W. D., Rose, Z. B. 1983. *Physiol. Chem. Phys.* In press
27. Gupta, R. K., Kostellow, A. B., Morrill, G. A. 1983. In *Water and Ions in Biological Systems*, ed. V. Vasilescu. London: Plenum. In press
28. Gupta, R. K., Kostellow, A. B., Morrill, G. A. 1983. *Biophys. J.* 41:128a (Abstr.)
29. Gupta, R. K., Mildvan, A. S. 1978. *Methods Enzymol.* 54:151–92
30. Gupta, R. K., Moore, R. D. 1980. *J. Biol. Chem.* 255:3987–93
31. Gupta, R. K., Wittenberg, B. A. 1983. *Fed. Proc.* 42:2065 (Abstr.)
32. Gupta, R. K., Yushok, W. D. 1980. *Proc. Natl. Acad. Sci. USA* 77:2487–91
33. Hamlyn, J. M., Ringel, R., Schaeffer, J., Levinson, P. D., Hamilton, B. P., Kowarski, A. A., Blaustein, M. P. 1982. *Nature* 300:650–53
33a. Hess, P., Weingart, R. 1981. *J. Physiol. London* 318:14–15
34. Ingwall, J. S. 1982. *Am. J. Physiol.* 242:H729–44
35. Jardetzky, O., Wertz, J. E. 1956. *Am. J. Physiol.* 187:608
36. Kaplan, J. G. 1978. *Ann. Rev. Physiol.* 40:19–41
36a. Ling, G. N., Walton, C., Ling, M. R. 1979. *J. Cell. Physiol.* 101:261–78
37. Maughan, D. 1983. *Biophys. J.* 43:75–80
38. Misawa, K., Lee, T. M., Ogawa, S. 1982. *Biochim. Biophys. Acta* 718:227–35
39. Moon, R. N., Richards, J. H. 1973. *J. Biol. Chem.* 248:7276–78
39a. Moore, R. D. 1983. *Biochim. Biophys. Acta* 737:1–30
40. Moore, R. D., Gupta, R. K. 1980. *Int. J. Quant. Chem. Quant. Biol. Symp.* 7:83–92
41. Moore, R. D., Munford, J. W., Pillsworth, T. J. 1983. *J. Physiol. London* 338:277–94
42. Morrill, G. A., Kostellow, A. B., Weinstein, S. P., Gupta, R. K. 1983. *Physiol. Chem. Phys.* In press
43. Morrill, G. A., Kostellow, A. B., Weinstein, S. P., Gupta, R. K. 1983. *Fed. Proc.* 42:1791 (Abstr.)
44. Nanninga, L. B. 1961. *Biochim. Biophys. Acta* 54:330–38
45. Nuccitelli, R., Deamer, D. W., eds. 1982. *Intracellular pH. Its Measurement, Regulation and Utilization in Cellular Functions.* New York: Liss. 594 pp.
46. Nunnally, R. L., Hollis, D. P. 1979. *Biochemistry* 18:3642–46
47. Odeblad, E., Bahr, B. N., Linstrom, G. 1956. *Arch. Biochem. Biophys.* 63:221–25
48. Ogino, T., den Hollander, J. A., Shulman, R. G. 1983. *Proc. Natl. Acad. Sci. USA* 80:5185–89
49. Redfield, A. G., Gupta, R. K. 1971. *Adv. Magn. Reson.* 5:81–115
50. Redfield, A. G., Gupta, R. K. 1971. *J. Chem. Phys.* 54:1418–19
51. Rink, T. J., Tsien, R. Y., Pozzan, T. 1982. *J. Cell Biol.* 95:189–96
52. Roberts, J. K. M., Jardetzky, O. 1981. *Biochim. Biophys. Acta* 639:53–76
53. Ross, B. D., Radda, G. K., Gadian, D. G.,

Rocker, G., Esiri, M., et al. 1981. *N. Engl. J. Med.* 304:1338–42
54. Rubin, H. 1975. *Proc. Natl. Acad. Sci. USA* 72:3551–55
55. Rubin, A. H., Terasaki, M., Sanui, H. 1979. *Proc. Natl. Acad. Sci. USA* 76:3917–21
56. Scarpa, A., Brinley, F. J., Dubyak, G. 1978. *Biochemistry* 17:1378–86
57. Shulman, R. G., Brown, T. R., Ugurbil, K., Ogawa, S., Cohen, S. M., den Hollander, J. A. 1979. *Science* 205:160–66
58. Smith, G. A., Hesketh, T. R., Metcalfe, J. C. 1982. See Ref. 7, pp. 65–75
59. Springer, C. S., Jr., Pike, M. M., Balschi, J. A. 1982. *Biophys. J.* 37:337a (Abstr.)
60. Tosteson, D. C. 1955. *J. Gen. Physiol.* 39:55–65
61. Tsien, R. Y. 1983. *Ann. Rev. Biophys. Bioeng.* 12:91–116
62. Vidair, C., Rubin, H. 1981. *J. Cell. Physiol.* 108:317–25
63. Wu, S. T., Pieper, G. M., Salhany, J. M., Eliot, R. S. 1981. *Biochemistry* 20:7399–7403
64. Wyrwicz, A. M., Schofield, J. C., Burt, C. T. 1982. In *Noninvasive Probes of Tissue Metabolism*, ed. J. S. Cohen, pp. 149–71. New York: Wiley Intersci.
65. Yeh, H. J. C., Brinley, F. J. Jr., Becker, E. D. 1973. *Biophys. J.* 13:56–71
66. Yushok, W. D., Gupta, R. K. 1980. *Biochem. Biophys. Res. Commun.* 95:73–81

Ann. Rev. Biophys. Bioeng. 1984. 13: 247–68

TOTAL INTERNAL REFLECTION FLUORESCENCE

Daniel Axelrod

Biophysics Research Division and Department of Physics, University of Michigan, Ann Arbor, Michigan 48109

Thomas P. Burghardt

Cardiovascular Research Institute, University of California, San Francisco, California 94143

Nancy L. Thompson

Department of Chemistry, Stanford University, Stanford, California 94305

INTRODUCTION

Total internal reflection fluorescence (TIRF) is an optical effect particularly well-suited to the study of molecular and cellular phenomena at liquid/solid interfaces. Such interfaces are central to a wide range of biochemical and biophysical processes: binding to and triggering of cells by hormones, neurotransmitters, and antigens; blood coagulation at foreign surfaces; electron transport in the mitochondrial membrane; adherence and mobility of bacteria, algae, and cultured animal cells to surfaces; and possible enhancement of reaction rates with cell surface receptors upon nonspecific adsorption and surface diffusion of an agonist. Liquid/solid interfaces also have important medical and industrial applications: e.g. detection of serum antibodies by surface immobilized antigens; and the manufacture of biochemical products by surface-immobilized enzymes.

Total internal reflection spectroscopy for optical absorption studies (called attenuated total reflection or ATR) was developed somewhat earlier than the fluorescence applications and has been widely used in surface chemistry studies (17). The basic book in this area is *Internal Reflection*

0084–6589/84/0615–0247$02.00

Spectroscopy by Harrick (18). Total internal reflection optics have also been combined with Raman spectroscopy (24, 32), X-ray fluorescence (2), infrared absorption spectroscopy (16), and light scattering (3, 33).

As a technique for selective surface illumination, total internal reflection fluorescence was first introduced by Hirschfeld (21) for solid/liquid interfaces, Tweet et al (39) for liquid/air interfaces, and Carniglia & Mandel (13) for high refractive index liquid/solid interfaces. TIRF has been combined with a variety of other conventional fluorescence techniques (polarization, Förster energy transfer, microscopy, spectral analysis, photobleaching recovery, and correlation spectroscopy) for a variety of purposes (detection of molecular adsorption; measurement of adsorption equilibrium constants, kinetic rates, surface diffusion, adsorbate conformation; and observation of cell/substrate contact regions). This review is organized according to those purposes. We begin with a summary of the relevant optical theory.

THEORY

Evanescent Intensity

When a light beam propagating through a transparent medium of high index of refraction (e.g. a solid glass prism) encounters an interface with a medium of a lower index of refraction (e.g. an aqueous solution), it undergoes total internal reflection for incidence angles (measured from the normal to the interface) greater than the "critical angle." The critical angle θ_c, is given by

$$\theta_c = \sin^{-1}(n_2/n_1), \qquad 1.$$

where n_2 and n_1 are the refractive indices of the liquid and the solid, respectively. Although the incident light beam totally internally reflects at the interface, an electromagnetic field called the "evanescent wave" penetrates a small distance into the liquid medium and propagates parallel to the surface in the plane of incidence. The evanescent wave is capable of exciting fluorescent molecules that might be present near the interface. This effect has been regarded as experimental proof of the existence of the evanescent wave (44).

The evanescent electric field intensity $I(z)$ decays exponentially with perpendicular distance z from the interface:

$$I(z) = I_0 e^{-z/d},$$

where

$$d = \frac{\lambda_0}{4\pi}[n_1^2 \sin^2\theta - n_2^2]^{-1/2} \qquad 2.$$

for angles of incidence $\theta > \theta_c$ and light wavelength in vacuum λ_0. Depth d is independent of the polarization of the incident light and decreases with increasing θ. Except for $\theta \simeq \theta_c$ (where $d \to \infty$), d is on the order of λ_0 or smaller.

The intensity at $z = 0$, I_0, depends on both the incidence angle θ and the incident beam polarization. I_0 is proportional to the square of the amplitude of the evanescent electric field $\mathbf{E}$ at $z = 0$.[1] (These expressions are given in the next subsection.) For incident electric field intensities $\mathscr{I}^{\parallel,\perp}$ with polarizations parallel and perpendicular, respectively, to the plane of incidence, the evanescent intensities $I_0^{\parallel,\perp}$ are

$$I_0^{\parallel} = \mathscr{I}^{\parallel} \cdot \frac{4\cos^2\theta(2\sin^2\theta - n^2)}{n^4\cos^2\theta + \sin^2\theta - n^2} \tag{3.}$$

and

$$I_0^{\perp} = \mathscr{I}^{\perp} \cdot \frac{4\cos^2\theta}{1 - n^2}, \tag{4.}$$

where

$$n = n_2/n_1 < 1. \tag{5.}$$

Figure 1 illustrates $I_0^{\parallel,\perp}$ as functions of θ. Note several features: (*a*) the evanescent intensity $I_0^{\parallel,\perp}$ is not weak and can be several times stronger than the incident intensity for angles of incidence within a few degrees of the critical angle; (*b*) the intensities of the two polarizations are different, with $I_0^{\parallel}$ somewhat more intense for all θ; (*c*) $I_0^{\parallel,\perp}$ both approach zero as $\theta \to 90°$.

Evanescent Fields

The phase behavior of the $\mathbf{E}$ field is quite remarkable (7). The field components are listed below, with incident electric field amplitudes $A^{\parallel,\perp}$ and phase factors relative to those of the incident $\mathbf{E}$ and $\mathbf{H}$ fields' phase at $z = 0$. (The coordinate system is chosen such that the x–z plane is the plane of incidence.)

$$E_x = \left[\frac{2\cos\theta\,(\sin^2\theta - n^2)^{1/2}}{(n^4\cos^2\theta + \sin^2\theta - n^2)^{1/2}}\right] A_{\parallel} e^{-i(\delta_{\parallel} + \pi/2)}, \tag{6.}$$

$$E_y = \left[\frac{2\cos\theta}{(1 - n^2)^{1/2}}\right] A_{\perp} e^{-i\delta_{\perp}}, \tag{7.}$$

[1] The evanescent intensity that excites fluorescence is given by $|\mathbf{E}|^2$ (12, 13). Generally, the energy flux of an electromagnetic field is given by the real part of the Poynting vector, $\mathbf{S} = (c/4\pi)\mathbf{E} \times \mathbf{H}$, where $\mathbf{H}$ is the magnetic field. For a transverse field, $|\mathbf{S}|$ is proportional to $|\mathbf{E}|^2$. Note that for an evanescent wave, $|\mathbf{S}|$ is not proportional to $|\mathbf{E}|^2$.

$$E_z = \left[\frac{2\cos\theta\sin\theta}{(n^4\cos^2\theta + \sin^2\theta - n^2)^{1/2}}\right] A_{\parallel} e^{-i\delta_{\parallel}}, \tag{8.}$$

where

$$\delta_{\parallel} \equiv \tan^{-1}\left[\frac{(\sin^2\theta - n^2)^{1/2}}{n^2\cos\theta}\right] \tag{9.}$$

and

$$\delta_{\perp} \equiv \tan^{-1}\left[\frac{(\sin^2\theta - n^2)^{1/2}}{\cos\theta}\right]. \tag{10.}$$

Note that the evanescent electric field is transverse to the propagation direction ($+x$) only for the $\perp$ polarization. The $\parallel$ polarization **E** field

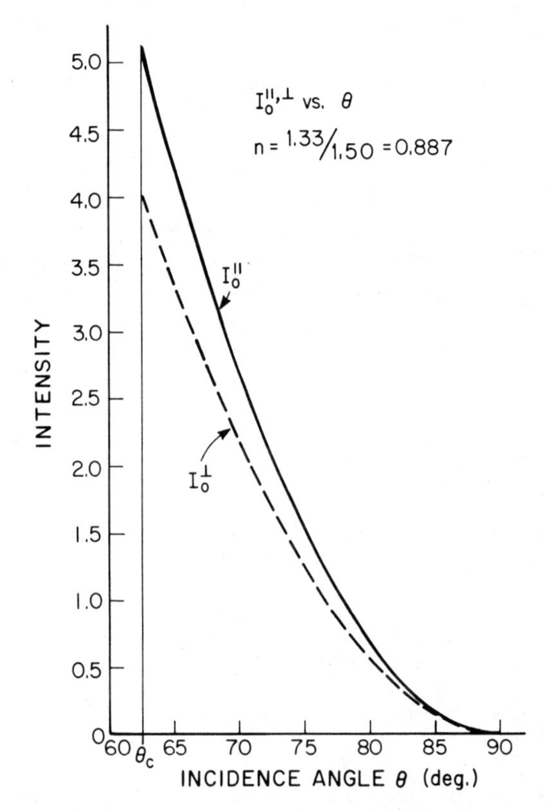

Figure 1 Intensities $I_0^{\parallel,\perp}$ vs incidence angle θ, for $n = 0.887$, corresponding to a critical angle of $\theta_c = 62.46°$. Intensity is expressed as the ratio of evanescent intensity at $z = 0$ to the incident intensity for each polarization.

"cartwheels" along the surface with a spatial period of $\lambda_0/(n_1 \sin \theta)$ as shown schematically in Figure 2.

For absorbers with magnetic dipole transitions, the evanescent magnetic field **H** is relevant. Assuming equal magnetic permeabilities at both sides of the interface, the components of the evanescent field **H** at $z = 0$ are

$$H_x = \left[\frac{2 \cos \theta \, (\sin^2 \theta - n^2)^{1/2}}{(1-n^2)^{1/2}}\right] A_\perp e^{-i(\delta_\perp - \pi)}, \tag{11}$$

$$H_y = \left[\frac{2n^2 \cos \theta}{(n^4 \cos^2 \theta + \sin^2 \theta - n^2)^{1/2}}\right] A_\parallel e^{-i(\delta_\parallel - \pi/2)}, \tag{12}$$

$$H_z = \left[\frac{2 \cos \theta \sin \theta}{(1-n^2)^{1/2}}\right] A_\perp e^{-i\delta_\perp}. \tag{13}$$

The angular dependence of the phase factors $\delta_\perp$ and $\delta_\parallel$ gives rise to a measurable longitudinal shift of a finite-sized incident beam, known as the Goos-Hanchen shift (29). Viewed physically, some of the energy of a finite-width beam crosses the interface into the lower refractive index material, skims along the surface for a Goos-Hanchen shift distance ranging from a fraction of a wavelength (at $\theta = 90°$) to infinite (at $\theta = \theta_c$), and then reenters the higher refractive index material.

In many TIRF experiments, an incident laser beam of Gaussian intensity profile passes through a converging lens, enters the side of a prism, and then focuses at a totally reflecting surface. The nature of the evanescent

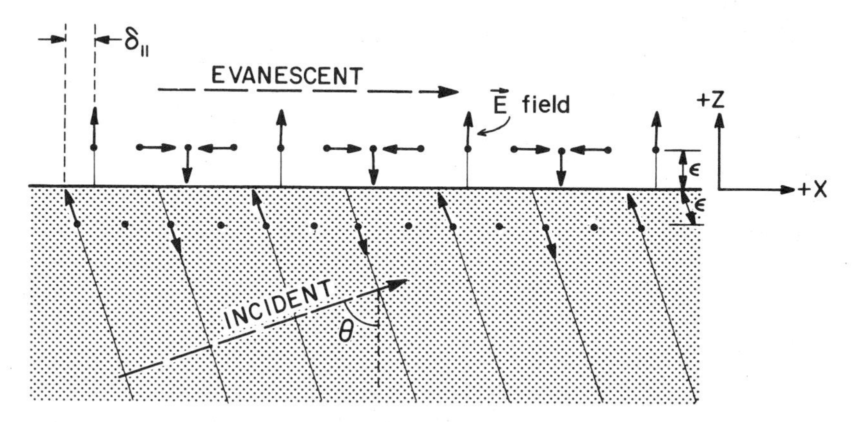

Figure 2 Electric field vectors of incident and evanescent light for the ∥ incident polarization, showing the phase lag $\delta_\parallel$ and the "cartwheel" or elliptical polarization of the evanescent field in the plane of propagation. Both the incident and evanescent vectors refer to the $z = 0$ position; they are schematically displaced a distance ε ($\rightarrow 0$) below and above the interface along their constant phase lines for pictorial clarity only.

illumination produced by such a focused finite beam geometry has been investigated in general (12). For typical experimental conditions, the evanescent illumination is of an elliptical Gaussian profile and the polarization and penetration depth are approximately equal to those of a plane wave.

The angular distribution of evanescent wave-excited fluorescence when viewed through the prism is anisotropic. A complete discussion appears in Lee et al (26).

Intermediate Layers

Thus far, the theoretical discussion has assumed that the medium in which the evanescent wave propagates contains no light absorbers, light scatterers, layers of unmatched refractive index, or electrical conductors. The effect of such intermediate layers is clearly relevant to TIRF.

The presence of absorbers perturbs the evenescent wave and thereby disturbs the linearity between actual surface concentration of absorber and observed fluorescence. (This effect is roughly analogous to the "inner filter" effect in conventional fluorimetry in which a high concentration of fluorophore attenuates the incident light.) Burghardt & Axelrod (10) have calculated the magnitude of this nonlinear effect as a function of surface concentration of absorbing material of arbitrary thickness. The result of this calculation shows that typical TIRF experiments are well within the linear range in which observed fluorescence and surface concentration are proportional. Harrick (18) displays results of an analogous calculation of the effect of absorbers on the intensity of the reflected laser beam.

Aside from the effect of absorbers, the presence of an intermediate layer of a homogeneous dielectric material of unmatched refractive index deposited on the totally reflecting surface can clearly affect the evanescent wave intensity and characteristic decay distance (31). Total internal reflection may even occur at the interface between the intermediate layer and the low refractive index medium. Regardless of their refractive indices, intermediate layers cannot thwart total internal reflection at some interface in the system if it would have occurred without the intermediate layers.

If a metallic conductor is deposited onto the interface, entirely new phenomena can be observed (15, 42). Rather than emitting fluorescence isotropically, an evanescent wave-excited dye molecule adsorbed to the intermediate metal layer will tend to transfer its energy to surface plasmons in the adjacent metal. If the metal film is sufficiently thin, the emission can then be observed as a hollow cone of light propagating back through the prism. If a semiconductor thin film is deposited on the interface, one might observe charge transfer between an excited dye molecule and the conduction band of the semiconductor (30).

The scattering of evanescent light by an intermediate layer has been treated theoretically (14). Evanescent light scattering is the basis of the dark field microscope introduced by Ambrose (3) for examining cultured biological cell-substrate contacts and also is the basis of a study of photoreceptor membrane attached to optical fibers (33). Whether scattering significantly increases the effective depth of light penetration into the low optical density medium has been tested in some of the experimental systems discussed in following sections.

In the next three sections, we review experimental results with TIRF optics.

ADSORPTION AT EQUILIBRIUM: DETECTION AND CALIBRATION

In an early published application of TIRF, Tweet et al (39) measured the emission spectrum and quenching behavior of chlorophyll *a* monolayers at an aqueous/air interface (see Figure 3*a*). TIRF provided a means of detecting the weak fluorescence from a very dilute monolayer with a high degree of exclusion of direct and scattered mercury arc excitation light.

A short note by Hirschfeld (21) introduced TIRF at a solid/liquid interface. A special cell (Figure 3*b*) designed to fit a commercial absorption spectrophotometer and providing for multiple total reflections of excitation light in a fused silica microscope slide was used to study the fluorescence of fluorescein in the 1–1000 ppm bulk concentration range. The goal here was to study bulk-dissolved dye in the vicinity of an interface rather than adsorbed dye. Hirschfeld lists several advantages of TIRF over conventional illumination for this purpose. (*a*) The small depth of penetration of the evanescent wave combined with observation of fluorescence emitted back into the prism could allow studies of turbid or highly adsorbing solutions. (*b*) The intensity/concentration relationship was linear at up to two orders of magnitude higher concentration for TIRF than for conventional illumination. (*c*) Effects of adsorption or the proximity of glass upon the fluorescence properties could be detected by changing the excitation incidence angle and thereby the penetration depth. (*d*) The possibility of multiple total reflections in the same prism allows the excitation to interact with the sample many times, thereby increasing sensitivity. (*e*) Many fluorescent materials (dyes, labeled proteins, etc) strongly adsorb onto solid surface from very dilute solutions, often unavoidably depleting the bulk concentration of the fluorophore and dramatically increasing the effective local concentration of the fluorophore in the proximity of the surface. Selective surface excitation by TIRF thereby leads to an effective enhancement of sensitivity to extremely low concentrations of fluorophore.

The sensitivity of TIRF to adsorbed films containing a fluorophore can be upgraded further, as described by Harrick & Loeb (19, 20). With a prism in which multiple total internal reflection of excitation light takes place and also in which much of the fluorescence emission is also trapped by total internal reflection, a large fluorophore-coated surface area can be observed with a large effective solid angle of light collection (Figure 3*c*). In one application of this sort of optics, a film of dansyl-labeled bovine serum albumin (BSA) was adsorbed onto a fused silica prism and dried in air. Note that the fluorescence here is not excited by the evanescent wave as

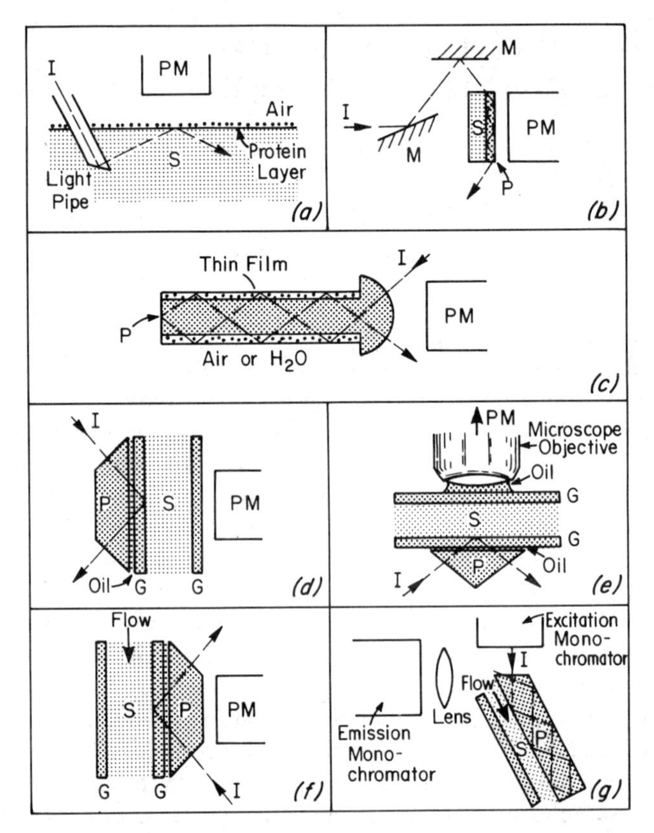

Figure 3 Optical designs for some TIRF experiments as described in the text. These are simplified drawings showing the general excitation geometries with most of the lenses in the excitation and emission pathways not shown. The following abbreviations are used: G, glass microscope slide; PM, photomultiplier (always assumed to be preceded by a colored optical filter to block scattered excitation light); P, prism (usually optical glass or fused silica); S, solution containing fluorophore (usually a fluorescence-labeled serum protein); M, mirror; I, incident light beam.

previously described, since total internal reflection occurs at the protein film/air interface instead of the silica prism/film interface. Rather, TIR serves to confine both the excitation and emission light on the film-coated prism, allowing entrance and exit of light only at one hemicylindrically shaped end.

Kronick & Little (25) have employed TIRF with an immunologically specific antigen-coated surface to assay for specific fluorescent-labeled antibodies in solution. The antigen complimentary to the antibody is conjugated to egg albumin and then is physically adsorbed onto a fused silica slide. The excitation light, from a helium cadmium laser, is introduced into the slide via an optical coupling with a trapezoidal shaped prism (Figure 3*d*). As fluorescent antibodies attach to the surface-immobilized antigens, the evanescent wave-excited fluorescence increases. The chief limit to the specific sensitivity of this assay is the nonspecific binding of antibodies to the egg albumin coated surface. This problem is less severe with strongly binding antigen-antibody combinations and specially treated surfaces that reduce nonspecific protein binding. With a silica slide coated by the antigen morphine, a concentration of 2×10^{-7} M of dissolved fluorescent labeled antimorphine could be detected specifically in their TIRF system. The system remains to be tested for success in assaying antibodies in blood serum.

A very elegant TIRF method and apparatus for assaying for viruses in human blood serum (called a "virometer") has been described by Hirschfeld and co-workers (22, 23) for use on the stage of an upright microscope (Figure 3*e*). A serum sample, mechanically filtered to eliminate particulates larger than virus particles, is treated with a nucleic-acid binding fluorophore and then observed by TIRF. As fluorescent-labeled constituents in the serum enter and leave the evanescent wave in the bulk, they cause visible fluorescence fluctuations. Because of their large size, the virions diffuse by Brownian motion more slowly than other fluorescent constituents and thereby cause slower fluorescence fluctuations. By autocorrelating the fluctuations, this slower component can be resolved. From the relative amplitude of this slow component, the absolute concentration of virions can be calculated. Mixtures of two fluorescent nucleic acid stains of widely different emission spectra could allow distinction between different types of virions. TIRF illumination, as opposed to conventional microscope epi-illumination, served mainly to (*a*) define a very small distance—the penetration depth d—through which virions can traverse in a short time, and (*b*) avoid background fluorescence from out-of-focus planes in the solution. The principles of the virometer are similar to some of those later applied in total internal reflection/fluorescence correlation spectroscopy (TIR/FCS) (38) for other purposes to be described later in the next section.

In TIRF experiments in which an adsorbed fluorophore is in reversible chemical equilibrium with bulk-dissolved fluorophore, it is desirable to determine what proportion of the observed fluorescence arises from actually adsorbed fluorophore versus bulk fluorophore merely close enough to the surface to be excited by the evanescent wave or scattered incident light. Given either (*a*) an independent calibration of adsorbed surface concentration (10), (*b*) an independent measurement of the total number of illuminated fluorescent molecules (36), or (*c*) an independent measurement of fluorescence from a nonadsorbant in a TIRF apparatus (27, 32a), the proportion of adsorbed vs bulk molecules can be calculated. Alternatively, the fraction of illuminated fluorophore that is adsorbed can be determined from the abscissa intercept of a plot of evanescent field depth (which can be varied with incidence angle; see Equation 2) vs measured fluorescence (corrected for the variation in evanescent intensity with incidence angle; see Figure 1). Another method of determining the adsorbed fraction of illuminated fluorophore is to interpolate the degree of polarization anisotropy of the sample's fluorescence (with adsorbed and bulk fluorophore in equilibrium) between that of purely adsorbed and purely solubilized fluorophore. (This method assumes slower rotational motion of adsorbed fluorophore compared to solubilized fluorophore.) Finally, the ratio r of fluorescent intensity to the intensity of the water solvent's Raman scattering peak can be compared between a surface under evanescent illumination and a bulk solution under conventional illumination. The relative values of ratio r are directly related to the fraction of illuminated fluorophore that is adsorbed. A large increase in ratio r (typically observed upon serum protein adsorption) demonstrates the dominance of surface adsorbed to bulk-dissolved fluorophore under TIRF illumination.

CHEMICAL KINETICS AND SURFACE DIFFUSION

It is often of interest to know how rapidly a solute adsorbs to a surface, how long it stays there before it desorbs, and whether it diffuses along the surface while adsorbed. Many of the studies in this area involve adsorption of blood serum proteins to nonbiological surfaces, partly because of the relevance to blood coagulation in medical prostheses.

The most direct approach to chemical kinetics in a TIRF system is a "concentration-jump": rapidly increasing bulk concentration from zero to a final value to measure adsorption rates, or the reverse to measure desorption rates. Employing this approach, Watkins & Robertson (41) measured the time course of uptake (as well as equilibrium adsorption isotherms) for fluorescein-labeled albumin, γ-globulin, and fibrinogen onto

Siliclad or silicone rubber-coated surfaces from static solution, and for γ-globulin onto silicone rubber from flowing solutions (Figure 3*f*). One qualitatively significant finding was a reversibly bound layer of γ-globulin constituting as much as 40% of the total surface bound protein.

Lok et al (27, 28) continued these type of studies with bovine albumin and fibrinogen adsorbing onto silicone rubber from flowing solutions. They find that the initial rate of adsorption for both proteins is diffusion limited (i.e. every solute molecule hitting an available site on surface sticks to it), and also that fibrinogen uptake does not plateau after several hours and may adsorb in multilayers. A particularly careful discussion of contributions from bulk-dissolved fluorophore and scattered incident light arising in artificial polymer coatings is presented in Lok et al (27).

The possibility exists that extrinsic fluorophores on serum proteins might affect the rates and amounts of physical adsorption. To avoid this possibility, Van Wagenen et al (40) studied the intrinsic tryptophan fluorescence of bovine serum albumin and γ-globulins adsorbed to fused silica following concentration jumps in a flushable cell. The general optical arrangement for TIRF was similar to that of Watkins & Robertson (41) (Figure 3*f*), but with the variations that the light source was an Hg-Xe arc instead of an argon laser and that both the excitation and emission wavelengths could be scanned by monochromators in order to obtain spectra. Both proteins studied exhibited a very rapid initial phase of adsorption from low bulk concentrations; γ-globulin also showed a prolonged slow uptake component continuing for at least 40 min after its introduction. Albumin adsorption continued to occur at concentrations exceeding those required for monolayer coverage, leading to the authors' inference of multilayers on the surface. No evidence is available as yet as to whether extrinsic probes affect the adsorption behavior seen with probe-free serum proteins.

Beissinger & Leonard (6) employed a multiple internal reflection prism adaptable to a commercial spectrofluorimeter to measure the concentration jump adsorption kinetics of fluorescein-labeled human γ-globulin to fused silica under flow conditions (Figure 3*g*). Again, both reversible and relatively irreversible binding was detected, with the reversible component increasing linearly with bulk concentration. Adsorption rates were found to be proportional to solution concentration.

Surface desorption rates and surface diffusion coefficients can be observed without perturbing the chemical equilibrium (as occurs in a concentration jump) by combining TIR with either fluorescence photobleaching recovery (TIR/FPR) or fluorescence correlation spectroscopy (TIR/FCS) (38). In TIR/FPR, adsorbed molecules are irreversibly photobleached by a flash of laser beam focused at a total internal reflection surface; subsequent fluorescence recovery vs time is monitored by an attentuated evanescent intensity as bleached molecules exchanged with unbleached ones from the solution or from surrounding nonilluminated regions of the surface (Figure 4). In TIR/FCS, the evanescent intensity is maintained at a constant and fairly dim level throughout, and spontaneous fluorescence fluctuations due to individual molecules entering and leaving a well-defined portion of the evanescent field are electronically autocorrelated. In general, the shape of the theoretical TIR/FPR and TIR/FCS curves depends in a complex manner upon the bulk and surface diffusion coefficients, the size of the illuminated or observed region, and the adsorption and desorption kinetic rates. However, under appropriate experimental conditions, the rate constants and surface diffusion coefficients can be readily obtained.

When the surface residency time of a reversibly bound adsorbant is much longer than the time necessary to enter or leave the vicinity of the observed surface area, the process is in the "reaction limit." Mathematically, this limit is described by

$$\frac{\beta}{k_2} \gg \frac{(\bar{C}/\bar{A})^2}{D_A} \qquad 14.$$

Surface residency time $\gg$ Bulk diffusion time

where

$$\beta = \begin{cases} 1 \text{ (for TIR/FPR), or} \\ \text{fraction of surface binding sites that are unoccupied} \\ \text{at equilibrium (for TIR/FCS),} \end{cases}$$

$\bar{A}$ = equilibrium concentration of bulk solute,

$\bar{C}$ = equilibrium concentration of surface adsorbed solute,

D_A = diffusion coefficient of solute in the bulk.

In the reaction limit for a large illuminated or observed area, the fluorescence recovery (for TIR/FPR) and the autocorrelation function (for TIR/FCS) depend only on k_2 and the shape is exponential. Thus, in the reaction limit the desorption rate k_2 can be obtained easily. Increasing the bulk concentration $\bar{A}$ or reducing the size of the observation area tends to drive the process toward the reaction limit.

If the opposite relationship to that of Equation 14 holds, for a large illuminated or observed area, the process is in the "bulk diffusion limit," in which TIR/FPR and TIR/FCS curves depend only on D_A and the equilibrium concentrations. Physically, this limit occurs when the surface behaves as a nearly perfect sink with respect to the adsorbing molecules.

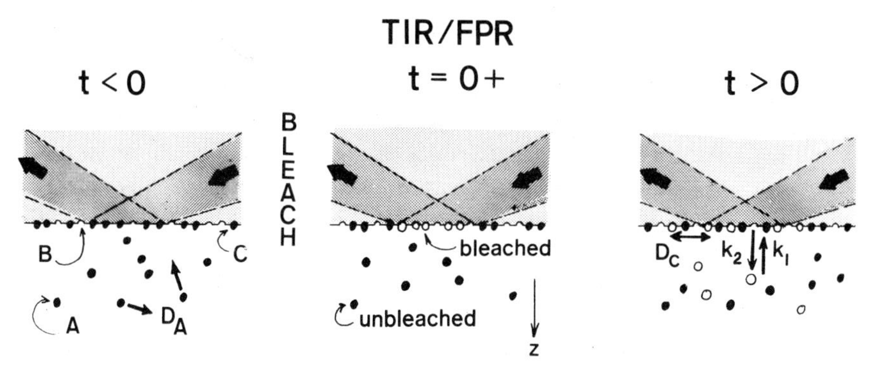

Figure 4 Exchange of surface-adsorbed bleached molecules with unbleached ones in TIR/FPR experiment. Symbols are as follows: A, unbleached solute molecules in the bulk solution, denoted by filled circles; B, unoccupied surface binding sites, denoted by hemicircular indentations in the surface; C, unbleached adsorbed molecules, which in the region of evanescent illumination, give rise to fluorescence; D_A and D_C, bulk and surface diffusion coefficients, respectively; k_1 and k_2, the adsorption and desorption kinetic rates, respectively. Bleached molecules, whether in the bulk solution or on the surface, are denoted by open circles.

Between the reaction and the bulk diffusion limits, the curves depend on both k_2 and D_A, $\bar{A}$, and $\bar{C}$. The shape of a TIR/FPR or TIR/FCS curve can thus distinguish between reaction limited, bulk diffusion limited, and intermediate processes. Thompson et al (38) provide a mathematical and physical description of these limits and intermediate cases; Thompson & Burghardt (37) further develop the meaning of these limits in terms of a microscopic model of reactions with target sites in one, two, or three dimensions.

If surface diffusion exists, its characteristic time clearly depends upon the size of the area across which an adsorbed molecule diffuses. Surface diffusion can thereby be distinguished from the adsorption/desorption process simply by changing the size of the illuminated region (for TIR/FPR) or observed region (for TIR/FCS). If surface diffusion occurs at a measurable rate, decreasing the illuminated or observed region size will increase the rapidity of fluorescence recovery or the autocorrelation function decay (38).

Both TIR/FPR and TIR/FCS have been confirmed as experimentally feasible (Figure 5). Using TIR/FPR, Burghardt & Axelrod (10) have measured a range of residency times of rhodamine-labeled BSA adsorbed to fused silica ranging from approximately 5 sec to at least several hours. The most "loosely" bound BSA molecules also appeared to surface diffuse with a coefficient of 5×10^{-9} cm^2/sec, fast enough to carry a BSA molecule several microns on the surface before desorption. This observation of surface diffusion of a biological molecule may be significant in view of the hypothesis of Adam & Delbruck (1) that nonspecific adsorption and diffusion on cell surfaces might enhance agonist/receptor reaction rates.

Using TIR/FCS, Thompson & Axelrod (36) have measured the nonspecific adsorption/desorption kinetics of rhodamine-labeled immunoglobulin and insulin on serum albumin coated fused silica. Rapidly reversible adsorption could be visualized under TIR as twinkling speckles of fluorescence as molecules enter and leave the evanescent wave region. Upon on-line autocorrelation of these fluorescence fluctuations, a range of desorption times was noted, with the shortest time less than 5 ms, limited by the rate of bulk diffusion. Antibody molecules appeared to bind specifically to antigen-coated fused silica, but this was accompanied by a large amount of reversible nonspecific binding. Such nonspecific binding is difficult to avoid and perhaps occurs on cell surfaces where it may be biochemically functional via the Adam & Delbruck (1) mechanism mentioned above. In suitable systems where nonspecific binding is low, TIR/FCS and TIR/FPR should prove useful for measuring specific solute-surface kinetic rates. In

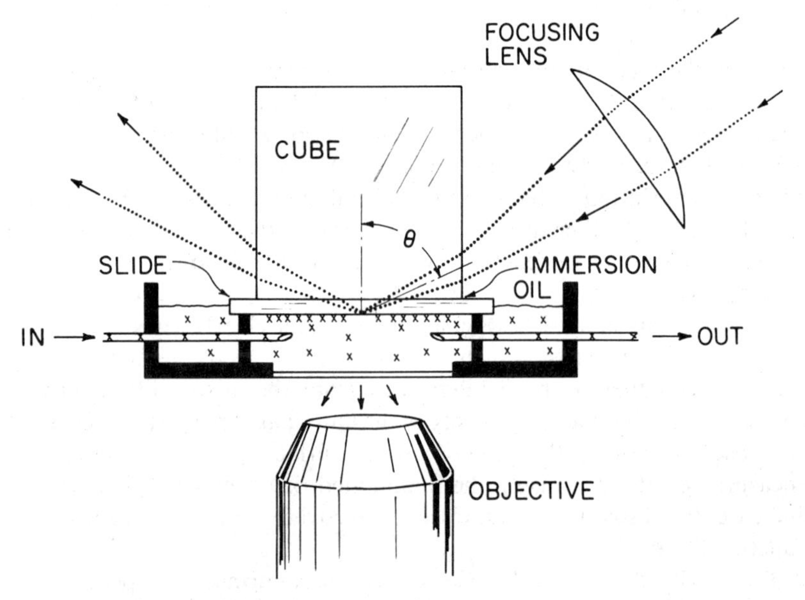

Figure 5 Experimental apparatus for TIR/FPR, for use on the stage of an inverted microscope. Fluorescent solute molecules are indicated by x's. The cubical prism and slide are made of fused silica. The drilled hole in the bottom of the dish is covered with a glass coverslip and sealed in with encapsulating resin. The apparatus for TIR/FCS is qualitatively similar, except that the inflow and outflow tubes are omitted and the spacing between the slide and the bottom coverslip is reduced to only 0.06 mm with a thin Teflon spacer to enable use of high aperture, short working distance oil immersion objectives.

principle, TIR/FCS can even measure specific solute/surface kinetic rates at equilibrium of nonfluorescent molecules by competing them with fluorescent analogs (35). In any case, TIR/FCS and TIR/FPR can be distinguished from the TIR concentration jump method by their ability to measure very rapid rates without necessitating macroscopic perturbation of the equilibrium.

MOLECULAR CONFORMATION OF ADSORBATES

The conformational and dynamical properties of adsorbed proteins can be studied by combining TIRF with conventional fluorescence spectroscopic techniques such as singlet-singlet (Förster) energy transfer and fluorescence polarization. For this purpose, a TIRF chamber easily fitted into a commercial spectrofluorimeter for spectra analysis has been designed (11) (Figure 6). Using this chamber, the authors observed a decrease in effective energy transfer rate upon adsorption in BSA multiply-labeled with the fluorophore donor/acceptor pairs of dansyl/eosin or 4-chloro-7-nitro-2,1,3-benzoxadiazole (NBD)/rhodamine. Under certain assumptions, this energy transfer change can be interpreted as a conformational change of BSA upon adsorption. The intrinsic tryptophan fluorescence of unlabeled BSA was also found to be less quenchable by iodide ion when adsorbed to fused silica than when dissolved in bulk (D. Axelrod, unpublished observation). This result indicates either a conformational change upon adsorption or reduced accessibility of iodide to tryptophan residues because of the proximity of the surface.

Using the same TIRF spectrofluorimeter chamber to measure steady state fluorescence polarization, Burghardt (8) has inferred a decreased rotational diffusion coefficient of NBD, rhodamine, or eosin fluorophores around their single covalent bonds binding them to BSA upon adsorption of the BSA to fused silica. The restricted rotational motion of the probe can be attributed to direct steric interference by the surface or to steric interference arising from a conformational change in BSA. (The same paper also contains a mathematical derivation of the expected fluorescence polarization anisotropy decay of a rotationally diffusing fluorophore while connected to an anisotropically rotationally diffusing protein.)

Preliminary experiments have been performed using TIRF with time-resolved fluorescence lifetime and anisotropy decay measurements to detect the fluorescence lifetimes and rotational motion of labeled protein at the solid/liquid interface (T. P. Burghardt and P. M. Torgerson, unpublished observations). These particular experiments utilize the unique polarization characteristics of the evanescent electric field to deduce the order of arrangement of labeled proteins in a biological structure (9).

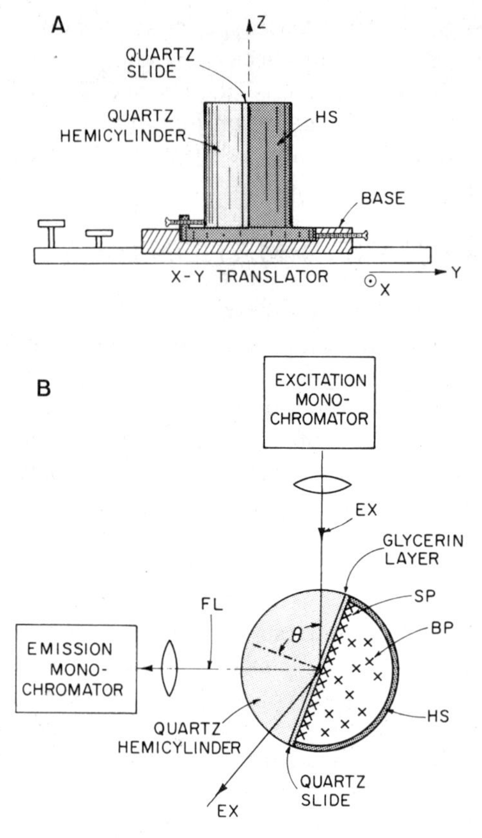

Figure 6 Two views of the total internal reflection fluorescence spectroscopy apparatus. (*A*) Side view shows the axis of rotation, *z*, about which the prism and slide and Plexiglass hemicylinder shell (HS) unit is rotated in the base to adjust the angle of incidence of the excitation light. The base is removably mounted on an *x-y* translator for lateral positioning of the unit. The apparatus fits inside the sample chamber of a commercial fluorescence spectrofluorometer. (*B*) Vertical view of the device shows the path of the excitation (EX) and the fluorescent emission (FL) through the quartz hemicylinder. Surface-adsorbed protein (SP) is illuminated by the evanescent field while in chemical equilibrium with the bulk-dissolved protein (BP). The fused silica hemicylinder has a radius of 1.3 cm and a height of 2.5 cm.

CELL/SUBSTRATE CONTACTS

TIRF can be used to excite fluorescence exclusively from regions of contact between living cultured cells and the substrate (usually plastic or glass) upon which they grow. These regions of contact are of considerable interest: they are anchors for cell motility, loci for aggregation of specific membrane proteins, and convergence points for cytoskeletal filaments. A recent review of TIRF microscopy has appeared (5).

Several experimental designs for TIRF microscopy are possible. Figure 7 (from 4) shows a design appropriate for an inverted fluorescence microscope. Phase contrast transmitted illumination and epi-illumination (i.e. through the objective) are easily interchanged with TIRF without changing the basic configuration (Figure 8). By varying the incidence angle of the totally reflecting laser beam, and thus the depth of evanescent illumination, details of the topography of the cell surface near the substrate can be revealed (4, 5).

Another TIRF design (Figure 9) (D. Axelrod, unpublished), used in an upright microscope, is particularly convenient for directly viewing cells growing in disposable plastic tissue culture dishes and works quite well despite the autofluorescence of tissue culture polystyrene plastic. (Corning brand dishes have significantly less autofluorescence than Falcon, Lux, or Nunc brands and are thereby more suitable.) Both this design and the one shown in Figure 7 permit the cells to be moved laterally while retaining TIR illumination centered in the microscope's field of view.

A third design (43) (for an inverted microscope) employs a triangular prism to guide a laser beam into a microscope slide where it undergoes multiple internal reflections (Figure 10). The cells may reside on top of the slide (for any kind of cells) (A. Brian and N. L. Thompson, unpublished) or underneath the slide (for tightly adhering cells). A novel feature is the use of intersecting laser beams (split from the same laser) at the total internal reflection surface to create a parallel line interference fringe pattern. This interference pattern can be seen on the fluorescing regions of cell/substrate contact, thereby confirming that the primary source of excitation is indeed the evanescent wave rather than light randomly scattered from the evanescent wave by the cells. Another potential application of these TIRF

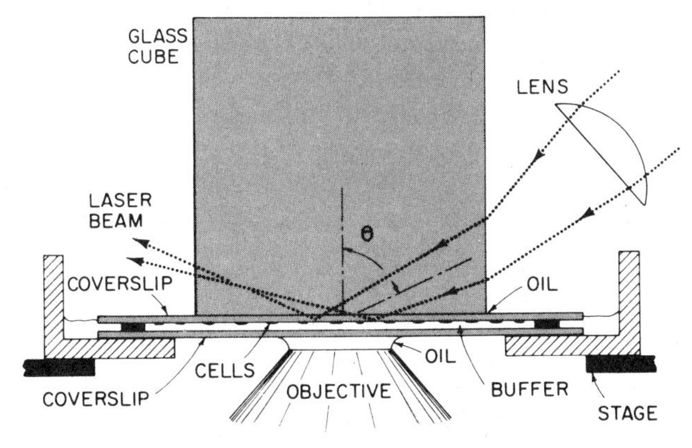

Figure 7 TIRF inverted microscope apparatus for viewing cells in culture. Cells are plated and grown on a standard glass coverslip, which is then inverted and placed in optical contact with the cubical prism via a layer of immersion oil or glycerin.

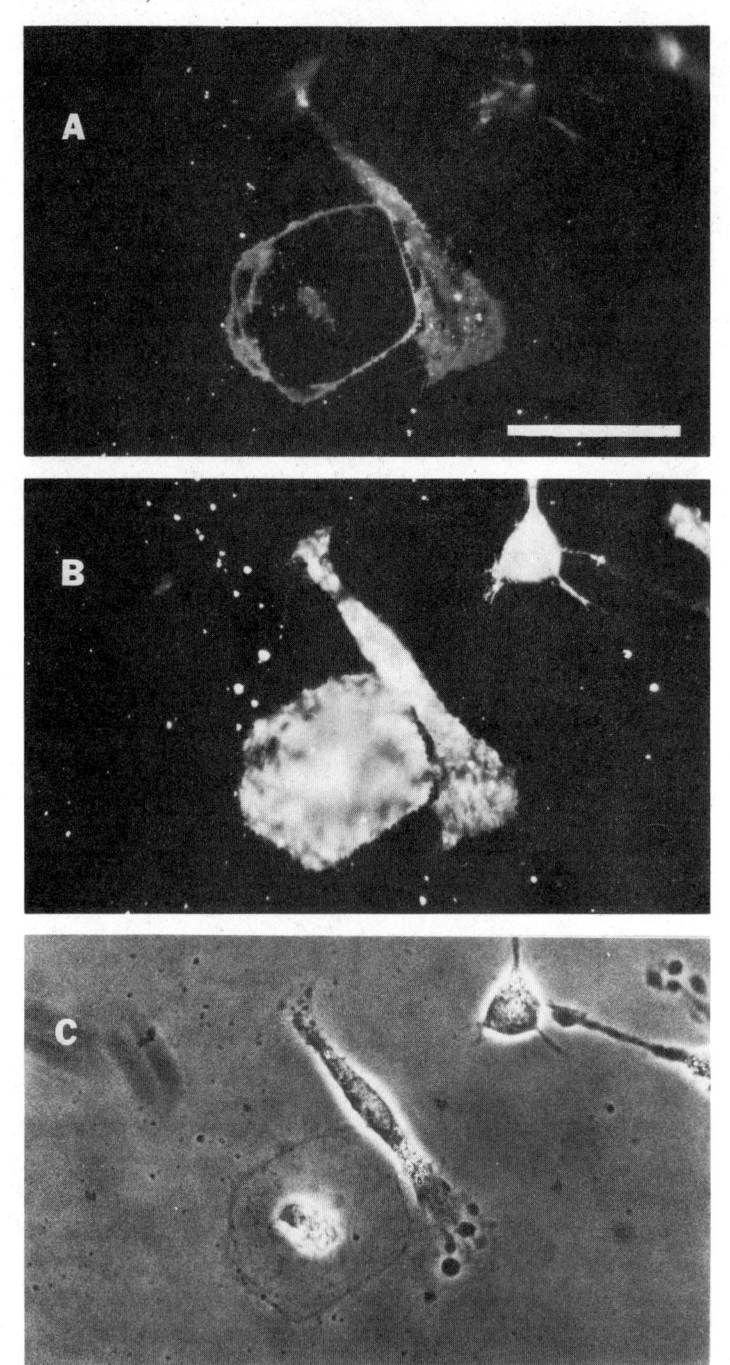
A
B
C

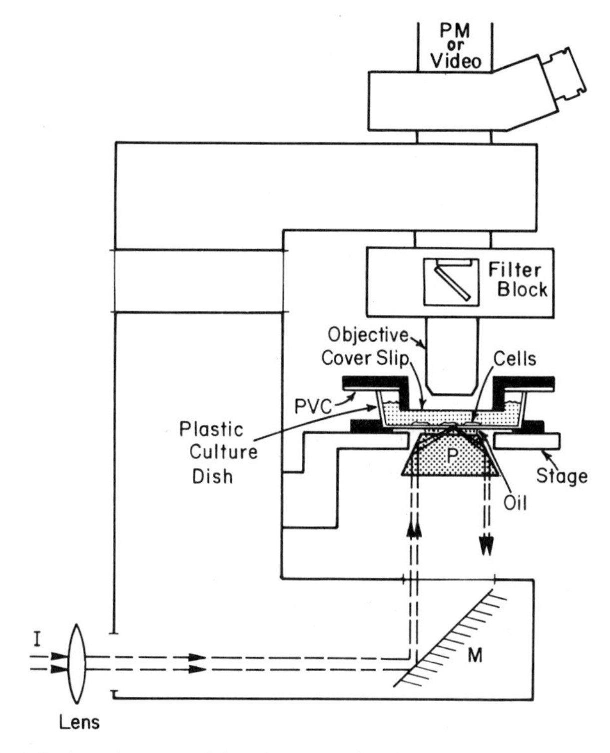

Figure 9 TIRF adapted to an upright microscope, for viewing cells in a plastic tissue culture dish. The prism is a truncated equilateral triangle. The region of the sample chamber is shown enlarged relative to the rest of the microscope for pictorial clarity. Abbreviations are as follows: I, incident light; M, mirror (part of microscope base); P, prism; PM, photomultiplier; PVC, polyvinylidene chloride ("Saran" film used to seal in a 10% CO_2 atmosphere over the tissue culture medium).

fringes is to measure lateral mobility of membrane components in cell/substrate contact regions by fluorescence redistribution after pattern photobleaching (34).

The features of TIRF on cells lead to potential applications as follows. (*a*) TIRF greatly reduces fluorescence from cytoplasmically internalized label, cellular debris, and autofluorescence in thick cells, relative to fluorescence from membrane regions close to the substrate. This feature may allow

Figure 8 Cells in a mixed fibroblast/myoblast primary culture of embryonic rat muscle, labeled with the membrane soluble dye, 3,3′-dioctadecylindocarbocyanine. (*A*) TIRF on inverted microscope. (*B*) Same field in epi-illumination. (*C*) Same field in phase contrast. TIRF shows that the large round cell is making contact with the substrate only around its periphery and in one small region near its center. A 40X, 0.75 N.A., water immersion objective was used. Space bar = 50 μm.

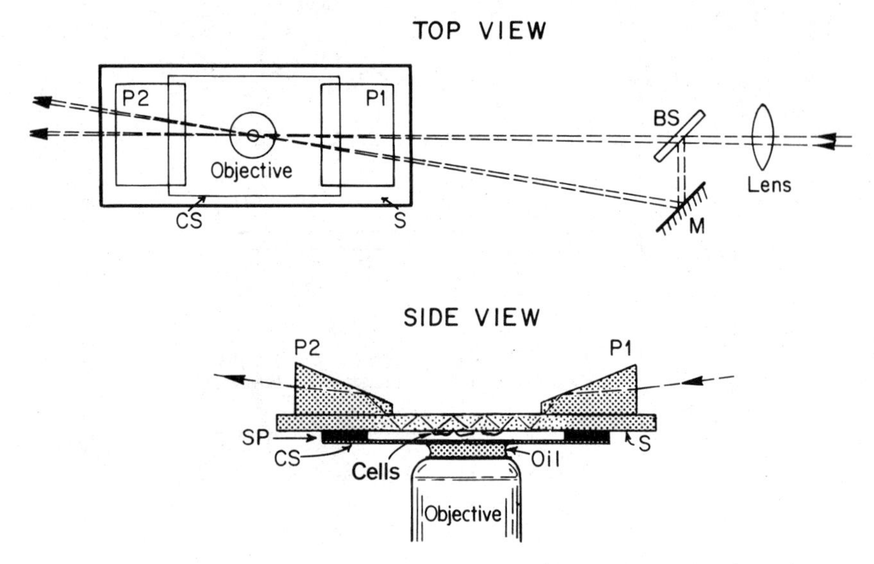

Figure 10 TIRF microscopy combined with interference fringes. The top view shows the two laser beams intersecting at the region under observation. The side view shows the path of a beam as it enters a prism P1, totally reflects multiple times in a glass slide to which cells are adhered (in this case on the lower surface), and exits via prism P2. Abbreviations are as follows: BS, beam splitter; CS, coverslip; L, lens; M, mirror; S, slide; SP, spacer.

detection of lower concentrations of fluorescence-marked membrane receptors than would otherwise be possible. (*b*) Study of the submembrane structure of the cell-substrate contact in thick cells can be facilitated. A fluorescently labeled cytoskeletal structure in the contact region can be visualized without interference from an out-of-focus background from fluorescently labeled cytoskeletal structure farther from the substrate. (*c*) By varying the incidence angle, the topography of the membrane facing the substrate can be mapped. (*d*) Reversibly bound fluorescent ligands on membrane receptors might be visualized without exciting background fluorescence from unbound ligand in the bulk solution. In this manner, certain cell surface receptors might be studied without the necessity of blocking them by irreversible antagonists.

EXPERIMENTAL SUGGESTIONS

In setting up a TIRF system, one may encounter a number of questions about design and materials. The following suggestions may be helpful.

1. The prism used to couple the light into the system and the (usually disposable) slide or coverslip in which the total reflection takes place need not be exactly matched in refractive index.

2. The prism and slide may be optically coupled with glycerin, cyclohexanol, or microscope immersion oil, among other liquids. Immersion oil has a higher index of refraction (thereby avoiding possible total internal reflection at the prism/glycerin interface for low incidence angles) but tends to be more autofluorescent (even the "extremely low" fluorescence types). This problem is usually not important to TIRF microscopy but can be serious in large area applications.
3. The prism and slide can both be made of ordinary optical glass for many applications, unless shorter penetration depths arising from higher refractive indices are desired. (More exotic high refractive index materials such as sapphire, titanium dioxide, and strontium titanate can yield penetration depths as low as $\lambda_0/20$.) However, optical glass does not transmit light below about 310 nm and also has a dim autoluminescence with a long (several hundred microsecond) decay time, which can be a problem in some photobleaching experiments. The autoluminescence of high quality fused silica (often called "quartz") is much lower.
4. The total internal reflection surface need not be polished to a higher degree than a standard commercial microscope slide.
5. Either a laser or conventional arc light source will suffice for study of macroscopic areas. But in TIRF microscopy, a conventional light source is difficult to focus to a small enough region at high enough intensity while still retaining sufficient collimation to avoid partial light transmission through the interface; a laser is more desirable here.
6. Illumination of surface-adsorbed proteins can lead to apparent photochemically-induced cross-linking and also to photobleaching at the higher range of intensities that might be used in TIRF studies. Apparent cross-linking, measured as a slow, continual, illumination-dependent increase in observed fluorescence, can be inhibited by 0.05 M cysteamine, among other substances; photobleaching can be reduced somewhat by deoxygenation or by 0.01 M sodium dithionite, among other substances.

ACKNOWLEDGMENTS

We thank Peter M. Torgerson of the University of California, San Francisco, and Adrienne Brian of Stanford University for permitting citation of some of their unpublished observations. This work was supported by USPHS NIH grant #14565 (to D.A.).

Literature Cited

1. Adam, G., Delbruck, M. 1968. In *Structural Chemistry and Molecular Biology*, ed. R. Alexander, D. Norman, pp. 198–215. San Francisco: Freeman
2. Aiginger, H., Wobrauschek, R. 1951. *J. Radioanal. Chem.* 61:281–93
3. Ambrose, E. J. 1961. *Exp. Cell Res. Suppl.* 8:54–73

4. Axelrod, D. 1981. *J. Cell Biol.* 89:141–45
5. Axelrod, D., Thompson, N. L., Burghardt, T. P. 1983. *J. Microsc.* 129:Pt. 1, pp. 19–28
6. Beissinger, R. L., Leonard, E. F. 1980. *Am. Soc. Artif. Internal Organs* 3:160–75
7. Born, M., Wolf, E. 1975. *Principles of Optics.* New York: Pergamon. 5th ed.
8. Burghardt, T. P. 1983. *J. Chem. Phys.* 78:5913–19
9. Burghardt, T. P. 1983. Submitted for publication
10. Burghardt, T. P., Axelrod, D. 1981. *Biophys. J.* 33:455–68
11. Burghardt, T. P., Axelrod, D. 1983. *Biochemistry* 22:979–85
12. Burghardt, T. P., Thompson, N. L. 1984. *Opt. Eng.* In press
13. Carniglia, C. K., Mandel, L., Drexgage, K. H. 1972. *J. Opt. Soc. Am.* 62:479–86
14. Chew, H., Wang, D., Kerker, M. 1979. *Appl. Opt.* 18:2679–87
15. Eagen, C. F., Weber, W. H., McCarthy, S. L., Terhune, R. W. 1980. *Chem. Phys. Lett.* 75:274–77
16. Fringeli, U. P., Gunthard, Hs. H. 1981. In *Membrane Spectroscopy*, pp. 270–332. New York: Springer-Verlag
17. Haller, G. L., Rice, R. W., Wan, Z. C. 1976. *Catal. Rev. Sci. Eng.* 13:259–84
18. Harrick, N. J. 1967. *Internal Reflection Spectroscopy.* New York: Wiley Intersc.
19. Harrick, N. J., Loeb, G. I. 1973. *Anal. Chem.* 45:687–91
20. Harrick, N. J., Loeb, G. I. 1982. *Mod. Fluorescence Spectrosc.* 1:211–25
21. Hirschfeld, T. 1965. *Can. Spectrosc.* 10:128
22. Hirschfeld, T., Block, M. J. 1977. *Opt. Eng.* 16:406–7
23. Hirschfeld, T., Block, M. J., Mueller, W. 1977. *J. Histochem. Cytochem.* 25:719–23
24. Iwamoto, R., Miya, M., Onta, K. 1981. *J. Chem. Phys.* 74:4780–90
25. Kronick, M. N., Little, W. A. 1975. *J. Immunol. Methods* 8:235–40
26. Lee, E.-H., Benner, R. E., Fen, J. B., Chang, R. K. 1979. *Appl. Opt.* 18:862–68
27. Lok, B. K., Cheng, Y.-L., Robertson, C. R. 1983. *J. Colloid Interface Sci.* 91:87–102
28. Lok, B. K., Cheng, Y.-L., Robertson, C. R. 1983. *J. Colloid Interface Sci.* 91:104–16
29. McGuirk, M., Carniglia, C. K. 1977. *J. Opt. Soc. Am.* 67:103–7
30. Memming, R. 1974. *Faraday Discuss. Chem. Soc.* 58:261–70
31. Palik, E. D., Holm, R. T. 1978. *Opt. Eng.* 17:512–24
32. Rabolt, J. F., Santo, R., Swalen, J. D. 1979. *Appl. Spectrosc.* 33:549–51
32a. Rockhold, S. A., Quinn, R. D., Van Wagenen, R. A., Andrade, J. D., Reichert, M. 1983. *J. Electroanal. Chem.* 150:261–75
33. Selser, J. C., Rothschild, K. J., Swalen, J. D., Rondelez, F. 1982. *Phys. Rev. Lett.* 48:1690–93
34. Smith, B. A., McConnell, H. M. 1978. *Proc. Natl. Acad. Sci. USA* 75:2759–63
35. Thompson, N. L. 1982. *Biophys. J.* 38:327–29
36. Thompson, N. L., Axelrod, D. 1983. *Biophys. J.* 43:103–14
37. Thompson, N. L., Burghardt, T. P. 1983. Submitted for publication
38. Thompson, N. L., Burghardt, T. P., Axelrod, D. 1981. *Biophys. J.* 33:435–54
39. Tweet, A. G., Gaines, G. L., Bellamy, W. D. 1964. *J. Chem. Phys.* 40:2596–2600
40. Van Wagenen, R. A., Rockhold, S., Andrade, J. D. 1982. In *Biomaterials: Interfacial Phenomena and Applications, Adv. Chem.*, ed. S. L. Cooper, N. A. Peppas, 199:351–70
41. Watkins, R. W., Robertson, C. R. 1977. *J. Biomed. Mater. Res.* 11:915–38
42. Weber, W. H., Eagen, C. F. 1979. *Opt. Lett.* 4:236–38
43. Weis, R. M., Balakrishnan, K., Smith, B. A., McConnell, H. M. 1982. *J. Biol. Chem.* 257:6440–45
44. Wood, R. W. 1934. *Physical Optics*, pp. 419–20. New York: Macmillan. 3rd ed.

Ann. Rev. Biophys. Bioeng. 1984. 13: 269–302

PATCH CLAMP STUDIES OF SINGLE IONIC CHANNELS

Anthony Auerbach and Frederick Sachs

Department of Biophysical Sciences, State University of New York, Buffalo, New York 14214

INTRODUCTION

The electrical activity of cells is controlled by transmembrane pores called ionic channels. Channels are extremely efficient enzymes that can increase the rate at which ions flow across cell membranes by more than fourteen orders of magnitude, with turnover numbers approaching 10^9/sec. This high rate of ion translocation means that individual channels conduct currents in the range of picoamperes, currents that can be readily measured with commercially available instrumentation.

Channels are allosteric enzymes that alter their conformation in response to ligand binding, the electric field in the membrane, and/or other stimuli. Some conformations conduct current and others do not; hence the flow of current through a single channel consists of a series of equal-amplitude pulses. The times between transitions are exponentially distributed random variables whose means can be related, often in simple ways, to the molecular rate constants governing state transitions of the channel.

The patch clamp is a recently developed electrophysiological technique that has sufficient resolution to record currents from a single ionic channel. The patch clamp has increased measurement sensitivity by three orders of magnitude over previous techniques. It is now routine to observe the behavior of one protein molecule with a time resolution approaching 10 μsec. Amazing!

The success of the patch clamp lies in the ability to form a high resistance seal between a glass micropipette and a patch of cell membrane (or lipid bilayer). The seal isolates the patch both electrically and chemically. Because of the electrical isolation and low resistance of the pipette relative to the membrane, a patch can be voltage clamped by simply applying a

0084–6589/84/0615–0269$02.00

voltage to the pipette. Chemical isolation allows the ionic environment of each face to be manipulated with little regard for composition or osmotic balance. Because the patch-pipette seal is mechanically stable, patches of membrane can be excised from the cell with either the intra- or extracellular aspect facing the bath and, because excised patches are small, solutions bathing the exposed surface can be exchanged within milliseconds.

A number of recent publications detail high resolution patch clamp techniques. The book *Single-Channel Recording*, edited by Sakmann & Neher (128), has the most complete and up-to-date discussion of single channel recording techniques and data analysis. Other methodological reviews include Neher (105), Hamill et al (62), Sachs & Auerbach (125), and Jackson et al (77). The mathematical basis for kinetic modeling of single channel data has been thoroughly developed by Colquhoun & Hawkes (29–31). Lecar et al (87) and Dionne & Leibowitz (42) have presented simplified analyses, while Horn & Lange (69) have presented a general method for comparing kinetic models to single channel data.

In this review we have confined our coverage almost entirely to single channel experiments. The results of whole-cell voltage clamps (cf 48), extracellular recording (51), and patch clamp experiments that have only recorded currents from ensembles of channels (9, 18, 41, 138) are not discussed. We first describe some of the physical characteristics of the patch and then some of the properties of single ionic channels revealed by patch clamp. Many channels cannot be cleanly classified according to gating or selectivity properties, and the organizational scheme we have used is, inevitably, oversimplified. Our emphasis on particular channels represents personal biases only.

PATCH CLAMP METHODOLOGY

The field of single channel electrophysiology in biological membranes has been dominated at every turn by the work originating in the laboratories of Neher & Sakmann in Göttingen, FRG. They were the first to publish records of single channel currents, obtained from acetylcholine receptors in extrajunctional membrane of denervated frog skeletal muscle (108). The unitary currents appeared as rectangular pulses superimposed on a background of Gaussian noise. Current levels that were integral multiples of the unitary current were adequately explained by the random overlap of currents from independent channels. Currents whose amplitude was less than the unitary current could be explained as arising from "rim" channels located in the sealing region between the pipette and the membrane (110).

These results in themselves did not constitute a conceptual breakthrough in the understanding of ionic channels, since single channels in artificial bilayers had provided prototypes of bistable channels (cf 46). However, the fact that the patch clamp permitted channel currents to be observed in native membranes, and that conclusions could be drawn from kinetic data with a new clarity (111), provided a powerful stimulus for development of the technique.

Patch clamp recording in biological membranes has merged with artificial bilayer techniques. The patch clamp has improved the signal-to-noise ratio of bilayer experiments by allowing smaller (lower capacity) membranes and by improved amplifier design. Taking advantage of the improved performance, Tank et al (141) developed a method whereby native membrane fragments are fused with large lipid vesicles that can be patch clamped. Coronado & Latorre (38) and Suarez et al (140) have developed a general technique for studying biological channels in bilayers using the patch clamp. A lipid monolayer is formed at the surface of a saline bath. Crude membrane vesicles are injected into the aqueous phase and then fuse with the monolayer, resulting in the incorporation of channels. A patch pipette is drawn upward through the monolayer and then brought down again, forming a bilayer across the tip of the pipette. Ion channels from heart muscle (38) and electroplax (140) have recently been studied with this method.

Properties of the Patch and the Seal

PATCH CHARACTERISTICS There is little data available to characterize the dimensions of a patch itself. Patch area can be estimated from the magnitude of capacity currents (~ 1 pA) arising from rapid changes in cell potential due to action potentials (~ 50 V/sec) or external voltage clamps (135). Assuming that the equivalent circuit for a sealed patch is a parallel RC network and that the specific membrane capacity is 1 $\mu F/cm^2$, capacity currents imply that a patch has an area between 2μm^2 (48) and 10 μm^2 (135; D. Nelson and F. Sachs, unpublished observations), an area that is equivalent to a hemisphere with a radius of 0.5–1.0 μm.

The patch area can be estimated indirectly from a comparison of the average number of channels in a patch compared to the number in a cell, assuming that the patches are representative of the whole cell. For potassium channels in tunicate eggs, Fukushima (54) reported that most patches contained only one channel. The channel density, computed from the ratio of the single channel current to the whole cell current, was 0.039/μm^2. Since the average patch contained one channel, the effective patch area was about 25 μm^2, equivalent to a hemisphere with a radius of 2 μm.

What is the conductance of the membrane patch? The resistivity of a pure lipid bilayer can be about 10^{10} Ω-cm^2 so that a 10-μm^2 patch of pure lipid would have an unmeasurably high resistance of 10^{17} Ω. For a typical biological membrane with a resistivity of 10^4 Ω-cm^2, a patch that is 2 μm in diameter would have a resistance of ~100 GΩ. Ayer et al (15) found that in tissue-cultured cells from chick heart patch resistances were >200 GΩ when the pipette contained normal (Na^+) saline and ~15 GΩ when it contained high K^+ saline. The excess conductance of the patch in high K^+ may be due to the presence of anomalous rectifier potassium channels that are continuously open in isotonic K^+ saline (54, 115). The presence of open channels can be tested by adding blockers such as Ba^{2+} or Cs^+. The high conductance of the patch in K^+ solutions can produce voltage errors in cell-attached recordings, since the patch impedance may approach that of a small cell (15, 48).

In chromaffin cells, Fenwick et al (48) observed an inward current of ~1 pA with cell-attached pipettes containing NaCl solutions. This current was driven by the membrane potential, but the origin of the leakage path (nonspecific damage or channel activity) was not determined.

Excised patches may not be representative of the normal cell membrane. There have been reports of significant kinetic changes following excision in sodium channels (23, 49), nicotinic acetylcholine-activated channels (143), calcium channels (49), and serotonin-activated channels (59). It is not known how much cytoplasmic material remains attached to membrane patches that have been excised. Guharay & Sachs (60) present indirect evidence that cytoskeletal elements remain attached to excised patches. Electron microscopic studies are clearly needed.

SEAL CHARACTERISTICS The nature of the seal between the membrane and the glass remains obscure. The seal resistance can be in the range of 100 GΩ, and if one assumes that the gap between the membrane and the glass is filled with normal saline and that the sealing region is a cylindrical annulus 1 μm long and 2 μm in diameter, the estimated thickness of the annulus must be ~0.16 nm (62). This is not a realistic value. Aside from the fact that membrane proteins can protrude from the plane of the membrane by 2 nm or more (80), there are membrane-bound polysaccharides that could hardly be compressed to the thickness of a carbon-carbon bond. Bilayers strongly repel each other at distances below ~2 nm (91), yet high resistance seals (although no better than biological membranes) can be formed with pure phospholipid bilayers (38, 141).

Rather than assume a short, or point, seal, it seems more reasonable to assume that the sealing region is distributed over several microns and that the distance between the glass and the insulating portion of the membrane is 1–2 nm. If the resistivity of the membrane-glass cleft were 5–10 times greater than that of normal saline, resistances in the range of 100 GΩ could be achieved with sealing regions of ~10 μm in length (F. Sachs and F. Guharay, unpublished calculations). This distributed-seal model predicts that the highest resistance seals will be obtained with membranes that are drawn far into the pipette so that the sealing region is longer than the space constant of the cleft between the glass and the membrane at the frequencies of interest. It is interesting that the distributed-seal model predicts that, as actually observed, the thermal noise-density of the patch will increase approximately linearly with frequency, since the input admittance of an RC cable falls as the square root of frequency. A detailed understanding of the properties of the seal-tip region awaits precise impedance measurements (F. Sachs and F. Guharay, in preparation) and electron microscopic studies.

The Limits of Patch Clamp Resolution

The lowest possible noise that could be attained in a patch clamp recording is the thermal noise of the patch/seal combination. A patch/seal of 100 GΩ would have an rms noise level of 0.13 pA, or a peak-to-peak level of about 0.8 pA, in a bandwidth of 10 kHz. Thus, for reliable identification of single channel currents, the smallest current that could be observed in a bandwidth of 10 kHz is about 1.6 pA. At present, in actual recordings the lowest noise in a bandwidth of 10 kHz is ~5 times the theoretical minimum. In the 10–500 Hz range the observed noise density is within a factor of two of the theoretical limit.

The excess noise at high frequencies can come from three sources: the membrane patch, the glass, or the seal. If the patch capacitance is not ideal and shows dielectric loss, then there will be an associated thermal noise. The noise from this source could be large. If the patch area is 10 μm^2 and the phase angle is 80° (rather than 90° for a perfect capacitor; see 26), at 10 kHz this source alone contributes noise equivalent to a 10-GΩ resistor. For a 70° phase angle the noise is equivalent to that of a 5-GΩ resistor. Dielectric loss in the glass of the pipette tip should be negligible compared to the loss in the membrane, because glass is a better dielectric than the membrane (57) and because the glass is about ten times thicker than the membrane. A distributed sealing region can produce noise equivalent to resistances far below that expected from the DC seal resistance, and it is likely that this will be the limiting noise source (F. Sachs and F. Guharay, in preparation).

The current amplifiers themselves produce excess noise that also has a density that increases approximately linearly with frequency. This noise may be due to uncharacterized losses in the structure of input transistors or to nonideal properties of the feedback resistors. Improvements in the electronics might be able to reduce the excess wideband noise by 30%, but more extensive reductions will require a better understanding of the patch itself.

Data Analysis and Kinetic Modeling

If the baseline and unitary current levels are well defined, the measurement of single channel amplitudes is straightforward; accurate estimates of channel current can be obtained by measuring a handful of events. For kinetic analysis, however, the exponential distribution of intervals makes it imperative that a great many events be collected in order to accurately estimate the population means. For example, with a well-separated, two-component, exponential distribution, at least 5,000 events need to be measured in order to estimate the population means with 90% confidence (125). Detailed kinetic studies are essentially impossible without computer-aided analysis.

An heuristic pattern recognition system was presented by Sachs et al (126). Additional discussion of data analysis is found in Sachs (124), Sachs & Auerbach (125), and Colquhoun & Sigworth (35).

Horn & Lange (69) have developed a method of analysis that compares single channel data to kinetic models. Given an idealized data trace (binary or n-ary currents), the probability of obtaining each interval is calculated based upon an assumed kinetic model and the number of channels in the patch. The effects of limited bandwidth can be included by estimating the probability of missed events in each interval. The likelihood for the entire interval sequence is maximized with respect to the rate constants and the number of channels by standard algorithms. Because the entire data set is fit as a whole rather than by combining parameters derived from regression to several different distributions, the analysis is more sensitive and stable than traditional eigenvector analyses. The computation time for this approach, however, can be substantial.

Regardless of the method of analysis, it is necessary to compute the observable rates from any kinetic scheme. Colquhoun & Hawkes (29–31) have developed a matrix formalism applicable to any stationary state model. However, the formalism does not deal cleanly with the case of multiple channel activity. Dionne & Liebowitz (42) and Leibowitz & Dionne (88) have analyzed three and four state linear models in which multiple channel activity is explicity included.

SINGLE ION-CHANNEL CURRENTS

Nicotinic Acetylcholine Receptor Channels

BACKGROUND Over twenty-five years of electrophysiological and pharmacological studies of nicotinic acetylcholine receptors have produced the following basic model of the channel activation sequence:

$$2A+R \underset{k_{-1}}{\overset{k_{+1}}{\rightleftharpoons}} A+AR \underset{k_{-2}}{\overset{k_{+2}}{\rightleftharpoons}} A_2R \underset{\alpha}{\overset{\beta}{\rightleftharpoons}} A_2R^*. \qquad 1.$$

Two molecules of agonist (A) bind to a closed channel (R) to form a doubly liganded, closed complex (A_2R), which can then undergo an isomerization to an open conformation (A_2R^*). In most preparations, power spectral analyses of agonist-induced current fluctuations and measurements of current relaxation rates following perturbations in agonist concentration or membrane potential indicate that the activation process behaves as a simple single-step process whose rate depends on temperature, membrane potential, and the agonist used to activate the channels (reviewed in 3, 139).

The suggestion of Magelby & Stevens (94) that the binding steps are fast compared to the isomerization has met widespread acceptance. According to this interpretation, at low agonist concentrations the macroscopic rate constant reflects the channel closing rate, α. Alternatively, if the binding steps are rate limiting and the isomerization fast (or perhaps there is no isomerization!), the apparent rate constant would represent the rate of agonist dissociation, k_{-2}. In this case, the elementary conductance determined from noise analysis would underestimate the true single channel conductance by a factor equal to the probability that a doubly liganded channel is in the open conformation. A third interpretation of the macroscopically derived rate constant is possible. Colquhoun & Hawkes (29) and Sakmann & Adams (127) pointed out that a single-step process would be apparent at low agonist concentrations if the lifetime of the doubly liganded form of the channel was brief compared to the open channel lifetime. In this "transient intermediate" case, the macroscopic relaxation rate constant, k_{obs}, is given by

$$k_{obs} = \alpha[k_{-2}/(k_{-2}+\beta)].$$

Regardless of the interpretation of the macroscopic rate constant, the sequential activation model predicts that a channel that has just closed and is in the A_2R state can either reopen or lose an agonist molecule. The number of reopenings before dissociation is given by β/k_{-2}. (Note that in the above model, k_{-2} is normalized on a per channel rather than a per site basis). The "fast binding" assumption predicts that reopenings are rare ($\beta \ll k_{-2}$), the "fast isomerization" assumption predicts that they are plentiful ($\beta \gg k_{-2}$), and the "transient intermediate" interpretation makes no predictions as to the number of reopenings—only that the lifetime of the A_2R state (given by $1/\beta + k_{-2}$) is brief.

The first single-channel recordings from acetylcholine receptors by Neher & Sakmann (108) in denervated frog skeletal muscle appeared to confirm the "fast binding" interpretation of the model. The currents were rectangular pulses that had lifetimes equivalent to those predicted from autocorrelation analysis of current noise (109). Moreover, the channel had a single open conformation with a conductance of $\sim$25 pS, similar to the value predicted from fluctuation analysis.

Nelson & Sachs (112) reported that at low concentrations of the agonist suberyldicholine (SubCh), channel openings in tissue-cultured chick muscle tended to occur in unit-amplitude bursts. The bursts clearly represented activity of a single channel, since openings from independent channels would have produced multi-unit events. This bursting pattern of single channel activity was predicted by the "transient intermediate" interpretation of Equation 1 (29), thus permitting estimates of channel opening, closing, and agonist dissociation rates.

HIGH RESOLUTION MEASUREMENTS Nicotinic channel activity occurs in bursts in all preparations. The detailed kinetic properties of the bursts, however, appear to vary substantially between preparations. Colquhoun & Sakmann (33) reported that, in tissue-cultured rat muscle, with low concentrations of SubCh, there are ~3 low conductance gaps/burst and the mean durations of the gaps within bursts is 45–70 μsec. In the mouse clonal cell line BC_3H1, bursts initiated by SubCh (outside-out patches) contain only ~1.2 gaps and the mean gap duration is ~75 μsec (137). Dionne & Leibowitz (42) and Leibowitz & Dionne (88) examined single nicotinic channels at the snake twitch muscle endplate and found that, with ACh as the agonist, there were only ~0.3 gaps/burst and the mean gap duration was ~200 μsec. Finally, Colquhoun & Sakmann (34) have given preliminary data indicating that at the frog motor endplate, nicotinic channels activated by ACh average ~3 gaps/burst with a mean burst duration of ~20 μsec.

Assuming that gaps within bursts represent sojourns of a channel in the A_2R state, the values given above have been used to estimate the channel opening and agonist dissociation rates (Table 1). The channel closing rate constant can be obtained directly as the inverse of the open period lifetime.

Table 1 Nicotinic acetylcholine receptor channel molecular rate constants[a]

Agonist	β	k_{-2}	α	Preparation
ACh	1,100	3,200	450	Snake endplate; 22°C, −90 mV (Ref. 88)
ACh	30,000	10,000	1,000[b]	Frog endplate; 12°C, −120 mV (Ref. 34)
SubCh	17,000	2,000	1,300[b]	
SubCh	12,000	4,000	400	Frog perisynaptic; 12°C, −120 mV (Ref. 33)
ACh	7,700	7,000	80	BC_3H1 cells; 10–22°C, outside-out patches (Ref. 137)
SubCh	4,600	12,000	60	
CCh	6,600	6,800	50	

[a] The rates of channel opening (β), closing (α), and agonist dissociation (per channel; k_{-2}) were estimated from kinetic analyses of bursts of single channel currents. The burst parameters were interpreted according to the activation sequence given in Equation 1, where short-lived gaps within bursts represent sojourns of a channel in the doubly liganded, closed state from which the channel opens.

[b] Calculated from the burst length (BL), where $\alpha = (k_{-2}+\beta)/k_{-2} \cdot \mathrm{BL}$.

While there is reasonable agreement as to the magnitude of k_{-2}, the estimates of β vary by a factor of 25.

The rise time and amplitude of miniature endplate currents (mepc) set certain constraints on the channel kinetic parameters and the number of molecules of transmitter in a quantum. If the density of postsynaptic channels is not limiting and assuming that the concentration of ACh is saturating and rises and falls rapidly in the cleft, the number of molecules in a quantum constrains the number of doubly liganded, closed channels that are formed. The channel opening rate and the agonist dissociation rate then determine how many of these channels will open within any specified time. In both snake (43) and frog (45) endplates, the mepc rise time is ~ 300 μsec and there are ~ 1500 channels open at the peak of the mepc. With the values given in Table 1, at snake endplates there must be at least 40,000 molecules of ACh in a quantum, whereas at the frog endplate there need only be 4000 molecules of ACh in a quantum. These values are lower limits, since the time required to open the channels is less than 300 μsec because diffusion and binding delays contribute substantially to the rising phase of the mepc (83, 84).

Leibowitz & Dionne (88) found that neither the number of gaps/burst nor the duration of the open period changed significantly with changes in the membrane potential; however, the mean duration of the gaps shortened with hyperpolarization. Their calculations indicate that β increases e-fold with a hyperpolarization of ~ 78 mV. In contrast, virtually all estimates of the voltage dependence of β from macroscopic measures indicate that the channel opening rate depends little on the membrane potential (2, 44, 56, 68, 107, 132). Another surprising result is that the channel closing rate, estimated from the mean open period, is faster in CCh-activated channel than it is in ACh-activated channels in the snake endplate (88), but the order is reversed in BC_3H1 cells (137).

Given the variety of preparations and experimental conditions from which the single nicotinic channel data were obtained, it is reasonable to expect substantial variation in channel kinetics. It is disquieting, however, to suppose that such elemental properties as the number of molecules of ACh in a quantum, or the channel opening rate, or the voltage and/or agonist dependence of the opening and closing rates could differ by an order of magnitude or more between vertebrate endplate channels.

MORE COMPLEX KINETIC MODELS ARE REQUIRED Colquhoun & Sakmann (33) noted that in tissue-cultured rat muscle, the distribution of durations of gaps within bursts could not be described by a single exponential. In addition to the 45–70 μsec population of gaps, there was a population with a mean duration of ~ 1 msec. Only about 2% of all gaps were from the slower population. This slower component of gaps has since been observed in other preparations (13, 137). Since a low concentration of agonist was used in these experiments, the relatively long-lived gaps cannot represent a cycle of agonist dissociation and binding. At 5 nM agonist, even with a diffusion-controlled association rate, hundreds of milliseconds are required to acquire a new agonist. Thus, the presence of "slow" gaps within bursts indicates that the channel can enter more than one low conductance, doubly liganded state.

Auerbach & Sachs (14) have found that nicotinic channels in tissue-cultured chick muscle can exist in at least four doubly-liganded, low conductance conformations. There is a population of gaps that is brief (<100 μsec), a population of intermediate duration (0.5–1.0 msec, depending on the agonist) with a conductance $\sim 12\%$ that of the open channel (see below), a population that is rather long lived (2–10 msec) with a conductance near zero, and a population of subconductance states through which bursts can terminate but which are shorter than the intra-burst subconductance gaps.

At low concentrations of agonist, it is not possible to distinguish unambiguously which of the gaps within bursts represent sojourns in the A_2R state. With increasing concentrations of agonist, the time required to re-bind an agonist will be reduced. At saturating concentrations of agonists there will be a component of the closed-time distribution with a characteristic time constant equal to $1/\beta$ (130). Sine & Steinbach (137) have measured gap durations over a wide range of concentrations of ACh and found no component that reaches the limiting value of $1/\beta$ predicted from kinetic analyses of the rapid component of gaps seen at low agonist concentrations (Table 1). They suggest that the fastest component of gaps within bursts does not exclusively represent sojourns in A_2R.

In addition to the multiple components of gaps within bursts, the distribution of open periods can often be described by the sum of two exponentials. The shorter component has a time constant of ~ 0.1 msec while the longer component has a time constant of ~ 1 msec. A number of studies indicate that these two populations do not arise from independent populations of channels. Rather, a single channel can open for either short or long periods (76, 78, 104, 137). In tissue-cultured rat muscle, a burst may contain both short and long openings even at low concentrations of agonist (78). In BC_3H1 cells, the short openings occur as isolated events at low concentrations of agonist (137). In chick muscle, the fraction of short openings is agonist dependent and is greater with 4-ketopentyltrimethylammonium than with ACh (12; A. Auerbach, unpublished observations).

The significance of the rapid component of open periods remains obscure. Colquhoun & Sakmann (33) suggested that short openings may represent open channels which are singly liganded. If so, short openings should become relatively less common at high agonist concentrations. Sine & Steinbach (137) have found that in BC_3 H1 cells the relative occurrence of short openings was similar at all concentrations of ACh (10 nM–300 μM), indicating that both long and short openings are likely to arise from channels that are doubly liganded.

It is uncertain whether the multitude of doubly liganded states represent kinetic "dead ends" or whether they can lead to unliganded or desensitized states without passing through the A_2R^* state. If it turns out that all the states can readily interconvert, state models of the burst may not be useful.

In summary, single channel recordings of ACh-activated channels have shown that models describing the channel kinetics need to be expanded to accommodate the number of new states that have been observed. In chick muscle, for example, there are at least six doubly-liganded forms of the channel. Snake endplate channels appear to have the simplest kinetics, and, perhaps, in this preparation the assumption that all brief closures are sojourns in A_2R is correct. The method employed by Sakmann et al (130) and Sine & Steinbach (137) in which the variation in gap durations is examined as a function of agonist concentration seems to be the most promising and least ambiguous approach.

It should be pointed out that in most recording situations, there is more than one channel present in the patch. Analyses of channel kinetics have assumed that all channels have identical behavior. Considering the opportunities for posttranslational modification of nicotinic channels (74) and for modulation of channel kinetics by the membrane environment (148), this assumption may not hold, and subsequent modelling may have to include a distribution of population means (hopefully not time dependent).

DESENSITIZATION In the continued presence of agonist, there is a progressive decline in the mean number of open channels because nicotinic channels enter inactivated, or desensitized, states. With single channel recording, there are usually only a few channels present, and it is often possible to drive all the channels in the patch to desensitized state. In this case, the random transition of a channel out of a desensitized state will give rise to a complex series of openings that can be attributed to an individual channel.

Sakmann et al (130) examined the kinetics of channels in denervated frog muscle at high (5–50 μM) concentrations of ACh. They found that the bursts of activity themselves occurred in groups. (Using the terminology of Sine & Steinbach (137), we refer to a series of closely spaced openings as a burst, a series of closely spaced bursts as a group, and a series of closely spaced groups as a cluster.) If bursts represent doubly liganded, non-desensitized forms of the channel (see above), then groups can be considered to be all non-desensitized forms, regardless of the extent of ligation. In the experiments of Sakmann et al (130), the mean group duration in 20 μM ACh was 420 msec. The gaps within groups grew shorter as the ACh concentration increased, but at 50 μM, the highest concentration tested, the gap duration had not reached a saturating value (equal to $1/\beta$). At 22 μM ACh, the burst duration was equal to the interburst duration, effectively defining the equilibrium constant. This value is in agreement with the value of ~30 μM estimated from macroscopic techniques (see 3, 139).

Groups tended to occur in clusters, with the gaps between clusters being much longer (34 sec) than the gaps between groups (180 msec at 20 μM ACh). The gaps between groups represent sojourns in one or more relatively short-lived desensitized states. The duration of cluster (4.8 sec) reflects the combined dwell-time of a channel in these short-lived desensitized and all activatable states. The gaps between clusters represent sojourns in long-lived desensitized states. Because there are usually many channels in the patch, these sojourns cannot be associated with an individual channel and thus are not easily interpreted. Sine & Steinbach (137) have observed a similar pattern of bursts, groups, and clusters at high concentrations of agonists in BC_3H1 cells.

While these studies do not suggest precise kinetic models joining open, activable, and desensitized forms of the channel, it seems clear that patch clamp experiments with high agonist concentrations will provide important information on the channel opening rate and on the short-lived component of desensitization. In particular, experiments with outside-out patches where the concentration of agonist can be jumped within a few tens of milliseconds should prove fruitful.

CHANNEL BLOCKADE A simple sequential model of open channel blockade has been remarkably successful in describing the interaction of local anesthetics and other ligands with nicotinic channels:

$$A_2R \underset{\alpha}{\overset{\beta}{\rightleftharpoons}} A_2R^* \underset{k_{-b}}{\overset{b \cdot k_{+b}}{\rightleftharpoons}} A_2R^*b$$

where A_2R^* is the conducting form of the channel and b is the concentration of the blocking ligand. There are many ligands capable of blocking open nicotinic channels, including uncharged (113) and charged (111) local anesthetics, the nicotinic antagonist *d*-tubocurarine (28, 111), and agonists such as ACh, carbamylcholine (CCh), and SubCh (137). Redmann (122) found that in BC_3H1 cells, high concentrations of permeant ions reduce k_{+b}, consistent with the notion that the ions and blockers compete for the same binding site in the channel (70). The conductance of the blocked state is unmeasurably small, less than 1% of the open-channel conductance (13, 14, 106).

The kinetics of channel blockade are quite variable. For example, at a membrane potential of -120 mV, k_{+b} is 1.8×10^6 sec^{-1} M^{-1} for benzocaine (113) and 2×10^7 sec^{-1} M^{-1} for QX-222 (111). For some ligands (e.g. QX-314, curare) the blocked state is so long that blockade appears as a reduced burst length. For others (e.g. QX-222, SubCh), the duration of the blocked state is short enough to be resolved as a brief interruption in the single channel current. In other cases (ACh, CCh), the block is too short lived to be resolved, and blockade is expressed as a decrease in the mean and an increase in the variance of the open-channel current. ACh has a rather low affinity for the channel with a K_D of 68 mM at -100 mV (137). Curare has a very high affinity for the channel with a K_D of 20 nM at -120 mV (28).

Neher & Steinbach's (111) analysis of local anesthetic blockade by QX-222 supported the simple sequential model shown above; however, recent results indicate that more complex models of channel block may be required. Neher (106) showed that QX-222 produces two populations of intraburst gaps rather than one predicted by the model. Additionally, in the sequential model a blocked channel can only lose a ligand by passing through the open state, so that the total time spent in the A_2R^* conformation is independent of the concentration of the blocker. In order to determine the total time spent in the conducting state without introducing errors that are due to a limited bandwidth, Neher measured the charge transferred per burst, which, according to the sequential model, must be independent of the concentration of blocker. Contrary to predictions of the model, the time spent in the open state decreased with increasing concentrations of QX-222. This result indicates that channels can enter some long-lived, nonconducting state directly from the blocked state. Lingle (89) reached similar conclusions in a study of blockade of crustacean nicotinic channels by chlorisondamine, and Connor et al (36) noted the existence of a long-lived, blocked state in their study of the effects of atropine on ganglionic nicotinic channels. It is unknown whether blockers or agonists can dissociate from these long-lived states.

The voltage dependence of the blocking and unblocking rates can provide insight into the structure of the channel lumen. For extracellularly applied cationic blockers, k_{+b} increases with hyperpolarization, whereas for uncharged blockers k_{+b} is voltage independent (1). This suggests that the field is acting on the blocking ligand rather than the channel protein. Gage et al (55), however, found that the voltage dependence of endplate channel blockade by procaine is the same whether the blocker is applied extracellularly or intracellularly, thus raising the possibility that the field is acting on the ligand-channel combination. This effect should be studied in excised patches where all solutions can be controlled.

If the field does act upon the ligand instead of the channel, the voltage sensitivity of the dissociation constant of the blocker provides a measure of how far through the membrane field the blocking site is located. Neher & Steinbach (111) estimated that the blocking site was located ~75% of the way through the field (from the extracellular surface). Sine & Steinbach (137) found similar values for block of channels in BC_3H1 cells by ACh and CCh, but the blocking site for SubCh was only ~40% through the field. Differences in site location may not reflect multiple ion-binding sites because steric and/or other interactions may determine where the blocker resides in the channel lumen. Again, assuming that the field acts upon the blocker, the forward blocking rate, k_{+b}, should increase exponentially with hyperpolarization until the rate becomes comparable to the diffusion controlled limit. Auerbach (11) found that the rate of blockade of chick nicotinic channels by 20 μM of benzyltrimethylammonium does not show a simple exponential dependence on voltage but saturates at hyperpolarized potentials. It is not clear whether k_{+b} is limited by voltage-independent diffusion of the blocker to the mouth of the channel (10) or whether a bound, but conducting, species of channel is formed.

Other Transmitter-Activated Channels

GLUTAMATE Glutamate opens large, cation-selective channels in locust muscle membranes. It has not been possible to form high resistance seals in this preparation, but the huge channel conductance (~130 pS) has enabled single channel recording to be done at bandwidths equivalent to those used to study nicotinic channels.

Single glutamate-activated channel currents have several characteristics in common with single nicotinic channel currents. The channel conductance does not vary with the membrane potential (119) or the agonist (39, 58). Openings tend to occur in bursts that contain a short-lived component of gaps, which have a lifetime of ~100 μsec (40). The lifetime of bursts is agonist dependent and is about 2.5 times longer with quisqualate than with glutamate (39, 58). Also, glutamate-activated channels can desensitize, although, in contrast to nicotinic channels, desensitization can be eliminated by treatment with the lectin Concanavalin A (ConA; 100).

Glutamate channels show spontaneous and marked changes in channel kinetics during the course of an experiment and shift between periods of low and high duty-cycle (119). In the high duty-cycle epochs, the channel lifetime is ~10 times that in the low duty-cycle epochs. This behavior is apparent even with low concentrations of glutamate and occurs both in ConA-treated and untreated preparations (58).

Because desensitization can be eliminated and because patches often contain only one channel, straightforward experiments at high agonist concentrations are possible. Cull-Candy et al (39) found that in extrajunctional channels during the low duty-cycle periods, the equilibrium constant for binding glutamate (defined by the concentration that produced a 50% probability of being open) was 300 μM. The linear sequential model (Equation 1) predicts that at agonist concentrations $\gg K_D$, low conductance gaps will have a mean duration equal to $1/\beta$. Cull-Candy et al (39) found the gap duration to be $\sim$500 μsec at 600 μM glutamate (2 times K_D), putting a lower limit of 2000 sec^{-1} on β. Cull-Candy & Parker (40) analyzed the kinetics of gaps within bursts according to the "transient intermediate" interpretation and obtained a value of 4000 sec^{-1}—a reasonable agreement.

Gration et al (58) examined the concentration dependence of the open channel lifetime in ConA-treated, denervated muscle fibers and found that, for concentration of glutamate between 0.1 and 10 mM, the lifetime increased from $\sim$3 to $\sim$800 msec. They concluded that, unlike nicotinic channels, the glutamate channels have at least two open states, both of which can bind agonists. In contrast to these results, Cull-Candy et al (39) found no increase in the lifetime in the range of 0.05 to 0.6 mM glutamate. The basis for this discrepancy may be that Gration et al included both low and high duty-cycle periods in their measurements, whereas Cull-Candy et al examined only low duty-cycle periods. Perhaps the channel can only bind an agonist during high duty-cycle periods.

MUSCARINIC CHANNELS Sakmann et al (129) studied acetylcholine-activated channels in dispersed cells from the S-A and A-V nodes of the rabbit heart. Presumably these channels were muscarinic, but the sensitivity of channel currents to muscarinic agonists and antagonists was not reported. The channels were predominantly K^+-selective, and the single channel currents showed a marked inward rectification, with a chord conductance of 25 pS and 5 pS for inward and outward currents (cell-attached patches, 20 mM KCl and 0.2 μM ACh in the pipette, 31°C). Channels in the patch could not be activated by ACh applied to the rest of the cell, implying that second messengers were not involved. Activity occurred in bursts that had slower kinetics than typical nicotinic channel currents. The distribution of burst durations was best fit with the sum of two exponentials, with the fast component predominating. The properties of these channels are drastically different from those of muscarinic channels from a variety of neuronal preparations, where application of muscarinic agonists shuts down a predominantly K^+-selective conductance, which is normally active at membrane potentials near threshold (4).

GABA (γ-AMINOBUTYRIC ACID) Jackson et al (76) examined Cl^--selective channels in tissue-cultured mouse spinal neurons activated by GABA (0.5–

1 μM), muscimol (0.3–1 μM) or (–)-pentobarbital (50 μM). The cells were impaled with KCl-filled conventional microelectrodes, and Cl^- was injected in order to raise the Cl^- equilibrium potential to ~ -20 mV. At 22°C, the channel conductance was ~23 pS for all three agonists. The channels showed bursting kinetics and the distribution of open times contained slow and fast components ($\tau = 8$ and 45 msec, -80 mV, muscimol).

ATP (ADENOSINE TRIPHOSPHATE) Single channels activated by ATP (1–100 μM) have been observed in tissue-cultured chick skeletal muscle (82). ADP, AMP, and adenosine were not effective. There were two types of ATP-activated channels (48 pS and 20 pS), neither differentiated between Na^+ and K^+, and neither was permeable to, or blocked by, Ca^{2+}. The 48-pS channel had a longer open time than the 20-pS channel (3.5 msec vs 1.3 msec, 22°, no applied potential), but both open times increased with hyperpolarization at a rate of 70 mV/*e*-fold change. The authors suggested that ATP activates nicotinic channels. This suggestion is particularly interesting since ATP is released from synaptic vesicles along with ACh (136) and since ATP can potentiate the action of ACh (47). More conclusive experiments (such as tests of cross-desensitization) are needed to validate the identity of the channels.

SEROTININ Guharay (59) studied single channels activated by serotinin in neuroblastoma cells. In cell-attached patches at 20°C, the channel conductance was 140 pS with a reversal potential near zero. The mean channel open time was ~2 msec and did not depend on the membrane potential. Like glutamate-activated channels, these channels spotaneously shifted between periods of high and low activity. In outside-out patches, only the low duty cycle periods were present.

Sodium Channels

CONDUCTANCE Sigworth & Neher (135) solved the technical difficulties of applying voltage steps to the patch electrode and recorded single sodium channel currents from tissue-cultured rat myotubes. The average current obtained from many single channel records resembled macroscopic sodium currents, supporting the notion that sodium channels function as independent units.

The conductance of single sodium channels at room temperature is consistently 10–15 pS from one preparation to another (23, 52, 73, 121, 135). The Q_{10} of the conductance is ~1.3, suggesting that the single channel current is not limited by high frequency conformational changes in the channel (73). Batrachotoxin, a toxin derived from poisonous frogs, decreases the conductance of sodium channels to 2 pS and greatly increases the open channel lifetime (121).

Sodium channel currents saturate at hyperpolarized potentials in some preparations (23, 49, 52). The sublinear current-voltage relationship may arise from saturation of a binding site within the channel (10), channel blockade by ions such as calcium (146), or artifactually, from band-limited recording. The correct explanation is not known.

Horn et al (72) probed the blocking site of sodium channels with the impermeant ion tetramethylammonium. The kinetics of the block were too fast to be resolved as individual gaps in the current, so that the blocking and unblocking rate constants had to be estimated from the mean open channel current. The voltage dependence of block indicated that the binding site was, surprisingly, ~90% of the way through the membrane field from the extracellular surface.

KINETIC MODELS OF ACTIVATION AND INACTIVATION There have been several attempts to understand sodium channel kinetics using single channel techniques. The basic kinetic model under consideration consists of three states:

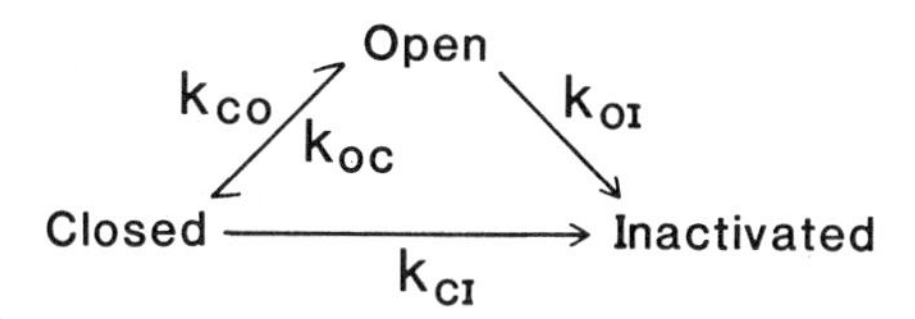

where each so-called state may actually consist of an ensemble of states. The essential difference between Closed and Inactivated is indicated by the irreversible reactions. Transitions out of the Inactivated state are much slower than the other rates.

The following issues have been raised: How many states are there in each ensemble? Can closed channels inactivate, and if so, at what rate? What proportion of open channels close by k_{OC} as opposed to k_{OI}? Does the decay of the sodium current reflect the lifetime of the open channel or do channels open more than once?

Published data are generally consistent with a single open state. With the exception of sodium channels found in tunicate eggs (52), the open channel lifetime has a slight (within a factor of 2.5), bell-shaped dependence on voltage (8, 23, 49, 73). In the scheme above, a weak voltage dependence of the open time implies that k_{OC} and k_{OI} are both voltage independent, or that k_{OC} and k_{OI} have equal but opposite voltage dependencies, or that one of the two rates is voltage independent and much larger than the other.

Activation kinetics have been studied by measuring the first latencies, which are the distribution of times from application of a depolarizing pulse to the appearance of the first opening. The distribution of first latencies is biphasic, which suggests that there are two or more states within the Closed ensemble (7, 71).

There are also at least two states in the Inactivated ensemble, one of which appears to be associated with resting inactivation (73). This desensitized, or hibernating, state appears in the data as runs of channel activity interspersed with runs of no activity. *N*-bromoacetamide (NBA) applied to the intracellular surface of the membrane exaggerates the development of this desensitized state, although similar behavior is occasionally present in untreated, cell-attached patches (C. Stevens, personal communication).

There is general agreement that in the tissue-cultured preparations, activation and inactivation processes significantly overlap in time (7, 73). The overlap is a reflection of the fact that the Closed ensemble has both a fast and a slow exit rate to the Open state, so that during the falling phase of the sodium current, some channels are opening long after others have inactivated. Over most of the activation curve, the rate of open channel inactivation is sufficiently fast that channels open only once during a sodium current.

Can Closed channels inactivate? In the sodium channel literature, distinctions have been made between coupled and uncoupled inactivation schemes. In the extreme case, the coupled scheme dictates that the Inactivated state(s) can be reached only from the Open state, i.e. that k_{CI} is zero. In the uncoupled scheme, typified by the Hodgkin-Huxley (67) model, Closed and Inactivated states can freely interconvert. The distinction between the two models is only of historical interest, since intermediate degrees of coupling (which appear to be the models of interest; 118) are obtained simply by varying k_{CI}.

Horn et al (71) tried to determine if Closed channels could inactivate. They applied depolarizing steps and selectively averaged either all records or only those records in which no channels opened for the first 4.3 msec. If Closed channels cannot inactivate during this silent period of 4.3 msec, all channels must still be in the Closed ensemble. After the silent period, the decay of the average current should have the same time course and probability of being open as the unconditional average, since both averages represent transitions from the same state(s). If Closed channels can inactivate, some of them will make transitions to the Inactivated state(s) during the 4.3-msec silent period. Then, after the delay, the probability of being open should then be less than the unconditioned average, because some channels are not available to make the opening transition. The time course of the average current should be essentially the same as the unconditional average. The data agreed with the latter case, indicating that k_{CI} is significant and that the extreme coupled scheme in which k_{CI} is zero can be excluded (118).

Aldrich & Stevens (8) found that k_{CI} is steeply voltage-dependent, varying from 0.015 $msec^{-1}$ at a membrane potential of about −60 mV to 0.59 $msec^{-1}$ near 0 mV. Following a depolarizing voltage step, the steep voltage dependence of k_{CI} drains channels from the Closed ensemble and hence determines the falling phase of the mean sodium current. The falling phase of the current is not a measure of the channel open time.

There is a controversy over the rate and voltage dependence of open channel inactivation (k_{OI}). The issues are not monumental, but the different ways in which the problem has been approached makes the controversy an interesting case study.

Aldrich et al (7, 8) estimated the relative probabilities of open channel inactivation and closing by counting the number of times a channel opens before inactivating. For a single channel, this value (K) is related to the probability of opening from a Closed state and inactivating from either the Open or Closed states:

$$K = k_{CO}/\{k_{CI}+k_{CO}[k_{OI}/(k_{OI}+k_{OC})]\}.$$

The opening rate, k_{CO}, was estimated from the first latency histogram and was found to be much greater than the rate of open channel inactivation, k_{CI}, which was estimated from the fraction of all records in which no openings occurred. When normalized by the number of channels in the patch (assumed to be equal to the maximum observed level), the data indicated that $k_{OI}/k_{OC} \simeq 6$. Thus, open channels close by inactivation about 85% of the time, and the voltage independence of the Open state lifetime indicates that k_{OI} is not strongly voltage-dependent (k_{OI} was ~ 2 $msec^{-1}$ at 5°C). Since k_{OI} dominates the channel closing rate, little can be said regarding the voltage dependence of k_{OC}.

Using a different approach, Horn et al (73) compared sodium channel kinetics with normal inactivation and with inactivation removed by NBA (116, 120). With inactivation intact, the open state lifetime is equal to $1/(k_{OC} + k_{OI})$ whereas with inactivation removed, the lifetime is $1/k_{OC}$. The lifetime of sodium channels in NBA-treated patches ($1/k_{OC}$) was voltage dependent, increasing *e*-fold with a hyperpolarization of 34 mV. Since the open state lifetime in untreated cells was voltage independent, Horn et al (73) concluded that k_{OI} had the opposite voltage dependence to k_{OC}. At a membrane potential of −20 mV and a temperature of 9°C, k_{OI} was ~ 0.2 $msec^{-1}$ and, according to their analysis, k_{OI}/k_{OC} varied from 5.7 at −60 mV to 0.6 at −10 mV. Thus, at potentials hyperpolarized from threshold, open channels preferentially close rather than inactivate.

It is important to point out that the conclusions of Horn et al and Aldrich et al only differ significantly for potentials negative to threshold. In strongly depolarized membranes, both groups agree that k_{OI} is much greater than

k_{OC}. Most errors in the analysis by Aldrich et al (7, 8) lead to underestimates of the k_{OI}, so we suspect that Horn et al (73) may have underestimated k_{OI} for cell-attached patches. If a preexisting, voltage dependent closing rate was increased about threefold by NBA, the conclusions of the two groups would be resolved. Alternatively, if the voltage dependence of inactivation were shifted by excision (23, 49), the technique of Horn et al, which combines data from excised and cell-attached patches, would lead to inappropriate comparisons of data at different effective potentials.

SOME SUGGESTIONS FOR IMPROVING DATA EXCHANGE: A MINOR DIGRESSION We have found that it is difficult to compare the results from different preparations and from different laboratories because of ambiguities in the membrane potential and the state of the channels. In cell-attached patches, the resting potential is not usually measured, so that the absolute membrane potential is uncertain by perhaps 30 mV. In order to make comparisons between different experiments, absolute potentials should be reported. While potentials are best measured with a separate intracellular electrode, it is simpler to calibrate the channel conductance and reversal potential and then convert channel current to potential. A running record of membrane potential is important, since drift in the resting potential during an experiment (particularly at low temperatures) can alter voltage dependent kinetics. As an alternative to direct measurement, the membrane potential could be clamped to zero by keeping the cells in a high potassium solution.

CAVEAT FREMULATOR (LET THE PATCHER BEWARE) Although the membrane potential is well defined for excised patches, the effects of excision on channel properties are not yet understood. Fenwick et al (49) have reported that excision produces a hyperpolarizing shift of 10–20 mV in both the activation and inactivation curves, and Cachelin et al (23) have reported a massive (calcium independent) hyperpolarizing shift of 50 mV. Related to excision artifacts, Trautmann & Siegelbaum (143) have found that the closing rate for nicotinic channels in outside-out patches is significantly faster than in cell-attached patches. Also, Guharay (59) found that excision altered the kinetics of serotinin-activated channels in neuroblastoma cells. It seems advisable, particularly for sodium channel experiments, to report steady state activation and inactivation curves so that offsets can be considered.

Potassium Channels

DELAYED RECTIFIER A delayed rectifier is a potassium-selective channel that produces an increase in membrane conductance with depolarization. Conti & Neher (37), in a technically demanding experiment, recorded currents from single potassium channels at the inner surface of a perfused squid axon. The single channel currents occurred in bursts, as was expected from macroscopic measurements that indicated the existence of multiple closed states (67). In contrast to predictions from the Hodgkin-Huxley n^4 independent-gating model (67), there were too many events per burst in the observed currents. Llano & Bezanilla (90) found an 18-pS delayed rectifier channel with complex kinetics in split-open squid axons. It appears likely, as with the sodium channel, that with the increase in resolution associated with the patch clamp kinetic models of this potassium channel will have to be revised.

In tissue-cultured heart cells, the delayed rectifier appears to have a conductance of 60 pS and bursting kinetics that are voltage dependent ($\sim$100 msec burst length at 27°C, -60 mV) (25, 50).

ANOMALOUS RECTIFIER An anomalous rectifier is a potassium-selective channel that produces an increase in membrane conductance with hyperpolarization. The anomalous rectifier is interesting in that there may be no conformational change associated with channel gating. Rather, the gating arises from blockade by diffusible ions.

In macroscopic relaxation experiments using tunicate eggs, the characteristic anomalous rectifier response to a hyperpolarizing step is a rapid (ohmic) increase in the current, followed by a slower, exponentially decaying phase called channel inactivation (114). The amplitude of the ohmic component and the time course of the decay depend upon the concentration and species of ion available to compete with K^+. In 100% K^+ saline there is virtually no inactivation, while in 100% Na^+, Cs^+, or Li^+ saline there is virtually no current (114).

Single channel experiments using tunicate eggs or rat myotubes show that in mixtures of K^+ and Na^+, Cs^+, Ba^{2+}, or Sr^{2+}, currents have the bursting pattern characteristic of channel blockade (53, 54, 115), although the concentration dependence of blockade has not been shown. The forward blocking rates are voltage dependent and can be much smaller than equivalent rates for local anesthetic blockade of nicotinic channels; at -130 mV, the association rate for Na^+ is 10^1 M^{-1} sec^{-1}. Cs^+ is a much more effective blocker, with an association rate of 3×10^6 M^{-1} sec^{-1} (54).

The dissociation, or unblocking, rates for extracellularly applied monovalent ions are nearly voltage independent, whereas the unblocking rates for divalent ions increase with hyperpolarization (54). This suggests that unblocking can occur by forcing the blocker completely through the channel. Because of their smaller charge, monovalent ions would not have sufficient electrostatic energy to overcome the barriers for a through passage.

The inward rectifier shows complex conductance/selectivity properties. The channel conductance in 150 mM KCl is about 10 pS for both tunicate eggs (53, 54) and rat myotubes (115). When 200 mM Na^+ is added to a K^+ saline, the channel currents show gaps that are due to Na^+ blockage, but most unexpectedly, the open channel current increases (54)! There is a clear bi-functional effect of Na^+: it blocks the channel on the time scale of milliseconds and facilitates K^+ transport on a much shorter time scale. This facilitation may be explained by ion-ion interactions within a single-file, multi-site pore (cf 66).

OTHER K^+ CHANNELS Single K^+-selective channels have been recorded in tissue-cultured smooth muscle (19), heart (129), glial (79), and HeLa (131) cells. The Ca^{2+}-activated potassium channel and the S channel of *Aplysia* are discussed elsewhere.

In tissue-cultured rabbit jejunum, there is a 50-pS, non-inactivating, K^+ channel activated by depolarization, which is insensitive to intracellular Ca^{2+} levels (19). The channel has both fast (160/sec) and slow (0.1/sec) opening rates. The slow component may play some role in slow wave generation. The channel exhibits to subconductance states with conductances 1/3 and 2/3 that of the open channel. All states have the same reversal potential.

In isolated rat heart cells there are three K^+-selective channels, one which is ACh-activated (see above) and two which are active at rest (129). With 70 mM K^+ saline on the extracellular face, the latter types both have conductances in the range of 35–40 pS. However, the "atrial" type found on beating cells has an open state lifetime of about 1.4 msec, whereas the "ventricular" type, found mostly in ventricular and non-beating atrial cells, has a lifetime of 48 msec.

In mouse oligodendrocytes there is a 70-pS, K^+-selective channel that has a high probability of being open at rest. The rates of opening and closing appear to be independent of both of the membrane potential and the concentration of Ca^{2+} on the cytoplasmic face (79). The channel shows outward rectification; in inside-out patches with high K^+ in the pipette and low K^+ in the bath, the conductance was over twice that when the solutions were reversed.

HeLa cells contain a potassium channel that has a conductance of 35–40 pS at potentials negative to E_K (131). The channel does not pass outward current but differs from the anomalous rectifier of tunicate eggs and rat muscle (54, 115) in that complex bursts of openings were apparent in 100% KCl solutions. The probability of finding an open channel increased with increasing concentration of KCl in the pipette (from 0.04 in 75 mM KCl to 0.32 in 300 mM KCl). The channel has a rather long-lived subconductance state(s) with an amplitude ~3/4 that of the main open state.

Calcium Channels

We use the term "calcium channel" to refer to a channel that is permeable to Ca^{2+}, as opposed to a channel whose gating is influenced by Ca^{2+}. Calcium channels are difficult to study because their conductance is small and the kinetics rapid, complicated, and variable. In addition, channel currents are labile. Ca^{2+}-selective channels have been recorded from snail neurons (21, 92), bovine chromaffin cells (49), heart cells (123), rat clonal pituitary cells (61), pheochromacytoma cells (61), and chick dorsal root ganglion cells (21).

Lux & Nagy (92) first recorded single calcium channels in snail neurons after ferocious enzymatic treatment to remove connective tissue. With potassium channels blocked by tetraethylammonium injection, calcium channel currents with a slope conductance of 5–15 pS were visible. With Ba^{2+} substitution for Ca^{2+}, the conductance increased to 10–30 pS. Fenwick et al (49) and Reuter et al (123) have obtained similar Ba^{2+} conductances in chromaffin cells and heart cells in culture. These conductances are much larger than the < 1 pS expected from the noise studies by Akaike et al (6). Why Akaike et al's estimate was so low is not clear, since noise studies by Lux & Nagy (92) and Fenwick et al (49) have yielded values for the unit conductance close to the single channel value.

As with most channels, calcium channels have multiple closed states that result in bursting behavior. Fenwick et al (49) have made extensive studies of the calcium channel kinetics using mostly whole-cell patch clamp recordings and have concluded that in chromaffin cells there is no voltage-dependent inactivation but there may be Ca^{2+}-dependent inactivation.

Interesting new results on the calcium channel relate to modulation of the channel by sympathomimetic amines and cyclic adenosine monophosphate (cAMP). Reuter et al (123) found that isoprenaline increased the probability of finding an open calcium channel by about a factor of ~2 (they did not report whether the drug was applied within the pipette or in the bath). The action of sympathomimetics has been linked to cAMP metabolism, so it was natural to look for the effects of cyclic nucleotides on channel kinetics. In cell-attached patches of tissue-cultured heart cells, bath applied bromo-cAMP increased the calcium channel opening rate within minutes (24). This effect does not appear to be dependent on intracellular Ca^{2+} levels, because intracellular dialysis with solutions containing EGTA did not produce inhibition. The single channel conductance and the number of channels in the patch did not change with bromo-cAMP application.

Calcium channels are under some additional form of metabolic control, since extensive dialysis or excision results in loss of activity; the loss for excised patches occurs within 10–20 sec (49). The ability to manipulate the chemical environment on both sides of excised patches should be of great benefit in revealing the basis for this metabolic control.

Channels Activated by Intracellular Ligands

CALCIUM-ACTIVATED POTASSIUM CHANNELS (Ca–K) The Ca–K channel is in many ways an ideal single-channel preparation. It is highly selective for potassium ($P_K/P_{Na} > 250$; 60). It has a conductance greater than 300 pS in isotonic KCl (17, 86), the largest of any channel observed in vivo, with the exception of the 430-pS chloride channel reported by Blatz & Magelby (20). The channel kinetics are strongly influenced by both membrane potential and the calcium activity on the cytoplasmic face of the membrane (17, 95, 101, 145). The channel is extremely stable, with successful recordings from excised patches lasting for days, even in the presence of 2% glutaraldehyde (J. Hidalgo, unpublished observations)! As well as being stable, the channel has many transition rates in the 1-kHz range, so that thousands of events may be easily collected (17).

The channel appears in many different preparations: tissue-cultured muscle cells (60, 117), chromaffin cells (95), clonal pituitary cells (145), neuroblastoma cells (121), T-tubules of rabbit muscle (103), sympathetic neurons (5), mouse parotic acinar cells (97), and snail neurons (93). With the exception of the last preparation, which has a relatively low conductance (19 pS), channels from the different sources have very much the same properties. The data of Moczydlowski & Latorre (103) and Vergara & Latorre (144) on the reconstituted T-tubule channel are the most detailed of the published results and, in view of the similarity to in vivo data, will serve here as a basis for discussion of the kinetics.

All the published data is consistent in requiring two calcium-binding sites. Moczydlowski & Latorre (103) present evidence that there are two open states that correspond to mono- and bi-liganded forms of the channel. The data are also consistent in requiring three or more closed states. Moczydlowski & Latorre (103) propose the following model:

$$
\begin{array}{lcl}
\mathbf{C} & & \\
\updownarrow & & \\
\mathbf{C}\text{–}Ca^{2+} & \leftrightarrow & \mathbf{O}\text{–}Ca^{2+} \\
\updownarrow & & \updownarrow \\
\mathbf{C}\text{–}Ca_2^{2+} & \leftrightarrow & \mathbf{O}\text{–}Ca_2^{2+}
\end{array}
$$

where **C** represents the closed channel and **O** the open channel. In the above scheme, the Ca^{2+} binding rates are voltage-dependent and the open-closed transitions rates are voltage-independent. Wong et al (145) have also proposed that the channel has voltage-dependent Ca^{2+}-binding rate constants.

There is some disagreement about the channel behavior at high calcium concentrations. Marty (95) reports that at 1 mM Ca^{2+}, the channel conductance is reduced, whereas Barrett et al (17) find that the channel conductance is independent of the calcium concentration up to 1 mM Ca^{2+}. Vergara & Latorre (144) show that, in the T-tubule-derived channel, Ca^{2+} and Ba^{2+} can both produce a long-lasting block of the channel. The time of block is too long to be confused with a drop in conductance; however, Vergara & Latorre also show that 10 mM Ca^{2+} produces an apparent drop in open channel current. They do not analyze the origin of the reduced conductance, but it may result from a competition between Ca^{2+} and K^+ in the wide antrum of the channel preceding the selectivity site (103), or from reductions in negative surface charge that may reduce the local K^+ activity.

CALCIUM-ACTIVATED, NONSELECTIVE CATION CHANNELS (Ca–N) Ca–N channels have been observed in tissue-cultured heart (32), neuroblastoma (147), and pancreatic acinar cells (98). The channels show bursting kinetics with burst durations measured in seconds. The kinetics of this channel have not been analyzed in detail, so that the analogy between the gating mechanism of the Ca–N channel and the Ca–K channel is unknown.

PROTON-ACTIVATED CHLORIDE CHANNELS A Cl^--selective channel from the noninnervated face of *Torpedo* electroplax cells consists of two identical, independently gated pores (dubbed "protochannels"; 102, 141). In reconstituted systems, the probability of finding a protochannel in the open state is voltage dependent and decreases with membrane hyperpolarization (*trans* side ground). Lowering the pH on the cis side of the bilayer increases the opening rate and decreases the closing rate of each protochannel (64). The voltage required to produce a half-saturation in the open time shifts ~60 mV with a unit change in pH. At a membrane potential of −100 mV, a drop in pH from 7.5 to 7.0 causes about a twofold increase in the probability of finding an open protochannel.

The effect of protons on the protochannel gate can be described by the reaction:

$$\begin{array}{ccc} \mathbf{C}+H^+ & \rightleftharpoons & \mathbf{O}+H^+ \\ \updownarrow & & \updownarrow \\ \mathbf{CH} & \rightleftharpoons & \mathbf{OH} \end{array}$$

where **C** and **O** represent the closed and open protochannel. The pH dependence of the probability of protochannel activation indicates that a single site on the protochannel binds H^+. The binding is not voltage sensitive, which indicates that the site is outside of the electric field of the membrane. The acid dissociation constant for this site decreases by three orders of magnitude (from $\sim 10^{-6}$ to $\sim 10^{-9}$ M) when the protochannel gate opens.

CHANNELS ACTIVATED BY SECOND MESSENGERS Nicotinic (62) or muscarinic (129) channels in cell-attached patches cannot be activated by applying ACh outside the patch. This not only indicates that the pipette/membrane seal acts as a barrier to diffusion but also that the receptor and channel moeties underlying the response are spatially linked. In other systems, stimulation of the cell surface adjacent to a gigaseal patch influences the activity of channels in the patch. In these cases, a diffusible, intracellular molecule (second messenger) mediates the response of the channel.

In certain *Aplysia* neurons, serotinin causes a reduction in a specific K^+ conductance (81). Patch clamp experiments of Siegelbaum et al (133) revealed a K^+-selective channel whose opening rate decreased when the bath contained 30 μM serotinin. There was no detectable change in channel selectivity, conductance, or open state lifetime. When the serotinin was washed out, channel activity returned. Several minutes were required for inactivation and reactivation. Ca^{2+} did not seem to be the intracellular mediator, since injection of EGTA into cells had no effect on channel kinetics, and exposure of inside-out patches to Ca^{2+}-free solutions did not alter channel gating. Intracellular injection of cAMP, however, caused a decrease in channel activity, suggesting that cAMP-dependent protein phosphorylation was responsible for the change in the opening rate.

The serotonin-sensitive channel (S channel) showed outward rectification, with chord conductances of 25, 55, and 100 pS at membrane potentials of -40, 0, and 20 mV (360 mM K^+ inside the pipette, 23°C). The probability of being open varied from 0.1 to 0.5 and was only weakly voltage dependent. There could be as many as five channels active in a patch. S channels, like glutamate-activated channels in locust muscle (119), shift spontaneously between periods of high and low activity. When more than one channel is active in the patch, the channels shift between high and low duty-cycle states independently. Serotonin closes all channels regardless of whether they are in the high or low duty-cycle mode.

Maruyama & Petersen (99) found that in cell-attached patches on mouse acinar cells, application of either ACh or the peptide hormone cholecystokinin to the surrounding membranes opened a Ca^{2+}-activated, cation-selective channel (Ca–N channel) of 35 pS. Agonists were applied by localized iontophoresis or diffusion from a micropipette, but activation did not occur for 10–40 sec. The lag between agonist application and channel activation is sufficiently long to allow for a complex sequence of metabolic events. When 1.5 mM cGMP was present in the bath, cholecystokinin failed to produce channel activation, although subsequent formation of inside-out patches showed that the channel could still be activated by Ca^{2+}. Ca^{2+} is an obvious candidate for the second messenger, but its source has not been identified.

Bacigalupo & Lisman (16) recorded single channel activity from the light-sensitive portion of *Limulus* photoreceptors. In ~5% of all patches, upon exposure to light, there was an increase in the activity of a 35-pS channel. The channel had a reversal potential near zero and a mean open time of 1–4 msec (20°C). The first latency for light-stimulated single channel activity lagged behind the macroscopic generator current by hundreds of milliseconds. Hence, it is doubtful that the single channel currents are responsible for generating the early light-response. The single channel activity (Ca–N channel?) may reflect delayed, light-induced changes in photoreceptor metabolism.

Anion-Selective Channels

Three anion-selective channels have been studied with the patch clamp technique: a GABA-activated channel in tissue-cultured mouse spinal neurons (76; discussed above in section on transmitter-activated channels), a pH-sensitive, double-barrelled channel in electroplax membranes (64, 102; discussed above in section on channels activated by intracellular ligands, and below under multiple conducting states), and a very high conductance voltage-activated channel found in tissue-cultured rat muscle (20). This last channel has a conductance of 430 pS in symmetrical 140 mM KCl solutions. The channel tends to be open at potentials near 0 mV and closed at potentials of either polarity. Channels entered a long-lived, inactivated state, but the inactivation could generally be removed by holding the potential at 0 mV for several hundred msec. The channel exhibited a variety of complex, voltage-dependent gating transitions. This channel is similar to the anion-selective channel found in mitochondrial outer membranes (27), and since it was present in only 5–10% of all patches, could conceivably be a mitochondrial escapee!

Stretch-Activated Channels

Guharay & Sachs (60) discovered a channel in tissue-cultured chick skeletal muscle that responded to membrane stretch produced by suction applied to the patch pipette. In 140 mM K^+ saline, the channel had a conductance of 70 pS with a P_K/P_{Na} of 4.0. The channel kinetics indicated one open and three closed states. All rates were voltage insensitive, and only the slowest opening rate was stretch sensitive. Cytochalasins increased the stretch sensitivity by a factor of thirty, probably by disrupting actin filaments that support most of the membrane tension. The probability of finding the channel open increased with increasing extracellular concentrations of K^+. The origin of this K^+ sensitivity is unknown. From an analysis of the stretch sensitivity, it is clear that the channel is able to gather its energy for gating from an area larger than 600 Å in diameter (perhaps using cytoskeletal components that are insensitive to cytochalasin).

Multiple Conductance States

Like many channels in artificial bilayers (22, 27; see 85), ion channels in biological membranes can adopt more than one conducting conformation. Some of these conformations are long lived and appear as rectangular current pulses whose mean current is different from the fully open channel, whereas others only appear as fluctuations in the mean current.

SUBCONDUCTANCE STATES OF NICOTINIC CHANNELS Hamill & Sakmann (63) found that at 8°C, nicotinic channels from tissue-cultured embryonic rat muscle had one of three conductances: 35 pS, 25 pS, and 10 pS. Based upon the probability of overlap, the 35- and 25-pS events were independent channels, but the 10-pS events occurred mainly as transitions from the other two levels and thus represented another conducting state of the channel. In one record (outside-out patch), the probability of being in the 35-pS, 25-pS, or 10-pS state was 0.15, 0.76, and 0.09. All three states had similar selectivity properties. The reversal potentials were similar, and the currents from each level remained in proportion when Na^+ was replaced with Li^+ or Cs^+. The conductance properties of the substate seemed similar to that of the main state, since both were weakly dependent on temperature (63, 142) and independent of agonist (14).

Trautmann (142) reported that in tissue-cultured rat muscle at 23°C, nicotinic channels activated by curare had conductances of 35 or 50 pS with a subconductance state of 10–13 pS. Tissue-cultured chick muscle at 22°C has main conductance states similar to those of Trautmann and a subconductance state of 5 pS (13, 14). Subconductance states are also present in the muscle-derived clonal cell-line L6 and in nicotinic channels reconstituted from *Torpedo* electroplax (D. Tank, personal communication).

Subconductance states may appear with variable frequency because the probability of adopting a substate is low, the duration of the substate is short, or the amplitude of the substate current is small. Therefore, it is not certain whether substates are a universal feature of nicotinic channels.

In contrast to the conductance, the lifetime of the substate depends upon both the temperature and the agonist. In tissue-cultured rat muscle, the substate lifetime was 24 msec at 8°C and <5 msec at 18°C (1–2 μM ACh; 63). In tissue-cultured chick muscle at room temperature, the substate lifetime was ~0.6 msec with ACh and ~1.1 msec with SubCh (14). Trautmann (142) observed substates with curare as an agonist but not with ACh. However, using the same preparation as Trautmann, Morris et al (104a) observed no substates with curare. Auerbach & Sachs (14) found that while the probability of entering the substate was variable, it was similar for channels activated by ACh and SubCh.

The simplest interpretation of a change in conductance with no change in selectivity is that a well in the energy profile for ion permeation is transiently deepened (65). This hypothesis makes two testable predictions. First, ions should bind more tightly to the subconductance conformation of the channel than to the main conductance conformation. The conductance-concentration curve for the substate should then saturate at lower ion concentrations than for the main state. Second, the rate of channel block by local anesthetics or other ligands should be the same for the main and subconductance states. While neither of these predictions has been explicitly tested, 50 μM curare appears to block the main conductance state, but not the substate (142). This suggests that the simple deeping of a well may not be sufficient to explain the subconductance state.

There is one report that transitions between the substate and the main state are not in equilibrium. Hamill & Sakmann (63) observed that sojourns in the substate followed, but did not precede, sojourns in the main state. In one experiment of 809 trials, there were 39 examples of transitions of 25–10–0 pS, but none of the reverse sequence. If we assume that channels that close through substates can reopen, then energy must be delivered to the system to maintain channel activity.

In contrast to Hamill & Sakmann's results, experiments by others at room temperature show no significant deviation from equilibrium. Trautmann (142) and Auerbach & Sachs (14) found that the probability of closed-sub-main and main-sub-closed transitions was equal within a factor of two. The results of Hamill & Sakmann (63) need to be repeated in order to verify the lack of equilibrium.

OPEN CHANNEL NOISE One of the more curious aspects of channel behavior revealed by the patch clamp is that open channel currents are noisier than the baseline. Excess white noise is expected from the increased conductance and the shot noise of ion transport itself. However, power spectra of the "excess" noise associated with the open nicotinic channels show that the noise consists of a Lorentzian component with a cutoff (3 db) frequency of about 300 Hz plus a white noise component (134). The cutoff frequency of the slow component of the noise is remarkably temperature sensitive but is not sensitive to the membrane potential. The amplitude distribution of the noise is not consistent with band-limited open-closed transitions, so that more complex kinetic models are necessary. The conformations associated with the Lorentzian component of the open channel noise are not correlated with gating transitions (F. Sigworth, personal communication).

The current during the subconductance state shows excess noise similar to that of open channel currents; however, the amplitudes of the fluctuations are larger (14). It seems reasonable to suppose that if small conformational changes can modulate the energy barriers for ion transport, then the low conductance states, which have higher barriers, will be more susceptible to modulation.

OTHER CHANNELS WITH SUBCONDUCTANCE STATES Subconductance states have been reported in Ca^{2+}-activated K^+ channels (103, 117), S channels of *Aplysia* (133), and potassium channels in HeLa cells (131).

A DOUBLE-BARRELLED CHLORIDE CHANNEL Cl^--selective channels reconstituted from *Torpedo* electroplax display a remarkable set of conductance states. On a time scale of seconds, the channel's kinetics can be modelled by a simple open-closed transition. Each "open" period is actually a burst of relatively rapid transitions (10–100 msec) between states having conductances of 0, 10, or 20 pS (102, 141). The fraction of time spent at each conductance state within a burst follows a binomial distribution, indicating that the transitions arise from the independent gating of two conducting pathways. Hence, the channel is composed of two identical (10 pS) barrels (dubbed protochannels), each with its own gate. The pair of protochannels is gated as a unit by a slow process:

$$\begin{array}{ccccccc} & & \mathrm{C_{co}} & \leftrightarrow & \mathrm{C_{oo}} & & \\ \mathrm{C_{cc}} & \leftrightarrow & & \mathrm{C_{oc}} & & & \\ & & \mathrm{O_{co}} & \leftrightarrow & \mathrm{O_{oo}} & & \\ \mathrm{O_{cc}} & \leftrightarrow & & \mathrm{O_{oc}} & & & \end{array}$$

O and **C** represent the open and closed state of the slow gate, and o and c represent the open and closed state of each protochannel gate. The **O**oo form of the channel has a conductance of 20 pS, the **O**co and **O**oc forms have a conductance of 10 pS, and all other forms are nonconducting. The three gates act completely independently. Bursts can start from either the 10-pS or 20-pS level, and the voltage dependence of the slow gating process is opposite that of the protochannel gating process.

CONCLUDING REMARKS

New information concerning ionic channels is rapidly becoming available, not only because of the great sensitivity of the patch clamp technique but also because it is now possible to patch clamp small cells that were previously inaccessible. Although we have concentrated upon single channel results, the patch clamp has much wider applicability. Marty & Neher (96), for example, have measured the capacitance, and hence area, of exocytotic vesicles of chromaffin cells. Almers et al (9) have measured the variation in sodium channel density across the surface of a muscle fiber and, by photobleaching the patch, have placed an upper limit on the diffusion constant of sodium channels in frog muscle membranes.

Because the patch clamp is so sensitive, we can see details of channel behavior that were never before visible. Consequently, models describing channel kinetics have become increasingly complex in order to accommodate a mushrooming number of states. Some of this complexity may be due to the fact that individual channels can differ from each other even though they represent the same gene product. Detailing all the observable properties of a channel may yield information on the structure and environment of the channel but may lead to kinetic models that are too complex to be useful. In the final analysis, our goal is not description, but understanding.

ACKNOWLEDGMENTS

We would like to thank the many people who have shared unpublished results and helped us through unfamiliar territory. In particular, we thank W. Almers, D. Colquhoun, L. DeFelice, R. DeHann, V. Dionne, F. Guharay, J. Hidalgo, R. Horn, H. Lecar, M. Leibowitz, C. Lewis, M. Montal, J. Rae, F. Sigworth, S. Sine, J. H. Steinbach, C. Stevens, and D. Tank. This work was supported by grant NS-13094 to FS from the USPHS.

Literature Cited

1. Adams, P. R. 1976. *J. Physiol. London* 260:532–52
2. Adams, P. R. 1977. *J. Physiol. London* 268:271–89
3. Adams, P. R. 1981. *J. Membr. Biol.* 58:161–74
4. Adams, P. R., Brown, D. A. 1982. *J. Physiol. London* 332:263–72
5. Adams, P. R., Constanti, A., Brown, D. A., Clark, R. B. 1982. *Nature* 296:746–49
6. Akaike, N., Fishman, H. M., Lee, K. S., Moore, L. C., Brown, A. M. 1978. *Nature* 274:379–82
7. Aldrich, R. W., Corey, D. P., Stevens, C. F. 1983. *Nature* 306:436–41
8. Aldrich, R. W., Stevens, C. F. 1983. *Cold Spring Harbor Symp.* 48: In press
9. Almers, W., Stanfield, P. R., Stuhmer, W. 1983. *J. Physiol. London* 336:261–84
10. Anderson, O. A. 1983. *Biophys. J.* 41:119–34
11. Auerbach, A. 1983. *Soc. Neurosci. Abstr.* 9:1137

12. Auerbach, A., del Castillo, J., Specht, P., Titmus, M. 1983. *J. Physiol. London* 343:551–68
13. Auerbach, A., Sachs, F. 1983. *Biophys. J.* 42:1–11
14. Auerbach, A., Sachs, F. 1984. *Biophys. J.* 45:187–98
15. Ayer, R. K. Jr., DeHaan, R. G., Fischmeister, R. 1983. *J. Physiol. London* 34:37P
16. Bacigalupo, J., Lisman, J. E. 1983. *Nature* 304:268–70
17. Barrett, J. N., Magleby, K. L., Pallotta, B. S. 1982. *J. Physiol. London* 331:211–30
18. Baylor, D. A., Lamb, T. D., Yau, K. W. 1979. *J. Physiol. London* 288:589–611
19. Benham, C. D., Bolton, T. B. 1983. *J. Physiol. London* 340:469–86
20. Blatz, A. L., Magleby, K. L. 1983. *Biophys. J.* 237–41
21. Brown, A. M., Camerer, H., Kunze, D. L., Lux, H. D. 1982. *Nature* 299:156–58
22. Busath, D., Szabo, G. 1981. *Nature* 294:371–73
23. Cachelin, A. B., dePeyer, J. E., Kokubun, S., Reuter, H. 1983. *J. Physiol. London* 340:389–401
24. Cachelin, A. B., dePeyer, J. E., Kokubun, S., Reuter, H. 1983. *Nature* 304:462–64
25. Clapham, D., DeFelice, L. 1984. *Biophys. J.* 45:40–42
26. Cole, K. C. 1968. *Membranes, Ions and Impulses.* Berkeley: Univ. Calif. Press
27. Colombini, M. 1979. *Nature* 279:643–45
28. Colquhoun, D., Dreyer, F., Sheridan, R. J. 1979. *J. Physiol. London* 293:247–84
29. Colquhoun, D., Hawkes, A. G. 1977. *Proc. R. Soc. London Ser. B* 199:231–62
30. Colquhoun, D., Hawkes, A. G. 1981. *Proc. R. Soc. London Ser. B* 211:205–35
31. Colquhoun, D., Hawkes, A. G. 1982. *Proc. R. Soc. London Ser. B* 300:1–59
32. Colquhoun, D., Neher, E., Reuter, H., Stevens, C. F. 1981. *Nature* 294:752–54
33. Colquhoun, D., Sakmann, B. 1981. *Nature* 294:464–66
34. Colquhoun, D., Sakmann, B. 1983. See Ref. 128, pp. 345–64
35. Colquhoun, D., Sigworth, F. J. 1983. See Ref. 128, pp. 191–263
36. Connor, E. A., Levy, S. M., Parsons, R. 1983. *J. Physiol. London* 337:137–58
37. Conti, F., Neher, E. 1980. *Nature* 285:140–43
38. Coronado, R., Latorre, R. 1983. *Biophys. J.* 43:231–36
39. Cull-Candy, S. G., Miledi, R., Parker, I. 1981. *J. Physiol. London* 321:195–210
40. Cull-Candy, S. G., Parker, I. 1982. *Nature* 294:464–66
41. Detwiler, P. B., Connors, J. D., Bodoia, R. D. 1982. *Nature* 300:59–61
42. Dionne, V. E., Leibowitz, M. D. 1982. *Biophys. J.* 39:253–61
43. Dionne, V. E., Parsons, R. 1981. *J. Physiol. London* 310:145–58
44. Dionne, V. E., Stevens, C. F. 1975. *J. Physiol. London* 251:245–70
45. Dwyer, T. 1981. *Biochim. Biophys. Acta* 646:51–60
46. Ehrenstein, G., Lecar, H., Nossal, R. 1970. *J. Gen. Physiol.* 55:119–33
47. Ewald, D. A. 1976. *J. Membr. Biol.* 29:47–65
48. Fenwick, E. M., Marty, A., Neher, E. 1982. *J. Physiol. London* 331:577–97
49. Fenwick, E. M., Marty, A., Neher, E. 1982. *J. Physiol. London* 331:599–635
50. Fischmiester, R., DeFelice, L. J., Ayer, R. K., Levi, R., DeHaan, R. L. 1983. Submitted for publication
51. Forda, S. R., Jessell, T. M., Kelly, J. S., Rand, R. P. 1982. *Brain Res.* 249:371–78
52. Fukushima, Y. 1981. *Proc. Natl. Acad. Sci. USA* 79:1274–77
53. Fukushima, Y. 1981. *Nature* 294:368–70
54. Fukushima, Y. 1982. *J. Physiol. London* 331:311–31
55. Gage, P., Hamill, O. P., Wachtel, R. E. 1983. *J. Physiol. London* 335:123–37
56. Gage, P., McBurney, R. 1975. *J. Physiol. London* 244:385–408
57. Giacoletto, L. J. 1977. In *Electronic Designers Handbook*, ed. L. J. Giacoletto, pp. 2-48–2-57. New York: McGraw-Hill
58. Gration, K. A. F., Lambert, J. J., Ramsey, R., Usherwood, P. N. R. 1981. *Nature* 291:423–25
59. Guharay, F. 1982. *Pharmacology and physiology of murine NIE-115 neuroblastoma cells.* PhD thesis. Univ Nottingham, England
60. Guharay, F., Sachs, F. 1984. *J. Physiol. London.* In press
61. Hagiwara, S., Ohmori, H. 1983. *J. Physiol. London* 336:649–61
62. Hamill, O. P., Marty, A., Neher, E., Sakmann, B., Sigworth, F. J. 1981. *Pflügers Arch. Eur. J. Physiol.* 391:85–100
63. Hamill, O. P., Sakmann, B. 1981. *Nature* 294:462–64
64. Hanke, W., Miller, C. 1983. *J. Gen. Physiol.* 82:25–45
65. Hille, B. 1975. In *Membranes*, ed. G. Eisenman, 3:255–323. New York: Dekker
66. Hille, B., Schwarz, W. 1978. *J. Gen. Physiol.* 72:409–42
67. Hodgkin, A. L., Huxley, A. F. 1952. *J. Physiol. London* 117:500–44

68. Horn, R., Brodwick, M. 1980. *J. Gen. Physiol.* 75:297–321
69. Horn, R., Lange, K. 1983. *Biophys. J.* 43:207–23
70. Horn, R., Patlak, J. 1980. *Proc. Natl. Acad. Sci. USA* 77:6930–34
71. Horn, R., Patlak, J., Stevens, C. F. 1981. *Nature* 291:426–27
72. Horn, R., Patlak, J., Stevens, C. F. 1981. *Biophys. J.* 36:321–27
73. Horn, R., Vandenberg, C. A., Lange, K. 1983. *Biophys. J.* 45:323–35
74. Huganir, R. L., Greengard, P. 1983. *Biophys. J.* 41:136a
75. Jackson, M. B., Lecar, H., Askanas, V., Engel, W. K. 1982. *J. Neurosci.* 2:1465–73
76. Jackson, M. B., Lecar, H., Mathers, D. A., Barker, J. L. 1982. *J. Neurosci.* 2:889–94
77. Jackson, M. B., Lecar, H., Morris, C. E., Wong, B. S. 1983. In *Current Methods in Cellular Neurobiology*, ed. J. L. Borden, J. F. McKelvy, pp. 61–99. New York: Wiley
78. Jackson, M. B., Wong, B. S., Morris, C. E., Lecar, H., Christian, C. N. 1983. *Biophys. J.* 42:109–14
79. Kettenmann, H., Okland, R. K., Lux, H. D., Schachner, M. 1982. *Neurosci. Lett.* 32:41–46
80. Kistler, J., Stroud, R. M., Klymkovsky, M. N., Lalancette, R. A., Fairclough, R. H. 1982. *Biophys. J.* 37:371–83
81. Klein, M., Kandel, E. 1978. *Proc. Natl. Acad. Sci. USA* 75:3512–16
82. Kolb, H., Wakelam, M. J. O. 1983. *Nature* 303:621–23
83. Land, B. R., Salpeter, E. E., Salpeter, M. M. 1980. *Proc. Natl. Acad. Sci. USA* 77:3736–40
84. Land, B. R., Salpeter, E. E., Salpeter, M. M. 1981. *Proc. Natl. Acad. Sci. USA* 78:7200–4
85. Latorre, R., Alvarez, O. 1981. *Physiol. Rev.* 61:78–150
86. Latorre, R., Miller, C. 1983. *J. Membr. Biol.* 71:11–30
87. Lecar, H., Morris, C. E., Wong, B. 1983. In *Structure and Function in Excitable Cells*, ed. D. Chang, I. Tasaki, W. Adelman, H. R. Leuchtag, pp. 159–73. New York: Plenum
88. Leibowitz, M. N., Dionne, V. E. 1984. *Biophys. J.* 45:153–63
89. Lingle, C. 1983. *J. Physiol. London* 339:395–418
90. Llano, I., Bezanilla, F. 1983. *Biophys. J.* 41:38a
91. Loosley-Millman, M. E., Rand, R. P., Parsegian, V. A. 1982. *Biophys. J.* 40:221–32
92. Lux, H. D., Nagy, K. 1981. *Pflügers Arch. Eur. J. Physiol.* 391:252–54
93. Lux, H. D., Neher, E., Marty, A. 1981. *Pflügers Arch. Eur. J. Physiol.* 389:293–95
94. Magleby, K. L., Stevens, C. F. 1972. *J. Physiol. London* 223:173–97
95. Marty, A. 1981. *Nature* 291:497–500
96. Marty, A., Neher, E. 1982. 79:6712–16
97. Maruyama, Y., Gallacher, D. V., Petersen, O. H. 1983. *Nature* 302:827–29
98. Maruyama, Y., Petersen, O. H. 1982. *Nature* 299:159–61
99. Maruyama, Y., Petersen, O. H. 1982. *Nature* 300:61–63
100. Mathers, D., Usherwood, P. N. R. 1976. *Nature* 259:409–11
101. Methfessel, C., Boheim, C. 1982. *Biophys. Struct. Mech.* 9:35–60
102. Miller, C. 1982. *Philos. Trans. R. Soc. London Ser. B* 299:401–11
103. Moczydlowski, E., Latorre, R. 1983. *J. Gen. Physiol.* 82:511–42
104. Montal, M., Labarca, P., Fredkin, D. R., Suarez-Isla, B. A., Lindstrom, J. M. 1984. *Biophys. J.* 45:165–74
104a. Morris, C. E., Wong, B. S., Jackson, M. B., Lecar, H. 1983. *J. Neurosci.* 3:2525–31
105. Neher, E. 1982. In *Techniques in Cellular Physiology*, ed. P. F. Baker, 112:1–16. Amsterdam: Elsevier
106. Neher, E. 1983. *J. Physiol. London* 339:663–78
107. Neher, E., Sakmann, B. 1975. *Proc. Natl. Acad. Sci. USA* 72:2140–43
108. Neher, E., Sakmann, B. 1976. *Nature* 260:799–802
109. Neher, E., Sakmann, B. 1976. *J. Physiol. London* 258:705–30
110. Neher, E., Sakmann, B., Steinbach, J. H. 1978. *Pflügers Arch. Eur. J. Physiol.* 375:219–28
111. Neher, E., Steinbach, J. H. 1978. *J. Physiol. London* 277:153–76
112. Nelson, D. J., Sachs, F. 1979. *Nature* 282:861–63
113. Ogden, D. C., Siegelbaum, S. A., Colquhoun, D. 1981. *Nature* 289:596–99
114. Ohmori, H. 1978. *J. Physiol. London* 281:77–99
115. Ohmori, H., Yoshida, S., Hagiwara, S. 1981. *Proc. Natl. Acad. Sci. USA* 78:4960–64
116. Oxford, G. S. 1981. *J. Gen. Physiol.* 77:1–22
117. Pallotta, B. S., Magleby, K. L., Barrett, J. N. 1981. *Nature* 293:471–74
118. Patlak, J. 1983. In *Physiology of Excitable Cells*, ed. A. Grinnel, W. Moody. New York: A. Lis
119. Patlak, J. B., Gration, K. A. F.,

Usherwood, P. N. R. 1979. *Nature* 278:643–45
120. Patlak, J., Horn, R. 1982. *J. Gen. Physiol.* 79:333–51
121. Quandt, F. N., Narahashi, T. 1982. *Proc. Natl. Acad. Sci. USA* 79:6732–36
122. Redmann, G. A. 1984. Submitted
123. Reuter, H., Stevens, C. F., Tsien, R. W., Yellen, G. 1982. *Nature* 297:501–4
124. Sachs, F. 1983. See Ref. 128, pp. 265–85
125. Sachs, F., Auerbach, A. 1983. In *Methods in Enzymology: Neuroendocrine Peptides*, ed. M. Conn, pp. 147–76. New York: Academic
126. Sachs, F., Neil, J., Barkakati, N. 1982. *Pflügers Arch. Eur. J. Physiol.* 395:331–40
127. Sakmann, B., Adams, P. 1978. In *Advances in Pharmacology and Therapeutics*, ed. J. Jacob, pp. 81–90. New York: Pergamon
128. Sakmann, B., Neher, E. 1983. In *Single Channel Recording*, ed. B. Sakmann, E. Neher, pp. 37–51. New York: Plenum
129. Sakmann, B., Noma, A., Trautwein, W. 1983. *Nature* 303:250–53
130. Sakmann, B., Patlak, J., Neher, E. 1980. *Nature* 286:71–73
131. Sauve, R., Roy, G., Payet, D. 1983. *J. Membrane Biol.* 74:41–49
132. Sheridan, R. E., Lester, H. 1977. *J. Gen. Physiol.* 70:187–219
133. Sieglebaum, S., Camardo, J. S., Kandel, E. R. 1982. *Nature* 299:413–17
134. Sigworth, F. J. 1982. *Biophys. J.* 37:309a
135. Sigworth, F. J., Neher, E. 1980. *Nature* 297:447–49
136. Silinsky, E. M. 1975. *J. Physiol. London* 247:145–62
137. Sine, S. M., Steinbach, J. H. 1984. *Biophys. J.* 45:175–85
138. Smith, T. G., Futamachi, K., Ehrenstein, G. 1982. *Brain Res.* 242:184–89
139. Steinbach, J. H. 1980. *Cell Surf. Rev.* 6:120–57
140. Suarez, B. A., Wan, K., Lindstrom, J., Montal, M. 1983. *Biochemistry* 22:2319–23
141. Tank, D., Miller, C., Webb, W. W. 1982. *Proc. Natl. Acad. Sci. USA* 79:7749–53
142. Trautmann, A. 1982. *Nature* 272–75
143. Trautmann, A., Sieglebaum, S. 1983. See Ref. 128, pp. 478–80
144. Vergara, C., Latorre, R. 1983. *J. Gen. Physiol.* 82:543–68
145. Wong, B. S., Lecar, H., Adler, M. 1982. *Biophys. J.* 39:313–17
146. Yamamoto, D., Yeh, J. Z., Narahashi, T. 1984. *Biophys. J.* 45:337–44
147. Yellen, G. 1982. *Nature* 296:357–59
148. Young, S. H., Poo, M. 1983. *Nature* 304:161–63

Ann. Rev. Biophys. Bioeng. 1984. 13: 303–30

IMMUNOELECTRON MICROSCOPY OF RIBOSOMES

Georg Stöffler and Marina Stöffler-Meilicke

Max-Planck-Institut für Molekulare Genetik, Abt. Wittmann, Ihnestrasse 63-73, D-1000 Berlin-3 (Dahlem), Germany

This article is dedicated to Prof. Dr. Herbert Braunsteiner, Innsbruck, on the occasion of his 60th birthday.

Introduction, Summary, and Perspectives

Immunoelectron microscopic techniques are applicable to many types of biological macromolecular assemblies and have attracted increasing attention in recent years. Immunoelectron microscopy (IEM) uses antibodies (IgG) as specific markers for the localization of components on the surface of multimeric macromolecular complexes. Basically, antisera are raised against a particular component of the complex. The whole complex is then incubated with antibodies, and the binding site of the antibody marks the position of the component in question.

The negative staining technique introduced by Brenner & Horne (6), together with the improved resolving power of present-day microscopes, first made possible the visualization of antibody molecules (93). Subsequently, antibody molecules have been visualized attached to virus surfaces (23, 109) and to ribosomes (98, 99). Because of their characteristic "Y" shape, the bound antibodies can be resolved without additional electron-dense markers. Asymmetric macromolecules (e.g. ribosomes) are better suited for this topographical technique than symmetric ones, because their distinctive structural features provide points of orientation and thus facilitate the localization of the antibody binding site. The resolution of the technique is about 3.5 nm; the limiting factor is the dimensions of the IgG molecule and in particular the width of the Fab-arm. IEM has been used extensively for elucidation of the structure of the *Escherichia coli* ribosome: antibodies specific for ribosomal proteins, nucleotides, or haptens have

0084–6589/84/0615–0303$02.00

been used to elucidate the spatial arrangement of the ribosomal proteins and rRNA within the ribosomal subunits and, recently, monoclonal antibodies against ribosomal proteins have also been employed. Stringent controls showing the specificity of the antibody preparations, absolutely necessary for the IEM approach, have been developed. Double antibody labelling studies have helped to determine the orientation of the subunits within the 70S monosome. Furthermore, the binding sites of ligands, which interact with the ribosome during protein biosynthesis, have been localized by IEM; this has made a considerable contribution to our knowledge of ribosome function. Ribosome structure and function have been studied by a variety of other approaches. It has thus been possible to correlate and compare the data obtained by IEM with results from other techniques.

This review is thus confined to a detailed description of the IEM methods used to study the *E. coli* ribosome, which is the most intensively studied structure to date. A detailed description of the work on this organelle demonstrates the various applications of this technique. The few reports in which IEM has been used to study the structure of other macromolecular complexes are summarized briefly. It is to be expected that in the future IEM will be applied to an increasing range of macromolecular structures.

Background

MOLECULAR COMPOSITION Bacterial ribosomes sediment with a sedimentation coefficient of 70S. When the magnesium ion concentration is lowered, 70S ribosomes dissociate into two subunits of unequal size: the small, 30S subunit and the large, 50S subunit. The 30S subunit consists of one 16S RNA molecule (1542 nucleotides) and 21 proteins numbered S1–S21. The 50S subunit contains two RNA molecules: the small, 5S RNA (120 nucleotides) and the large, 23S RNA (2904 nucleotides), along with 32 proteins designated L1–L34. There are four copies of L7 plus L12 on the ribosome; each of the other proteins obeys a stoichiometry of one or less (22, 84). The rRNA molecules represent two-thirds of the mass of the ribosome. The complete primary structures of the RNA and protein components of *E. coli* ribosomes have been determined [for reviews see (7, 105)].

SHAPE OF RIBOSOMES AND SUBUNITS The first electron micrographs of *E. coli* ribosomes were obtained by Hall & Slayter (21), who used shadow-casting with platinum. Negative contrasting with phosphotungstic acid was introduced by Huxley & Zubay (24) and, later, the single carbon layer technique (94) was frequently employed for the preparation of ribosomal particles. A double carbon layer technique was found to give a more uniform stain distribution and optimal structural preservation of the specimen (98). Tischendorf et al (86, 87) developed a new preparation technique, in which double-layer regions can be produced in a controlled way (71).

70S ribosomes and 30S and 50S ribosomal subunits, negatively stained with uranyl acetate, are shown in Figure 1. The 30S subunits show elongated contours with a one-third/two-thirds partition and appear to be randomly oriented on the carbon film with respect to rotation about their long axis (Figure 1). 50S subunits are observed in two characteristic views: the crown view and the kidney view. In contrast to ribosomal subunits, electron micrographs of 70S ribosomes lack such distinctive structural features (Figure 1). The nomenclature used for the typical structural features of the ribosomal subunits is given in Figure 2.

By assuming that the different views of a given particle are various projections of a unique three-dimensional structure, it has been possible to construct three-dimensional models. A reasonable model should be consistent with the majority of projections observed on the electron micrographs. The three-dimensional models derived by our group for the ribosomal subunits are shown in Figure 5 and that for the 70S monosome in Figure 9. The models derived from electron microscopy are compatible with models obtained by small angle scattering studies (46a, 48a, 63b, 82a); artifacts due to specimen preparation are thus negligible.

Three-dimensional models have also been proposed by other groups. With few exceptions, the ribosome models, after a number of years, have converged and received general acceptance (see 106). An exception is the 30S model of Lake (32), which differs in its extreme flatness [see Figure 5 and (106)]. The 50S model of Lake (32) has received almost general acceptance, and the 70S models of Stöffler (26, 28, 29) and Vasiliev (95) are in reasonable agreement with a model proposed by Lake (32; see Figure 9*b*). Only the 50S and 70S models of Boublik (3, 4) deviate markedly from the general picture [see Figure 9 and (106)]. A more detailed description of the differences between the models has been given elsewhere (39, 106).

ANTISERA AGAINST INDIVIDUAL RIBOSOMAL PROTEINS Precipitating antisera against all 21 30S- and the 32 50S-proteins have been obtained primarily from rabbits and also from sheep (64, 74–76). No immunological cross-reaction among the 21 proteins of the 30S subunits of *E. coli* could be detected by any of a variety of immunochemical methods (75). With the exception of proteins L7 and L12, the proteins of the 50S subunit were also immunologically distinct (64, 74). It was then found that the proteins of the two ribosomal subunits were immunologically distinct except for anti-S20 and anti-L26, which cross-reacted with the mixture of proteins from the other subunit (64). We concluded that, with the exception of the protein pairs L7/L12 and S20/L26, ribosomal proteins generally had distinct primary structures. The protein-specific antisera could thus be used for

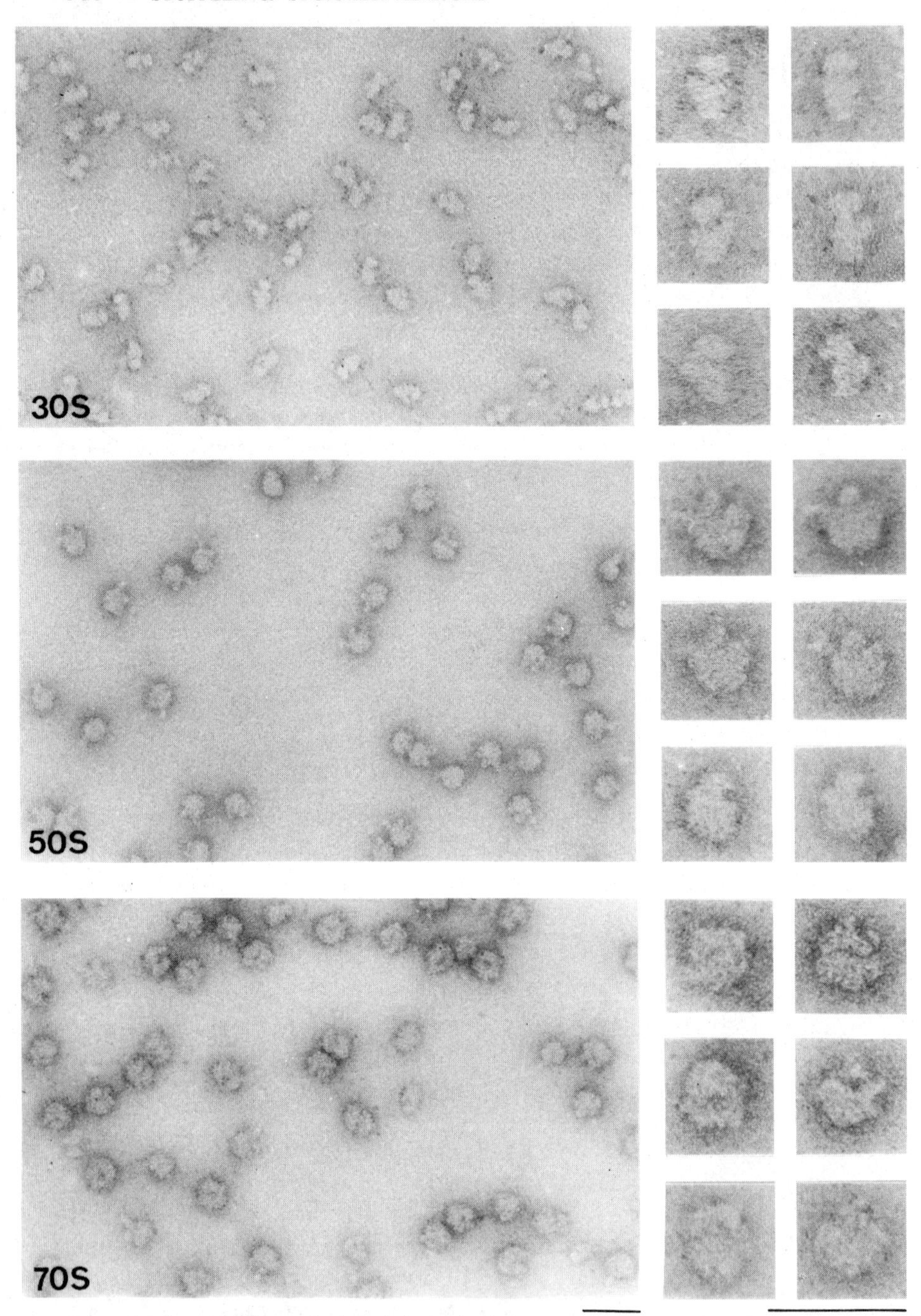

Figure 1 General fields and selected electron micrographs of 30S, 50S, and 70S ribosomal particles, negatively contrasted with uranyl acetate. Bar = 50 nm.

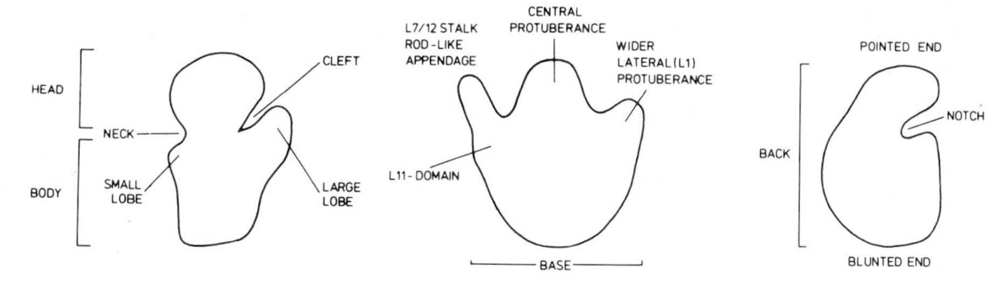

Figure 2 Diagrammatic representation of the 30S subunit, shown in the cloven asymmetric projection (*left*), and the 50S subunit seen in the crown view (*center*) and the kidney view (*right*). The nomenclature of the characteristic features, as used in the text, is given.

IEM. Meanwhile, the amino acid sequences of all 53 ribosomal proteins have been elucidated (107) and have confirmed the immunological results: there are no obvious sequence homologies between the ribosomal proteins; proteins S20 and L26 are identical, and protein L7 is the acetylated form of protein L12 (105).

ANTIGENIC DETERMINANTS ON THE SURFACE OF RIBOSOMAL PARTICLES A prerequisite for the use of antibodies against single ribosomal proteins in many instances is that the protein should have at least one antigenic determinant accessible in the intact ribosomal subunit. In these studies it is important to recognize that a protein antigen, when injected into an animal, gives rise to a population of individual species of antibodies that are specific for various antigenic determinants distributed over the surface of the antigen protein. The majority of these antibody species is reactive towards the isolated antigen, but only a minority will react with the few antigenic determinants that are accessible for antibody binding in intact subunits.

In 1973 we showed, with a variety of different experimental approaches such as ultracentrifugation, gel filtration, double-antibody precipitation, and functional inhibition tests, that ribosomal proteins are accessible for antibody binding *in situ*. Most of the 30S and many of the 50S ribosomal proteins have at least one determinant that is accessible in the intact ribosomal subunit (49, 64, 68, 76). An example of such an accessibility experiment is shown in Figure 3*a–d*. Upon incubation of 50S subunits with IgGs specific for protein L11, dimeric immunocomplexes are formed, and these remain stable during sucrose density gradient centrifugation. At high antibody concentrations, 60–80% of the subunits are dimerized by the antibody (Figure 3*d*). If such a large fraction is reactive with antibodies, it can also be concluded that the determinant to be mapped is present on functionally active ribosomes, since in standard ribosome preparations half of the particle population is functionally active (2a).

The reactivity with 50S subunits on sucrose gradients has been plotted

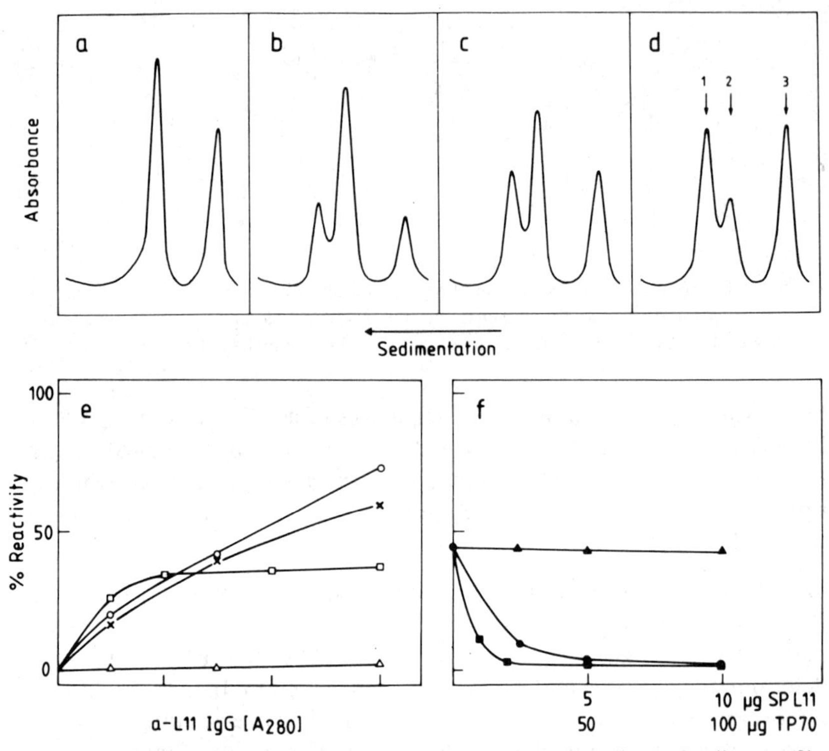

Figure 3 Antibody-ribosome complex formation in sucrose gradients. Gradient profiles obtained with (*a*) 170-fold excess of preimmune IgG; (*b–d*) 30- to 170-fold excess of anti-L11. Peak 1 = 50S · IgG · 50S complexes; peak 2 = 50S subunits; peak 3 = IgG. (*e*) Reactivity of anti-L11 IgG preparations obtained from different animals, plotted against antibody concentration. (*f*) Inhibition of anti-L11 · ribosome complex formation by preincubation of the antibody with increasing amounts of single protein L11 (■——■), total ribosomal proteins (●——●), and total ribosomal proteins that lack protein L11 (▲——▲). Reactivity of 50S subunits was determined by planimetry as described by Stöffler et al (68).

against the antibody concentration for anti-L11 IgG obtained from different animals (Figure 3*e*). Only two out of four IgG preparations dimerize more than 50% of the subunits, and one IgG preparation does not form dimeric immunocomplexes at all. This result illustrates that it is impossible to conclude from negative results that a particular nonreactive protein is completely buried. This general rule, which has been critically considered for ribosomes by Cantor (10), has not always been appreciated by other investigators (102).

Protein Topography in Ribosomal Particles

A key principle of immunoelectron microscopy is that a purified IgG-antibody, specific for a single ribosomal protein, becomes bound to the appropriate ribosomal subunit; the divalent antibody dimerizes two subunits, which can then be examined under the electron microscope. The location of the bound antibody on the subunit surface can be determined, and in this way the position of an antigenic determinant of the particular protein is found. For electron microscopy, the immunocomplexes are separated from unreacted ribosomal subunits and, especially, from unbound IgG molecules by sucrose density gradient centrifugation, and the dimeric immunocomplexes (Figure 3*b–d*, peak 1) are used directly for specimen preparation.

THE 30S SUBUNIT Figure 4 shows examples of electron micrographs of small subunits reacted with protein-specific antibodies. Images of subunits connected by an IgG-molecule are easily recognized, and the location of the antibody attachment site can be described readily in two dimensions. For example, antibodies against protein S13 bind at the head of the small subunit, anti-S11 binds at the large lobe, anti-S5 at the small lobe, and anti-S17 binds near the lower pole (Figure 4).

THE 50S SUBUNIT Altogether, 14 of the 32 ribosomal proteins of the 50S subunit have been mapped in comprehensive studies (13, 79). Figure 4 shows general fields of electron micrographs obtained from 50S subunits that had been incubated with protein-specific antibodies. Since 50S subunits are observed predominantly in the crown view, a two-dimensional localization of the antibody binding site on this projection is easily achieved. For example, anti-L1 binds to the broad lateral protuberance, anti-L18 to the central protuberance, anti-L7/L12 to the rod-like appendage, and anti-L23 to the base (Figure 4).

THREE-DIMENSIONAL LOCALIZATION In order to determine the three-dimensional location of a given protein, it is necessary to locate the antibody attachment site on subunits seen in different orientations. Thus on 50S subunits, antibody binding to the crown and kidney projection has to be determined. 50S·IgG·50S complexes, in which a crown view is combined through one and the same IgG molecule with a kidney projection, demonstrate the identity of the epitopes on the two views, thus allowing an unambiguous localization of the antibody binding site in three dimensions. An example of such an immunocomplex is seen for anti-L23 in Figure 4 (⟵); a detailed description has been given elsewhere [see Figure 10 in (71)].

For the three-dimensional localization of the proteins of the 30S ribosomal subunit, the antibody attachment sites on three projections have been used [these projections have been described in detail in (78)]. The Berlin group has localized 15 of the 21 ribosomal proteins at distinct sites on the surface of the 30S subunit of the *E. coli* ribosome (Figure 5*a*; 28, 71,

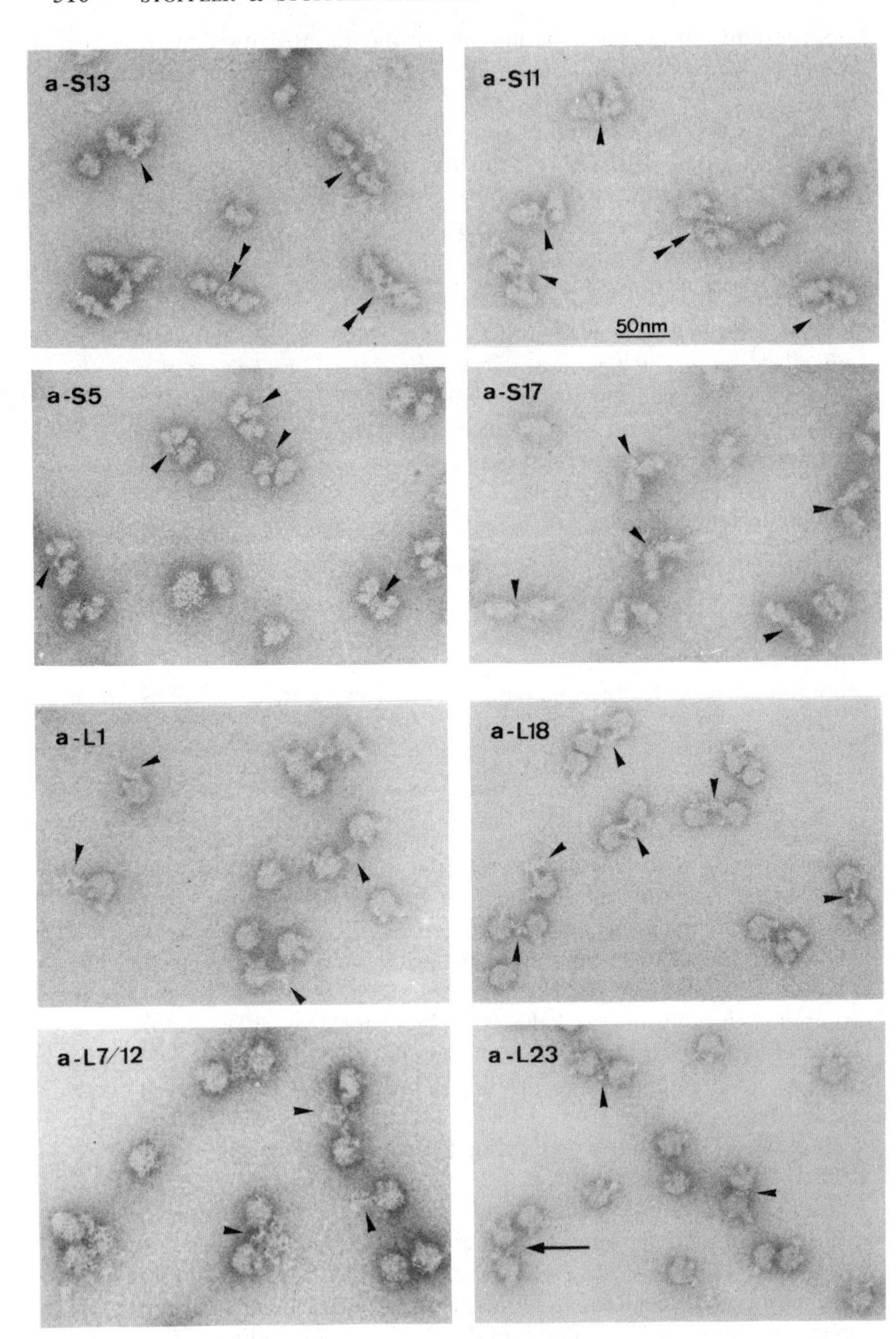
a-S13
a-S11
50nm
a-S5
a-S17
a-L1
a-L18
a-L7/12
a-L23

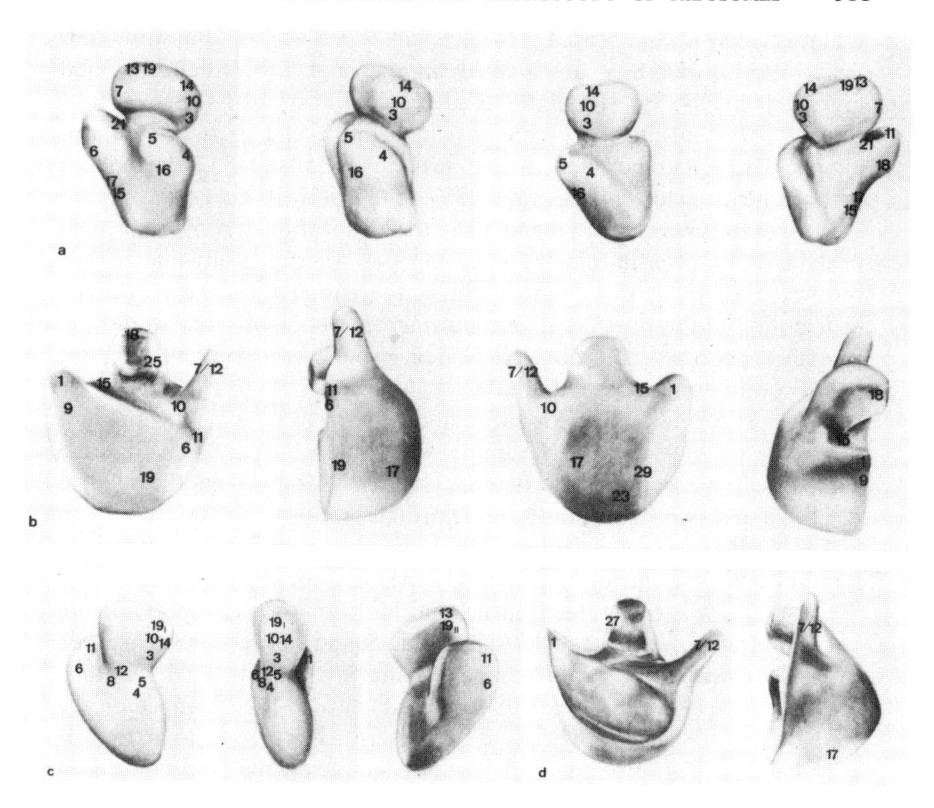

Figure 5 Three-dimensional models of the 30S (*a*, *c*) and 50S (*b*, *d*) ribosomal subunits. The locations are indicated for the centers of antibody binding sites for individual ribosomal proteins, as determined in Berlin (*a*, *b*) (see 13, 28, 40, 51, 69, 71, 77–79) and Los Angeles (*c*, *d*) (see 37, 104). Proteins S3 and S7 have been mapped in a comparative study with monoclonal and conventional antibodies (G. Breitenreuter, M. Lotti, M. Stöffler-Meilicke, G. Stöffler, manuscript in preparation).

77, 78), and Lake and collaborators have localized 12 proteins (Figure 5*c*; 104). To date, a total of 17 proteins have been located (Figure 5).

The 30S proteins are localized in distinctive domains. There are two domains on the head of the subunit. One comprises proteins S3, S10, and S14, and the second comprises proteins S7, S13, and S19. There is evidence

Figure 4 Electron micrographs of 30S and 50S subunits, respectively, reacted with antibodies specific for individual ribosomal proteins. Arrowheads mark characteristic immunocomplexes; double arrowheads indicate immunocomplexes in which a pair of subunits is simultaneously connected by two IgG molecules. ⟵ indicates a 50S · IgG · 50S complex, in which a crown projection is connected by one and the same IgG molecule with a kidney projection.

that protein S9 is also located at the head of the 30S subunit. Four proteins, S6, S11, S18, and S21, are located on the large lobe (platform), and four proteins, S4, S5, S12, and S16, are located at or near the small lobe (Figure 5). According to preliminary data, protein S20 is also in this region. The antigenic sites of two proteins, S15 and S17, have been mapped near the lower pole of the 30S subunit body (Figure 5).

Most of the previous discrepancies in the localization of some ribosomal proteins by the Berlin and the Madison/Los Angeles groups have been resolved recently [for a comparison of the earlier data see (18)]. The distribution of the proteins in domains agrees well with other studies on protein topography (39, 50), e.g. protein-protein cross-linking (90) and neutron scattering (56). This finding is noteworthy, since, owing to the intrinsic properties of the various methods, complete agreement would not necessarily be expected. For example, while the protein sites in Figure 5 depict surface locations, the neutron scattering data yield the relative positions of the centers of mass of the proteins (56).

The binding sites in both crown and kidney views have been determined for antibodies specific for 14 proteins of the 50S subunit (13, 28, 51, 69, 77, 79), and the resulting three-dimensional locations of these proteins are shown in Figure 5*b*. Proteins L7/L12, L1, L17, and L27 have been mapped by Lake and collaborators (37, 82; see Figure 5*d*). Proteins L7/L12, which are present in four copies per 50S subunit, have also been mapped on the stalk by Boublik et al (4). These authors described a second site on the L1-protuberance. This discrepancy is still not completely resolved; a location of two copies of L7/L12 on the central protuberance has been proposed from other studies by Möller and his colleagues (48, 110).

As in the 30S subunit, the proteins of the 50S subunit are also clustered in domains. The central protuberance contains antigenic determinants of proteins L18 and L25, both of which bind independently to 5S RNA. This data is in agreement with the placing of the 3′ end of 5S RNA (see Figure 10). According to Lake (37), protein L27 is also located on this protuberance (Figure 5*d*). Another domain is found on the broad lateral protuberance and comprises proteins L1 and L9. The stalk contains several antigenic determinants of the proteins L7/L12, whereas the protuberance from which the stalk originates contains proteins L6, L10, and L11. A determinant of protein L19 was found slightly separated from the latter domain (Figure 5). Another domain is on the concave surface of the 50S subunit, below proteins L1 and L9, and comprises determinants of proteins L23 and L29. Protein L17 has been mapped at a unique position. The domain structure as found by IEM also agrees in principle with other topographical studies as described above (39, 50, 106).

THE STRUCTURE OF THE 70S RIBOSOME Electron micrographs of 70S monomeric ribosomes show a large number of different projections. 70S ribosomes have globular, polygonal, or slightly prolate contours and varying regions of accumulated stain. In contrast to ribosomal subunits, electron micrographs of 70S ribosomes lack highly distinctive structural features (Figure 1). Despite the many 70S projections, only two forms have been described thus far (3, 32, 88). In one projection the contours of both subunits can be discerned (non-overlap view), whereas the other projection shows the contours of the 30S subunit superimposed on the 50S subunit or vice versa (overlap view).

It is thus not surprising that the three-dimensional models of the 70S monosome differ in the relative orientation of the ribosomal subunits within the 70S ribosome [28, 35; for a detailed discussion see (39, 106)].

In order to develop an accurate three-dimensional model that is based on experimental data and not only on interpretation of images, we developed two approaches to determine the relative orientation of the subunits within the 70S ribosome and to define most of the various projectional forms seen on electron micrographs.

Double antibody labelling experiments The first approach was based on the development of a double-antibody labelling technique (26–29). Two antibodies, one directed against a 30S protein and another directed against a 50S protein, were incubated with a mixture of 70S ribosomes and 30S and 50S subunits. Electron micrographs were screened for 30S-70S-50S triples connected by two antibody molecules. The examination of such complexes revealed the orientation of the subunits within the 70S ribosome, since distances between several pairs of antigenic determinants could be measured. The subunit orientation determined by this approach was independent of any interpretation of the various 70S projectional forms.

Using this approach, Stöffler and co-workers in 1980 showed that ribosomal proteins L1 and S13 are close together and that proteins S13 and L7/L12 are far apart on the 70S ribosomal surface (27, 28). Examples of these experiments are shown in Figure 6. We concluded that these data are incompatible with the subunit arrangement proposed by Boublik (3) but in agreement with the other 70S models (Figure 9). Later these studies were extended to distance measurements between several pairs of antigenic determinants [see (26, 29) and Table 1]. It appears that the comparison of distances between the central protuberance (the marker was a fluorescence label at the 3′ end of 5S RNA) and the determinants of proteins S5 and S6, respectively, provides a means of discriminating between the model proposed by Lake and that which we have put forward. The finding that S5 and S6 were both far from the central protuberance and the distances measured between proteins L7/L12 or L11 of the large subunit and several proteins of the small subunit (see Figures 6, 9 and Table 1) are all incompatible with Lake's model. Together with results obtained by binding a monovalent Fab against proteins L7/L12 to 30S · 70S · 50S triples as a third marker, that is distinguishable from an IgG (Figure 6*f*, *g*), we were able

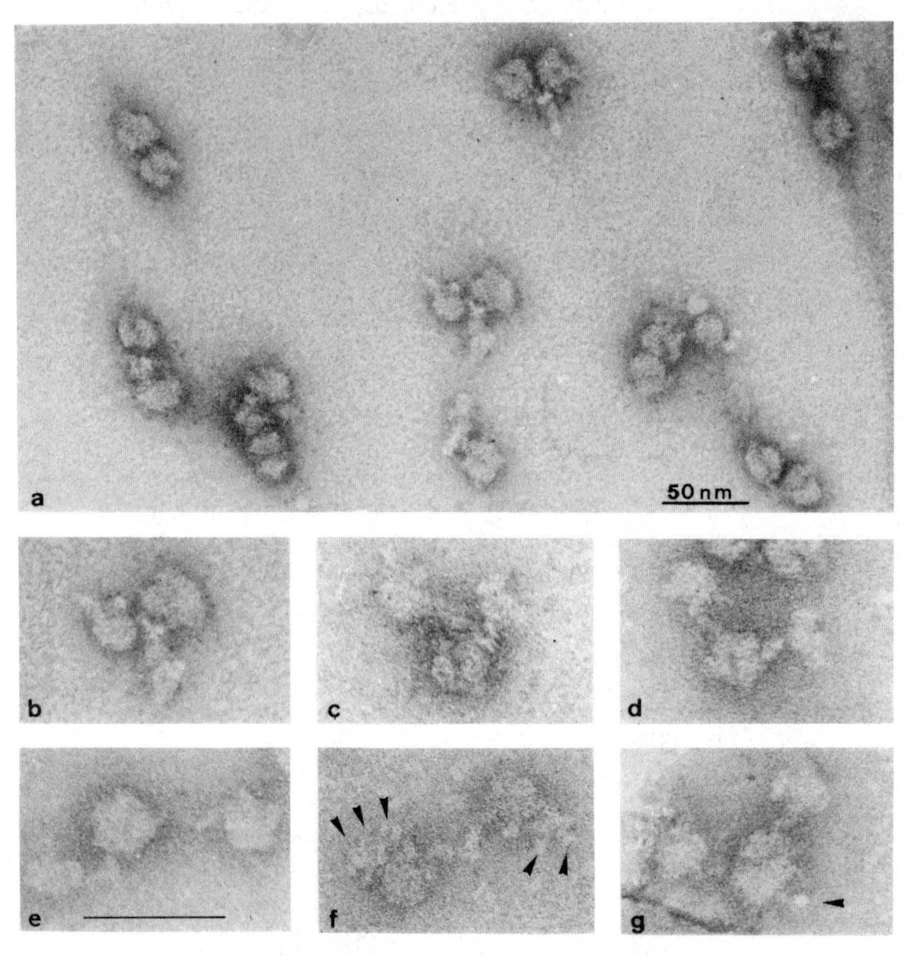

Figure 6 General field of a double labelling experiment with anti-S13 and anti-L1 (*a*). A typical 30S · 70S · 50S triple is shown in the center of the field. Dimeric complexes of ribosomal particles in different combinations (70S · 30S, 70S · 50S, 70S · 70S) are also seen. Selected electron micrographs of 30S · 70S · 50S triples, obtained with (*b*) anti-S13/anti-L1 (very close); (*c*) anti-S6/anti-L1 (close); (*d*) anti-S6/5S RNA (far); (*e*) anti-S13/anti-L7/L12 (very far). (*f*, *g*) Monovalent Fab fragments against L7/L12 as additional marker (*arrowed*). The 3′ end of 5S RNA was labelled with antibodies to fluorescein (80).

Table 1 Relative distances between characteristic subunit features within the 70S ribosome[a]

	S5	S6	S13	S14
L1	Far	Close	Very close	Close
L7/L12	n.d.	n.d.	Very far	Very far
5S RNA	Very far	Very far	Close	Close
L11	Far	Far	n.d.	n.d.

[a] The distances between the various pairs of antibody binding sites have been measured and histograms have been drawn. Parameters: very close = <10 nm (too close for measurement); close = 10–13 nm; far = 14–16 nm; very far = >16 nm. n.d. = not determined.

to show that the L7/L12 stalk protrudes from the 70S particle and does not join the 30S subunit as proposed by Lake (34). Vasiliev and co-workers (95) recently came to the same conclusion (Figure 9).

Single antibody labelling experiments In the second approach, antibodies were selected that bound to one of the several distinctive regions of the ribosomal subunits. Dimeric immunocomplexes, in which an antibody molecule combined a 70S ribosome with either a 30S or a 50S subunit, were examined under the electron microscope (Figure 7). Several distinctive subunit features could thus be located on the numerous 70S projections. In this way it was possible to characterize nine different projectional forms, of which the most completely characterized forms are shown in Figure 8. In contrast to the first method, this approach allowed an interpretation of the contours and the internal regions in which stain was accumulated (26–29). The three-dimensional locations of ribosomal proteins on the 70S monosome as shown in Figure 9 have thus been assigned not merely by transferring their locations (which had been obtained experimentally on subunits) onto the 70S model but by determining the antibody binding sites on 70S ribosomes directly.

DETERMINATION OF ANTIBODY SPECIFICITY It is a general working principle in immunochemistry that an antigen which is pure by biochemical criteria should still be assumed to be impure as an immunogen. Consequently, double immunodiffusion, quantitative immunoprecipitation, immunoelectrophoresis, etc, have been employed to assess the purity of the various antisera against pure ribosomal proteins (64). Impure antisera have been purified when necessary by immunoaffinity chromatography (85). The purity, as established by these criteria, was sufficient for most applications (64).

When studying the reaction of antibodies that have been raised against

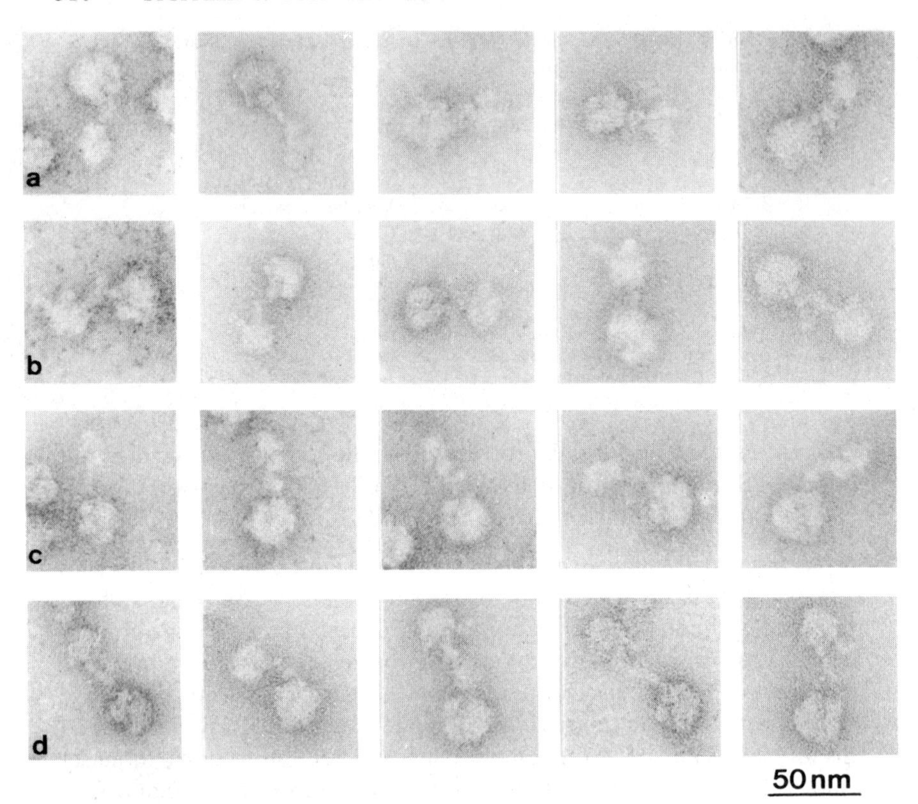

Figure 7 30S·70S and 50S·70S antibody complexes. (*a*) 30S·IgG·70S and (*b*) 50S·IgG·70S complexes, illustrating the locations of typical structural features of the subunits on the 70S ribosomes. (*c*) 30S·anti-S13·70S and (*d*) 50S·anti-fluorescein·70S complexes, which show antibody binding to different projectional forms of 70S ribosomes (see Figure 8). Anti-fluorescein marks the 3′ end of 5S RNA (80).

isolated ribosomal proteins with this protein *in situ*, only a minor fraction of the antibody population is reactive towards antigenic determinants that are accessible for antibody binding in the intact particle. If this fraction contains contaminating antibodies in small amounts that possess a high reactivity towards ribosomes, then these are likely to escape detection by the standard immunochemical techniques mentioned above.

Various attempts have been made to demonstrate the specificity of antibodies used for immunoelectron microscopy, but none of these has so far been sufficient to prove the specificity of the antibodies used. Purification of antibodies specific for *E. coli* ribosomal protein S4 by immunoaffinity chromatography did not eliminate contaminating antibodies (85). Similarly, substitution of *E. coli* S4 with its *B. stearothermo-*

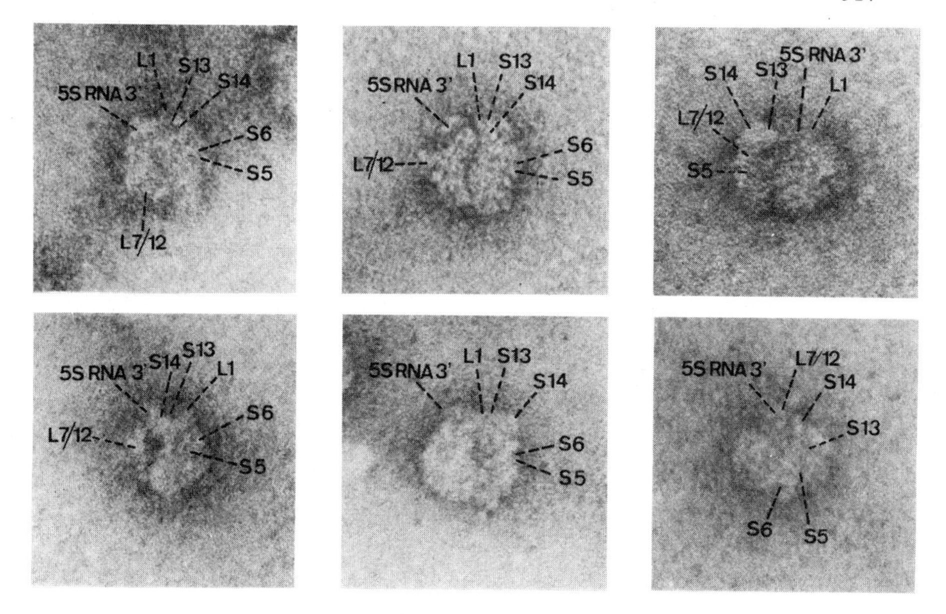

Figure 8 Electron micrographs of 6 different projectional forms of 70S ribosomes on which several antibody binding sites, characterizing typical structural features of ribosomal subunits, have been localized.

philus analogue did not prove to be an adequate criterion for the absence of contaminating antibody (36, 102). Furthermore, the latter control cannot be used for antisera against *E. coli* ribosomal proteins, which strongly cross-react with their *B. stearothermophilus* analogue, e.g. anti-S11 and anti-S19 (35a).

In view of all these technical difficulties, several control experiments intended to eliminate effects of both contaminating and/or cross-reacting antibodies (13) have been developed recently. First of all, dimeric immunocomplexes must be abolished completely by preincubation of the antibody with stoichiometric amounts of the antigen protein (Figure 3*f*). This experiment shows that the determinant to which anti-L11 binds is contained in protein L11. The double antibody assay that has been performed by Lake & Strycharz (37) in order to demonstrate the specificity of the antibodies used for the localization of three proteins of the 50S subunit has the disadvantage that it is performed under equilibrium conditions, whereas the immunocomplexes used for electron microscopy are prepared under nonequilibrium conditions (25).

The absorption experiment described above, however, does not exclude the possibility that the reactive determinant on the ribosome is present on another protein that has the same antigenic determinant. This is at least

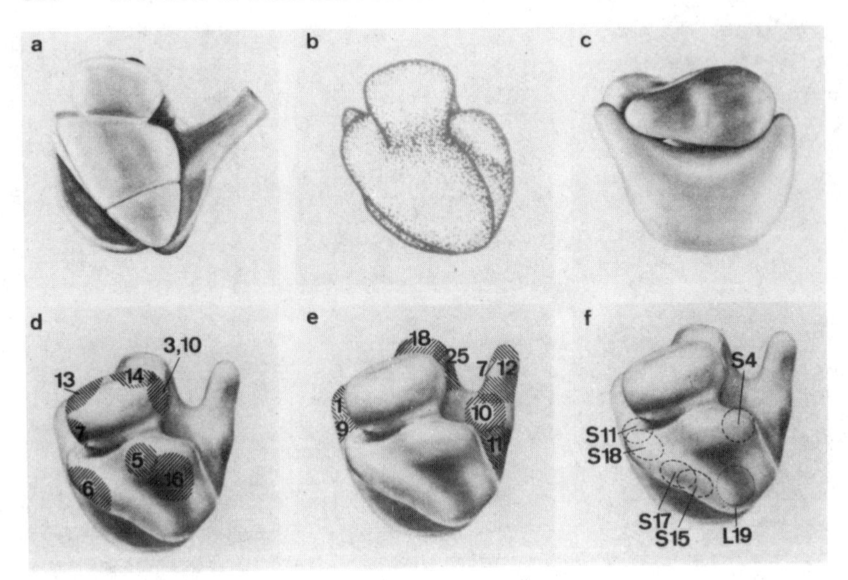

Figure 9 Three-dimensional model of the 70S ribosome as proposed by (*a*) Vasiliev (95); (*b*) Lake (34); (*c*) Boublik (3); and (*d–f*) Stöffler (28, 29). The locations of small subunit proteins are shown in (*d*) and of large subunit proteins in (*e*). The dotted circles in (*f*) indicate the location of ribosomal proteins, which are not accessible for antibody binding in the 70S monosome. Lake has also proposed an alternative model with the stalk sticking out (32).

conceivable, since an antibody to protein L11 may, for example, bind to protein L16, because these two proteins contain a common hexapeptide (Thr-Phe-Val-Thr-Lys-Thr). In principle, such cases of cross-reactivity could occur frequently, since common tetrapeptides have been found with a high frequency (reviewed in 105) and a tetrapeptide has the minimal size for an antigenic determinant (58).

Cross-reacting antibodies can be excluded if the formation of dimeric immunocomplexes is not inhibited by absorption with a mixture of all ribosomal proteins that lacks only the protein to be mapped (Figure 3*f*). In contrast, the formation of dimeric immunocomplexes is abolished by absorption with total ribosomal protein (Figure 3*f*). A mixture lacking a single ribosomal protein can be prepared from purified individual proteins or by immunoaffinity chromatography of total ribosomal proteins on columns to which IgG antibodies against the particular protein have been bound (G. Stöffler, R. Hasenbank, and W. Tate, unpublished data). The locations of the proteins mapped in this laboratory, shown in Figure 5, have all been determined with the complete set of control experiments.

Mutants lacking a single ribosomal protein are particularly useful for demonstrating antibody specificity; several such mutants have been isolated by Dabbs (12, 14). Proteins S17, L1, L11, L15, L19, and L29 have been mapped on such mutant ribosomes after reconstitution with the respective protein from wildtype (13, 40, 67, 69).

SHAPE OF RIBOSOMAL PROTEINS IN THE RIBOSOME In early work, well-separated antigenic determinants were found for several proteins (33, 76). These data suggested that some ribosomal proteins were highly elongated. At present, it is quite likely that all these results, perhaps with a few exceptions, were due to impure antibodies.

It has, however, been shown in rigorously controlled experiments that antibodies against a number of proteins (S4, S5, S6, S7, S11, S13, S14, S17, S18) frequently led to the formation of dimeric complexes in which the two subunits are linked simultaneously by two antibody molecules (see 88). In each case, the two sites are close to one another. These results suggest that some ribosomal proteins may have axial ratios up to 5:1; that is, they are globular, but not spherical. Since negative immunochemical results cannot rule out the possibility that elongated proteins exist in the ribosome, a search for additional antigenic sites is still necessary.

Selective labelling of amino-acid side chains with haptens, as performed for localizing the cysteine residues of proteins S4, S17, S18, and L6 (77, 78; L. Schuster, R. Lührmann, M. Stöffler-Meilicke, and G. Stöffler, manuscript in preparation) may help to solve this question and should serve as a refined mapping technique in the future. Interesting results are also emerging from the use of antibodies against fragments of ribosomal proteins as recently exemplified for protein L11 (51, 83).

Monoclonal antibodies, with the exception of a feasibility study (62), have not yet been used. Recently the locations of proteins S3 and S7 were determined in a comparative study with monoclonal and conventional antibodies (G. Breitenreuter, M. Lotti, M. Stöffler-Meilicke, G. Stöffler, manuscript in preparation; see Figure 5). This technique, now in use, should contribute considerably to our knowledge of the topography of ribosomal proteins and RNA. It should be emphasized that specificity controls, which exclude any cross-reactivity with other ribosomal proteins, also need to be performed for monoclonal antibodies (see above).

Topography of Ribosomal RNA

Two-thirds of the ribosomal mass is contributed by RNA. There is evidence that 16S and 23S RNA molecules are important for providing the overall structure of ribosomal subunits (96, 97), and more recent data indicate that the RNA molecules play important roles in ribosome function (52). Despite these facts, very little has been learned about the location of specific RNA

Table 2 Modified nucleotides at known positions of *E. coli* 16S and 23S RNA[a]

	Position	
	16S RNA	23S RNA
N^6-Dimethyladenosine	1518, 1519	—
7-Methylguanosine	526	2069
N^2-Methylguanosine	966, 1207, 1516	—
5-Methylcytidine	1407 (+1)	2394
N^6-Methyladenosine	—	1618
N^2-Methylguanosine	—	2251
1-Methylguanosine	—	745
Pseudouridine	—	746, 1911, 1917

[a] See (8, 9).

sites by IEM. Data on the location of specific RNA sites have been obtained in principle by two approaches.

LOCALIZATION OF MODIFIED NUCLEOTIDES Nucleotides themselves are not immunogenic, but they are reasonable haptens when coupled to carrier proteins. The technique of Erlanger & Beiser (15) for coupling nucleosides to proteins is almost generally applicable for this purpose, and antibodies can be produced against each of the four common nucleosides and nucleotides and, more important, against modified nucleosides. Antibodies against several modified nucleotides that occur at known positions within the 16S and 23S rRNA (7) are available in several laboratories [for references see (81)] and can be used for electron microscopy (Table 2).

The first initiative to localize specific RNA sites on the ribosome by IEM came from Glitz and collaborators (45, 54, 55, 91, 92). They localized the two adjacent dimethyladenosine residues at the junction between the large lobe and the body of the small subunit (54), a result in agreement with our studies (65; Figure 10). The one m^7G (at position 526 of 16S RNA) was mapped in the vicinity of the small lobe, close to the site of protein S4 (92). It is noteworthy that the m^7G at position 526 is within the binding site for ribosomal proteins S4 and S20 (cf 5, 10, 111).

LOCALIZATION OF HAPTENATED NUCLEOTIDES Antibodies to chemical modification reagents also can be used to localize specific reaction sites on RNA molecules. The ends of ribosomal RNAs have been localized in this manner.

The location of the 3′ and 5′ ends of 16S RNA A 2,4-dinitrophenyl-residue (DNP) was attached to the 3′ terminal adenosine of intact 30S subunits

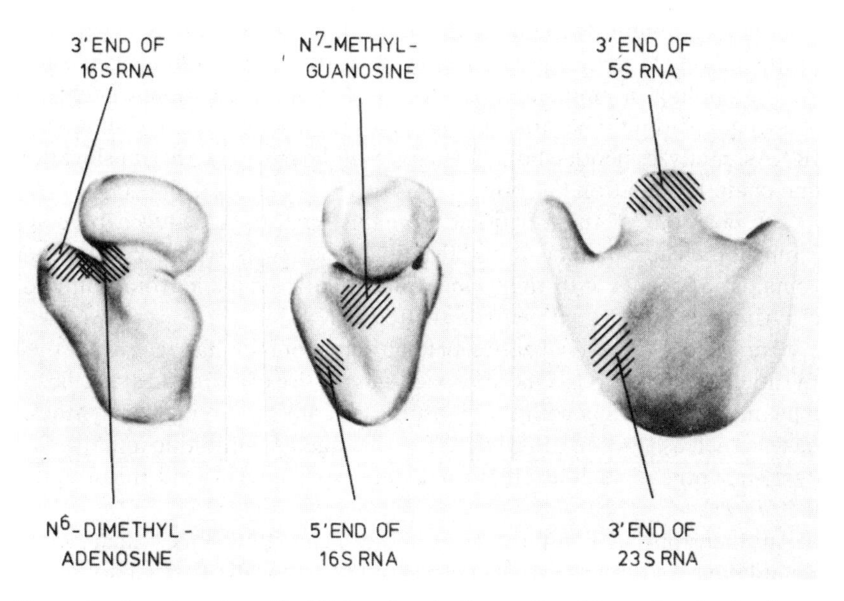

Figure 10 Location of specific RNA regions in the small and large *E. coli* subunits, as determined by IEM.

(45) and of isolated 16S RNA (DNP-16S rRNA). In the latter case, 30S ribosomal subunits were reconstituted from DNP-16S rRNA and total 30S ribosomal proteins (43). The reconstituted subunits were functionally active. Antibodies to DNP were allowed to react with these subunits so as to form ribosome-antibody complexes; these were purified and used for electron microscopy. A single binding site was found and was located at the inner side of the large lobe of the 30S subunit (43; Figure 10). The same location was also obtained when the 3′ end was labelled with phenyl-β-D-lactosides (59) or with fluoresceinthiosemicarbazide (80). The 5′ end of 16S RNA has been mapped by Mochalova et al (47; Figure 10) with 2,4-dinitrophenylethylenediamine as the modification reagent and hapten.

The location of the 3′ ends of 5S and 23S RNA Using the same approach, the 3′ ends of 5S and 23S RNA on the 50S subunit have been mapped (60, 61, 80; Figure 10).

Location of Functional Domains

The translation of mRNA into protein occurs on ribosomes with the participation of macromolecules such as tRNAs and protein synthesis factors. The factors play important roles during the three main steps of the cycle of protein synthesis, namely initiation, elongation, and termination.

The small subunit recognizes the initiation site on mRNA with the participation of three initiation factors and initiator tRNA. It is responsible for the binding of aminoacyl tRNAs and for the translational fidelity of messenger reading. The large subunit binds the acceptor stem of aminoacyl tRNAs, catalyzes peptide bond formation, and participates in translocation and chain termination (39, 50).

The positions of the functional domains have been deduced by combining biochemical data about the functional roles of individual ribosomal components with the topographical data available from immunoelectron microscopy and neutron scattering (39, 76).

More recently, the localization of functional domains has been attempted by more direct methods. One approach was the use of antibodies specific for a ligand that interacts with the ribosome (e.g. antisera to protein synthesis factors such as IF-3 or EF-G) or antibodies against antibiotic inhibitors of protein synthesis such as chloramphenicol, puromycin, and thiostrepton (65, 70, 71).

Another approach employed antibodies against haptens, which can be bound covalently to the ribosomal ligand to be localized, e.g. haptenated mRNA (65, 70, 71). These experiments have been described in detail elsewhere (71); a short summary of them is given below.

FUNCTIONAL DOMAINS ON THE 30S SUBUNIT The 3′ terminal region of 16S rRNA participates in the correct alignment of natural mRNAs during formation of the initiation complex. A purine-rich sequence of the mRNA, upstream of the initiation codon, binds to a conserved pyrimidine-rich sequence at the 3′ end of 16S RNA (63). The 3′ end of the 16S rRNA and the two N^6,N^6-dimethyladenosines, found respectively 24 and 25 nucleotides away from this 3′ end, have been located by IEM; they are found on the large lobe of the 30S subunit (Figure 10). The anticodon of the tRNA is also closely associated with the junction between the large lobe and the head, as determined by affinity IEM (30).

The mRNA binding domain was determined using a synthetic photoaffinity analogue of mRNA, poly(4-thiouridylic acid), which was haptenated with DNP, covalently bound to 30S subunits, and mapped with DNP-specific antibodies. The analogue was located at the junction between the large lobe and the body, with additional sites found within the neck region of the 30S subunit (Figure 11*a*; 42, 65, 70). In a recent experiment, poly(U) with an average chain length of 40–70 nucleotides was labelled at its 5′ or 3′ terminus with 2,4-dinitrophenyl residues. Both ends of the synthetic mRNA were located with DNP-specific antibodies on the 30S subunit and, in addition to the previous study (42), on 70S ribosomes (16). These authors suggest that the mRNA forms a loop around the large lobe of the 30S subunit as it passes through the ribosome (16). These data support conclusions drawn from an exceptional electron micrograph of polysomes from *Chironomus tentans* in which ribosomal subunits, mRNA, and the nascent polypeptide chains are directly seen (Figure 12; 17).

The binding site of initiation factor IF-3 on 30S subunits has been determined with antibodies against IF-3 (Figure 11*b*). First, the factor was cross-linked to the small subunit with suberimidate; the 30S·IF-3

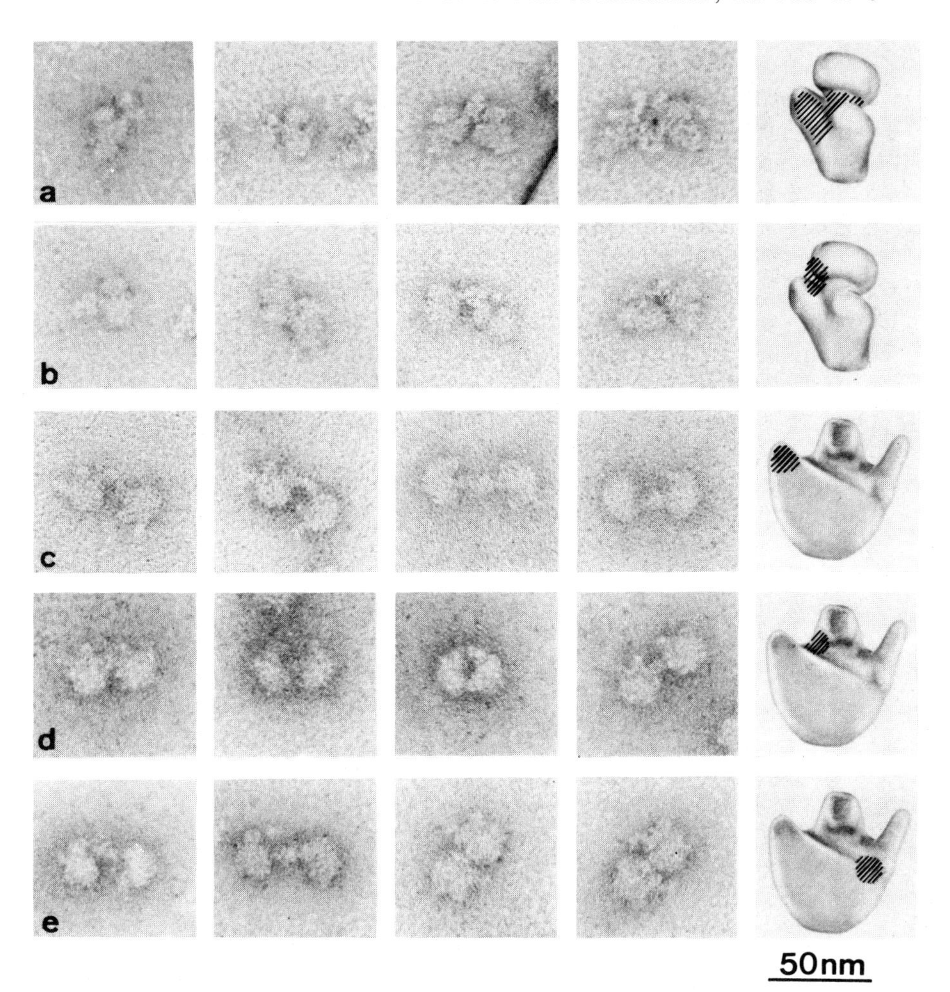

Figure 11 Localization of functional domains on ribosomal subunits, as determined by IEM: Binding site of (*a*) mRNA and (*b*) initiation-factor IF-3 on the 30S subunit. The mRNA binding sites could be localized in two dimensions only. Furthermore, the binding sites of (*c*) puromycin, (*d*) chloramphenicol, and (*e*) thiostrepton have been localized on the 50S subunit.

complexes were then allowed to react with antibodies against IF-3. The IF-3 binding site occupies a region extending from the lower part of the head to the large lobe (Figures 11*b*, 13; 42, 65, 70), and this region is at the subunit interface (Figures 9, 13).

Thus, functional sites on the 30S subunit involved in initiation, in binding of both synthetic and natural mRNA and in the ribosomal decoding site (30), have been located within a confined area in the 30S subunit around the one-third/two-thirds partition (see Figures 11, 13).

FUNCTIONAL DOMAINS AND ANTIBIOTIC BINDING SITES ON THE 50S SUBUNIT

Topography of the ribosomal peptidyl transferase center The binding sites of the peptidyl transferase inhibitor drugs puromycin and chloramphenicol on 50S subunits were localized by IEM. Iodo- and bromacetyl derivatives of these two antibiotics were used as affinity analogues for covalent labelling of ribosomal subunits, which then were incubated with antibodies specific for each of the antibiotics. Electron microscopy revealed a single antibody attachment site at the L1-protuberance for puromycin (Figure 11*c*) and in the angle between the lateral and the central protuberance for chloramphenicol (41, 65, 66, 70; Figure 11*d*). A similar experiment has been performed with ribosomes that were photoaffinity labelled with puromycin in the presence of tetracycline (46). Antibody binding was observed at three sites, and it has been concluded that one of them is the specific drug binding site that is located between the central and the lateral protuberances. All these data suggest that the peptidyl transferase center may be located in the angle formed by the central and lateral protuberances, opposite the 50S appendage (Figure 13).

The exit site of the nascent polypeptide chain has been located by electron microscopy (17; Figure 12) and by immunoelectron microscopy (1)

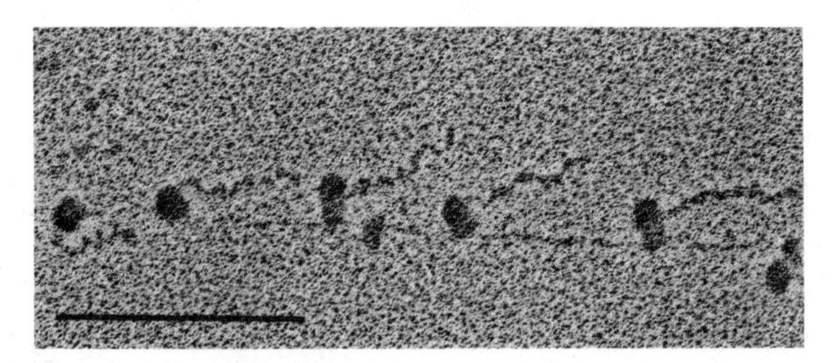

Figure 12 High magnification power view of an exceptional preparation in which the large polysomes were stretched during specimen preparation (from 17). The two ribosomal subunits and their relationship to mRNA and the nascent protein chain can be seen. Bar = 0.25 μm. (With kind permission from J.-E. Eström.)

at the back of the 50S subunit, i.e. 100–150 Å away from the presumed site of peptidyl transfer (see Figure 13).

The binding site of elongation and termination factors and of the antibiotic thiostrepton The four copies of proteins L7/L12 are involved in the binding of the elongation factors EF-G and EF-Tu and the termination

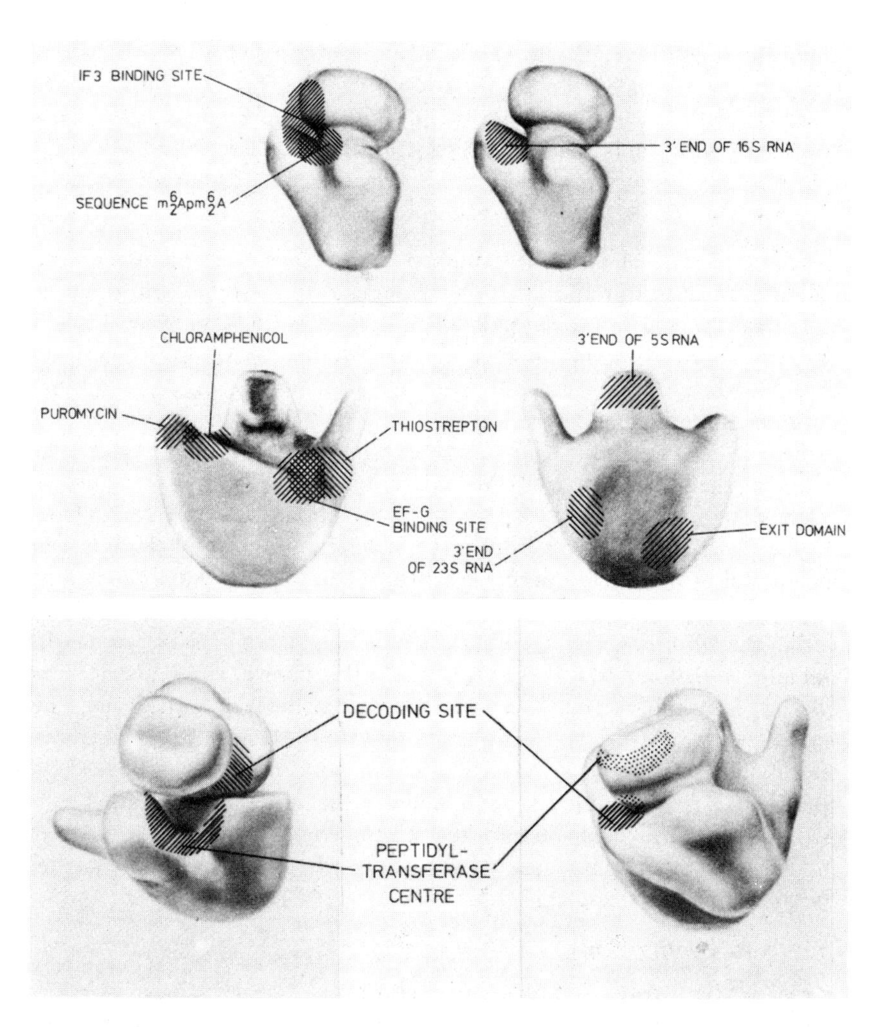

Figure 13 (*Top*) Functional domains on *E. coli* ribosomal subunits as determined by IEM [see Figure 11 and (41–43, 65, 66)]. The location of the EF-G binding site is from (19) and that of the exit domain from (1). (*Bottom*) Location of the ribosomal decoding site and the peptidyl transferase center in the 70S ribosome. The dotted areas indicate location at the interface.

factors RF-1 and RF-2 to ribosomes (see 39, 73, 76). Proteins L6, L10, L11 contribute to a domain that is located at the protuberance from which the L7/L12 stalk originates (see Figure 5). The binding site of elongation factor EF-G has been mapped in this region with EF-G-specific antibodies (19; Figure 13).

The thiostrepton binding site has been localized with antibodies against thiostrepton (65, 66, 70). These experiments were performed without an affinity analogue. A single site was found at the origin of the L7/L12 stalk, at the body of the 50S subunit (Figures 11*e*, 13). The drug-binding site coincides with the antigenic sites of protein L11 (cf Figures 5, 13). This correlates with the finding that protein L11 is essential for binding thiostrepton to ribosomes, and that ribosomes lacking L11 are resistant to this drug (11, 67, 101).

A MODEL OF THE FUNCTIONING RIBOSOME A reasonable picture of the location of structural and functional domains on the *E. coli* ribosome is emerging. Distances between pairs of ribosomal proteins, between specific regions of rRNAs, and between proteins and rRNA have also been determined by other methods, notably protein-protein cross-linking, fluorescence energy transfer, and neutron scattering between deuterated protein pairs. The results from all these techniques are in good agreement with the results obtained by IEM as described above.

The data on the relative orientation of the subunits in the 70S ribosome, on the location of the 30S ribosomal decoding site, and on the possible location of the peptidyl transferase center, taken together, have allowed us to propose a model of how mRNA and tRNAs are likely to be arranged within the 70S ribosome (71; Figure 13). The data available at present do not allow a definite location of the peptidyl transferase center, nor do they allow a discrimination between a tRNA bound to the A or to the P site. The model of the functioning ribosome as depicted in Figure 13 is thus still speculative. An alternative model, which has recently been proposed by Spirin (63a), is also compatible with much of the topographical data. It can be hoped that the localization of further ribosomal proteins and of other functional sites on the ribosome will soon give a unified picture of the functioning 70S ribosome.

Further Applications

COMPARATIVE STUDY OF RIBOSOMES FROM DIFFERENT ORGANISMS The ribosome is the physical site of protein synthesis in all cells and cellular organelles. Comparative IEM studies, aimed at mapping specific structural or functional elements on ribosomes from different organisms, may provide us with information about the extent of evolutionary conservation.

The locations of the two dimethyladenosines and of m^7G (position 526) in the 16S RNA are identical in *E. coli* and pea chloroplast ribosomes (91, 92). Ribosomes from *E. coli* mutants lacking protein L1 have been reconstituted with protein L1 from *Bacillus stearothermophilus*; mapping was then performed with antibodies to the *Bacillus* protein (M. Stöffler-Meilicke, J. Dijk, and G. Stöffler, unpublished results). The epitopes of the *Bacillus* protein mapped at the same position as those from *E. coli*. It is gratifying that some proteins that have been mapped on *E. coli* and *B. stearothermophilus*, such as S4, S5, L7/L12, L6, and L9, have been found at identical positions (72, 78). It appears from these data that the three-dimensional organization of ribosomes has been conserved at least during the evolution of prokaryotic organisms.

Some studies on topographical mapping of ribosomal proteins have been described for eukaryotic ribosomes (2, 44). Bielka and co-workers have shown that the technique can be applied to eukaryotic ribosomes (2). However, in view of the lessons learned from *E. coli*, their results should be considered with caution until control experiments are performed. Monoclonal antibodies have been obtained against several ribosomal proteins from chicken liver (89). None of the clones was reactive towards intact subunits (89), and they were, therefore, of no value for IEM.

MACROMOLECULAR STRUCTURES OTHER THAN RIBOSOMES The IEM methods, whose application to the *E. coli* ribosome were described in detail in this article, have been used for the elucidation of the structure and function of a variety of macromolecular structures (20, 31, 38, 53, 57, 100, 108). Following the pioneering work of Yanagida & Ahmad-Zadeh (109), Schwarz et al could assign the serological cross-reactivity and host specificity of T-even type *E. coli* phages to distinct stretches on the distal parts of the tail fibers (57). A number of investigators have used antibodies specific for subunits or defined portions of the macromolecule to determine their distribution *in situ*. In this way, the structure of yeast fatty acid synthetase (100), of isolated hemagglutinin and neuraminidase molecules from influenza virus (108), and of fibrinogen (53) and clathrin (31) have been studied. IEM has also successfully been applied to the study of the subunit arrangement of hemocyanin from *Androctonus australis* (38), an investigation that was performed with monovalent Fab fragments. In an extensive study on the axonally transported fodrin molecule, Glenney et al (20) have demonstrated how a large set of monoclonal antibodies with differing specificities can be used for a detailed topographical analysis of macromolecules.

For structures that have not yet been crystallized or where crystals yield diffractograms that are difficult to evaluate because of the complexity of the data, IEM offers a large field of applications. The theoretical possibility thus exists that immunoelectron microscopy, perhaps in combination with three-dimensional electron microscopy, may close the gap between the two realms of structural investigation—X-ray crystallography and conventional electron microscopy.

ACKNOWLEDGMENTS

The authors are pleased to have the opportunity to thank Berthold Kastner, Steve White, and Paul Woolley for many stimulating discussions and for their help in preparing the manuscript. The constant interest and encouragement of Heinz-Günter Wittmann is gratefully acknowledged. We thank Marina Lotti and Berthold Kastner for providing unpublished electron micrographs, Renate Albrecht-Ehrlich, Ulla Liebing, Eva Philippi, and Inge Popella for their help in preparation of the figures, and Johanna Belart for typing the manuscript. The work was supported by the Deutsche Forschungsgemeinschaft (Sfb 9).

Literature Cited

1. Bernabeu, C., Lake, J. A. 1982. *Proc. Natl. Acad. Sci. USA* 79:3111–15
2. Bielka, H. 1982. *The Eukaryotic Ribosome.* Berlin/Heidelberg/New York: Springer-Verlag
2a. Bodley, J. W., Zieve, F. J., Lin, L., Zieve, S. T. 1970. *J. Biol. Chem.* 245:5656–61
3. Boublik, M., Hellmann, W., Kleinschmidt, A. K. 1977. *Cytobiologie* 14:293–300
4. Boublik, M., Hellmann, W., Roth, H. E. 1976. *J. Mol. Biol.* 107:479–90
5. Deleted in proof
6. Brenner, S., Horne, R. W. 1959. *Biochim. Biophys. Acta* 34:103–10
7. Brimacombe, R., Maly, P., Zwieb, C. 1983. *Prog. Nucleic Acid Res. Mol. Biol.* 28:1–48
8. Brosius, J., Dull, T. J., Noller, H. F. 1980. *Proc. Natl. Acad. Sci. USA* 77:201–4
9. Brosius, J., Palmer, M. L., Kennedy, P. J., Noller, H. F. 1978. *Proc. Natl. Acad. Sci. USA* 75:4801–5
10. Cantor, C. R. 1979. In *Ribosomes, Structure, Function and Genetics*, ed. G. Chambliss, G. R. Craven, J. Davies, K. Davis, L. Kahan, M. Nomura, pp. 23–49. Baltimore: Univ. Park Press
11. Cundliffe, E., Dixon, P., Stark, M., Stöffler, G., Ehrlich, R., Stöffler-Meilicke, M. 1979. *J. Mol. Biol.* 132:235–52
12. Dabbs, E. R. 1979. *J. Bacteriol.* 140:736–37
13. Dabbs, E. R., Ehrlich, R., Hasenbank, R., Schroeter, B. H., Stöffler-Meilicke, M., Stöffler, G. 1981. *J. Mol. Biol.* 149:553–78
14. Dabbs, E. R., Hasenbank, R., Kastner, B., Rak, K.-H., Wartusch, B., Stöffler, G. 1983. *Mol. Gen. Genet.* 192:301–8
15. Erlanger, B., Beiser, S. 1964. *Proc. Natl. Acad. Sci. USA* 52:68–74
16. Evstafieva, A. G., Shatsky, I. N., Bogdanov, A. A., Semenkov, Y. P., Vasiliev, V. D. 1983. *EMBO J.* 2:799–804
17. Francke, C., Edström, J.-E., McDowall, A. W., Miller, O. L. Jr. 1982. *EMBO J.* 1:59–62
18. Gaffney, G., Craven, G. R. 1979. See Ref. 10, pp. 237–65
19. Girshovich, A., Kurtskhalia, T. V., Ovchinnikov, Yu. A., Vasiliev, V. D. 1981. *FEBS Lett.* 130:54–59
20. Glenney, J. R., Glenney, P., Weber, K. 1983. *J. Mol. Biol.* 167:275–93
21. Hall, C. E., Slayter, H. S. 1959. *J. Mol. Biol.* 1:329–32
22. Hardy, S. J. S. 1975. *Mol. Gen. Genet.* 140:253–74
23. Höglund, S. 1967. *Virology* 32:602–77
24. Huxley, H. E., Zubay, G. 1960. *J. Mol. Biol.* 2:10–18
25. Kahan, L., Winkelmann, D. A., Lake, J. A. 1981. *J. Mol. Biol.* 145:193–214

26. Kastner, B. 1982. PhD thesis. Tech. Univ., Berlin
27. Kastner, B., Stöffler-Meilicke, M., Stöffler, G. 1980. *Electron Microscopy 1980, 7th Eur. Congr. Electron Microsc. Found., Leiden*, pp. 564–65
28. Kastner, B., Stöffler-Meilicke, M., Stöffler, G. 1981. *Proc. Natl. Acad. Sci. USA* 78:6652–56
29. Kastner, B., Stöffler-Meilicke, M., Stöffler, G. 1982. *Proc. 10th Int. Congr. Electron Microsc., Hamburg*, 3:105–6
30. Keren-Zur, M., Boublik, M., Ofengand, J. 1979. *Proc. Natl. Acad. Sci. USA* 76:1054–58
31. Kirchhausen, T., Harrison, S. C., Perham, P., Brodsky, F. M. 1983. *Proc. Natl. Acad. Sci. USA* 80:2481–85
32. Lake, J. A. 1976. *J. Mol. Biol.* 105:131–60
33. Lake, J. A. 1978. In *Advanced Techniques in Biological Electron Microscopy II*, ed. J. K. Koehler, pp. 173–211. Berlin/Heidelberg/New York: Springer-Verlag
34. Lake, J. A. 1979. See Ref. 10, pp. 207–36
35. Lake, J. A. 1982. *J. Mol. Biol.* 161:89–106
35a. Lake, J. A., Kahan, L. 1975. *J. Mol. Biol.* 99:631–44
36. Lake, J. A., Pendergast, M., Kahan, L., Nomura, M. 1974. *Proc. Natl. Acad. Sci. USA* 71:4688–92
37. Lake, J. A., Strycharz, W. A. 1981. *J. Mol. Biol.* 153:979–92
38. Lamy, J., Bijlholt, M. M. C., Sizaret, P.-Y., Lamy, J., van Bruggen, E. F. J. 1981. *Biochemistry* 20:1849–56
39. Liljas, A. 1982. *Prog. Biophys. Mol. Biol.* 40:161–228
40. Lotti, M., Dabbs, E. R., Hasenbank, R., Stöffler-Meilicke, M., Stöffler, G. 1983. *Mol. Gen. Genet.* 192:295–300
41. Lührmann, R., Bald, R., Stöffler-Meilicke, M., Stöffler, G. 1981. *Proc. Natl. Acad. Sci. USA* 78:7276–80
42. Lührmann, R., Stöffler-Meilicke, M., Dieckhoff, J., Tischendorf, G., Stöffler, G. 1980. See Ref. 27, pp. 568–69
43. Lührmann, R., Stöffler-Meilicke, M., Stöffler, G. 1981. *Mol. Gen. Genet.* 182:369–76
44. Lutsch, G., Noll, F., Theise, H., Enzmann, G., Bielka, H. 1979. *Mol. Gen. Genet.* 176:281–91
45. McKuskie Olson, H., Glitz, D. G. 1979. *Proc. Natl. Acad. Sci. USA* 76:3769–73
46. McKuskie Olson, H., Grant, P. G., Cooperman, B., Glitz, D. G. 1982. *J. Biol. Chem.* 257:2649–56
46a. Meisenberger, O., Pilz, I., Stöffler-Meilicke, M., Stöffler, G. 1984. *Biochim. Biophys. Acta*. In press
47. Mochalova, L., Shatsky, I., Bogdanov, A., Vasiliev, V. 1982. *J. Mol. Biol.* 159:637–50
48. Möller, W., Schrier, P. I., Maassen, J. A., Zantema, A., Schop, E., et al. 1983. *J. Mol. Biol.* 163:553–73
48a. Moore, P. B. 1979. See Ref. 10, pp. 111–33
49. Morrison, C. A., Tischendorf, G., Stöffler, G., Garrett, R. A. 1977. *Mol. Gen. Genet.* 151:245–52
50. Nierhaus, K. H. 1982. *Curr. Top. Microbiol. Immunol.* 97:81–155
51. Noah, M. 1982. PhD thesis. Tech. Univ., Berlin
52. Noller, H. F., Woese, C. R. 1981. *Science* 212:403–11
53. Norton, P. A., Slayter, H. S. 1981. *Proc. Natl. Acad. Sci. USA* 78:1661–65
54. Politz, S. M., Glitz, D. G. 1977. *Proc. Natl. Acad. Sci. USA* 74:1468–72
55. Politz, S. M., Glitz, D. G. 1980. *Biochemistry* 19:3786–91
56. Ramakrishnan, V. R., Yabuki, S., Sillers, I.-Y., Schindler, D. G., Engelmann, D. M., Moore, P. B. 1981. *J. Mol. Biol.* 153:739–60
57. Schwarz, H., Riede, I., Sonntag, I., Hennig, U. 1983. *EMBO J.* 2:375–80
58. Sela, M. 1969. *Science* 166:1365–74
59. Shatsky, I. N., Evstafieva, A., Bystrova, T., Bogdanov, A., Vasiliev, V. 1980. *FEBS Lett.* 121:97–100
60. Shatsky, I. N., Evstafieva, A., Bystrova, T., Bogdanov, A., Vasiliev, V. 1980. *FEBS Lett.* 122:251–55
61. Shatsky, I., Mochalova, L., Kojouharova, M., Bogdanov, A., Vasiliev, V. 1979. *J. Mol. Biol.* 133:501–15
62. Shen, V., King, T. C., Kumar, V., Daugherty, B. 1980. *Nucleic Acids Res.* 8:4639–49
63. Shine, J., Dalgarno, L. 1975. *Nature* 254:34–38
63a. Spirin, A. S. 1983. *FEBS Lett.* 156:217–21
63b. Spirin, A. S., Serdyuk, I. N., Shpungin, J. L., Vasiliev, V. D. 1979. *Proc. Natl. Acad. Sci. USA* 76:4867–71
64. Stöffler, G. 1974. In *Ribosomes*, ed. M. Nomura, A. Tissières, P. Lengyel, pp. 615–67. Long Island, NY: Cold Spring Harbor Lab.
65. Stöffler, G., Bald, R., Kastner, B., Lührmann, R., Stöffler-Meilicke, M., et al. 1979. See Ref. 10, pp. 171–205
66. Stöffler, G., Bald, R., Lührmann, R., Tischendorf, G., Stöffler-Meilicke, M. 1980. See Ref. 27, pp. 566–67
67. Stöffler, G., Cundliffe, E., Stöffler-Meilicke, M., Dabbs, E. R. 1980. *J. Biol. Chem.* 255:10517–22
68. Stöffler, G., Hasenbank, R., Lütgehaus,

M., Maschler, R., Morrison, C. A., et al. 1973. *Mol. Gen. Genet.* 127:89–110
69. Stöffler, G., Noah, M., Stöffler-Meilicke, M., Dabbs, E. R. 1984. *J. Biol. Chem.* In press
70. Stöffler, G., Stöffler-Meilicke, M. 1981. *International Cell Biology 1980–1981*, pp. 93–102. Berlin/Heidelberg/New York: Springer-Verlag
71. Stöffler, G., Stöffler-Meilicke, M. 1983. In *Modern Methods in Protein Chemistry*, ed. H. Tschesche, pp. 409–457. Berlin/New York: de Gruyter
72. Stöffler, G., Stöffler-Meilicke, M., Littlechild, J. 1982. See Ref. 29, pp. 103–4
73. Stöffler, G., Tate, W., Caskey, C. T. 1982. *J. Biol. Chem.* 257:4203–6
74. Stöffler, G., Wittmann, H. G. 1971. *J. Mol. Biol.* 62:407–9
75. Stöffler, G., Wittmann, H. G. 1971. *Proc. Natl. Acad. Sci. USA* 68:2283–87
76. Stöffler, G., Wittmann, H. G. 1977. *Molecular Mechanisms of Protein Biosynthesis*, pp. 117–202. New York/San Francisco/London: Academic
77. Stöffler-Meilicke, M., Epe, B., Steinhäuser, K. G., Woolley, P., Stöffler, G. 1983. *FEBS Lett.* 163:94–98
78. Stöffler-Meilicke, M., Epe, B., Woolley, P., Lotti, M., Littlechild, J., Stöffler, G. 1983. *Mol. Gen. Genet.* Submitted for publication
79. Stöffler-Meilicke, M., Noah, M., Stöffler, G. 1983. *Proc. Natl. Acad. Sci. USA* 80:6780–84
80. Stöffler-Meilicke, M., Stöffler, G., Odom, O. W., Zinn, A., Kramer, G., Hardesty, B. 1981. *Proc. Natl. Acad. Sci. USA* 78:5538–42
81. Stollar, B. D. 1973. *The Antigens*, 1:1–85. New York/London: Academic
82. Strycharz, W. A., Nomura, M., Lake, J. A. 1978. *J. Mol. Biol.* 126:123–40
82a. Stuhrmann, H. B., Koch, M. J., Parfait, R., Haas, J., Ibel, K., Crichton, R. R. 1977. *Proc. Natl. Acad. Sci. USA* 74:2316–20
83. Tate, W. P., Dognin, M., Noah, M., Stöffler-Meilicke, M., Stöffler, G. 1983. *J. Biol. Chem.* In press
84. Thammana, P., Kurland, C. G., Deusser, E., Weber, J., Maschler, R., et al. 1973. *Nature* 242:47–49
85. Tischendorf, G. W., Stöffler, G. 1975. *Mol. Gen. Genet.* 142:193–208
86. Tischendorf, G. W., Zeichhardt, H., Stöffler, G. 1974. *Mol. Gen. Genet.* 134:187–208
87. Tischendorf, G. W., Zeichhardt, H., Stöffler, G. 1974. *Mol. Gen. Genet.* 134:209–23
88. Tischendorf, G. W., Zeichhardt, H., Stöffler, G. 1975. *Proc. Natl. Acad. Sci. USA* 72:4820–24
89. Towbin, H., Ramjoué, H.-P., Kuster, H., Liverani, D., Gordon, J. 1982. *J. Biol. Chem.* 257:12709–15
90. Traut, R. R., Lambert, J. M., Boileau, G., Kenny, J. W. 1979. See Ref. 10, pp. 89–110
91. Trempe, M. R., Glitz, D. G. 1981. *J. Biol. Chem.* 256:11873–79
92. Trempe, M. R., Ohigi, K., Glitz, D. G. 1982. *J. Biol. Chem.* 257:9822–29
93. Valentine, R. C., Green, N. M. 1967. *J. Mol. Biol.* 27:615–17
94. Valentine, R. C. Shapiro, B. M., Stadtman, E. R. 1968. *Biochemistry* 7:2143–52
95. Vasiliev, V. D., Selivanova, O. M., Baranov, V. I., Spirin, A. S. 1983. *FEBS Lett.* 155:167–72
96. Vasiliev, V. D., Selivanova, O. M., Koteliansky, V. E. 1978. *FEBS Lett.* 95:273–76
97. Vasiliev, V. D., Zalite, O. M. 1980. *FEBS Lett.* 121:101–4
98. Wabl, M. R. 1973. PhD thesis. Freie Univ., Berlin
99. Wabl, M. R. 1974. *J. Mol. Biol.* 84:241–47
100. Wieland, F., Siess, E. A., Renner, L., Verfürth, C., Lynen, F. 1978. *Proc. Natl. Acad. Sci. USA* 75:5792–96
101. Wienen, B., Ehrlich, R., Stöffler-Meilicke, M., Stöffler, G., Smith, I., et al. 1979. *J. Biol. Chem.* 254:8031–41
102. Winkelmann, D., Kahan, L. 1979. *J. Supramol. Struct.* 10:443–55
103. Winkelmann, D. A., Kahan, L. 1983. *J. Mol. Biol.* 165:357–74
104. Winkelmann, D. A., Kahan, L., Lake, J. A. 1982. *Proc. Natl. Acad. Sci. USA* 79:5184–88
105. Wittmann, H. G. 1982. *Ann. Rev. Biochem.* 51:155–83
106. Wittmann, H. G. 1983. *Ann. Rev. Biochem.* 52:35–65
107. Wittmann-Liebold, B. 1980. In *Genetics and Evolution of RNA Polymerase, tRNA and Ribosomes*, ed. S. Osawa, H. Zozaki, H. Uchida, T. Yura, pp. 639–54. Amsterdam: Elsevier-North Holland Biomed.
108. Wrigley, N. G., Laver, W. G., Downie, J. C. 1977. *J. Mol. Biol.* 109:405–21
109. Yanagida, M., Ahmad-Zadeh, C. 1970. *J. Mol. Biol.* 51:411–21
110. Zantema, A., Maasen, J. A., Kriek, J., Möller, W. 1982. *Biochemistry* 21:3077–82
111. Zimmermann, R. A. 1979. See Ref. 10, pp. 135–69

Ann. Rev. Biophys. Bioeng. 1984. 13: 331–71

FLUCTUATIONS IN PROTEIN STRUCTURE FROM X-RAY DIFFRACTION

Gregory A. Petsko and Dagmar Ringe

Department of Chemistry, Massachusetts Institute of Technology, Cambridge, Massachusetts 02139

INTRODUCTION

It is only recently that the extraordinary variety of motions possible for proteins and nucleic acids has been appreciated. In addition, the development of new physical techniques for probing the structure and behavior of macromolecules has allowed characterization of their dynamics for the first time.

It seems fitting that X-ray crystallography, which unintentionally contributed to the old "rigid molecule" view, should make an equal contribution to the new picture of the flexible biopolymer. No other physical method presents such a high resolution view of the non-hydrogen atoms in a macromolecule. It has long been held that this view is without any dynamic information and is therefore inherently inferior to information from techniques such as NMR. However, during the past few years it has become clear that much valuable information about the spatial distribution of atomic fluctuations can be obtained by careful analysis of crystal structures at high resolution. X-ray diffraction is thus complementary to spectroscopic and other techniques, which are low in spatial resolution but rich in temporal information. When used in concert, the modern arsenal of physical methods provides not only a general notion of how fast overall processes occur for a given biopolymer but also a detailed map of the various motions undergone by different regions of the molecule in different time regimes.

This review focuses on the application of single crystal X-ray diffraction techniques to the study of protein dynamics. First, an overview is presented

0084–6589/84/0615–0331$02.00

of the nature and types of internal motions that are found in proteins. The methods currently used to extract dynamic information from crystallographic results are outlined, and the results obtained to date are reviewed. These data are compared with information from other techniques, and some suggestions for the functional role of the motions observed are offered. Finally, an outlook for future applications of X-ray crystallography to this field is presented. No attempt is made to review exhaustively the literature of protein dynamics or of protein crystallography. Many good reviews of both subjects have been published recently (21, 26, 28, 39, 46, 47, 104; 30, 60, 60a). Rather, this article concentrates on results that allow a critical appraisal of the present status of research in this subject.

Other types of diffraction experiments, such as small-angle scattering and neutron diffraction, can also provide dynamic information. A review of neutron protein crystallography has just appeared (51), and small-angle scattering of both neutrons and X rays is covered elsewhere (86). Finally, the focus of this review is on proteins, although the techniques described here are being applied to single crystals of transfer RNA (75) and oligonucleotide DNAs (29) as well.

OVERVIEW

Dynamics is a term that covers all of the intramolecular motions of proteins. These may be the motions of individual atoms, groups of atoms, or whole sections of the protein. These motions can be divided into three broad categories, characterized by the extent and time scale of the motion and by the method that is used to study it (Table 1).

The first category contains atomic fluctuations, such as vibrations. These motions are random, very fast (21), and rarely cover more than 0.5 Å. The energy for these motions comes from the kinetic energy inherent in the protein as a function of temperature. Because of the short time scale (picosecond), these motions are studied by both molecular dynamics and X-ray diffraction. Molecular dynamics is a method of mathematical modelling of the distances that an atom can traverse under the constraints placed on it by bonding and energy considerations (46, 47, 54).

The second category contains collective motions, such as the movements of groups of atoms that are covalently linked in such a way that the group moves as a unit. The size of the group ranges from a few atoms to many hundreds of atoms. Entire structural domains may be involved, as in the case of the flexible Fc portion of immunoglobulins, where rigid-body motion of a 50,000 dalton unit occurs (25). There are two types of these motions: those that occur fast but infrequently (internal tyrosine ring flips belong to this category), and those that occur slowly (the cis-trans isomerization of proline is one example). The energy for collective motions also derives from the thermal energy inherent in a protein as a function of temperature. The time scale of these motions [from picoseconds to nanoseconds or slower (21)] allows some of them to be studied by such techniques as NMR (99) and fluorescence spectroscopy (63).

The third category contains motions that can be described as triggered conformational changes. These are the motions of groups of atoms (i.e. individual side-chains) or whole sections of a protein (i.e. loops of chain, domains of secondary and tertiary structure, or subunits), which occur as a response to a specific stimulus. The distance moved can be as much as 10 Å or more. The time scale can be estimated by studies of the rate of binding or turnover reactions. The energy for triggered conformational change comes from specific interactions, such as electrostatic attractions or hydrogen bonding interactions. The best-known example of a triggered conformational change is the transition in tertiary structure that occurs when ligands bind to the iron atoms of hemoglobin (70, 71).

X-ray diffraction can be used easily to study motions in categories one and three. The electron density of any atom in a structure derived from an X-ray diffraction study is an average of several components. It reflects the position of that atom averaged over every unit cell of the crystal. It also reflects the possible positions in which an atom can exist within the unit cell

Table 1 Types of motion found in proteins

Motion	Spatial displacement (Å)	Characteristic time (sec)	Energy source	Method of observation
Fluctuations (e.g. atomic vibrations)	0.01 to 1	10^{-15} to 10^{-11}	$k_B T$	Computer simulation, X-ray diffraction
Collective motions (*A*) fast, infrequent (e.g. Tyr ring flip) (*B*) slow (e.g. domain hinge-bending)	0.01 to >5	10^{-12} to 10^{-3}	$k_B T$	NMR, fluorescence, hydrogen exchange, simulation, X ray
Triggered conformational change	0.5 to >10	10^{-9} to 10^{3}	Interactions	X ray, spectroscopy

and the amount of time the atom spends in each. Both effects result in a spread of electron density over a range of space. For instance, the vibrations defined in category one would result in a distribution of the electron density about a most probable position. The most probable position is the point of highest electron density and can be used to define the amplitude of the vibration. These vibrations are frequent but limited in range. By contrast, the motions described as category three occur when triggered and cover a wider range of space. The distinguishing feature of these motions, which makes it possible to observe their effect, is the stability of discrete positions, which are highly populated relative to all possible positions. These highly populated positions are the ones observed in an X-ray crystallographic study. The absence of this feature, i.e. discrete states that are highly populated, is the reason that motions described in category two are difficult to observe by X-ray methods. If a motion occurs too infrequently or too slowly to produce distinguishable states, the electron density of the atoms involved is too low to be observed.

X-RAY DIFFRACTION TECHNIQUES

The Nature of Protein Crystals

Macromolecules are irregular in shape and do not pack tightly into lattices. In a protein crystal, the different neighboring molecules make contact with one another at only a few points on their surface. There are large open spaces in the lattice that must be filled with crystallization solvent to prevent the crystal from collapsing. A typical protein crystal is 40–60% solvent by volume (60). The average surface amino acid residue in a crystalline protein is in contact not with another protein molecule but with solvent. There is ample space in the lattice for most atoms to move. Contacts between neighboring molecules are expected to influence only the mobility of those few atoms in the contact region.

The nature of protein crystals suggests that both the structure and the dynamics of proteins in the "solid" state will be relevant to these properties in aqueous solution. There is now abundant evidence that the time-averaged protein structure is normally not changed by crystallization (74, 99; for reviews, see 57, 81). Dynamic properties are more difficult to assess. Before considering the way in which dynamic information is reflected in the diffraction pattern, it is useful to consider the question of whether such information is pertinent to the behavior of proteins under physiological conditions.

The Flexibility of Proteins in the Crystalline State

Indirect evidence for the flexibility of crystalline proteins has existed for many years. A number of crystalline enzymes possess catalytic activity comparable to that found in solution (57), so, if flexibility is important for function, at least a large measure of flexibility must be retained in the crystal. Oxygen-binding proteins such as hemoglobin and myoglobin, which are known to require flexibility of structure for oxygen binding (see below), reversibly bind ligands in the crystalline state (23, 24).

Recently, powerful direct evidence for the internal mobility of crystalline proteins has come from observations of hydrogen/deuterium and hydrogen/tritium exchange in crystals. Normally, one is interested in the exchange of peptide amide hydrogens, which reflect the secondary and tertiary structure of the protein (14, 32, 108, 110). Two extreme models have been proposed to explain the phenomenon that even amide groups buried deep within the protein interior will exchange with deuterium or tritium, albeit slowly. One is the "local unfolding" model, which postulates that regions of the structure move away from the center of mass of the protein and therefore transiently expose internal hydrogens to bulk solvent (31). The other is the "penetration" model, which speculates that small random fluctuations in the structure can create channels that allow solvent molecules to reach the interior of the folded protein, where the chemical exchange step occurs (109). For our purposes, the importance of both models is that they require protein flexibility for exchange to take place.

Neutron diffraction can be used to observe hydrogens and deuteriums directly in crystalline macromolecules. Studies have now been completed on several protein crystals (40, 50, 107) and show that if these crystals are suspended in deuterated mother liquor for extended periods of time, nearly all of the amide protons are exchanged. This is true even for amides that are completely inaccessible to solvent in the static structure. Internal mobility must exist even in the crystalline state.

Even more compelling evidence has been obtained from a pioneering series of studies by Tüchsen & Ottesen on the rates of amide hydrogen exchange in protein crystals. The total tritium-hydrogen back exchange kinetics for hen egg white lysozyme are nearly identical in cross-linked crystals and in solution (94, 95). The slowest exchanging 16 protons show identical exchange kinetics for cross-linked crystals and dissolved lysozyme in long time back-exchange experiments (E. Tüchsen, unpublished). In the case of insulin, where zinc-dependent hexamerization occurs in the crystals and leads to a substantial conformational change over the solution structure (111), the total tritium exchange in the cross-linked crystal is slower than for dissolved insulin [see discussion following (35)].

The Debye-Waller Factor

Accurate determination of atomic fluctuations by crystallography requires measurement of high-resolution data. Unfortunately, these data are low in intensity and hard to measure. For this reason, the study of protein dynamics by X-ray diffraction can be divided into two broad categories: qualitative studies at low resolution, where the absence of significant electron density for a group of atoms is taken as evidence for disorder in their positions, and studies at high resolution, where quantitative estimate of individual atomic motions is attempted.

For quantitative work a physical model is needed. Small-molecule crystallography has developed a number of models to deal with such effects as anharmonic motion, libration, and anisotropic vibration. For the most part, these have not been applied to proteins. The simple Debye-Waller model (105) has been widely used for macromolecular crystal structure analysis. In this treatment of atomic motion, the probability of finding an atom a given distance x from its equilibrium position x_0 is Gaussian. If it is assumed that the motion is isotropic, the model states that the motion in any direction can be characterized in terms of a mean-square vibration amplitude, $\langle x^2 \rangle$, also termed the mean square displacement. The X-ray scattering from each atom is modified by a Gaussian function that is related to the mean-square displacement of that atom. The form of the Gaussian is

$$\exp\left(-B \sin^2 \theta/\lambda^2\right)$$

where θ is the Bragg angle, λ is the wavelength of the incident radiation, and B is related to the mean-square displacement by

$$B = 8\pi^2 \langle x^2 \rangle.$$

B is called the atomic temperature factor or Debye-Waller factor.

It is important to state explicitly the assumptions in the Debye-Waller model of individual atomic motion. It is assumed that the potential in which each atom moves is harmonic, or at least has a very large harmonic component. It is further assumed that the motion is isotropic. The validity of these assumptions in the case of protein dynamics is considered below.

Determination of $\langle x^2 \rangle$

Each atom in a crystalline protein may be assigned an individual isotropic temperature factor, B, related to its mean-square displacement, $\langle x^2 \rangle$. Once a protein crystal structure has been solved and approximate atomic coordinates have been measured for most of the non-hydrogen atoms in the molecule, the preliminary atomic model may be improved by refinement. If only low-resolution data have been measured for the protein, it is not appropriate to include the mobility of the atoms in the model. If high-resolution data are available, then the X-ray scattering contribution from each atom can be modified by the exponential term containing the Debye-Waller B-factor, and B becomes an additional variable parameter for each atom in the least-squares refinement. Additional information is inserted in the form of known bond lengths, bond angles, planarity, and chirality. This structural information is assigned weights relative to that given to the X-ray data, and by appropriate adjustment of these weights a distribution of bond lengths, angles, etc. about the ideal values is obtained. Because the interatomic distances are not constrained to exact numbers but are allowed to range about the expected quantities, this modified least-squares process is called restrained refinement.

The use of restraints in protein structure refinement is open to criticism on the grounds of insufficient knowledge of the actual distribution of bond lengths and angles in real proteins, but it is only by this method that most crystalline proteins can yield any dynamic information. Very few protein crystals diffract to high enough resolution (beyond 1.5 Å) to yield enough data to permit unrestrained least-squares refinement with individual temperature factors. Evidence discussed below suggests that the restraints do not conceal meaningful variations in individual atomic mobilities. However, although mean-square displacements obtained from restrained refinements appear to have considerable relative accuracy, their absolute accuracy is open to doubt. One reason for this is the imposition of restraints on *B*-factors as well as on interatomic distances.

The technical aspects of the imposition of relative *B*-factor restraints have been discussed by Konnert & Hendrickson (48), whose method is the most widely used. To allow refinement of individual isotropic *B*-factors for all atoms in a protein with relatively low-resolution data (i.e. 2-Å resolution or even lower) it is assumed that the positional disorder of two covalently bonded atoms is highly correlated. In other words, two bonded atoms are restrained to have mean-square displacements that do not differ by a large amount. Intuition suggests that this is physically reasonable, but it must be kept in mind that, reasonable or not, restraints represent the imposition of preconceived notions about structure and dynamics on what would otherwise be an objective physical technique.

Factors Contributing to $\langle x^2 \rangle$

The *B*-factor calculated for each atom as the result of a restrained least-squares refinement may be viewed as the attempt to fit a Gaussian to the spread of electron density about the average position of that atom. Anything that produces a spreading of electron density will contribute to the *B*-factor, and consequently to the estimated mean-square displacement. Atomic motion, whether individual or part of a collective mode, will contribute. The crystal may be heterogenous with respect to the position of an atom: different molecules in the crystal may have the atom in different places, and these positions may not interconvert because of large potential energy barriers. This is equivalent to saying that the protein has folded into a number of distinct conformations, at least at the site in question. This static disorder will also cause a spreading of the electron density if the different positions are closer together than the resolution of the structure refinement, because the measured X-ray data represent an average over all unit cells in the lattice. It is important to realize that there may be some higher temperature at which enough kinetic energy exists to convert this static disorder into a dynamic one.

If the protein in question were in solution instead of in the crystalline state, the diffusional motion of the entire molecule would contribute to the observed displacements. In the crystal, lattice forces prevent such motion and no diffusional term need be considered. However, overall lattice vibration of the entire molecule as a rigid body is a possibility. Thus far, attempts to identify such vibrations in protein crystal structure analyses have not given any definitive results (84).

Finally, there is a contribution to the mean-square displacement of every atom that is directly due to the crystalline nature of the system. If the crystal lattice is not perfect, that is, if every molecule in the crystal is not in exactly the same position relative to the origin of its unit cell as every other molecule, there will be a spreading of the electron density that is due to static lattice disorder. Unlike the "frozen-in" conformational disorder discussed earlier, this lattice disorder is a temperature-independent property of the particular crystal being measured. It is possible to separate the lattice disorder into two components: translational and rotational. Rotational disorder should have a magnitude that increases with increasing distance from the axis of rotation; translational lattice disorder should affect all atoms in the protein equally.

Thus, the following expression can be written for every individual atomic mean-square displacement in the protein

$$\langle x^2 \rangle = \langle x^2 \rangle_v + \langle x^2 \rangle_c + \langle x^2 \rangle_{ld},$$

where $\langle x^2 \rangle_v$ is the contribution due to thermal vibrations, $\langle x^2 \rangle_c$ is the contribution due to motions that have a larger potential energy barrier than simple vibrations (and are therefore slower and may be static at some temperature), and $\langle x^2 \rangle_{ld}$ is the contribution due to lattice disorder (36). The crystallographer who wishes to extract the maximum amount of dynamic information from a refined crystal structure must separate the effects of lattice disorder from the other two quantities and must estimate the relative contributions of $\langle x^2 \rangle_v$ and $\langle x^2 \rangle_c$ for each atom or residue.

The Problem of Lattice Disorder

Thus far, only one attempt has been made to determine the magnitude of $\langle x^2 \rangle_{ld}$ in a protein crystal. Frauenfelder et al (36) made use of the insensitivity of the Mössbauer effect to lattice disorder: every Mössbauer-active nucleus absorbs randomly, and the recoilless fraction of the Mössbauer effect is not decreased by stationary disorder. At some temperature above the characteristic temperature of the sample (the temperature at which the mean time of the relaxation in question is equal to the nuclear lifetime),

$$\langle x^2 \rangle_M = \langle x^2 \rangle_v + \langle x^2 \rangle_c.$$

Comparison of this formula with that given above for $\langle x^2 \rangle$ from X-ray diffraction suggests a direct method of determining $\langle x^2 \rangle_{ld}$. One measures $\langle x^2 \rangle$ for some atom in a protein by both Mössbauer absorption and X-ray crystallography, on the same crystal if possible, and subtracts the Mössbauer value from the X-ray value, yielding the lattice disorder at that atom. Unfortunately, only a few nuclei are Mössbauer active. Frauenfelder et al (36) subtracted Parak & Formanek's (68) measurements of the $\langle x^2 \rangle_M$ for ^{57}Fe-doped iron in the heme of crystalline metmyoglobin from their own $\langle x^2 \rangle$ for the iron from a crystallographic refinement at 1.5 Å. The resulting $\langle x^2 \rangle_{ld}$ at the iron position was 0.045 Å^2. Use of this value to correct every atom in the protein for the effect of lattice disorder required the assumption that the lattice disorder was purely translational, an assumption based on no evidence but which yielded sensible results. Comparison of this value with the overall $\langle x^2 \rangle$ for the protein, 0.175 Å^2 at room temperature, suggests that lattice disorder contributes less than 30% of the total $\langle x^2 \rangle$.

Recently, the myoglobin lattice disorder determination has been re-evaluated in the light of new theories about the Mössbauer effect in crystalline proteins (42). Any motion having a characteristic time slower than 0.1 ms is now known not to contribute to $\langle x^2 \rangle_M$. Therefore, if motions on this time scale contribute to $\langle x^2 \rangle$ as determined by X-ray diffraction, subtracting $\langle x^2 \rangle_M$ with its underestimation of the dynamic contribution will lead to an overestimation of the lattice disorder term. It is now believed that $\langle x^2 \rangle_{ld}$ is approximately 0.025 Å^2 for metmyoglobin crystals, or less than 15% of the total mean-square displacement.

The problems encountered in the interpretation of the Mössbauer effect in protein crystals, plus the difficulty of the technique and the scarcity of Mössbauer active nuclei in most proteins, limit the use of this method. Its historical importance was to establish that, contrary to the belief of many scientists at the time, lattice disorder was not dominant in the myoglobin crystal. This result is in accord with the observation that myoglobin is a very well-diffracting protein, yielding strong X-ray reflections beyond a resolution of 1.2 Å. Other protein crystals that diffract to very high resolution should have small lattice disorder contributions.

Although it is very difficult to measure $\langle x^2 \rangle_{ld}$ accurately, it is also unnecessary. The important question is whether lattice disorder dominates the measured $\langle x^2 \rangle$ values. If it does not, the relative $\langle x^2 \rangle$ values will be meaningful, even though $\langle x^2 \rangle_{ld}$ cannot be subtracted from them. In their pioneering study of the dynamics of hen-egg-white and human lysozymes in the crystalline state, Phillips and co-workers pointed out that if the same protein crystallizes in two different crystal forms, comparison of the relative $\langle x^2 \rangle$ values for the atoms in each structure, refined independently, will reveal the effect of the lattices (10). If the structural correlations of $\langle x^2 \rangle$ are the same in the two distinct crystals, lattice disorder cannot be dominant. This was observed for the two lysozyme crystals, one of which was in a tetragonal space-group and the other in an orthorhombic space-group. The only areas of the two lysozyme structures that differed significantly in relative $\langle x^2 \rangle$ values were those residues involved directly in intermolecular contacts in the two lattices (Figure 1). This important result shows that the dynamics of proteins in the crystalline state are not disturbed by crystal

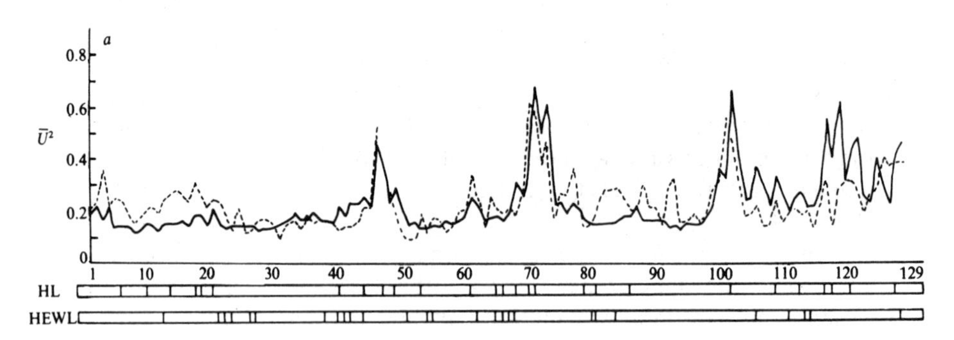

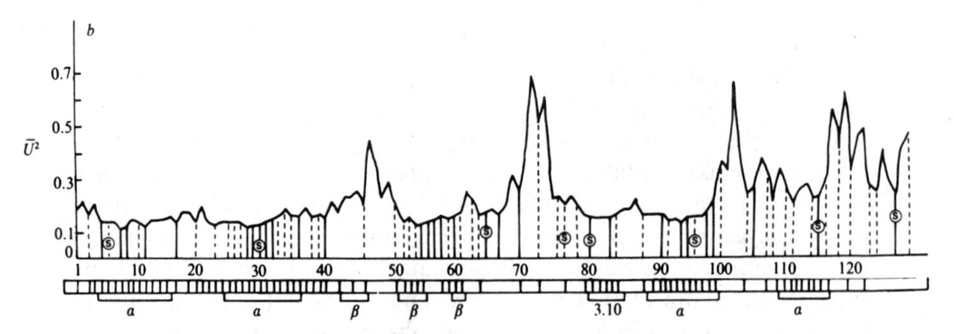

Figure 1 Plot of the mean main-chain U^2 $(= \langle x^2 \rangle)$ values against residue number for orthorhombic human lysozyme (HL, *full line*) and tetragonal hen egg white lysozyme (HEWL, *broken line*). Residues involved in intermolecular contacts in the two crystals are indicated in the lower rectangles. [Reproduced with permission from (10).]

contacts, except in very restricted regions. Extrapolating to other protein crystals, the $\langle x^2 \rangle$ values determined by X-ray crystallography should reflect the intrinsic dynamic behavior of the protein structure and will therefore be relevant to the dynamics of these molecules in solution.

The inability to determine the magnitude of $\langle x^2 \rangle_{ld}$ is serious when two structures are to be compared for changes in the mobility of residues or regions of structure. For example, evidence from hydrogen-deuterium exchange measurements in solution suggests that an enzyme becomes more rigid when inhibitor is bound. Restrained least-squares refinement of the crystal structures of the native enzyme and the enzyme-inhibitor complex yields two sets of $\langle x^2 \rangle$ values, but how are these data to be compared? If the inhibited crystal shows lower overall $\langle x^2 \rangle$s than the native, it is possible that the difference arises not from a change in the mobility of the protein on ligand binding but from a difference in the lattice disorder of the particular crystals.

There are several ways to deal with this problem. Ideally, both data sets should be measured from the same crystal, but the approach that is the simplest to carry out and that is the most widely used is the minimum function method (35), which involves an assumption about the level of $\langle x^2 \rangle$ values found in crystals. Values of $\langle x^2 \rangle$ smaller than the zero-point vibrational limit are not to be expected in any structure, and atoms that have $\langle x^2 \rangle$s near to this limit (0.01 Å^2) should never show a decrease in those $\langle x^2 \rangle$s. Therefore, if one takes two independently determined sets of mean-square displacements and adjusts the one with the higher overall $\langle x^2 \rangle$ so that its lowest values are equal to those in the other structure, the two data sets will have been placed on an approximately identical scale and any lattice disorder difference will have been subtracted out. A corollary of this reasoning is that if two sets of $\langle x^2 \rangle$s have roughly equal values for the smallest displacements, the lattice disorder in the two structures may be taken as approximately equal. This approach has been used in a comparison of crystal data from nine different complexes of ribonuclease A (37a).

In conclusion, it is probably more important to be aware of the possible effects of lattice disorder and to try to ascertain its relative effect on a given system than to try to measure its magnitude exactly.

Temperature Dependence of $\langle x^2 \rangle$

To separate the effects of static and dynamic disorder, and to obtain an assessment of the height of the potential barrier that is involved in a particular $\langle x^2 \rangle$, it is necessary to vary the temperature. A static disorder will be temperature independent, whereas a dynamic disorder will have a temperature dependence related to the shape of the potential well in which the atom moves and to the height of any barriers it must cross (36). Simple harmonic thermal vibration is expected to decrease linearly with temperature until the Debye temperature T_D; below T_D the mean-square displacement due to vibration is temperature independent and equal to the zero-point vibrational $\langle x^2 \rangle$. The high-temperature portion of a curve of $\langle x^2 \rangle$ vs T will therefore extrapolate smoothly to 0 at $T = 0$ K if the dominant contribution is $\langle x^2 \rangle_v$. In such a plot the low-temperature limb is expected to have values of $\langle x^2 \rangle$ equal to about 0.01 Å^2 (105). Departures from this behavior are characteristic of more complex motion and/or static disorder.

Proteins undergo other motions than just vibration in a harmonic potential with weak restoring forces. An example of a more complicated motion would be large-scale librations of aromatic rings. Any motion with a collective character will have a potential well different from a simple thermal vibration and will require more energy. There will thus be some temperature of relaxation, T_R, below which the motion will be "frozen-out" and a static distribution of conformations will exist. In the nomenclature of Frauenfelder et al (36), each member of this distribution is called a conformational substate. It has been shown that, if the barriers between substates are small, the distribution can still be dynamic even at temperature as low as 80 K (42). However, evidence from Mössbauer scattering suggests that the average value of T_R for several proteins is about 180 K (69). Thus, measurements of $\langle x^2 \rangle$ at temperatures below this value should show a much less steep temperature dependence than measurements above, if $\langle x^2 \rangle_c$ is a significant component of the total $\langle x^2 \rangle$.

Measurements of $\langle x^2 \rangle$ as a function of temperature also establish whether lattice disorder dominates the apparent dynamics. If the $\langle x^2 \rangle_{ld}$ term is the major contributor to the observed $\langle x^2 \rangle$ for the protein, there should be little or no temperature dependence for the individual atomic $\langle x^2 \rangle$ values.

Small-molecule crystallographers are familiar with these concepts, since it is routine to measure data at low temperature to improve precision by reduction of thermal motion. Albertsson et al have reported the crystal structure of D(+)-tartaric acid at 295 K, 160 K, and 35 K (7). Figure 2 shows the thermal elipsoids for the structure at each of these temperatures: the smooth variation of B with T is apparent.

Accuracy and Precision of $\langle x^2 \rangle$ Values

The precision of isotropic B-factors ordinarily can be calculated from the inverse of the least-squares matrix used in the refinement, but application of restraints in the procedure makes this impossible. Several small proteins have been refined by unrestrained least-squares at very high resolution; from these structures and from comparison of B-factors from the same

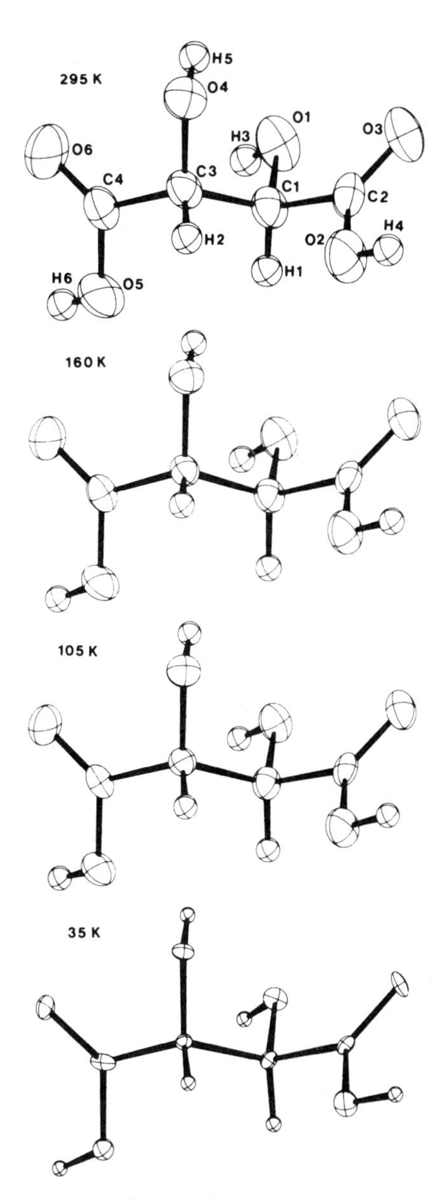

Figure 2 Molecular geometry and thermal ellipsoids (75% probability) of D(+)-tartaric acid at 295, 160, 105, and 35 K. [Reproduced with permission from (7).]

protein refined in different crystals, with different data sets and refinement methods (J. Kuriyan and G. A. Petsko, unpublished), a consensus has emerged about the precision of Debye-Waller factors. If the resolution of the structure is 2 Å or higher and if the atomic coordinates are well-determined (R-factor of about 20% or less), then the standard deviation of an individual B is about 10% of its value. A B of 20 $Å^2$ corresponds to an $\langle x^2 \rangle$ of 0.25 ± 0.025 $Å^2$.

Systematic errors have not yet been considered in this treatment. In comparing different data sets it is assumed that these errors are small. Unfortunately, they can be quite large. Any errors in atomic coordinates will be reflected in B-factors that are larger than their true values. Errors in correcting for radiation damage will also cause a systematic shift in B: overcorrection of the high resolution reflections will make B smaller than it should be, whereas undercorrection will cause B to be too large. Absorption of X-rays by the crystal, solvent, and capillary tube affects the relative measured intensities; if not properly corrected for, absorption will affect the refined B-factors in unpredictable ways. Finally, the resolution of the data set is directly reflected in the overall B-factor level. If the analysis is performed at too low a resolution, the B-factors will be systematically too low. This is particularly important when different data sets are being compared. If the two crystals were not measured to the same resolution, one data set should be adjusted to make them of equal resolution before refinement. If this is not done there could be a systematic difference in the B-factors that does not arise from differences in dynamics or lattice disorder. Experience suggests that a resolution of 2 Å is the lowest at which this effect may disappear; thus, comparison of a 1.9-Å resolution structure with a 1.5-Å resolution structure may be free of systematic differences that are due to resolution if both structures are well refined, whereas comparison of a 2.5-Å resolution structure with a 1.7-Å resolution structure would be subject to major difficulties (J. Kuriyan and D. Ringe, unpublished).

The final problem that affects the accuracy of crystallographically derived atomic displacements is the failure of low occupancy states to contribute to electron density above the noise level. An atom whose positional distribution has a shape with long tails will yield a B-factor reflecting only the highly populated center of the distribution. Consequently, the refined $\langle x^2 \rangle$ will be an underestimate of the actual displacement. In general, measurements of atomic motion by protein crystallography will be smaller than values determined by techniques such as NMR and protein dynamics, which can observe states of low probability.

In conclusion, careful refinement can yield $\langle x^2 \rangle$ values of good precision, but their absolute accuracy is difficult to determine and is likely to be much worse. Unless some independent measurement of one or more $\langle x^2 \rangle$ values is available for a cross-check, it is unwise to place much faith in the absolute values of individual atomic (or even residue averaged) $\langle x^2 \rangle$s. Their relative values, however, have been shown to be meaningful (see below).

RESULTS

Atomic Fluctuations

OVERALL AND INDIVIDUAL MAGNITUDES In their pioneering studies of rubredoxin, Jensen and his co-workers demonstrated that it was possible to refine protein crystal structures by least-squares methods if very high-resolution data were available (101). These investigators were the first to obtain individual isotropic *B*-factors for every non-hydrogen atom in a protein. Their analysis has now progressed to a resolution of 1.2 Å with a crystallographic *R*-factor of 12.8% (102). The overall *B* for rubredoxin is 12 Å^2, corresponding to an $\langle x^2 \rangle$, uncorrected for lattice disorder, of 0.15 Å^2. Individual *B*-factors vary from 7 Å^2 (0.088 Å^2 for $\langle x^2 \rangle$) to over 50 Å^2 (0.63 Å^2). Even the smallest of these values are larger than those normally observed for atoms in small organic crystal structures at room temperature ($B = 3$–4 Å^2, $\langle x^2 \rangle = 0.04$ Å^2). Other small, well-diffracting proteins give similar numbers to rubredoxin. The large magnitudes suggest that, compared to crystals of simple organic substances, protein crystals have either much larger lattice disorder or much larger $\langle x^2 \rangle_c + \langle x^2 \rangle_v$ values, or both. Atoms or residues with *B* values larger than 50 Å^2 are nearly invisible in the electron density map and either occupy many distinct conformations or have large amplitudes of motion.

There are a few small proteins that diffract to resolutions comparable to that of small molecule crystals, i.e. resolution of 1 Å or higher. These proteins include ribonuclease A (73), crambin (90), and avian pancreatic polypeptide (37b). In the refined crystal structures of these proteins a number of atoms have *B*-factors of 3–4 Å^2. The presence in the list of *B* values of numbers this small implies that lattice disorder is very small in these crystals, as suggested by their diffracting power, and also implies that these particular atoms or residues undergo predominantly simple atomic vibrations.

The resolution of the rubredoxin structure refinement allowed assignment of individual isotropic *B*-factors to every atom without the imposition of restraints on the difference in *B* between bonded atoms. Nonetheless, such atoms were found not to differ excessively in temperature factor. This observation suggests that the restraints imposed in nearly all other protein structure refinements are reasonable.

CORRELATION OF $\langle x^2 \rangle$ WITH STRUCTURE The individual atomic $\langle x^2 \rangle$ values determined for a number of crystalline proteins vary widely depending on their position in the tertiary structure of the protein. Their relative magnitudes correlate with structural features, with smaller *B*-factors occurring where atoms appear to be more rigidly bound by covalent and non-covalent interactions. It is this sensible behavior of $\langle x^2 \rangle$ values that, more than any theoretical consideration, is persuasive of their relative accuracy.

Consider for example the average *B*-factors for some side chains in rubredoxin. The average *B* for Val 24, which is internal, is 9.6 Å^2. The corresponding value for the surface Val 52 is 22.7 Å^2. Similar behavior is observed for Ile residues 33 and 12: the buried side-chain of Ile 33 has average $B = 10.9$ Å^2, but the value for the external Ile 12 is 35.7 Å^2 (102). Correspondence of average *B* with location has also been observed in myoglobin (36) and lysozyme (10), among other proteins. Residues that are involved in strong covalent interactions, such as cysteines in disulfide bridges or ligands to metal atoms, have significantly smaller *B*-factors than the same residue type when not so involved. Residues in loops that project away from the globular mass of the protein tend to have relatively high *B*-factors. Residues involved in α-helices and β-pleated sheets have, on average, lower mean-square displacements than residues in irregular structure. The termini of the polypeptide chain, unless involved in strong intra- or intermolecular interactions, have very high mean-square displacements, in keeping with the observation that the electron density for these regions is often very weak or absent altogether (78).

Figure 3*a* and *b* shows the variation in average main-chain atomic displacement for myoglobin (36) and lysozyme (10). In myoglobin (Figure 3*a*), the regions that form the heme binding pocket are significantly more rigid than the external interhelix loops. In lysozyme (Figure 3*b*), the active site cleft is particularly mobile. In both cases, the presence of secondary structure is correlated with relative rigidity.

Individual atomic $\langle x^2 \rangle$ values in other proteins show the same sensible correlation with structure. An example is given in Table 2, taken from the work of Baker on actinidin (12). Lysine 145 in actinidin projects away from the surface of the molecule, and the *B*-factors for the side-chain atoms increase steadily along the chain. The objection has been raised that this behavior is caused by the restraints imposed during the refinement, which decrease as one nears the terminal atoms in the side chain [see discussion following (35)]. However, examination of the other lysine residues in Table 2 shows that this is not the case. Lysine 17 and 181 are internal, and their *B* values are, as expected, low and nearly uniform along the side chain. Lysine 39 is confined in a surface pocket and Lys 106 and 217 are exposed, but unlike Lys 145 they do make some interaction with other protein atoms. The tabulated *B*-factors are perfectly in accord with the qualitative predictions made on the basis of these locations. Similar sensible behavior has been demonstrated for the *B*-factors of internal and external side chains in sperm whale metmyoglobin (36), lysozyme (10), and rubredoxin (102).

One interesting common feature in all of the careful analyses of protein $\langle x^2 \rangle$ values carried out to date is that the atoms of the backbone of a particular residue all have very similar *B*-factors, but the carbonyl oxygen atom of the peptide bond often has a slightly higher *B*. This may indicate torsional flexibility of the carbonyl unit. Because of this systematic effect, most workers use only the N, C_{alpha}, and carbonyl carbon atoms of the backbone for computing average main-chain *B*-factors.

Plots of individual $\langle x^2 \rangle$ values vs distance from the center of mass of the protein for a number of different proteins all show a general increase in the magnitude of $\langle x^2 \rangle$ with increasing distance (10, 36). Frauenfelder et al pointed out that, in terms of X-ray determined mobility, proteins can be considered to have a condensed core and a semi-liquid outer shell (36).

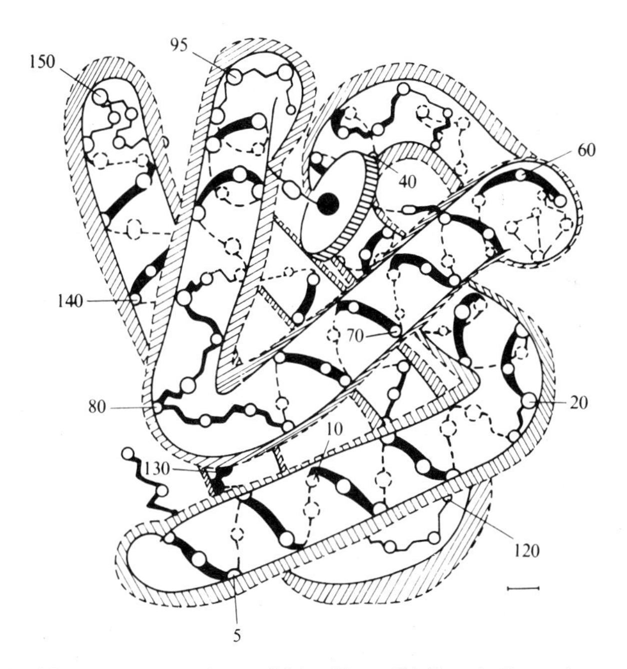

Figure 3a Backbone structure of myoglobin. The solid lines indicate the path of the polypeptide chain; alpha carbons are circles. Shaded areas indicate regions in which the main chain can move, with 99% probability, assuming the measured $\langle x^2 \rangle$ pertain to a Gaussian distribution. The scale bar is 2 Å. [Reproduced with permission from (36).]

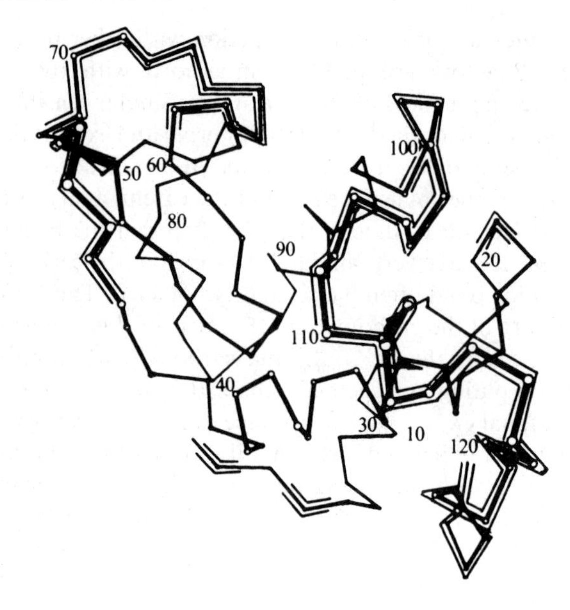

Figure 3b Perspective drawing of the main chain of lysozyme with residues having $\langle x^2 \rangle$ values > 0.2 Å^2 in the refined human lysozyme structure outlined by parallel lines. The active site cleft runs almost vertically down the front of the drawing. [Reproduced with permission from (10).]

Table 2 B values for all Lys and Phe side chains in actinidin[a]

Residue	Location	Side-chain atom and *B*-factor (Å^2)						
Lys		CB	CG	CD	CE	NZ		
17	Internal	6.6	7.8	6.0	5.8	7.6		
39	Confined in pocket	8.0	5.2	8.7	7.1	9.2		
106	Exposed	10.2	12.4	13.4	24.2	21.7		
145	Projects into solvent	16.3	22.8	33.8	50.0	48.3		
181	Internal	4.5	7.2	9.8	12.8	8.0		
217	Exposed	12.7	16.4	18.7	15.4	14.0		
Phe		CB	CG	CD1	CE1	CZ	CE2	CD2
28	Internal	6.8	5.4	8.4	9.5	8.5	6.1	6.3
74	Internal	5.0	5.8	5.2	8.7	6.3	5.3	8.5
76	One exposed edge	4.8	6.7	9.1	9.0	7.0	9.2	9.6
144	Internal	9.9	9.0	13.2	15.6	13.2	18.9	13.7
152	One exposed edge	12.5	10.4	10.0	9.3	15.2	12.8	11.3

[a] From (12).

However, within each of these regions there is considerable local variation, suggesting an inhomogeneous system. Mobility of the core is therefore likened to an aperiodic solid, whereas the term "semi-liquid" as applied to the outer shell refers to the large but bounded $\langle x^2 \rangle$ values possible for surface atoms. Even on the surface there are differences in mobility assignable to chemical properties. Charged side chains have larger $\langle x^2 \rangle$ than uncharged side chains, even when the two are exposed to the same degree (36).

TEMPERATURE DEPENDENCE OF $\langle x^2 \rangle$ To date, only two studies of protein crystal structures at very low temperatures have reported refined *B*-factors. Huber and associates have refined the structure of trypsinogen in methanol-water, initially at 213 K (83) and subsequently at 173 and 103 K using X-ray data collected with synchrotron radiation (100). The purpose of this study was to ascertain if the "activation domain," a region of 15% of the molecule, which is disordered in the native crystal structure at room temperature (43), became ordered at low temperature. Cooling to 103 K reduced the overall *B*-factor of the trypsinogen structure from 16.1 to 11.5 Å^2, and some residues at the N-terminal region became more clearly visible, but the majority of the activation domain remained disordered. There are three possible explanations for this effect: either the domain is still mobile at 100 K, the disorder is static at all temperatures, or a dynamic disorder has been frozen into a static disorder in which a number of conformations of nearly equal energy are nearly equally populated, but none to such an extent that it is visible above the noise level in the electron density map. A number of elegant chemical labelling (100) and physical measurements (20) have been carried out by Huber and co-workers, who conclude that the disorder in the activation domain of trypsinogen is dynamic in solution at room temperature with conformational transitions occurring with a reorientation correlation time of 11 ns (20). Stroud and associates, working with essentially the same crystal form of trypsinogen, have reported reasonably clear electron density for much of the activation domain (52). Differences of this kind may reflect differences in sample preparation, crystallization procedures, and methods of refinement, as well as the effect of omitting low resolution data that are sensitive to the presence of highly mobile regions.

One interesting aspect of the trypsinogen low temperature study is that no change in overall *B* occurred on cooling from 173 to 103 K (100). This may reflect the relaxation temperature believed to exist for proteins at around 180 K, below which many of the collective motions are frozen out, or it may indicate that increasing lattice disorder, after extreme cooling, compensated for reduction in mobility.

The structure of metmyoglobin has been determined to a resolution of 2 Å at 80 K (42). Data collection at this temperature was achieved by a "flash-freezing" technique that did not involve the use of a cryoprotective mother liquor. The overall *B*-factor decreased from 14 Å^2 at 300 K ($\langle x^2 \rangle$ = 0.175 Å^2) to 5 Å^2 at 80 K ($\langle x^2 \rangle$ = 0.063 Å^2). The drop in $\langle x^2 \rangle$ with cooling observed in this study and that of Huber's group on trypsinogen are the most convincing evidence that proteins are flexible in the crystalline state and that static disorder and experimental errors do not dominate the observed *B*-factors. When the 80 K data were compared with those observed earlier over a more restricted range of temperatures (36), it was seen that only 46 of the 153 residues in myoglobin had average *B*-factors that extrapolated to 0 at 0 K. The temperature dependence of the remainder of the protein was inconsistent with simple harmonic vibration but consistent with the notion that conformational substates could be frozen out at sufficiently low temperatures. Fifty-one residues can be modelled with a linear temperature dependence, with $\langle x^2 \rangle$ = approximately 0.04 Å^2 at 0 K. The remaining 56 amino acids in myoglobin do not show a good linear fit of $\langle x^2 \rangle$ vs *T*.

Figure 4 shows a plot of the average *B*-factor at 300 and 80 K for the backbone atoms in myoglobin vs residue number. From these data it can be seen that, despite differences in crystal, data collection method, absorption correction and other parameters, the overall correlation of $\langle x^2 \rangle$ with structure is the same at the two temperatures. The two curves also show

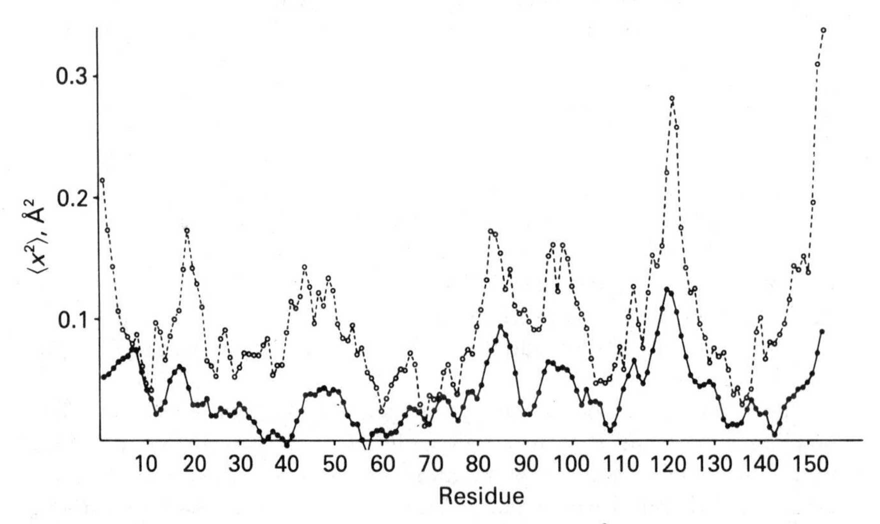

Figure 4 Average backbone (N, alpha C, and carbonyl C) $\langle x^2 \rangle$ values for myoglobin vs residue number: ●, 80 K; ○, 300 K. An $\langle x^2 \rangle_{ld}$ of 0.045 Å^2 has been subtracted from the observed $\langle x^2 \rangle$ values in both structures. [Reproduced with permission from (42).]

that residues with large displacements tend to have large temperature dependence, whereas residues with small *B*-factors show little change on cooling. Although the observed data do not fit a simple harmonic motion model, this behavior is quasi-harmonic. Huber has noted similar results for trypsinogen (100).

As mentioned above, spectroscopic data on a number of proteins suggest that there is a relaxation temperature at about 180 K. Unfortunately, the present data for myoglobin cluster in points far removed from this value and could be fit by a single straight line with high scatter or by two lines, one nearly temperature-independent, intersecting just below 200 K. It is important to determine the exact form of the temperature dependence experimentally. Parak and associates (personal communication) are currently measuring myoglobin data at 165 K and at near 4 K; these two points should allow a proper determination.

One remarkable observation from the comparison of the myoglobin structures at 300 and 80 K was an apparent reduction in the volume of the protein at low temperature. The unit cell volume of the crystal is 4.5% smaller at 80 K than at room temperature (42), reflecting a general shrinkage of the molecule by approximately the same amount (34). Because of the nonuniformity of interactions within the protein, the thermal expansion coefficient might be expected to be anisotropic. Analysis of the spatial distribution of the shrinkage of myoglobin on cooling has shown that some regions of the molecule move much more than others. In particular, an external loop of charged amino acids, the CD corner, moves in toward the center of the protein considerably more than any other region (34). Taken together with the temperature dependence of the $\langle x^2 \rangle$ values, this study of the thermal contraction of myoglobin provides the most detailed picture yet available of the physical chemistry of a protein. It should be noted that Huber observed a change in the center of mass of trypsinogen, on cooling to 100 K, consistent with an anisotropic contraction of the structure (100). Since myoglobin is somewhat atypical, being an all-helical protein with a large prosthetic group, analysis of the thermal contraction of other proteins is awaited with interest.

Low temperature crystallographic studies have been carried out on one nucleic acid, a B-DNA dodecamer (29). Refinement at 16 K revealed a large overall drop in *B*, but some of the atoms in the molecule still had large *B*-factors even at this very low temperature. These large residual mean-square displacements were interpreted as demonstrating static disorder; however, by analogy with the protein studies discussed above, a dynamic disorder even at 16 K cannot be ruled out, and the freezing-in of a disorder among conformational substates, which is dynamic at room temperature, is also consistent with the data.

COMPARISON WITH MOLECULAR DYNAMICS Since X-ray diffraction is only able to visualize the highly occupied states of a molecule (in an average over many molecules and many days of data collection), it is expected that high frequency, low amplitude motions will be the major contributors to the observed $\langle x^2 \rangle$. Molecular dynamics can simulate motions on this time scale. Given the atomic coordinates of a protein from an X-ray structure, an empirical potential energy function may be written as a function of the coordinate set. It is then possible to derivatize this function to obtain the force on each atom, and to solve the Newtonian equations of motion for a small time interval, usually a fraction of a picosecond (61, 62). The importance of the potential energy function chosen, the limitations of the method, and general strategies in its use have been discussed recently (54). The complexity of the calculation is such that, with currently available computers, simulations may be run for total time periods of only a hundred picoseconds or so. Nevertheless, calculations have been carried out for a number of proteins, and the results have been compared with the dynamic picture obtained from X-ray diffraction analysis of $\langle x^2 \rangle$ values.

Remarkably, the agreement between the theoretical calculation and the experimental data is good. To compare X-ray-derived $\langle x^2 \rangle$ values with those obtained from the mean-square displacement of atoms during the molecular dynamics simulation, one must allow for the assumed isotropy of the Debye-Waller model and remove the lattice disorder term. Thus,

$$\langle x^2 \rangle_{\text{mol dyn}} = \frac{3B}{8\pi^2} - \langle x^2 \rangle_{\text{ld}}$$

where the equal sign represents "should equal." The earliest molecular dynamics simulations were performed by McCammon, Karplus and associates, who studied the basic trypsin inhibitor of bovine pancreas. They found that after 9 ps of simulation the structure had altered in some regions but generally remained close to the X-ray structure (61). The same conclusions were reached after 96 ps. However, comparison of the mean-square displacements with X-ray $\langle x^2 \rangle$ values was difficult, since the refinement method employed for trypsin inhibitor was not optimum for the determination of individual B-factors. Such a comparison was performed for reduced cytochrome c (64), where the crystal structure had been refined by restrained least-squares (89). The lattice disorder correction was made by the minimum-function method and was assumed to be 0.07 Å^2. Figure 5 shows the comparison of the calculated and experimental root-mean-square fluctuations vs residue number. There is excellent agreement between the two in terms of regions of high and low displacement.

In the simulations performed to date, the atomic motions have been found to be highly anisotropic and somewhat anharmonic. The rms fluctuation of an atom in its direction of largest displacement is often twice that observed in its direction of smallest displacement and may be even larger (98). There is often very little correlation between the preferred direction and local bonding (65, 88). Anharmonicity predominates, but a quasi-harmonic model in which the potential of mean force for an atomic displacement is itself temperature dependent appears to fit the data reasonably well (58, 98).

The size of the molecular dynamics calculation necessitates that most simulations are run for an isolated protein in vacuum. Lack of good models for the solvent continuum has also been a contributing factor. Therefore, there should be discrepancies between theory and experiment in those regions of the protein involved in intermolecular contacts in the crystal lattice and in some of the external side-chains. For cytochrome *c*, the largest differences are found in surface charged residues, particularly lysines (47). Omission of the solvent and crystal contacts is not found to affect the dynamics of the interior of the protein significantly. Simulation of pancreatic trypsin inhibitor in a Lennard-Jones solvent or with crystal contacts improved the agreement between the average structure from simulation and the X-ray structure; the rms deviation was reduced from 2.2 to 1.35 Å in the "solvent" and to 1.52 Å in the "crystal" (97, 98).

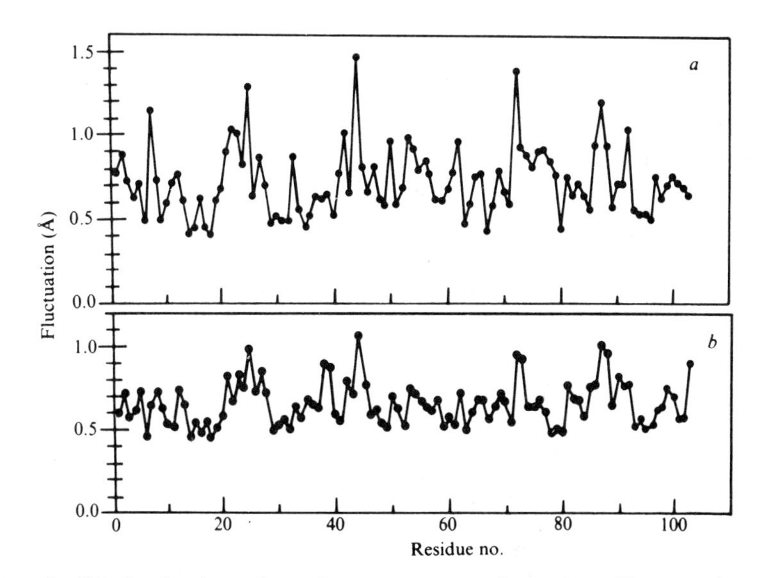

Figure 5 Calculated and experimental root-mean-square fluctuations of ferrocytochrome *c* averaged over each residue plotted vs residue number. (*a*) molecular dynamics simulation. (*b*) X-ray temperature factor data corrected for an estimated lattice disorder of (3 × 0.073 Å^2) by the minimum function method. [Reproduced with permission from (64).]

The effect of neglect of solvent and crystal contacts in the dynamics calculations has also been shown in simulations of myoglobin (J. Kuriyan, R. M. Levy, S. Swaminathan, and M. Karplus, unpublished). Although overall agreement between the X-ray $\langle x^2 \rangle$ and the fluctuations from molecular dynamics is good, one region of the protein shows much greater mobility in the simulation than indicated by its *B*-factors. This region is the CD corner, which is also observed to undergo the largest change in conformation on cooling of the crystalline protein to 80 K. As this region is highly charged, it is possible that electrostatic effects are not properly treated in the simulation. Nevertheless, even vacuum simulations appear to give an excellent representation of the relative mobilities of internal residues in proteins.

Recently, a 12-ps molecular dynamics simulation has been carried out on the full unit cell of trypsin inhibitor containing 4 protein molecules and 560 water molecules, the latter generated theoretically (96). The magnitude of the fluctuations corresponds to those found in the earlier simulations of the isolated protein, whereas the structure averaged over the full simulation was found to have an rms discrepancy of only 1.2 Å from the X-ray structure. The difference between the average structures of the four molecules in the unit cell was found to be of the same magnitude as the difference between each of the molecules and the X-ray structure. Averaging over the four molecules decreased the difference from the X-ray structure, suggesting that no systematic deviation from the X-ray structure occurred. Comparison between the rms fluctuations from this simulation and the $\langle x^2 \rangle$ values from refinement of the trypsin inhibitor crystal structure shows better agreement than seen in earlier simulations of the same protein, but still poorer than observed for myoglobin, lysozyme, or cytochrome *c*. Forty-seven bound water molecules have been located in the crystal structure of trypsin inhibitor, but only 9 of these are reproduced within 1 Å in the simulation.

Van Gunsteren and associates have tried to use the molecular dynamics results from this simulation to overcome the limitations of isotropy and harmonicity imposed on the X-ray analysis by the Debye-Waller model. Their method is as follows: molecular dynamics produces a large number of configurations, which together describe the trajectories of all atoms in the unit cell. From each configuration, a model electron density map can be generated and the resulting set of maps may be combined to produce an average map containing electron density distributions for each atom that reflect its motion throughout the simulation, regardless of the complexity of that motion. Fourier transformation of this average map yields structure factors that can be compared with the measured X-ray data. The 1 Å discrepancy between the dynamics structure and the X-ray structure is reflected in a 52% R-factor for this comparison. However, the R-factor for low resolution data is actually better than obtained from crystallographic refinement, indicating better treatment of the bulk solvent in the crystal interstices. Even more encouraging, when the average molecular dynamics positions were shifted back to the X-ray positions, but the *B*-factors were taken from the molecular dynamics simulation, the R-factor for data between resolutions of 6.65 and 1.5 Å was 29%, compared with 25.8% when the X-ray temperature factors were used (96).

To date, only one molecular dynamics simulation has been carried out at low temperatures and compared with corresponding X-ray data. Calculation was performed on a decaglycine α-helix as a function of temperature between 5 and 300 K (55). Below 100 K the harmonic approximation is valid for the motions of this helix, but above that value the average $\langle x^2 \rangle$ for the alpha carbon atoms is more than twice that found in the harmonic model. These data were compared with the observed $\langle x^2 \rangle$ for the helices in myoglobin at 300 and 80 K (42). Correcting the X-ray data by an assumed lattice disorder of 0.025 Å^2, the root-mean-square displacements of the alpha carbon atoms of all the helical regions of myoglobin are 0.48 Å at 300 K and 0.28 Å at 80 K. The simulation values are 0.52 Å at 300 K and 0.22 Å at 80 K for the decaglycine helix. Once again, the agreement between theory and experiment is excellent, giving confidence in both approaches.

COMPARISON WITH OTHER MEASUREMENTS X-ray diffraction provides information about the spatial distribution of protein fluctuations but not about their time scales. Techniques such as NMR (74, 99), fluorescence spectroscopy (13, 53, 63), and Mössbauer spectroscopy (28, 68, 69) are complementary. The time domains of most of these methods are much longer than that of molecular dynamics, and it is likely that X-ray diffraction results are most appropriate to the high frequency regime. Although one might expect from theoretical considerations that regions of a protein undergoing large rapid fluctuations also demonstrate considerable mobility on a longer time scale, this has never been proven experimentally. Fluorescence and NMR spectroscopy can identify the presence of a highly mobile region within a protein (13, 79), which can be compared with the absence of strong electron density in a crystal structure determination of the same molecule. The strength of such an approach is the complementary nature of the information provided. Spectroscopy can indicate that a region of the polypeptide chain containing, for example, a tryptophan and a histidine is unusually mobile. X-ray diffraction can indicate to which segment of the amino acid sequence these residues belong.

One can then return to the spectroscopic approach to examine the time dependence of this motion and its alteration upon substrate binding in light of the location of the region in the three-dimensional structure.

One exciting advance in NMR technology that promises to allow direct comparison between some spectroscopic and X-ray data is solid-state NMR. It is possible to measure the electric quadrupole splitting of a specifically deuterium-labelled group in a magnetically ordered suspension of small, paramagnetic protein crystals (66, 80). Further developments in this area are awaited with interest.

Mössbauer scattering can give direct information about the mean-square displacement of atoms in proteins (28). Myoglobin and hemoglobin have been studied extensively by this method, and the results at the iron position have been compared with data from X-ray diffraction (28, 68, 69). The agreement is quite good. The temperature dependence of the $\langle x^2 \rangle$ from the myoglobin iron Mössbauer effect reveals several different dynamic processes (69). Below 160 K, $\langle x^2 \rangle$ increases linearly with T, as expected for $\langle x^2 \rangle_v$ in a harmonic system. Above 160 K, collective conformational motions become important, and $\langle x^2 \rangle_c + \langle x^2 \rangle_v$ deduced from Mössbauer data on myoglobin crystals are compatible with X-ray diffraction data. Correlation of the mean-square displacements in myoglobin as measured from crystallography and Mössbauer scattering has led to a new understanding of the molecular events responsible for certain effects in the Mössbauer spectra of atoms in proteins (42, 69, and F. Parak, personal communication).

One interesting technique that can be correlated directly with both neutron and X-ray diffraction is hydrogen exchange. Several different classes of peptide amide hydrogens can be seen on the basis of their exchange rates, with the slowest requiring days or longer for complete exchange (110). Neutron diffraction can indicate which protons in the crystalline molecule are exchanged completely after a given time of soaking the crystal in deuterated mother liquor. Several classes of hydrogens are also observed by this method (40, 50, 51, 107), but even in the crystal nearly all amide protons are able to exchange eventually. These observations indicate that the motions which allow buried protons to exchange with solvent occur within the crystal lattice. One can then ask whether there is any correlation between the slowest exchanging amide protons and regions of the molecule with very low $\langle x^2 \rangle$. The answer in the case of bovine trypsin is that no correlation is found (50). Motions that permit solvent to reach protons that are inaccessible in the static structure are apparently too slow (i.e. generate conformations too low in occupancy) to make a major contribution to individual atomic B-factors.

CAVITIES INSIDE PROTEINS Both experimental and theoretical studies have confirmed the mobility of atoms inside globular proteins. A pertinent question is where these atoms can move to, given the tightly packed nature of most macromolecules. Richards pointed out that the packing of atoms in the interior of proteins is not perfect and that small cavities are found (76). It is these cavities that provide the room for internal residues to move. Presumably, as the structure fluctuates, new cavities are formed as atoms move to fill old cavities. This suggests that internal motions are highly collective in character, even over nonbonded neighbors (61a).

The assumption has been made for some years that large empty spaces do not exist inside globular proteins and that any big cavity will be occupied by one or more water molecules. However, there are a number of large cavities inside myoglobin that appear to be highly apolar. Recently, an investigation of the binding of xenon in those cavities by both NMR (91) and high-resolution X-ray diffraction (92) has shown that xenon will occupy four of the internal cavities in myoglobin with only slight alteration in the position of surrounding amino acids. Filling these cavities has a small effect on the overall $\langle x^2 \rangle$ for the protein, which is reduced by 15% relative to the xenon-free structure at the same temperature (the minimum-function method was used to correct the two data sets for lattice disorder). Larger reductions in $\langle x^2 \rangle$ are observed for some of the side chains in van der Waals contact with the xenon atoms. This study establishes the existence of cavities inside myoglobin and implies that reduction in cavity volume is correlated with reduction in protein flexibility. It is interesting to note that there is no obvious pathway from outside the protein to any of the internal cavities occupied by xenon in the static structure of myoglobin. The ability of xenon to reach these sites, even in the crystalline protein, is further evidence that much of the solution flexibility of macromolecules is retained in the crystal state.

CHANGES IN MOBILITY Assuming that lattice disorder differences can be estimated or are unimportant, protein crystallography provides an ideal method for determining changes in the mobility of groups of atoms (or the entire molecule) when a protein binds ligands or is otherwise altered. The effects of Xenon binding on the $\langle x^2 \rangle$ values of myoglobin have already been noted. Other studies have focused on the changes in B-factors that occur when inhibitors bind to ribonuclease A (106), *Streptomyces griseus* serine protease A (44), and other enzymes. Figure 6 shows the difference in B-factor plotted as a function of residue number for the serine protease; the reduction in mobility by those residues that make contact with the inhibitor is specific and dramatic. Individual changes in mobility have been seen for

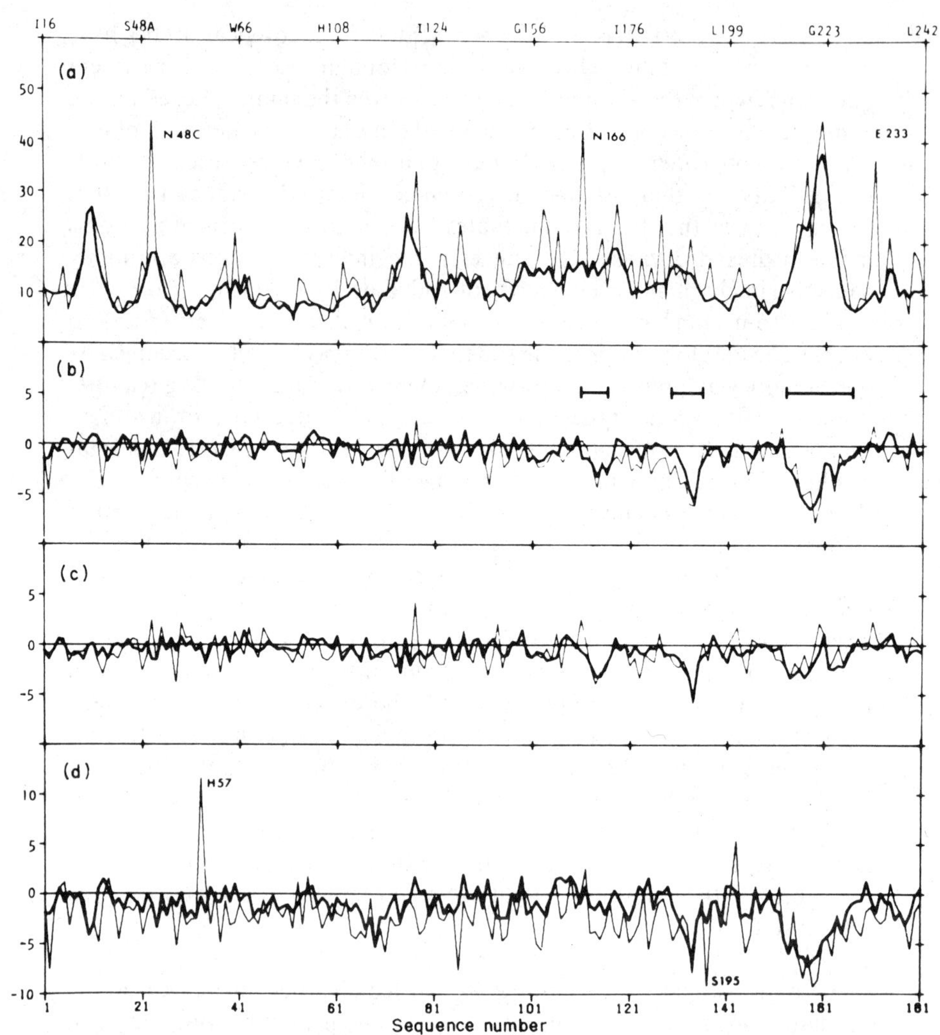

Figure 6 Average *B*-factors ($Å^2$) for main-chain (*thick line*) and side-chain (*thin line*) atoms in *Streptomyces griseus* Protease A vs residue number. (*a*) Average *B* values for the enzyme in its complex with a tetrapeptide product inhibitor. The residues underlined by horizontal bars form the binding site. The three labelled residues are all external. (*b*) Difference in average *B* between the atoms of the enzyme-product complex and those of the native enzyme. The regions showing the largest differences are the substrate binding sites, which become more ordered. (*c*) As for (*b*), but with a different product. (*d*) as for (*b*) and (*c*), but the complex is with a peptide aldehyde that forms a covalent hemiacetyl bond to Ser 195. His 57 has undergone a triggered conformational change and its *B*-factor increases by 12 $Å^2$. Formation of the covalent bond from the gamma oxygen of the serine to the carbonyl carbon of the aldehyde is accompanied by a large decrease in average *B* for Ser 195 (-9 $Å^2$). [Reproduced with permission from (44).]

residues of myoglobin on going from the met to the oxy form (35). Overall and individual changes in $\langle x^2 \rangle$ are observed for ribonuclease A when the inhibitory sulfate ion is removed from the active site (20a).

It is not necessary to carry out complete refinement of a structure to detect changes in the disorder of a residue or group of residues. Huber and associates have solved the structures of several complexes of trypsinogen with peptides and inhibitors and have noted a marked increase in the clarity of the electron density in the activation domain prior to any refinement (17, 18, 43). Alber and co-workers have noted similar behavior for a ten-residue flexible loop in the structure of triose phosphate isomerase (5). In the native enzyme this loop is almost invisible in the electron density map, apparently because of high thermal motion (2, 3). On binding of substrate, the loop adopts a different average conformation and becomes ordered (2–5). It is expected that the qualitative assessment of changes in mobility will be one of the most important uses of protein crystallography in the area of protein dynamics.

Collective Motions

Evidence that collective motions occur in proteins comes predominantly from molecular dynamics simulations. Analysis of the time dependence of the atomic fluctuations in the trypsin inhibitor simulations demonstrated that motions contributing to the mean-square displacement of any atom could be separated into local oscillations superimposed on collective motions (88). Individual atomic fluctuations have a subpicosecond time scale; the collective motions have time scales of at least 1 ps and often longer. These motions may involve all the atoms in a residue, an atom plus its nonbonded neighbors, or groups of linked atoms such as an alpha helix. Interestingly, over a time period of 25 ps, the individual fluctuations contributed only 40% to the average $\langle x^2 \rangle$ of the main-chain atoms. Further, these motions tended to be of uniform magnitude over the whole structure. The low-frequency, collective motions provided the variation of $\langle x^2 \rangle$ with structure (88).

Examination of a plot of $\langle x^2 \rangle$ vs residue number from X-ray diffraction such as shown in Figure 4 is intriguing in the light of these findings. The observed structural dependence of such a plot must come from collective modes of motion to a large extent. But because individual isotropic *B*-factors are used to fit the X-ray data, detailed information about the direction and magnitudes of these collective motions is lost. Consider the data for individual side-chain atomic displacements in actinidin shown in Table 2. Internal lysines show $\langle x^2 \rangle$ values that are uniform over the residue, suggesting collective motion. This is further indicated by the *B* values in the table for the phenylalanine residues, all of which are internal.

With the exception of Phe 144 (which is adjacent to the highly mobile Lys 145 and may be undergoing a collective motion with that residue), the phenylalanines have *B*-factors that are nearly equal within a side chain, assuming the 10% experimental precision. Aromatic rings are expected to have highly correlated displacements, and these data are consistent with expectation.

Collective motions destroy the correlation of anisotropy with chemical bonding because of their relatively large amplitudes (62a). Their large contribution to the $\langle x^2 \rangle$ determined from crystallography means that any attempt to fit an anisotropic model to protein fluctuations in an X-ray analysis must not assume that the principal directions of motion will lie along bond directions. Although such correspondence may occur, it is not predominant.

Normal mode calculations, though difficult for large proteins, are expected to be useful in understanding collective fluctuations in structure (19a, 38). Calculation of the normal modes of low-frequency vibrations in bovine pancreatic trypsin inhibitor revealed a rich variety of motions that are due to concerted variations in soft variables, i.e. dihedral angles (38). Most modes with frequencies above 50 cm^{-1} (time periods of less than 0.7 ps) behaved harmonically at room temperature. As expected, the mean-square displacements in atomic positions were found to be determined mainly by collective modes with frequencies below 30 cm^{-1} (19a). Comparison of the main-chain average $\langle x^2 \rangle$ computed from the normal mode analysis with values from X-ray crystallography and molecular dynamics revealed good agreement.

If a large-scale collective motion is important, its presence may be detected by an X-ray diffraction study, since the electron density for the particular group of atoms may vanish altogether (3, 25, 78). Quantitative information about such motions cannot be obtained, since there is no electron density to fit, but a qualitative picture that can aid in the interpretation of spectroscopic data is possible. Another class of motions that cannot be seen by crystallography are those that are fast but infrequent. Tyrosine ring flips are one example. The actual flipping of even a buried ring occurs in less than 1 ps, but the event is very improbable (47, 99). Consequently, the intermediate states in the rotation are very low in occupancy. X-ray results confirm this picture: although ring flips are known to occur, the electron density for aromatic rings in protein structures is disk shaped, not spherical (37, 102).

Refinement methods that make collective motion analysis possible have been described for small molecule crystallography. Anisotropic temperature factor analysis should aid in the identification of collective modes, since groups of atoms with the same principal direction of motion could be presumed to have correlated mobility. Some protein refinement schemes treat certain covalently bonded atoms as a group (42a). This approach has promise but neglects the noncovalent interactions that also lead to large collective modes in dynamics simulations. Clearly, extracting information about collective motions from a protein crystal structure is very difficult. What is exciting is that the information is present.

Triggered Conformational Change

When a ligand binds to a protein, the protein may adjust to accommodate the presence of the new molecule. This new structure has a lifetime as long as the ligand remains bound, or as long as the structure of the ligand remains unchanged. Such changes in protein structure have long been observed by spectroscopic methods but until recently could not be defined. The availability of structural data has made it possible to observe the end effect of such movements. These motions are not random but rather are the result of a specific stimulus. Motion is inferred because a portion of the protein occupies a different position in the structure with bound ligand compared to the native structure. For instance, when the substrate dihydroxyacetone phosphate binds to the glycolytic enzyme triose phosphate isomerase, the section of polypeptide chain bounded by Trp 168 and Thr 177 moves more than 10 Å from a relatively external, disordered position to a well-defined one covering the active site pocket (2, 3). A similar movement of about 4.5 Å has been observed in the acid protease penicillopepsin when an inhibitor is bound (43a).

X-ray diffraction can observe the initial and final states of such a movement but gives no information about the path or time of the motion involved. In the case of triose phosphate isomerase, the loop that moves goes from a relatively disordered state to an ordered one. In terms of dynamics this means that the individual atoms of the loop have a high mobility in the native protein and a low mobility in the bound form. This is not always necessarily true. Since the energy for triggered conformational change comes from the energy of specific interactions, such as breaking and remaking hydrogen bonds and/or salt bridges, not from thermal energy, it does not follow that a segment involved in such a movement has to be initially mobile in a dynamic sense. Hemoglobin is one example of this: the quarternary structural change on oxygenation is a transformation between two ordered states (70, 71).

X-ray diffraction studies cannot give information on the speed of such a movement. A lower limit can be obtained for the time scale from well-defined kinetic studies only if the rate-limiting step can be associated with a binding step. For instance, the rate-limiting step in the reaction catalyzed by triose phosphate isomerase is the dissociation of product, which occurs with a microscopic rate constant of $4 \times 10^3\ s^{-1}$ (8). Therefore, it can be inferred that the opening of the flexible loop, which must take place to allow product to diffuse away from the active site, occurs in less than 1 ms.

The types of substances whose binding can trigger a specific conformational change include substrates [dihydroxyacetone phosphate in triose phosphate isomerase (3)], coenzymes [NAD binding to lactate dehydrogenase (1)], allosteric effectors [CTP and ATP in aspartic transcarbamylase (56)], inhibitors [statine-containing polypeptide inhibitors of acid proteases (19, 43a)] and even protons [the Bohr effect in hemoglobin (71); pH-dependent conformational changes in adenylate kinase (67)]. Despite its limitations, X-ray diffraction is the only method that can give detailed information on the initial and final states and, possibly, stabilized intermediate states (6). By combining this information with spectroscopic data and computer simulations (if these can be extended to such large, slow moving systems), it should eventually be possible to define both the pathway and time scale of triggered conformational changes.

FUNCTIONAL CONSEQUENCES OF PROTEIN MOTION

There is little direct evidence that the flexibility of proteins is essential for their biological function, but there is considerable indirect evidence. Assuming that flexibility and function are related, one question is whether the entire range of motions that occur are involved. It seems likely that the answer is yes. Even small amplitude individual atomic motions are probably important; they provide the means by which the structure can relax during larger scale movements. Intuition suggests that collective modes are likely to be particularly important, since they could lead to the displacement of individual loops of polypeptide chain or a critical side chain. There are three areas where X-ray diffraction suggests that protein dynamics and protein function are interrelated: ligand binding, enzymatic catalysis, and regulation.

Ligand Binding

Kendrew, Perutz, Watson, and their colleagues noted over twenty years ago that there was no obvious pathway in the static structure of myoglobin or hemoglobin for oxygen or other ligands to reach the iron atom from outside the protein (72). The observed on-rate for ligands is sufficiently fast that any process effecting entry must itself be very rapid. The logical conclusion is that fluctuations in the protein structure open a transient channel for ligand penetration. Frauenfelder and associates have studied the ligand binding process in myoglobin by spectroscopy at multiple temperatures, pressures, and viscosities (11, 15). They observed non-exponential time dependence of the binding of both CO and O_2 at temperatures < 160 K. From this they concluded that the protein must exist in a number of conformational substates at low temperature and that diffusion of the ligand through the protein matrix must occur. Case & Karplus analyzed this problem theoretically. Empirical energy function calculations showed that the rigid protein would have barriers to ligand entrance into the heme pocket on the order of 100 kcal/mol (22). Trajectory calculations of the escape of model ligands from the pocket showed a primary escape route between the distal histidine 64 (F7) and the distal valine 68 (E11). The energy barrier could be reduced to reasonable magnitude (5 kcal/mol) by small rotational motions of these side chains.

Inspection of the X-ray derived mean-square displacements for myoglobin (36) reveals that the distal side of the heme pocket is, surprisingly, more rigid overall than the proximal side. Nonetheless, there is a region of relatively high $\langle x^2 \rangle$ values that leads from the binding site at the iron past His 64 and outside the protein past Arg 45, which forms a salt bridge with a propionic acid side-chain of the heme. The general direction of this mobile region is in agreement with the prediction of Case & Karplus.

Direct evidence for the existence of a channel and its probable location has now come from crystallographic studies of the binding of a large ligand to metmyoglobin. Ringe and co-workers diffused phenylhydrazine into the crystal and determined the structure of the resulting aryl heme complex at a resolution of 1.5 Å (77). The phenyl ring is bound end-on, and its steric interactions prevent the closing of the channel, which had to open to admit its entrance into the distal pocket. The channel is formed by side-chain rotations of Val 68 and His 64; the histidine moves by over 5 Å, folding into a cavity in the protein interior. Movement of the histidine displaces a sulfate ion, which occupies part of this cavity in the met structure, and is accompanied by a movement of Arg 45. Rotation of this side chain about two carbon-carbon bonds ruptures the salt bridge to the heme propionic acid, which in turn becomes highly disordered (large increase in $\langle x^2 \rangle$). The net effect of these movements is to create a channel that static accessibility calculations show is wide enough to accommodate an oxygen molecule. Ringe has referred to the phenyl group as a "molecular doorstop" that holds open a passage normally open only transiently (77).

Myoglobin (and, by implication, hemoglobin) is not the only protein where buried regions become accessible, but it is the best studied. Oxygen quenching of aromatic side-chain fluorescence is observed with other proteins (53), implying similar diffusion is possible. The functional correlation of mobility is obvious for the heme proteins. Protection of a buried reactive group from access by other than the desired ligands while retaining rapid binding is one important role for protein flexibility.

Recognition and Regulation

Huber (43) and Alber (2) have discussed the role of protein dynamics in the regulation of enzymatic activity and the recognition of one molecule by another. The trypsinogen-trypsin system is an example of regulation by flexibility. The rigid enzyme is the active species; the proenzyme with its disordered activation domain is unable to bind substrate tightly. Covalent modification of trypsinogen produces a triggered conformational change that leads to activation (17, 18, 43).

Antibodies provide striking examples of large-scale collective motion in proteins. The structure of the intact IgG molecule Kol showed no significant electron density for the entire 50,000 dalton Fc portion of the immunoglobulin (25). This domain is connected to the Fab portion of the antibody by a hinge region, which is thought to be flexible. The Fab "arms" of the protein also have some flexibility (59, 82). It has been speculated that the functional significance of independent arm and stem movement may lie in the ability to reach antigenic determinants in different arrangements (9, 43). This is an example of the importance of a large-scale correlated motion for recognition.

Koshland proposed how triggered conformational changes could provide recognition and regulation in the induced-fit hypothesis (49). As originally stated, ligand binding stabilized the active form of a conformational equilibrium. Inhibitors lacked some essential interactions needed to stabilize this form. Extension to a disordered region in a protein only requires that the correct allosteric effector or substrate order the flexible segment, while inhibitors are unable to do so. A particularly elegant example of the role of mobility in recognition comes from crystallographic studies of viral proteins and intact viruses (16, 41, 85). The viral proteins that interact with the viral nucleic acid to build the stable virus particle are all found to have at least one disordered region. The mobility of this segment of the protein is believed to allow bonding to differently arranged RNA strands.

Catalysis

The most controversial role for atomic fluctuations in proteins is in enzymatic catalysis. Huber has pointed out that a collective motion of the side chain of His 57 appears to be essential for serine protease catalysis (43). The initial structure in which this ring is hydrogen bonded to Ser 195 must rearrange to one in which the imidazole hydrogen bonds to the amide of the leaving-group peptide. An aromatic ring flip would accomplish this. Gilbert & Petsko postulate that flexibility of the side chain of Lys 41 is essential for ribonuclease A catalysis of RNA hydrolysis (37). The substrate undergoes large changes in stereochemistry and charge configuration during the reaction, and a flexible lysine is needed to follow these changes without interfering with them sterically. Thus one function for flexibility in catalysis is to provide an active site capable of responding to alterations in the substrate as the reaction proceeds.

Speculation has been offered that the role of a flexible segment in citrate synthase may be to transfer a reactive intermediate between active sites (43, 103). A similar role for a disordered region has been proposed for the E2 component of the multienzyme complex 2-oxo acid dehydrogenase (79), where a lipoyl-lysine is thought to move rapidly over considerable distances. Thus, a large-scale collective motion may function to provide a mechanism for active site coupling.

Alber has pointed out that highly mobile regions of an enzyme may play a role in facilitating catalysis by weakening substrate and product binding (2). If the interactions between enzyme and substrate produce very tight binding, product dissociation is likely to be very difficult. Alber suggests that the observation that several crystalline enzymes contain disordered segments of polypeptide chain that become ordered on substrate or product complexation may be explained by the advantage of entropy loss. If some of the intrinsic binding energy of substrate with enzyme is expended to order a mobile region of the protein, the net binding strength will be weaker than the sum of the interactions would suggest. The enzyme may thus have specificity without excessively tight binding. In triose phosphate isomerase a ten amino acid disordered loop folds over the substrate and becomes highly ordered (3). For product release to occur, the loop must move out of the way. A thermally driven fluctuation in atomic positions can break transiently the bonds between the loop and the substrate, but the tendency of the loop to resume its open, native conformation when this happens will be increased by the gain in entropy that results from regenerating the disordered form. Flexible segments may contribute to the optimization of enzymatic reaction rates.

Finally, there remains a possibility that the protein structure may favor those modes of motion that lie along the reaction coordinate at the expense of those perpendicular to it. This is the so-called "directed fluctuations" hypothesis (37). In this view, the intramolecular contacts and arrangement of secondary structural features would lead to a distribution of vibrational modes in the active site that tended to favor the movements of atoms required during catalysis. There is no evidence in favor of this hypothesis, or against it. Analysis of the anisotropies of protein $\langle x^2 \rangle$ values in enzyme-substrate complexes stabilized at subzero temperatures (6) might provide some clues.

PERSPECTIVE AND OUTLOOK

Crystallographic studies of protein dynamics have, to date, yielded information rich in structural detail. Properly carried out, a high-resolution X-ray diffraction investigation followed by careful least-squares refinement can give the spatial distribution of the high-frequency mean-square displacements in a protein. These displacements reflect both individual atomic fluctuations in hard variables (bond lengths and bond angles) and collective motions involving soft variables (torsion angles, nonbonded interactions). Lower frequency, large-amplitude motions may not be measurable by an X-ray study, but they may also lead to such complete disorder that their existence can at least be inferred from the absence of interpretable electron density for some sections of the structure. Results obtained thus far may be summarized as follows.

1. A good general picture is available from the combined efforts of protein crystallography and molecular dynamics of how proteins can move in a range from subpicoseconds to tens of picoseconds. Interior residues are more rigid than groups on the surface, and structural constraints are reflected in restricted motion even for surface residues. Amplitudes of motion of 0.5 Å or greater are not uncommon.
2. The temperature dependence of these fast motions varies considerably over the structure. In general, large $\langle x^2 \rangle$ values have large temperature dependence, whereas small displacements are less affected by temperature; however, exceptions are common. Significant reduction in $\langle x^2 \rangle$ on cooling establishes that proteins are mobile even in the crystalline state and that static disorder is not the dominant contributor to the individual *B*-factors.
3. Some evidence for collective motions can be seen with crystallographic studies, but many such modes can only be observed spectroscopically.
4. Disordered regions in electron density maps are no longer automatically taken as signs of errors in structure determination. It is now recognized that the absence of strong electron density is often an indicator of considerable flexibility. Such flexibility may have functional significance for the protein in question. Some of the functional roles for protein dynamics are beginning to be understood.
5. The end result of a large-scale motion can be seen when it forms a stable state. This often occurs in the case of triggered conformational change. Transitions from initially disordered to ordered states are also observed for some systems.

6. The overall picture of protein motion obtained from X-ray diffraction analyses of crystalline proteins, and even many of the details, agrees well with information obtained from solution spectroscopy, other physical measurements on protein crystals, and theoretical calculations.

Missing from these results are the details that can be extracted from thermal motion analysis of small-molecule crystal structures. Admittedly, it may be overinterpretation to apply these methods to protein data, but it is well to remember that just over ten years ago it was commonly felt that protein structures could not even be refined. Certainly some small, well-diffracting proteins such as crambin and avian pancreatic polypeptide should be amenable to many of the sophisticated small-molecule methods, since they diffract to resolutions comparable to simple organic crystals.

The most important type of analysis that should be done is anisotropic *B*-factor refinement, which would give the principal directions of motion along with the amplitude information now obtained. The problem is that unrestrained anisotropic thermal elipsoids of the type shown in Figure 2 require six parameters for each atom instead of the single isotropic parameter, and even data of 1.5 Å resolution do not provide enough overdeterminacy. Anisotropic refinement of some aromatic residues in rubredoxin (102) and lysozyme (10) has been carried out by holding the rest of the structure fixed and appears to give sensible results (Figure 7). This approach should be pursued further. Konnert & Hendrickson have described a procedure for implementing restraints based on bond direc-

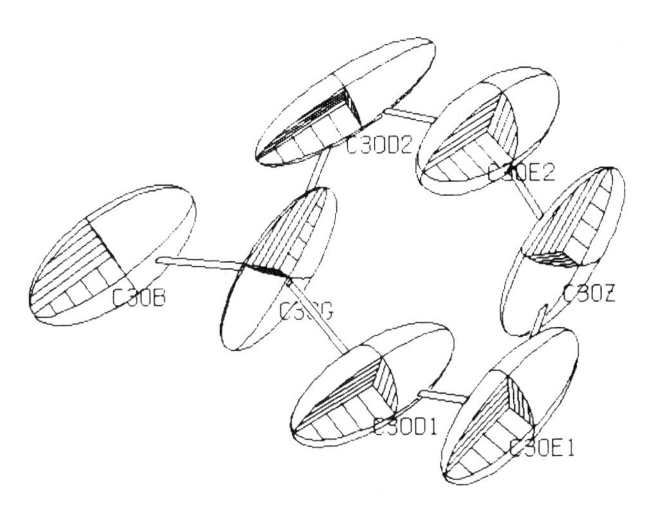

Figure 7 Thermal ellipsoids for the side chain of Phe 30 in rubredoxin after 3 cycles of anisotropic refinement of this residue alone. The ring appears to move as a rigid body, as expected. [Reproduced with permission from (102).]

tions in anisotropic refinement (48); unfortunately, molecular dynamics results show that collective motions destroy this correlation, so it is very dangerous to use this approach. Optimally, very high resolution data should be measured, and unrestrained anisotropic refinement should be carried out on at least a few proteins, so that development of restrained or group-atom methods can be undertaken in the light of unbiased information. Blundell and associates have undertaken precisely this study on avian pancreatic polypeptide, measuring data beyond a resolution of 1 Å and refining the structure by small-molecule procedures (37b). The results of this pioneering work are awaited with great interest.

Anharmonicity of protein motions has been demonstrated by both temperature dependence and theoretical calculations. Methods of reformulating the temperature factor expression to take this into account have been described by Dawson (27) and Johnson (45). However, both experiment and theory suggest that the low-amplitude high-frequency fluctuations in protein structure are quasi-harmonic, so the necessity for more complex treatment is a moot point.

Van Gunsteren and colleagues have shown that combination of molecular dynamics descriptions of anharmonic and anisotropic fluctuations in atomic position with the average positions determined from X-ray refinement is possible and gives results in reasonable agreement with measured X-ray data (96). Such a partnership may eventually produce an improved protein model for the application of sophisticated refinement methods.

Trueblood & Dunitz have applied simple physical models to analysis of small-molecule temperature factors; these models have yielded information about force constants, frequencies of motion, and energy barriers (93). This exciting feasibility study opens up the possibility of extracting these parameters for protein motions, given anisotropic refinement at very high resolution. Efforts are now underway to evaluate a Trueblood-Dunitz analysis on myoglobin (J. Kuriyan, unpublished observations).

X-ray diffraction can provide atomic-resolution information about the initial and final states of a process such as ligand binding or triggered conformational change, but the pathway is much more difficult to unravel. Ringe has shown that a bulky ligand can be used to prevent the return of a channel to its resting state after binding, which allows the residues that move to form the channel to be identified (77). Ligands with chemical "tails" might be used to investigate other channels in proteins. Intermediate states in triggered conformational changes may be trapped by low temperatures (6, 33) or careful choice of inhibitors or by examination of abortive complexes in multi-step reactions.

Proteins are not static systems. They undergo a great variety of dynamic processes at ordinary temperatures. X-ray diffraction, which was once held to be a static technique, can provide information about the extent and direction of some of these motions. In the future, application to problems in protein dynamics should form one of the major uses of protein crystallography.

ACKNOWLEDGMENTS

The authors have benefited greatly from discussions with J. Kuriyan, T. Alber, M. Karplus, H. Frauenfelder, I. D. Kuntz, R. Tilton, J. A. McCammon, F. Parak, R. M. Levy, D. C. Phillips, and E. Pai. The help of many colleagues whose work is referred to in the text is also gratefully acknowledged. The authors' research is supported by the National Institutes of Health.

Literature Cited

1. Adams, M. J., Bühner, M., Chandrasekhar, K., Ford, G. C., Hackert, M. L., et al. 1973. *Proc. Natl. Acad. Sci. USA* 70: 1968–72
2. Alber, T. 1981. *Structural origins of the catalytic power of triose phosphate isomerase.* PhD thesis. Mass. Inst. Technol., Cambridge. 242 pp.
3. Alber, T., Banner, D. W., Bloomer, A. C., Petsko, G. A., Phillips, D. C., Rivers, P. S., Wilson, I. A. 1981. *Philos. Trans. R. Soc. London Ser. B* 293: 159–71
4. Alber, T., Gilbert, W. A., Ringe Ponzi, D., Petsko, G. A. 1982. In *Mobility and Function in Proteins and Nucleic Acids*, ed. R. Porter, M. O'Connor, J. Whelan, pp. 1–24. London: Pitman. 357 pp.
5. Alber, T., Petsko, G. A., Rose, D. R. 1984. *J. Mol. Biol.* In press
6. Alber, T., Petsko, G. A., Tsernoglou, D. 1976. *Nature* 263: 297–300
7. Albertsson, J., Oskarsson, A., Stahl, K. 1979. *J. Appl. Crystallogr.* 12: 537–44
8. Albery, W. J., Knowles, J. R. 1976. *Biochemistry* 15: 5627–31
9. Amzel, L. M., Poljak, R. J. 1979. *Ann. Rev. Biochem.* 48: 961–97
10. Artymiuk, P. J., Blake, C. C. F., Grace, D. E. P., Oatley, S. J., Phillips, D. C., Sternberg, M. J. E. 1979. *Nature* 280: 563–68
11. Austin, R. H., Belson, K. W., Eisenstein, L., Frauenfelder, H., Gunsalus, I. C. 1975. *Biochemistry* 14: 5355–73
12. Baker, E. N. 1980. *J. Mol. Biol.* 141: 441–84
13. Bandyopadhyay, P. K., Wu, F. Y.-H., Wu, C.-W. 1981. *J. Mol. Biol.* 145: 363–73
14. Barksdale, A. D., Rosenberg, A. 1982. *Methods Biochem. Anal.* 28: 1–113
15. Beece, D., Eisenstein, L., Frauenfelder, H., Good, D., Marden, M. C., et al. 1980. *Biochemistry* 19: 5147–57
16. Bloomer, A. C., Champness, J. N., Bricogne, G., Staden, R., Klug, A. 1978. *Nature* 276: 362–68
17. Bode, W. 1979. *J. Mol. Biol.* 127: 357–74
18. Bode, W., Schwager, P., Huber, R. 1978. *J. Mol. Biol.* 118: 99–112
19. Bott, R., Subramanian, E., Davies, D. R. 1982. *Biochemistry* 21: 6956–62
19a. Brooks, B., Karplus, M. 1983. *Proc. Natl. Acad. Sci. USA* 80: 6571–75
20. Butz, T., Lerf, A., Huber, R. 1982. *Phys. Rev. Lett.* 48: 890–93
20a. Campbell, R. L., Petsko, G. A. 1984. Submitted for publication
21. Careri, G., Fasella, P., Gratton, E. 1975. *CRC Crit. Rev. Biochem.* 3: 141–64
22. Case, D. A., Karplus, M. 1979. *J. Mol. Biol.* 132: 343–68
23. Chance, B., Ravilly, A. 1966. *J. Mol. Biol.* 21: 195–98
24. Chance, B., Ravilly, A., Rumen, N. 1966. *J. Mol. Biol.* 17: 525–34
25. Colman, P. M., Deisenhofer, J., Huber, R., Palm, W. 1976. *J. Mol. Biol.* 100: 257–82
26. Cooper, A. 1981. *Sci. Prog. Oxford* 66: 473–97
27. Dawson, B. 1967. *Proc. R. Soc. London Ser. A* 298: 255–63
28. Debrunner, P. G., Frauenfelder, H. 1982. *Ann. Rev. Phys. Chem.* 33: 283–99
29. Drew, H. R., Samson, S., Dickerson, R.

E. 1982. *Proc. Natl. Acad. Sci. USA* 79:4040–44
30. Eisenberg, D. 1970. In *The Enzymes*, ed. P. Boyer, 1:1–89. New York: Academic. 559 pp.
31. Englander, S. W. 1975. *Ann. NY Acad. Sci.* 244:10–27
32. Englander, S. W., Englander, J. J. 1978. *Methods Enzymol.* XLIX:24–39
33. Fink, A. L., Petsko, G. A. 1981. *Adv. Enzymol.* 52:177–246
34. Frauenfelder, H., Hartmann, H., Karplus, M., Kuntz, I. D. Jr., Kuriyan, J., Parak, F., et al. 1984. *Science*. In press
35. Frauenfelder, H., Petsko, G. A. 1980. *Biophys. J.* 32:465–83
36. Frauenfelder, H., Petsko, G. A., Tsernoglou, D. 1979. *Nature* 280:558–63
37. Gilbert, W. A., Kuriyan, J., Petsko, G. A., Ringe Ponzi, D. 1983. In *Structure and Dynamics: Nucleic Acids and Proteins*, ed. E. Clementi, R. H. Sarma, pp. 405–20. Guilderland, NY: Adenine Press. 487 pp.
37a. Gilbert, W. A., Petsko, G. A. 1984. Submitted for publication
37b. Glover, I., Haneef, I., Pitts, J. E., Wood, S., Moss, D. S., Tickle, I., Blundell, T. L. 1983. *Biopolymers* 22:293–304
38. Gō, N., Noguti, T., Nishikawa, T. 1983. *Proc. Natl. Acad. Sci. USA* 80:3696–3700
39. Gurd, F. R. N., Rothgeb, T. M. 1979. *Adv. Protein Chem.* 33:73–165
40. Hanson, J., Schoenborn, B. 1981. *J. Mol. Biol.* 153:117–46
41. Harrison, S. C., Olsen, A. J., Schutt, C. E., Winkler, F. K., Bricogne, G. 1978. *Nature* 276:368–73
42. Hartmann, H., Parak, F., Steigemann, W., Petsko, G. A., Ringe Ponzi, D., Frauenfelder, H. 1982. *Proc. Natl. Acad. Sci. USA* 79:4967–71
42a. Holbrook, S., Kim, S.-H. 1983. Submitted for publication
43. Huber, R. 1979. *Trends Biochem. Sci.* 4:271–76
43a. James, M. N. G., Sielecki, A. R. 1983. *J. Mol. Biol.* 163:299–361
44. James, M. N. G., Sielecki, A. R., Brayer, G. D., Delbaere, L. T. J., Bauer, C.-A. 1980. *J. Mol. Biol.* 144:43–88
45. Johnson, C. K. 1969. *Acta Crystallogr. Sect. A* 25:187–94
46. Karplus, M., McCammon, J. A. 1981. *CRC Crit. Rev. Biochem.* 9:293–349
47. Karplus, M., McCammon, J. A. 1983. *Ann. Rev. Biochem.* 53:263–300
48. Konnert, J. H., Hendrickson, W. A. 1980. *Acta Crystallogr. Sect. A* 36:344–49
49. Koshland, D. E. Jr. 1958. *Proc. Natl. Acad. Sci. USA* 44:98–104
50. Kossiakoff, A. A. 1982. *Nature* 296:713–21
51. Kossiakoff, A. A. 1983. *Ann. Rev. Biophys. Bioeng.* 12:159–82
52. Kossiakoff, A. A., Chambers, J. L., Kay, L. M., Stroud, R. M. 1977. *Biochemistry* 16:654–64
53. Lakowicz, J. R., Weber, G. 1973. *Biochemistry* 12:4171–79
54. Levitt, M. 1983. *J. Mol. Biol.* 168:595–620
55. Levy, R. M., Perahia, D., Karplus, M. 1982. *Proc. Natl. Acad. Sci. USA* 79:1346–50
56. Lipscomb, W. N. 1983. *Ann. Rev. Biochem.* 52:17–34
57. Makinen, M. W., Fink, A. L. 1977. *Ann. Rev. Biophys. Bioeng.* 6:301–43
58. Mao, B., Pear, M. R., McCammon, J. A., Northrup, S. H. 1982. *Biopolymers* 21:1979–89
59. Matsushima, M., Marquart, M., Jones, T. A., Colman, P. M., Bartels, K., Huber, R., Palm, W. 1978. *J. Mol. Biol.* 121:441–59
60. Matthews, B. 1975. In *The Proteins*, ed. H. Neurath, R. L. Hill, C. L. Bolder, 3:403–590. New York: Academic. 663 pp.
60a. McCammon, J. A. 1984. *Rep. Prog. Phys.* 47:1–46
61. McCammon, J. A., Gelin, B. R., Karplus, M. 1977. *Nature* 267:585–90
61a. McCammon, J. A., Lee, C. Y., Northrup, S. H. 1983. *J. Am. Chem. Soc.* 105:2232–37
62. McCammon, J. A., Wolynes, P. G., Karplus, M. 1979. *Biochemistry* 18:927–42
62a. Morgan, J. D., McCammon, J. A., Northrup, S. H. 1983. *Biopolymers* 22:1579–93
63. Munro, I., Pecht, I., Stryer, L. 1979. *Proc. Natl. Acad. Sci. USA* 76:56–60
64. Northrup, S. H., Pear, M. R., McCammon, J. A., Karplus, M., Takano, T. 1980. *Nature* 287:659–60
65. Northrup, S. H., Pear, M. R., Morgan, J. D., McCammon, J. A., Karplus, M. 1981. *J. Mol. Biol.* 153:1087–1109
66. Oldfield, E., Rothgeb, T. M. 1980. *J. Am. Chem. Soc.* 102:3635–37
67. Pai, E. F., Sachsenheimer, W., Schirmer, R. H., Schulz, G. E. 1977. *J. Mol. Biol.* 114:37–45
68. Parak, F., Formanek, H. 1971. *Acta Crystallogr. Sect. A* 27:573–78
69. Parak, F., Frolov, E. N., Mössbauer, R. L., Goldanskii, V. I. 1981. *J. Mol. Biol.* 145:825–33
70. Perutz, M. F. 1970. *Nature* 228:726–34

71. Perutz, M. F. 1979. *Ann. Rev. Biochem.* 48:327–86
72. Perutz, M. F., Mathews, F. S. 1966. *J. Mol. Biol.* 21:199–202
73. Petsko, G. A. 1975. *J. Mol. Biol.* 96:381–92
74. Poulsen, F. M., Hoch, J. C., Dobson, C. M. 1980. *Biochemistry* 19:2597–607
75. Quigley, G. J., Teeter, M. M., Rich, A. 1978. *Proc. Natl. Acad. Sci. USA* 75:64–68
76. Richards, F. M. 1979. *Carlsberg Res. Commun.* 44:47–63
77. Ringe, D., Petsko, G. A., Kerr, D., Ortiz de Montellano, P. R. 1984. *Biochemistry* 23:2–4
78. Ringe, D., Petsko, G. A., Yamakura, F., Suzuki, K., Ohmori, D. 1983. *Proc. Natl. Acad. Sci. USA* 80:3879–83
79. Roberts, G. C. K., Duckworth, H. W., Packman, L. C., Perham, R. N. 1982. See Ref. 4, pp. 47–71
80. Rothgeb, T. M., Oldfield, E. 1981. *J. Biol. Chem.* 256:1432–46
81. Rupley, J. A. 1969. In *Structure and Stability of Biological Macromolecules*, ed. S. N. Timasheff, G. D. Fasman, pp. 291–93. New York: Dekker. 694 pp.
82. Segal, D. M., Padlan, E. A., Cohen, G. H., Rudikoff, S., Potter, M., Davies, D. R. 1974. *Proc. Natl. Acad. Sci. USA* 71:4298–4302
83. Singh, T. P., Bode, W., Huber, R. 1980. *Acta Crystallogr. B* 36:621–27
84. Sternberg, M. J. E., Grace, D. E. P., Phillips, D. C. 1979. *J. Mol. Biol.* 130:231–45
85. Stubbs, G., Warren, S., Holmes, K. 1977. *Nature* 267:216–21
86. Stuhrmann, H. B., Miller, A. 1978. *J. Appl. Crystallogr.* 11:325–45
87. Sussman, J. H., Holbrook, S. R., Church, G. M., Kim, S.-H. 1977. *Acta Crystallogr. Sect. A* 33:800–4
88. Swaminathan, S., Ichiye, T., van Gunsteren, W., Karplus, M. 1982. *Biochemistry* 21:5230–41
89. Takano, T., Dickerson, R. E. 1980. *Proc. Natl. Acad. Sci. USA* 77:6371–75
90. Teeter, M. M., Hendrickson, W. A. 1979. *J. Mol. Biol.* 127:219–24
91. Tilton, R. F. Jr., Kuntz, I. D. Jr. 1982. *Biochemistry* 21:6850–57
92. Tilton, R. F. Jr., Kuntz, I. D. Jr., Petsko, G. A. 1983. *Biochemistry*. In press
93. Trueblood, K. N., Dunitz, J. D. 1983. *Acta Crystallogr. Sect. B* 39:120–33
94. Tüchsen, E., Hvidt, A., Ottesen, M. 1980. *Biochimie* 62:563–66
95. Tüchsen, E., Ottesen, M. 1979. *Carlsberg Res. Commun.* 44:1–10
96. van Gunsteren, W. F., Berendsen, H. J. C., Hermans, J., Hol, W. G. J., Postma, J. P. M. 1983. *Proc. Natl. Acad. Sci. USA* 80:4315–19
97. van Gunsteren, W. F., Karplus, M. 1981. *Nature* 293:677–78
98. van Gunsteren, W. F., Karplus, M. 1982. *Biochemistry* 21:2259–74
99. Wagner, G., Demarco, A., Wüthrich, K. 1976. *Biophys. Struct. Mech.* 2:139–58
100. Walter, J., Steigemann, W., Singh, T. P., Bartunik, H. D., Bode, W., Huber, R. 1982. *Acta Crystallogr. B* 38:1462–72
101. Watenpaugh, K. D., Sieker, L. C., Herriott, J. R., Jensen, L. H. 1973. *Acta Crystallogr. B* 29:943–56
102. Watenpaugh, K. D., Sieker, L. C., Jensen, L. H. 1980. *J. Mol. Biol.* 138:615–33
103. Wiegand, G., Kukla, D., Scholze, H., Jones, T. A., Huber, R. 1979. *Eur. J. Biochem.* 93:41–50
104. Williams, R. J. P. 1979. *Biol. Rev.* 54:389–420
105. Willis, B. T. M., Pryor, A. W. 1975. *Thermal Vibrations in Crystallography*. Cambridge: Cambridge Univ. Press. 280 pp.
106. Wlodawer, A., Miller, M., Sjölin, L. 1983. *Proc. Natl. Acad. Sci. USA* 80:3628–31
107. Wlodawer, A., Sjölin, L. 1982. *Proc. Natl. Acad. Sci. USA* 79:1418–22
108. Woodward, C. K., Hilton, B. D. 1979. *Ann. Rev. Biophys. Bioeng.* 8:99–127
109. Woodward, C. K., Hilton, B. D. 1980. *Biophys. J.* 32:561–75
110. Woodward, C. K., Simon, I., Tüchsen, E. 1982. *Mol. Cell. Biochem.* 48:135–60
111. Yu, N.-T., Jo, B. H. 1973. *Arch. Biochem. Biophys.* 160:614–22

Ann. Rev. Biophys. Bioeng. 1984. 13: 373–98

THE Na/K PUMP OF CARDIAC CELLS

David C. Gadsby

Laboratory of Cardiac Physiology, The Rockefeller University, 1230 York Avenue, New York, New York 10021

Perspectives and Overview

The heart is a rhythmically active syncytium throughout which action potentials propagate more than once each second. Because the passive ion fluxes underlying this electrical activity tend to dissipate the electrochemical potential gradients that drive them, active ion transport systems must continually work to maintain the ideal ionic environment inside the cells. The necessary Na extrusion and K uptake is accomplished in cardiac cells, as in most other cells, primarily by the Na/K pump, a membrane-bound Na,K-stimulated ATPase. From studies in red blood cells and nerve and muscle, it is clear that the Na/K pump transports more Na than K; most evidence indicates that about 3 Na ions are pumped out and 2 K ions pumped in for each molecule of ATP that is split. In other words, the Na/K pump continuously generates an outwardly directed, i.e. hyperpolarizing, current across the cell membrane. The net ionic conductance of the membrane determines the size of the resulting pump potential: because the Na/K pump makes this direct contribution to the membrane potential, it is commonly referred to as an electrogenic (electricity-generating) pump and less commonly, though arguably more correctly, as a rheogenic (current-generating) pump. Changes in Na/K pump current not only cause important modifications of the electrical activity of excitable cells but also provide a convenient signal for monitoring Na/K pump rate in voltage-clamped preparations. A number of thorough reviews of the literature on the Na/K pump and its electrogenic nature, in heart as well as in other tissues have been published (5, 26, 45, 46, 50, 89). The aim of this article is to summarize present knowledge on the Na/K pump of cardiac cells, focussing on recent electrophysiological data, and to identify remaining areas of uncertainty and other pressing topics for future investigation.

0084–6589/84/0615–0373$02.00

The Na/K Pump of Cardiac Cells is Electrogenic

It is easily shown that enhanced activity of the Na/K pump increases the membrane potential of cardiac cells, but it has been more difficult to prove that the hyperpolarization is caused directly by an increase in pump current. The problem is that the hyperpolarization could arise indirectly, as a result of local depletion of K just outside the cell due to the increased pumping. Local depletion of extracellular K is likely to accompany net uptake of K in most cardiac preparations because they contain restricted extracellular spaces, such as intercellular clefts, T-tubules, membrane infoldings, and narrow intercellular pathways across the endothelial cell sheath. In the steady state, intracellular K concentration ($[K]_i$) is constant and there is no net movement of K, so that accumulation or depletion of K ions cannot occur. Pump activity is usually studied, however, in cells loaded with Na to stimulate the pump. The increased pump rate causes net loss of Na and net uptake of K and, hence, almost certain depletion of extracellular K. For this reason, one must view with considerable caution claims that membrane potentials recorded during enhanced pump activity exceed estimates of the electrochemical equilibrium potential for K ions, E_K, calculated from the Nernst equation, $E_K = (RT/F) \ln ([K]_o/[K]_i)$, using measured values of $[K]_i$ and assuming that $[K]_o$, the K concentration just outside the cells, equals the bulk fluid K concentration.

By careful design of experiments, the problem of K depletion can be largely avoided. Vassalle (93) subjected spontaneously beating Purkinje fibers to periods of rapid stimulation and showed (Figure 1) that, during the stimulation, the diastolic membrane potential first decreased and then slowly increased with respect to its initial control value. After stopping stimulation, there was a period of temporary quiescence before spon-

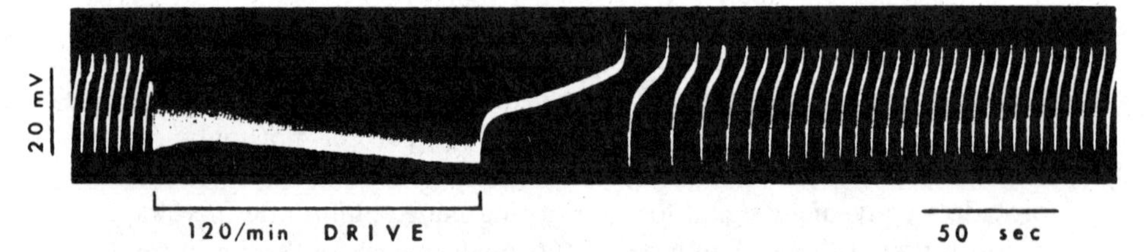

Figure 1 Effect of a 2-min period of rapid drive (2 Hz; indicated by bar) on diastolic potentials of a spontaneously beating Purkinje fiber bathed in 2.7 mM K Tyrode's solution. Only diastolic potentials and the most negative portions of action potentials are shown. Note that maximum diastolic potential becomes more positive during the first 20–30 sec of overdrive but then slowly increases; at the end of the drive it is clearly more negative than the initial control level. From Vassalle (93).

taneous activity resumed. By analogy with the earlier work of Connelly (19) and Rang & Ritchie (85) on nerve fibers, Vassalle argued that the extra influx of Na during the rapid drive increased intracellular Na concentration, $[Na]_i$, and so stimulated electrogenic extrusion of Na, giving rise to the hyperpolarization, both during the drive and after it, and to the temporary abolition of spontaneous activity. These effects were not seen in Li-substituted solution or in the presence of DNP, which suggested that they did indeed reflect enhanced Na/K pump activity. Moreover, since net uptake of K cannot be expected during the overdrive (65, 67), the hyperpolarization observed then cannot be attributed to extracellular depletion of K and so must result from increased pump current (cf 19). Pump-induced depletion of K is known to occur after periods of rapid drive (65, 67) but cannot by itself account for the concomitant suppression of automaticity, because reducing $[K]_o$ causes an increase in spontaneous frequency (92).

Instead of increasing Na influx, another way to raise $[Na]_i$ is to temporarily reduce Na efflux by briefly inhibiting the pump via exposure to K-free solution. Restoring extracellular K results in pump stimulation and causes hyperpolarization and temporary abolition of spontaneous action potentials in small preparations from mammalian sinoatrial node (79, 80) and in small Purkinje fibers (42). These results can be attributed to enhanced pump current rather than depletion of K, because lowering $[K]_o$ is known to cause depolarization and an increase in spontaneous rate. A transient increase of pump current in Purkinje fibers following brief periods of K-free superfusion is corroborated by two other results: (*a*) in regularly driven fibers, restoration of $[K]_o$ is followed by a temporary reduction in action potential duration, an effect opposite to that caused by lowering $[K]_o$ (77, 92, 98); (*b*) in Purkinje fibers with resting potentials at the "lower" level (40), a reduction in $[K]_o$ causes monotonic depolarization, whereas brief exposures to K-free solution are followed by temporary hyperpolarization (42).

Another way to study electrophysiological effects of pump activity is to inhibit the pump rather than stimulate it. Because Na influx occurs even in resting cardiac cells (e.g. 36), in the steady state, there is an equivalent pumped efflux of Na and therefore a steady pump current. Isenberg & Trautwein (64) blocked the Na/K pump in Purkinje fibers with the fast-acting cardiotonic steroid, dihydro-ouabain (DHO). They showed that, within the first minute of DHO action, there was a marked prolongation of the action potential in fibers driven at a low rate, and, in voltage-clamped fibers, there was a parallel inward shift of the membrane current-voltage relation over a wide voltage range. Both results are consistent with abolition of pump current by DHO, but not with any extracellular accumulation of K that might result from loss of cellular K. Similarly, rapid pump inhibition by DHO causes a parallel inward shift of the membrane current-voltage relationship in small, voltage-clamped, ventricular trabeculae from guinea pigs (22).

By combining the procedure for rapid Na loading (brief exposures to K-free fluid) with the voltage-clamp technique, it has been possible to determine directly, in small Purkinje fibers, changes in pump current under a variety of conditions (30, 32, 33, 38, 41, 48). As an example, Figure 2*b* shows changes in holding current caused by three brief exposures to K-free fluid, in the absence or presence of the cardiotonic steroid acetylstrophanthidin (2 μM). The transient outward current normally seen on returning to 4 mM $[K]_o$ is abolished by acetylstrophanthidin, despite the fact that a similar increment in $[Na]_i$ must have occurred during all three periods of K-free superfusion. That transient outward current can thus be attributed to enhanced pump activity rather than to other possible consequences of a rise in $[Na]_i$, for example, reduction of inward Na current, or increase in outward current generated by electrogenic Na/Ca exchange (75), or increase in Ca-activated K current (e.g. 63). Pump-induced depletion of K can also be ruled out as an explanation for the outward current shift, because the same records show that reducing $[K]_o$ causes a monotonic inward shift of holding current. The transient outward current can therefore be attributed to an increment in the current generated directly by the Na/K pump. The record in Figure 2*a* shows the time course of the inward shift of holding current caused by inhibiting the Na/K pump, reversibly, with acetylstrophanthidin. The inward shift reflects abolition of steady-state outward pump current; pump inhibition is complete within a couple of minutes. The current change cannot be attributed to extracellular accumulation of K because, at the low holding potential used in this experiment (see Figure 4 for explanation), increasing $[K]_o$ causes an outward, not inward, current shift (cf 38, 40, 77). Since the inward current caused by acetylstrophanthidin indicates the magnitude of the steady-state pump current, the records in Figure 2*b* suggest that the pump current is, transiently, roughly doubled in size following the 2-min exposure to K-free fluid (cf 38).

Dependence of Na/K Pump Rate on $[Na]_i$

In cardiac cells, as in other cell types, the pumped rate of Na extrusion varies with $[Na]_i$ (for review see 45, 46, 89). An attempt to determine the relationship between these variables was made by Glitsch et al (49) in a difficult series of experiments, using flame photometric determination of changes in cellular Na content of guinea-pig auricles during the first 10 min of recovery from Na loading. Although the considerable scatter of the data precluded detailed kinetic analysis, estimated Na efflux increased with

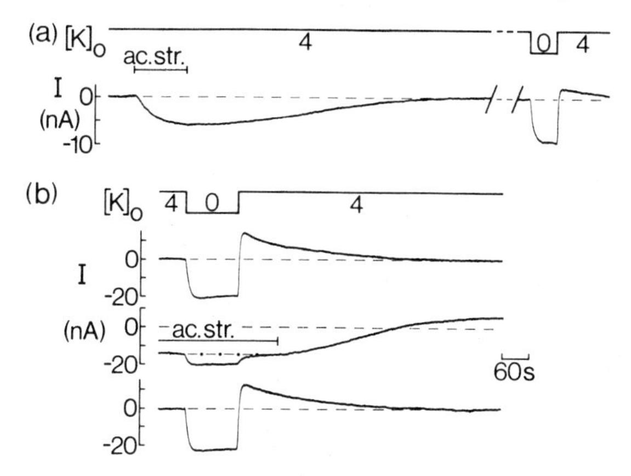

Figure 2 Effects of micromolar concentrations of acetylstrophanthidin (bars labeled ac. str.) on net membrane current (I) of Purkinje fibers voltage clamped at the lower resting potential in 4 mM K, low-chloride (isethionate) solution. (*a*) 2-min application of 5 μM acetylstrophanthidin at a holding potential (V_H) of -32 mV (subsequent break indicates omission of 5-min section of record), followed by 1-min exposure to zero $[K]_o$ as indicated by top line. (*b*) Net current changes caused by three consecutive 2-min exposures to K-free fluid as indicated by upper line; $V_H = -40$ mV. 2 μM acetylstrophanthidin was added 3 min before the start of the second record and was washed out at the end of the bar. The 60-sec time calibration applies to both (*a*) and (*b*). The broken lines mark zero current; the dot-dash line in (*b*) marks the current level in 4 mM $[K]_o$ plus acetylstrophanthidin, which was less inward than in zero $[K]_o$, with or without acetylstrophanthidin, presumably because outward K current is diminished on switching from 4 mM to zero $[K]_o$. From Gadsby & Cranefield (41).

$[Na]_i$ and showed a tendency towards saturation at very high $[Na]_i$ (i.e. >40 mM).

The introduction of fine-tipped, Na-sensitive microelectrodes (e.g. 36, 88) made reliable measurements of intracellular Na activity (a^i_{Na}) more accessible. Deitmer & Ellis (23) soon demonstrated that, in sheep Purkinje fibers, a^i_{Na} declined with an exponential time course on restoration of $[K]_o$ following brief periods of Na loading in K-free solution. If the net loss of cell Na can be attributed solely to increased Na efflux, then the exponential time course implies that, over the 5–15 mM a^i_{Na} range investigated, the increment in Na efflux is proportional to the increment in a^i_{Na}. Deitmer & Ellis were cautious about drawing such a conclusion (*a*) because the observed changes in membrane potential could have caused Na influx to vary, and (*b*) because of possible contamination from Na movements associated with Na/Ca exchange, although they pointed out precedents for a linear dependence of Na efflux on $[Na]_i$ over a limited concentration range, e.g. in snail neurones (88) and squid giant axons (11, 59).

Subsequent experiments using voltage-clamped Purkinje fibers showed that, on reactivating the Na/K pump with K or Rb ions after brief periods of

Na loading, not only does the increment in pump current decay with a single exponential time course (30, 38, 41) but the increment in a^i_{Na} decays simultaneously with the same time course (32, 33, 48). The importance of these findings is their demonstration that the rate of net Na extrusion and the rate of electrogenic Na extrusion are both linearly related to a^i_{Na}, at least over the range tested (3–17 mM a^i_{Na}, equivalent to 4–22 mM $[Na]_i$ assuming an intracellular Na activity coefficient of 0.75) (32). The findings also indicate that, at a fixed level of $[K]_o$, the rate constant for Na extrusion does not vary with a^i_{Na}. The maximal rate constant for pump current decay at high $[K]_o$ averaged 0.88 min^{-1} in canine Purkinje fibers (38).

This apparently linear dependence of pump rate on $[Na]_i$ is somewhat puzzling because, like any other enzyme system, the Na/K pump is expected to show saturation kinetics at high enough levels of $[Na]_i$ (as it does at high levels of $[K]_o$; see below). For red blood cells (86), and recently for squid giant axons (66, but cf 11, 59), a sigmoid relationship has been demonstrated between strophanthidin-sensitive Na efflux rate and $[Na]_i$, with half-maximal activation occurring at 20–30 mM $[Na]_i$ (for review, see 50, but cf 10). It is not yet possible to rule out a similar relationship for cardiac cells since, over a limited range, a sigmoid relationship can be difficult to distinguish from a linear one. An important test for a nonlinear "foot" to the relationship at very low $[Na]_i$ would be to compare the size of the steady-state pump current, obtained from the rapid current shift on adding strophanthidin (e.g. Figure 2, above), with its expected size (assuming that pump current is always proportional to a^i_{Na}), estimated by extrapolating back to zero a^i_{Na} the linear relationship between pump current increment and a^i_{Na} that is determined during recovery from Na loading (see Figure 4*B* in 32). A systematic comparison of this kind has not yet been made, but the results of Eisner et al (32, 35) indicate that there might be a nonlinear foot to the relationship between the pump rate and a^i_{Na}. Although the precise dependence of pump rate on a^i_{Na} remains to be determined, it seems fair to state that, over the physiological range of a^i_{Na}, the relationship is, to a good approximation, linear and that any nonlinearity at very low or very high levels of a^i_{Na} has yet to be confirmed experimentally.

Dependence of Na/K Pump Rate on [K]$_o$

Early experiments indicated that, as in other tissues, the Na/K pump in cardiac cells is inhibited by a reduction, and stimulated by an increase, in $[K]_o$ (for review, see 45). Subsequently, use of the voltage clamp and/or Na-sensitive microelectrodes has greatly facilitated detailed investigation of the $[K]_o$ dependence of pump activity. Deitmer & Ellis (23) showed that, in sheep Purkinje fibers, the rate constant for exponential recovery of a^i_{Na} after Na loading increased with $[K]_o$ over the range 1–12 mM. They plotted, as a function of $[K]_o$, the maximum rate of a^i_{Na} decline, calculated by multiplying the rate constant by the initial size of the increment in a^i_{Na}, and found a half-maximal response at a $[K]_o$ of 10 mM. However, because the steady-state level of a^i_{Na} varies with $[K]_o$, the increment in a^i_{Na} is itself a function of $[K]_o$, so that the $[K]_o$ dependence of the maximum rate of a^i_{Na} recovery does not accurately reflect the affinity of the pump for external K. The rate constant of a^i_{Na} decline, on the other hand, provides a measure of pump activation independent of the level of a^i_{Na}: Deitmer & Ellis have recently reported that the rate constant in their experiments was half-maximal at a $[K]_o$ of 3 mM (personal communication in 33). Glitsch et al (47) found a similar value for half-maximal activation of a^i_{Na} decay by either $[K]_o$ or $[Rb]_o$ in sheep Purkinje fibers.

Eisner et al (32, 33) used voltage-clamped sheep Purkinje fibers and measured simultaneously changes in a^i_{Na}, tension, and pump current. They confirmed that K and Rb are equipotent activators of the rate constant for a^i_{Na} decay (32) and showed that Rb ions activate the identical rate constants for pump current decay and for a^i_{Na} decay with Michaelis-Menten kinetics and that half-maximal activation occurs at 4 mM $[Rb]_o$ (33). This value compares well with their previous estimate of 6 mM $[Rb]_o$ for the concentration yielding half-maximal activation ($K_{0.5}$) of the rate constant for pump current decay (30). For reasons not fully understood, a somewhat lower value, of 1 mM, was found for half-maximal activation of the rate constant for pump current decay by $[K]_o$ in small, voltage-clamped Purkinje fibers from dog hearts (38).

Since all of these measurements were made during recovery from Na loading, i.e. under conditions of cellular Na loss and K gain, a certain degree of extracellular depletion of K might be expected, and some evidence of this in sheep Purkinje fibers has been reported (30). Any such K depletion would tend to cause $K_{0.5}$ values to be overestimated, and the effect should be the same for Rb as for K because they are equipotent activators of the pump. One possible explanation for the difference between the $K_{0.5}$ values of 3–6 mM for sheep, and 1 mM for dog, Purkinje fibers might be that extracellular depletion is smaller in the latter preparation (33, 38) because of the greater average width of the spaces between cells (29). If that explanation is correct then, as Eisner et al (33) point out, the rather good exponential fits to the decay of pump current and of a^i_{Na} (and, hence, the linear relationship between pump current increment and a^i_{Na}) are difficult to understand, because the decay rate constants are sensitive to $[K]_o$ (or $[Rb]_o$) and so should gradually increase as recovery of a^i_{Na} proceeds, since, when the steady state has been regained, there can be no extracellular ion depletion.

On the other hand, some suggestion that extracellular depletion causes overestimation of the $K_{0.5}$ values for K and Rb obtained from the kinetic data in sheep Purkinje fibers derives from the $[K]_o$ ($[Rb]_o$) dependence of steady-state a^i_{Na} measurements made in the same experiments (23, 33, 47). In the steady state, Na influx and efflux must be equivalent and, if Na influx is independent of $[K]_o$ ($[Rb]_o$) (as expected in a voltage-clamped preparation; 38), then steady-state Na efflux should remain constant as $[K]_o$ ($[Rb]_o$) varies. Na efflux, A_{Na} (mol. s^{-1}), shows first order dependence on $[Na]_i$ (see above) and we can write $A_{Na} = vk\,[Na]_i$, where k (s^{-1}) is the $[K]_o$-($[Rb]_o$)-dependent rate constant and v (cm^3) is the intracellular volume of the preparation. If the steady-state Na efflux remains unchanged, then at each level of $[K]_o$ ($[Rb]_o$) the steady-state level of Na_i is expected to vary inversely with the rate constant, k, appropriate for that $[K]_o$ (33, 38). Plots of steady-state a^i_{Na} against $[K]_o$ ($[Rb]_o$) yield apparent $K_{0.5}$ values of about 1 mM $[K]_o$ (23), about 2 mM $[K]_o$ or $[Rb]_o$ (47) and, in voltage-clamped fibers, ≤ 1 mM $[Rb]_o$ (33). These are close to $K_{0.5}$ values for $[K]_o$ activation of the Na/K pump in red blood cells as well as in other cells (for review, see 50). They therefore tend to support the suggestion that $K_{0.5}$ estimates obtained during enhanced pump activity might be too high, presumably because of extracellular depletion.

Stoichiometry of Na/K Exchange

By analogy with work on red blood cells, it might be assumed that the most straightforward way to determine the stoichiometric ratio of pumped Na and K fluxes would be to measure radioactive tracer Na efflux and K influx, before and after suddenly blocking the pump with ouabain, and then to take the ratio of the ouabain-sensitive fluxes. However, this approach is likely to yield misleading information if the pump is electrogenic and makes a substantial contribution to the resting potential, because the depolarization resulting from pump inhibition is expected to change passive fluxes to compensate for loss of pump current. In other words, the ouabain-sensitive fluxes (especially K influx) include a component reflecting the change in passive flux. This error will be smaller for cells in which the Na/K pump, although electrogenic, makes only a small contribution to the membrane potential because of a relatively high membrane permeability to other ions, such as Cl; then changes in pump current can be offset easily by changes in Cl current with little change in membrane potential and, consequently, in passive K or Na fluxes. Such conditions apply to red blood cells (e.g. 61) and perhaps account for their crucial role in early successful determinations of Na/K pump stoichiometry (see 50).

In cardiac cells, the resting Cl conductance is relatively small and cannot "clamp" the membrane potential during pump inhibition. Measurement of tracer fluxes in cardiac preparations voltage-clamped electronically is technically demanding (e.g. 95), and determinations of ouabain-sensitive

Na and K fluxes under voltage clamp have yet to be made. Nevertheless, the stoichiometry of the Na/K pump in voltage-clamped Purkinje cells has been estimated on the basis of measures (*a*) of the extra charge associated with electrogenic extrusion of a given increment in $[Na]_i$ following a brief exposure to K(Rb)-free solution and (*b*) of the corresponding increment in Na efflux (31, 32, 48). The extra charge was determined as the time integral of the exponentially decaying increment in pump current and is unlikely to be much in error; e.g. errors from any extracellular ion depletion should have been small because the sensitivity of membrane current to changes in extracellular K (Rb) concentration was diminished by substitution of Rb for K (31, 32) or by addition of 0.5–2 mM Ba (48). The increment in net Na efflux underlying the increment in pump current was calculated using the corresponding change in a^i_{Na}, from which the change in $[Na]_i$ was estimated, assuming that the intracellular activity coefficient for Na is 0.75 (32, 48). To calculate the extruded quantity of Na underlying the change in $[Na]_i$, that concentration change must be multiplied by the volume in which the Na is distributed, i.e. the cytoplasmic volume of the preparation under study, and this is probably where errors most easily arise. To determine the volume of Purkinje strand occupied by cells, Eisner et al (32) assumed cylindrical geometry and made measurements of length and diameter with a dissecting microscope; they calculated the electrogenic fraction of extruded Na to be 0.26 (SEM = 0.06; $n = 8$). Cell volume so determined might systematically overestimate the cytoplasmic Na space by inappropriate inclusion of extracellular space (intercellular clefts, caveolae) and any intracellular space from which Na is excluded. The estimated electrogenic fraction varies inversely with cytoplasmic volume and would be increased by correcting for any such overestimate: a 30% overestimate would be required to bring the electrogenic fraction to 0.33, the value expected if the $[Na]_i$ increment is extruded only by the Na/K pump and if 3 Na are pumped out for every 2 K (Rb) pumped in. Glitsch et al (48) calculated cellular volume using a surface/volume ratio of 0.4 μm^{-1} and an estimate of the total surface area of the preparation obtained from measurements of capacitative current assuming a specific membrane capacitance of 1 $\mu F\ cm^{-2}$. They then calculated the electrogenic fraction to be 0.39 (SD = 0.18; $n = 14$); correcting the cellular volume for any space that excludes Na would, again, increase that fraction.

More recently, Eisner et al (31) estimated the electrogenic fraction without making any assumptions about cytoplasmic volume or the activity coefficient of intracellular Na. Instead, they had to make an assumption about the action of the local anesthetic, lidocaine, which causes an outward shift of holding current and, simultaneously, initiates a fall in a^i_{Na}, in voltage-clamped sheep Purkinje fibers; the assumption is that the outward current shift accurately reflects the reduction in steady Na current, which is the sole cause of the fall in a^i_{Na}. They were then able to compare the change in current and initial rate of change of a^i_{Na} on adding lidocaine, with a pump current transient and associated fall in a^i_{Na} (all recorded in the same preparation), thereby eliminating cytoplasmic volume and Na activity coefficient from the relationships and allowing estimation of the electrogenic fraction, which turned out to be 0.38 (SEM $= 0.07$; $n = 4$).

In spite of the uncertainties, these results from cardiac Purkinje fibers tend to support the general consensus that the Na/K pump in a wide variety of cells extrudes roughly 3 Na ions for each 2 K (Rb) ions taken up (for review, see 89), although few would argue that the data presently at hand for cardiac cells preclude a coupling ratio of 4 Na:3 K or of 2 Na:1 K.

Whatever its absolute value, however, it has recently become clear that the Na:K coupling ratio remains constant when the pump rate is changed over a fairly wide range by variations in $[Na]_i$ and/or $[K]_o$ ($[Rb]_o$) (30, 32, 33, 38, 41, 48). The results and arguments supporting this claim are largely based on those of Thomas (88, 90), who recorded changes in pump current and in a^i_{Na} in snail neurones and showed that, as described above for Purkinje fibers (e.g. 32), both parameters declined with identical exponential time courses during recovery from Na loading. These results indicate that the rate of loss of Na from the cells, i.e. the increment in Na efflux, is proportional to the increment in $[Na]_i$, as is the increment in pump current. If net Na extrusion occurs only via the Na/K pump under these conditions, then the linear relationship between the increments in Na efflux and pump current demonstrates that a constant fraction of pumped Na efflux appears as pump current, in other words that the pump coupling ratio does not vary with $[Na]_i$.

The coupling ratio also seems to be independent of $[K]_o$. As already mentioned, the rate constant, k, of the exponential decay of increments in $[Na]_i$ and in pump current increases with $[K]_o$ (or concentration of other activator cations) according to simple Michaelis-Menten kinetics (e.g. 33). During recovery from a given increment in $[Na]_i$, the peak amplitude of the increment in pump current $[\Delta I_p(0)]$ also increases with $[K]_o$, with identical kinetics to the $[K]_o$ activation of k (30, 38). Since the area under the exponentially decaying increment in pump current, i.e. the total additional charge (Q) accompanying recovery from Na loading, is given by $Q = \Delta I_p(0)/k$, it is clear that, for a given increment in $[Na]_i$, Q remains constant as $[K]_o$ is varied (38). Similarly, Eisner & Lederer (30) demonstrated that k is linearly related to $\Delta I_p(0)$ for pump current transients recorded at various $[Rb]_o$ or $[Cs]_o$; the same line fits Rb and Cs data. Subsequently, these authors have shown that $Q/\Delta a^i_{Na}$ (where Δa^i_{Na} is the overall change in a^i_{Na} and provides a measure of the total increment in Na extrusion) remains constant as $[Rb]_o$ varies (32, 33). These results all argue persuasively that a constant fraction of extruded Na appears as charge, and hence that the Na : K coupling ratio of the pump is independent of changes in $[Na]_i$, $[K]_o$, $[Rb]_o$ or $[Cs]_o$ and, further, that the coupling ratio is the same for Rb transport as it is for Cs transport. These conclusions apply over the entire range of pump activation by extracellular cations and over a several-fold range of activation by $[Na]_i$. There is, as yet, no good reason for believing that the coupling ratio of the Na/K pump in cardiac cells is variable, even though its precise value remains to be pinpointed.

Dependence of Na/K Pump Rate on Membrane Potential

Because the Na/K pump normally uses the free energy liberated by hydrolysis of ATP to effect the stoichiometric transfer of Na and K ions across the cell membrane against their electrochemical potential gradients, it is, in principle, possible to adjust the conditions (ionic concentrations or membrane potential) such that the electro-osmotic work required is equal to, or exceeds, the free energy available from splitting ATP, so that the pump either stops or runs backwards. For a Na/K pump that extrudes 3 Na ions and takes up 2 K ions per molecule of ATP hydrolyzed, the reversal potential of the pump (V_{rev}) is defined as the membrane voltage at which the free energy per molecule of ATP is equal to the osmotic work needed to move 3 Na and 2 K ions up their respective chemical potential gradients together with that needed to move the single excess charge up the transmembrane potential gradient, V_{rev} (e.g. 12, 27). From this equality, V_{rev} can be estimated to be roughly -160 mV from the difference between the negative free energy of ATP hydrolysis, 13–15 kcal/mol or about -600 meV/molecule, and the sum of the osmotic work in transporting 3 Na and 2 K ions, approximately 240 meV and 200 meV, respectively (for healthy mammalian cardiac cells with concentration gradients of $\sim$20-fold for Na and $\sim$30–40 fold for K). Hence, there would appear to be little reason for expecting the pump to run backwards over the physiological voltage range unless, for example, the energy available from ATP should be drastically reduced (but cf 12).

Since there is good reason to expect the pump current to decline towards zero as the membrane is hyperpolarized towards V_{rev}, it would be interesting to know just how the pump current varies with membrane potential, i.e. the shape of the instantaneous pump current-voltage relationship. In this context, "instantaneous" means before a change can occur in any of the concentrations that specify V_{rev}. The little information available suggests that the pump current in cardiac cells is practically independent of voltage, at least over the range -90 mV to about -20 mV. However, reliable instantaneous current-voltage relationships are notoriously difficult to determine in multicellular cardiac preparations, partly because of the apparent plethora of time-dependent currents. As already mentioned, Isenberg & Trautwein (64) measured "late currents" (which appeared reasonably steady between 1 and 2 sec after each voltage step) in sheep Purkinje fibers before, and 2 min after, inhibiting the Na/K pump with 10 μM dihydro-ouabain (DHO); they found that between -90 and -30 mV the current-voltage curve was simply shifted downward by DHO with no obvious change in shape.

More recently, Eisner & Lederer (30) used the same preparation to obtain current-voltage relationships in 10 mM $[Rb]_o$ during, and after recovery from, transiently enhanced pump activity; they also found a roughly parallel shift of the current-voltage curve between -80 and -20 mV. These authors, however, measured "steady-state" currents which, reportedly, required voltage steps of 2–4 sec at potentials positive to -50 mV and often longer at more negative potentials because of slow decay of pacemaker current. In both studies, the implicit assumption is that any time-dependent currents inadvertently included in the measurements should be similar in control and test conditions, and should therefore cancel in the subtraction, and that time-independent currents should be identical apart from the changes in pump current. However, there is good reason to expect differences in ionic concentrations between test and control conditions in both cases. After 2 min of pump inhibition by DHO in sheep Purkinje fibers, extracellular accumulation of K can be expected but probably only a very small increase in $[Na]_i$ (see Figure 9 in 23), whereas during enhanced pump activity in 10 mM $[Rb]_o$ following a 20-min period of Na loading, some depletion of extracellular Rb can be expected in addition to a marked increase in $[Na]_i$ and hence, presumably, in $[Ca]_i$ (e.g. 34). It is not at all clear whether those altered concentrations result in changes in time-dependent and/or time-independent currents that are significant with respect to the concomitant changes in pump current.

An attempt to determine instantaneous current levels following step changes in membrane potential was made by Glitsch et al (48). These authors voltage clamped sheep Purkinje fibers exposed to 13.5 mM K solution containing 0.5 mM Ba (to diminish K currents) and reported that the increment in pump current following a 15-min period of Na loading in K-free solution showed little instantaneous dependence on voltage between -30 and -100 mV, although they did mention that a small decrease was occasionally observed at potentials between -100 and -120 mV. They showed, moreover, that this apparent voltage independence was maintained throughout the decay of the pump current increment, indicating that the size and shape of the pump current transient were independent of voltage; this suggests that the pump-coupling ratio is also independent of membrane potential. Unfortunately, neither of these later reports (30, 48) includes suitable records of the current changes elicited by the clamp steps, from which the relative importance of contaminating, time-dependent components might be judged.

Given the general consistency of the results obtained in these different studies, however, it is probably safe to conclude that the Na/K pump current in cardiac Purkinje fibers shows no marked potential dependence over the voltage range between the plateau of the action potential and the normal resting, or diastolic potential. Consistent with that statement, Figure 3 shows that the abbreviation of the action potential, and increase in resting potential, observed in a canine Purkinje fiber during enhanced pump activity can be closely mimicked by intracellular injection of a small, steady (i.e. time- and voltage-independent), hyperpolarizing current (43).

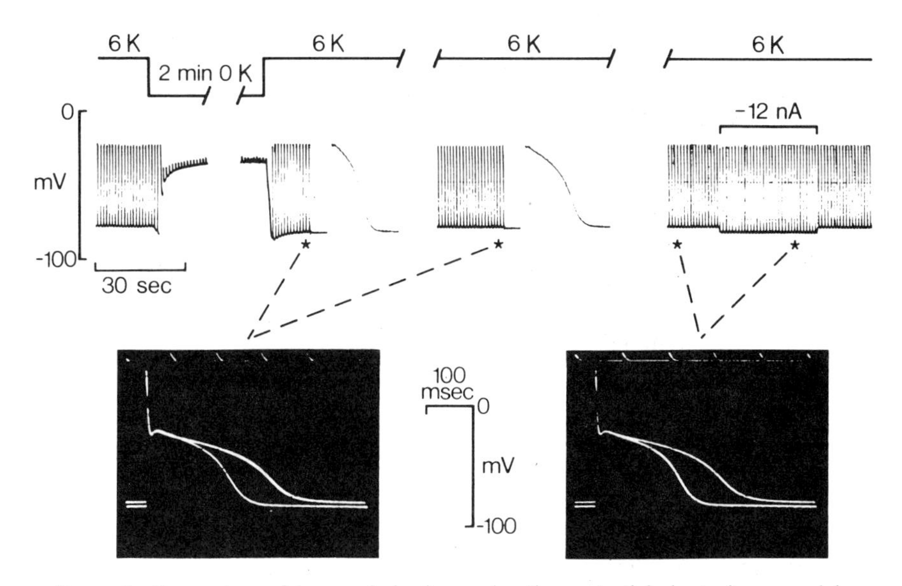

Figure 3 Comparison of hyperpolarization and action potential shortening caused by enhanced pump activity with that caused by current injection via a second microelectrode, in a Purkinje fiber driven at 1 Hz in 6 mM K Tyrode's solution. The top line shows the 2-min exposure to K-free fluid and indicates omission of three sections of record of duration 90 sec, 10 min, and 3 min, respectively. Action potential upstrokes are attenuated by the pen recorder, but the two high-speed chart records (for which the time calibration represents 500 msec) clearly reveal the temporary reduction in action potential duration following the exposure to zero $[K]_o$. The asterisks indicate when the action potentials, superimposed below, were displayed on the storage oscilloscope. The pump-induced hyperpolarization and shortening of the action potential are both closely approximated by injection of a steady (i.e. time- and voltage-independent) 12-nA hyperpolarizing current (bar over chart record). From Gadsby & Cranefield (43).

Even if the pump current shows no instantaneous dependence on membrane potential, its size must be expected to vary with voltage in the steady state. This is because, in the steady state, pumped Na efflux must equal Na influx, so that if the coupling ratio is independent of voltage (see above), the pump current must show the same steady-state voltage dependence as Na influx. The question then is how does Na influx vary with membrane potential? Prolonged depolarization of voltage-clamped sheep Purkinje fibers, from around -70 mV to about -20 mV at 4 mM $[K]_o$ (or $[Rb]_o$), has been shown to cause a measurable fall in a^i_{Na}, which recovers on repolarization (33). In addition, on suddenly inhibiting the Na/K pump with K-free, Rb-free solution, the rate of rise of a^i_{Na} is greater at the more negative potential (33). A rise in a^i_{Na} was also found on hyperpolarization from -30 to -110 mV at 4 mM $[K]_o$, and all these results were interpreted as indicating that passive Na influx increases with the inward driving force on Na ions (33). Although the results are also in the direction expected from effects on the pump of K accumulation and depletion associated with the voltage clamp steps, Eisner et al (33) argue that this is unlikely to be an adequate explanation because a similar voltage-dependence of a^i_{Na} is seen after pump inhibition by 10 μM strophanthidin, when other Na extrusion systems, such as Na–Ca exchange, are presumed to regulate a^i_{Na} (37). A small surprise is that the Na influx, as reflected in the changes in a^i_{Na}, reveals no indication of the TTX-sensitive "window" Na current which is known to have a bell-shaped voltage dependence and to peak at -50 to -60 mV (4, 18); the influence of the TTX-sensitive component might be slight if it represents less than half the total Na influx (24, 33).

Figure 4 shows a result suggesting that the voltage dependence of Na influx in dog Purkinje fibers might be different. Ignoring the brief applications of isoprenaline, the figure shows the current changes resulting from 3-min exposures to K-free solution, first at a holding potential of -40 mV, then at -90 mV, and then again at -40 mV. Both records at -40 mV show a monotonic, inward current shift on switching to K-free fluid that reflects abolition of pump current as well as reduction of outward K current (see Figure 2*b*, above; 38, 41, 77); on switching back to 4 mM $[K]_o$, the usual, exponentially decaying increment in pump current is recorded. At -90 mV however, the switch to K-free solution causes a complex, biphasic current change lasting about 15 sec, the time required for K ion equilibration in the extracellular space. Due to inward rectification and the resulting crossover of K current-voltage relationships at different $[K]_o$ levels (77), the outward K current at -90 mV first increases and then declines, as $[K]_o$ falls from 4 mM to zero; that current change is superimposed on a simultaneous, presumably monotonic, decline in pump current. Of course, the biphasic change in passive K current also occurs on switching back to 4 mM $[K]_o$ and tends to partially obscure the transient increment in pump current (cf p. 1784 in 41). Nevertheless, after that rapid current change is complete, a slowly decaying increment in pump current can be seen at −90 mV, although it is considerably smaller than those recorded in the same fiber at −40 mV. The simplest explanation for this result is that the Na load gained during 3 min of K-free superfusion is smaller at −90 mV than at −40 mV—in other words, that the passive Na influx is smaller at −90 mV.

Plainly, measurements of a^i_{Na} in voltage-clamped dog Purkinje fibers under various conditions would help to solve this puzzle, but, whatever the outcome, it is clear that any intervention that alters membrane potential is

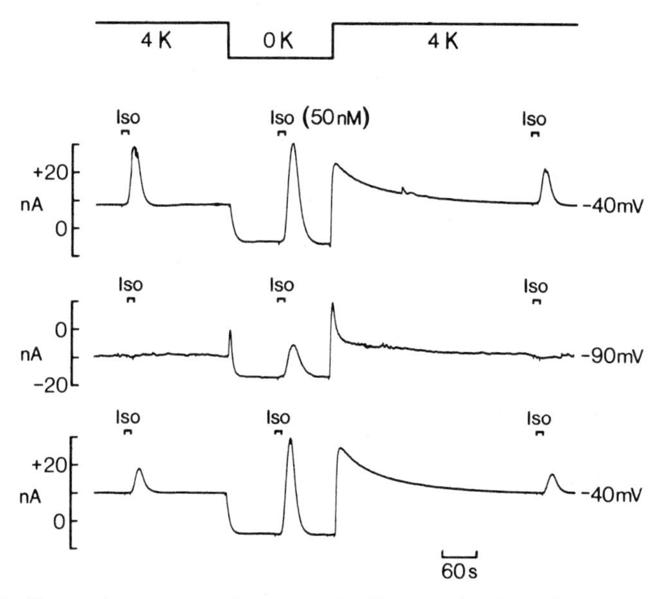

Figure 4 Changes in net current in response to 10-sec applications of 50 nM isoprenaline (bars labeled Iso) in 4 mM K, and in K-free, Tyrode's solution (as indicated by top line) at the holding potentials, −40 mV, then −90 mV, and then again −40 mV, indicated beside each record. The intervals between the records were 11 min and 5 min, respectively. The usual, exponentially declining increment in pump current was recorded on switching from zero to 4 mM $[K]_o$ at −40 mV. The more rapid transients at −90 mV on switching to and from K-free fluid are due to changes in outward K current, as $[K]_o$ changes, and reflect "crossing over" of the steady-state, inwardly rectifying, K current-voltage relationships at different $[K]_o$ levels so that, under these conditions, K current is transiently increased and then decreased when $[K]_o$ falls. An additional, slowly decaying increment in pump current is seen on switching back to 4 mM $[K]_o$ at −90 mV, but it is smaller than that recorded at −40 mV, presumably because the Na load (determined by the rate of Na influx) is smaller at −90 mV than at −40 mV. Note that the isoprenaline-induced current can be increased either by depolarization or by lowering $[K]_o$, i.e. by increasing the outward K ion driving force, V_H–E_K. From Gadsby (39).

likely to simultaneously change Na influx which, in turn, will lead to an adjustment of $[Na]_i$ until, in the steady state, Na efflux and influx are once again in balance (33, 38, 43). Even in the absence of a change in membrane potential, interventions that directly or indirectly modify Na influx (e.g. variation of $[Na]_o$, application of TTX or local anesthetics, changes in external divalent cation concentration) will result in changes in $[Na]_i$, and hence in pump current.

Na/K Pump Stimulation by Low Concentrations of Cardiotonic Steroids or by β-Catecholamines?

Evidence has accumulated over the years to suggest that low (nanomolar) concentrations of cardiotonic steroids might stimulate the Na/K pump, whereas high (micromolar) concentrations undoubtedly inhibit it (for review, see 78). The importance of such findings is that they complicate explanation of the clinically relevant, positive inotropic effect of therapeutic doses of these agents. It is clear that micromolar, and in some cases submicromolar, concentrations of cardiotonic steroids inhibit the Na/K pump and cause a rise in $[Na]_i$, which, in turn, via Na–Ca exchange, causes an increase in $[Ca]_i$ and, hence, an increase in twitch tension (e.g. see 71). It seems likely, however, that micromolar plasma concentrations of these steroids elicit toxic effects, such as arrhythmias, and that therapeutic plasma concentrations are in the nanomolar range. The great debate, then, centers on whether "therapeutic" concentrations of cardiotonic steroids cause inhibition or stimulation of the Na/K pump and whether the therapeutic positive inotropic effect is causally related to either; a further complication has been introduced by the recent results of Hart et al (55) showing that low concentrations of strophanthidin (5–500 nM) can cause either positive or negative inotropic effects, depending on experimental conditions. Since there is some evidence that low concentrations of strophanthidin can cause a^i_{Na} to fall (e.g. 23, 51, 87), then a fall in $[Ca]_i$, mediated by Na–Ca exchange, would provide a possible explanation for the negative inotropism reported by Hart et al (55). Whatever the mechanism for the inotropic effects of low doses of cardiotonic steroids turns out to be, the evidence for stimulation of the Na/K pump by such low doses is worth discussing.

The phenomenon is by no means confined to cardiac cells. A small, variable increase in Na efflux by 10 nM of strophanthidin has been reported for squid giant axons (25), an occasional transient stimulation of K influx has been found in red blood cells on addition of 0.1 or 1 μM ouabain (57), and stimulation of the Na,K-ATPase by low concentrations of ouabain has been demonstrated in preparations from rabbit brain and chicken kidney (81), from guinea-pig and calf ventricular muscle (82), and from dog kidney (54). The evidence from cardiac tissue is largely indirect. Godfraind & Ghysel-Burton (51) reported that incubating guinea-pig atria in 1 to 10 nM ouabain for 3 hr resulted in a significant increase in cell K content and decrease in cell Na content; they also showed that this effect is not mimicked by dihydro-ouabain and so argued that unsaturation of the lactone ring is a prerequisite for pump stimulation (44). These results were obtained with atria electrically driven at 3.3 Hz, but no information on possible changes in diastolic membrane potential or in action potential is presented. Voltage changes are likely to be important, since they can be expected to result in changes in passive Na influx (as already discussed) and, if depolarization were to markedly diminish Na influx, it is even conceivable that a slight reduction in pump rate might lead to a fall in $[Na]_i$ (15; but seen discussion of voltage dependence of Na influx, above). However, Ellis (23, 36) showed that low concentrations of ouabain or strophanthidin can cause a^i_{Na} to decline in quiescent sheep Purkinje fibers, even when there is no change in resting potential, and Sheu et al (87) sometimes observe a small hyperpolarization associated with the fall in a^i_{Na} caused, in driven Purkinje fibers, by 1–10 nM of strophanthidin.

In voltage-clamped sheep Purkinje fibers, 50–500 nM ouabain was found to cause an outward shift of the current-voltage relationship at negative potentials, and negative shift of the apparent reversal potential for the pacemaker current. Both effects were attributed to extracellular K depletion consequent to ouabain-induced stimulation of the Na/K pump (16). Figure 5 shows an outward shift of holding current in a voltage-clamped dog Purkinje fiber on application of 1 or 10 nM acetylstrophanthidin; this result cannot be attributed to extracellular K depletion secondary to pump stimulation because, as illustrated, reduction of $[K]_o$ at such low holding potentials causes an inward, not outward, current shift. The inward current shift caused by 1 μM acetylstrophanthidin (*bottom, right*) reflects pump inhibition and consequent reduction of outward pump current. The outward current shift caused by 1 or 10 nM acetylstrophanthidin might thus reflect an increase in pump current that is due to stimulation of the Na/K pump.

More direct evidence comes from the experiments of Hamlyn et al (54) on partially purified Na,K-ATPase from dog kidney, which is stimulated ~20% by low concentrations of cardiotonic steroids; higher concentrations cause inhibition along the usual, sigmoid dose-effect curve. Washing the Na,K-ATPase with 100 mM buffered salts of K, Rb, NH_4, or Cs activated it by ~20% and, under those conditions, cardiotonic steroids caused only inhibition of the Na,K-ATPase along the dose-effect curve usually obtained at higher concentrations of cardiotonic steroids. These results suggest strongly that the apparent stimulation reflects displacement

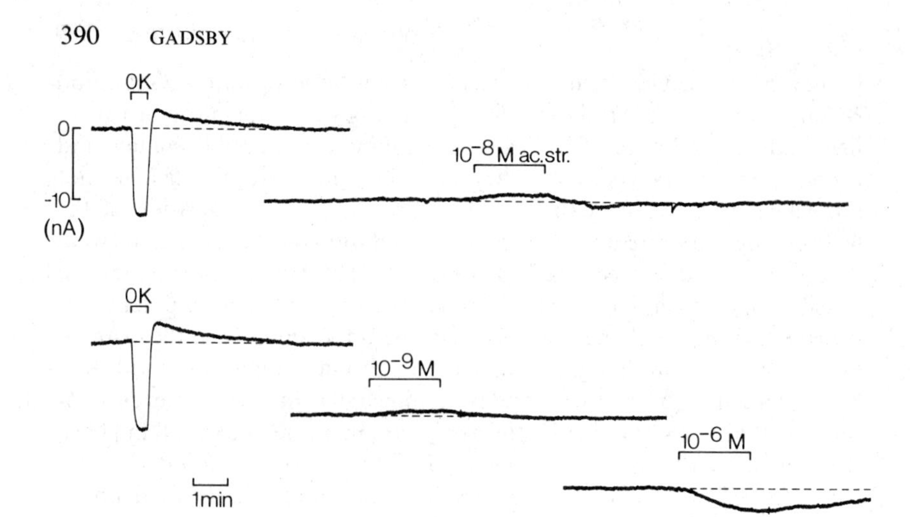

Figure 5 Outward shift of net current caused by nanomolar concentrations of acetylstrophanthidin (bars labeled ac. str.) in a Purkinje fiber held at −33 mV in 4 mM K, low Cl solution. The 30-sec exposures to K-free fluid (bars labeled 0K) were followed by the usual transient increments in pump current; 1 μM acetylstrophanthidin (*bottom right*) caused an inward current shift, presumably because of reduction of pump current. A possible explanation for the small outward current shifts shown at the upper right is that nanomolar acetylstrophanthidin might stimulate, rather than inhibit, the Na/K pump. From Gadsby & Cranefield (43).

of an inhibitor from the partially purified Na,K-ATPase rather than direct stimulation of the pure enzyme (54; cf 78).

An alternative explanation for some of the above effects of low concentrations of cardiotonic steroids in intact tissue (but not for the results obtained with red cells, squid axons, or Na,K-ATPase preparations) has been proposed by Hougen et al (62), who argue that the pump stimulation is, in fact, caused by endogenous β-catecholamines that are somehow released from cellular (presumably neuronal) stores. They found that 3 nM ouabain, in quiescent pieces of guinea-pig atria, caused an increase in ouabain-inhibitable Rb uptake; that effect could be blocked by the β-antagonist propranolol, was mimicked by nanomolar concentrations of norepinephrine or isoproterenol, and was absent after in vivo depletion of endogenous myocardial catecholamines by either reserpine or 6-hydroxydopamine. It is not clear by what mechanism ouabain releases endogenous catecholamines, nor is the time course of that release known, so that it is difficult to assess the plausibility of such an explanation for either the rapid outward current shift caused by acetylstrophanthidin in Figure 5 or the fall in a^{i}_{Na} at low strophanthidin concentrations observed by Deitmer & Ellis (23) (both results were obtained in small, continuously superfused, Purkinje fibers). Hart et al (55) argue that the outward current shifts they record at relatively negative potentials, when low concentrations of strophanthidin are applied to small, voltage-clamped, segments of Purkinje fibers, are unlikely to reflect activation of β-adrenoceptors. The reason is that β-catecholamines are expected to increase inward currents via the pacemaker system (56) and, negative to E_K, to increase inward K current via an increase in steady-state K conductance (see below; 21, 39). Therefore, any outward current that is due to a catecholamine-induced increase in pump activity and hence, possibly, K depletion, would have to be larger than those inward current shifts in order to cause a net outward current. Nevertheless, it would clearly be prudent in future experiments to rule out possible complications from β-catecholamine effects by including low concentrations of propranolol in superfusion solutions.

The mechanism of the Na/K pump stimulation by β-catecholamines is itself a topic of considerable interest. Partially purified Na,K-ATPase preparations from rat brain or from rat skeletal muscle can apparently be stimulated by both D- and L-isomers of some β-catecholamines such as isoprenaline or adrenaline; but half-maximally effective concentrations are in the micromolar range, and only the effects of the L-isomers are blocked by propranolol (1, 13; for review, see 83) or, in Na,K-ATPase from brain, even by the α-antagonist, pentolamine (83). Such non-stereospecific, low affinity effects are unlikely to be involved in the enhanced Rb uptake observed by Hougen et al (62) in response to nanomolar concentrations of β-catecholamines. Two recent studies have shown that catecholamines can cause a decrease in a^i_{Na}, an effect attributed to pump stimulation, although only high concentrations of the catecholamines, 0.2–1 μM, were tested (72, 97). Wasserstrom et al (97) found that isoprenaline or norepinephrine caused a^i_{Na} to fall, after an unexplained delay of some 10 min, in both quiescent and driven preparations, and that the effect was abolished by propranolol. They also reported hyperpolarization of 1–2 mV in driven and resting preparations and suggested that the hyperpolarization and fall of a^i_{Na} both reflect enhanced electrogenic Na extrusion, an argument that would seem difficult to sustain. Presumably, the preparations are in Na balance before exposure to catecholamines, so that Na influx and efflux are equivalent. Pump stimulation then enhances efflux (and, hence, pump current), and as long as efflux exceeds influx a^i_{Na} will fall. As a^i_{Na} falls, so will pumped Na efflux, until it again balances influx. By then, pump current will have returned to its initial level, but a^i_{Na} will now be at a lower, steady level. In other words, the increment in pump current, and therefore the pump hyperpolarization, are expected to be only transient, whereas the reduced a^i_{Na} should be maintained (7). These arguments should certainly hold for the quiescent preparations, in which Wasserstrom et al (97) reported the larger hyperpolarization, but they may be complicated by other current changes in the driven preparations. Lee & Vassalle (72) were able to record a^i_{Na} continuously in driven Purkinje fibers by using low-pass filters (70). They showed that norepinephrine caused a^i_{Na} to fall after a delay of about 1 min and that the decline in a^i_{Na} did not occur after inhibiting the Na/K pump with strophanthidin, which strongly suggests that it was due to pump stimulation.

A number of other studies have yielded results that suggest that β-catecholamines can stimulate the Na/K pump. For example, β-catecholamines have been shown to increase both K influx and K efflux in quiescent (96) as well as beating (6, 94, 96) cardiac preparations and to cause a net uptake of K (6, 94, 96), which was diminished by strophanthidin (6). They can also increase the resting, or maximum diastolic, potential in Purkinje fibers and atrial preparations (28, 60, 91, 99), and this hyperpolarization has been attributed to enhanced electrogenic Na extrusion. A more detailed proposal has been made by Akasu et al (2, 3), who found that 10 μM adrenaline increased the size of the ouabain-sensitive hyperpolarization obtained on brief application of 2 mM $[K]_o$ to trabeculae from bullfrog atria kept in K-free solution (2); under sucrose-gap voltage clamp, adrenaline shifted to lower $[K]_o$ values the $[K]_o$-dependence of the ouabain-sensitive outward current recorded during similar brief restoration of various levels of $[K]_o$ (3). The authors reasonably interpreted these results as suggesting that adrenaline increased the affinity of the Na/K pump for extracellular K ions; half-maximal activation occurred at 0.7 mM $[K]_o$ in the presence of adrenaline and at 1.3 mM $[K]_o$ in its absence.

Some studies indicating pump stimulation by catecholamines might bear further analysis, however, in light of the recent demonstration that β-catecholamines can increase background K permeability in atrial cells of the dog coronary sinus (7) and in dog Purkinje fibers (21, 39). In quiescent coronary sinus strips, noradrenaline caused a marked hyperpolarization, which was abolished by propranolol but not by cardiotonic steroids and which was augmented in K-free solution; conductance measurements revealed a reversal potential for the noradrenaline effect that varied with $[K]_o$ approximately as did estimated E_K (7). The results from voltage-clamped Purkinje fibers were similar: outward current induced by isoprenaline was abolished by propranolol but not by acetylstrophanthidin and was augmented in K-free solution (see Figure 4, above). The current declined as the holding potential approached E_K and reversed approximately at E_K (21, 39). Clearly, these new results per se do not rule out the possibility that β-catecholamines can directly stimulate the Na/K pump, but they do suggest alternative mechanisms that should be considered before such a conclusion is drawn. Thus, an increase in K permeability provides a reasonable explanation for membrane hyperpolarization towards E_K; if small cardiac cells (with a large surface-to-volume ratio) are exposed to ouabain or K-free solution for long periods of time, it is even possible that the resulting large fall of $[K]_i$, and hence of E_K, might conspire to confer an apparent "ouabain sensitivity" on a hyperpolarization caused solely by an increase in K permeability (cf 7). Certainly, such an increase in K permeability is expected to contribute to the increases in K efflux caused by catecholamines (e.g. 94, 96). If the enhanced K efflux leads to extracellular accumulation of K this would be expected to stimulate the Na/K pump.

Pump stimulation by β-catecholamines under some conditions, therefore, might be indirect, secondary to extracellular K accumulation caused by the catecholamine-induced increase in K permeability. That indirect stimulation of the pump would still be sensitive to cardiotonic steroids. Catecholamine-induced K accumulation could conceivably occur (at least transiently) in preparations that are not voltage clamped, but it is more likely in preparations voltage clamped at potentials positive to E_K, and so could provide an explanation for results such as those of Akasu et al (3). In fact, these authors (74) subsequently obtained experimental support for such a mechanism by showing that acetylcholine, which is known to increase K permeability in atria, induced a partially ouabain-sensitive outward current when applied to voltage-clamped bullfrog atrial trabeculae kept in K-free solution. They suggested that the increased K efflux caused K accumulation and, hence, stimulation of the electrogenic Na/K pump. If adrenaline also increases K permeability in frog atria, then a similar indirect stimulation of the pump should occur, but only if the test $[K]_o$ level is below that for maximal activation of the pump; the result would be an apparent shift to lower $[K]_o$ levels of the saturable $[K]_o$-dependence of pump activation (cf 3). It seems possible that a similar kind of explanation might apply to other demonstrations of catecholamine-induced increases of pump activity [for example, those reported by Hougen et al (62) and by Lee & Vassalle (72)], but, obviously, it would not apply to any results obtained either with preparations already exposed to maximally activating $[K]_o$ levels or with Na,K-ATPase preparations.

Effects of the Na/K Pump on Electrical Activity of Cardiac Cells

Mullins & Noda (76) showed that, in theory, the contribution of the electrogenic Na/K pump to the steady resting potential can be calculated if the stoichiometric ratio, r, of pumped Na and K fluxes is known. The reason is that, when the cell is in a steady state with respect to Na and K, the influx and efflux of each ion must be equivalent, so that if the active fluxes of Na and K are in the ratio r, so will be the passive fluxes. This ratio of the passive fluxes can then be incorporated into the Goldman (52), Hodgkin & Katz (58) equation for the resting potential, as described in detail by Thomas (89); he also quotes a personal communication by P. Ascher showing that the pump contribution to the steady-state resting potential cannot exceed $(RT/F)\ln(1/r)$, i.e. about -10 mV for $r = 3/2$. An important additional assumption in these calculations is that membrane permeabilities to Na and K should be voltage independent. That assumption is reasonable for voltage changes of a millivolt or two, but it is unlikely to hold for larger voltage changes, particularly in cardiac cells, because of the marked inwardly-rectifying K current and the voltage-dependent, TTX-sensitive, steady Na current, which combine to produce the N-shaped, steady-state current-voltage relationship that is characteristic of most cardiac cells. In that case, sudden inhibition of a 3:2 Na/K pump could cause a depolarization much greater than 10 mV. Such voltage-dependent permeabilities are likely to underlie the large potential changes reported on pump inhibition or mild stimulation in cardiac, as well as other, cell types (e.g. see Figure 2 in 42, and Figure 2 in 43; for review, see 89).

In general, the voltage change resulting from a given change in pump current is determined by the membrane slope conductance. In Purkinje fibers, the slope conductance is considerably reduced during the plateau of the action potential so that changes in pump current have more pronounced effects there than at diastolic potentials (42, 64). This is illustrated in Figure 3, where enhanced pump activity following brief exposure to K-free solution caused a marked reduction in the duration of the action potential plateau, but only a small increase in resting potential. Similar, substantial changes in action potential duration occur slowly following sudden alterations of the drive rate; they probably also reflect changes in pump current (8, 17, 43) that are due to changes of $[Na]_i$ (14) caused by changes of Na influx, although such effects have sometimes been attributed to variations in $[Ca]_i$ and hence in $[Ca]_i$-activated K conductance (for review, see 9). In Purkinje fibers driven at low rates, the abbreviation of the action potential plateau, following either periods of rapid drive (43, 65) or brief exposures to K-free fluid (Figure 3; 42), is readily attributed to increased Na/K pump current; extracellular K depletion has been measured in large fibers using K-sensitive microelectrodes (65), but cannot be the underlying mechanism because reduction of $[K]_o$ is known to prolong, not shorten, Purkinje fiber action potentials (e.g. 77, 92, 98).

A form of self-sustaining, repetitive electrical activity, called triggered activity (20), can be recorded in cardiac preparations under various conditions. During triggered activity, each action potential gives rise to a delayed afterdepolarization that exceeds threshold and so initiates the next action potential, and so on (for review see 20, 100). Such bursts of triggered activity are readily elicited by appropriate electrical stimulation of preparations from the canine coronary sinus exposed to moderately low concentrations of norepinephrine; during each burst, however, there is generally a gradual slowing of the activity, associated with a progressive hyperpolarization, until, usually within a few minutes, the activity stops (101). The gradual hyperpolarization and slowing during the triggered burst can be countered by inhibition of the Na/K pump (by prolonged K-free superfusion or application of acetylstrophanthidin) and enhanced by stimulation of the pump (by brief, temporary withdrawal of external K, followed by its return, or by brief periods of rapid overdrive). This strongly suggests that the bursts of triggered activity are normally terminated as a result of enhanced electrogenic Na extrusion, owing to the increase in Na influx and, hence, in $[Na]_i$, associated with the bursts of action potentials (101). A transient increase in pump current, caused by brief exposure to K-free solution, can also abolish, either temporarily or permanently, spontaneous action potentials arising in partially depolarized Purkinje fibers (42) and can temporarily abolish pacemaking activity in normally polarized Purkinje fibers (42) or in cells of the sinoatrial node (79) or atrioventricular node (68); these latter effects closely resemble the temporary, post-drive suppression of pacemaker activity in Purkinje fibers previously reported by Vassalle (93) and attributed to enhanced electrogenic pump current.

Since there is a possibility that bursts of triggered action potentials, as well as abnormal repetitive activity in depolarized cardiac cells, might be involved in the initiation of certain cardiac arrhythmias, it appears that enhanced electrogenic Na extrusion can have an antiarrhythmic effect via membrane hyperpolarization and abolition of the abnormal automaticity. For this reason, any agent that could be shown to specifically stimulate the Na/K pump would be of therapeutic interest.

Present Conclusions and Future Directions

In the past decade, considerable advances have been made in understanding Na/K pump function in cardiac cells. It has been demonstrated, unequivocally, that the Na/K pump is electrogenic, and direct measurements of changes in pump current have been made under voltage clamp. Substitution of Rb for K, and addition of Ba have been introduced as techniques for diminishing electrical effects of the extracellular K depletion caused by changes in pump rate that result in cellular K uptake. The $[K]_o$-dependence of the cardiac cell Na/K pump has been shown to be similar to that for other cell types, e.g. red blood cells, and its dependence on $[Na]_i$ has been found to be roughly linear over a limited concentration range. Within a moderate, physiological range of membrane potentials, the pump shows little, if any, instantaneous voltage dependence and so constitutes an approximately constant current source. The stoichiometric ratio of pumped Na:K fluxes could be 2:1, 3:2, or 4:3, with the ratio 3:2 appearing most probable, as in other cell types. Whatever its absolute value, there is reasonable evidence that the Na:K coupling ratio is unaffected by moderate changes in $[K]_o$, $[Na]_i$, or membrane potential. Relatively small changes in pump current have been shown to have pronounced effects on the electrical activity of cardiac cells and can be expected to occur, for example, as a result of changes in heart rate.

Despite these important additions to our knowledge, there are many outstanding questions. Some of these amount to extensions of range: for instance, is the $[K]_o$ dependence of pump rate sigmoidal or parabolic at low levels of $[K]_o$, and does the $[Na]_i$ dependence of pump rate show a sigmoidal "foot" at very low $[Na]_i$ and does it saturate at very high $[Na]_i$? What is the reversal potential for pump current and what is the shape of the instantaneous pump current-voltage relationship? What is the precise value of the pump coupling ratio and can it be varied experimentally? The intriguing effects of catecholamines and of low concentrations of cardiotonic steroids remain to be fully explained, and there are many steroid and polypeptide hormones whose reputed modulatory effects on the Na/K pump, in cardiac as well as other cells, deserve further investigation. The recent establishment of reliable procedures for obtaining suspensions of isolated cardiac cells or of cardiac sarcolemmal vesicles might encourage a renewed interest in flux studies, perhaps combined with biochemical or pharmacological interventions, while voltage-sensitive dyes might provide information on concomitant changes in membrane potential that would be essential in interpreting the results of such studies. Isolated cells, voltage-clamped using either the whole-cell recording (patch-clamp) technique of Hamill et al (53) or perhaps even wider-tipped suction pipettes that allow intracellular dialysis (73), should prove useful in answering some of the questions. Intracellular dialysis could greatly facilitate biochemical manipulation of Na/K pump activity in voltage-clamped cardiac cells but seems unlikely to provide quite the same opportunity for simultaneous measurement of changes in pump current and pump flux under voltage clamp as does the squid giant axon (84) or barnacle muscle fiber (69).

The known electrical and chemical effects on the heart of changes in Na/K pump activity, mediated for example by variations in heart rate, by ischemia, by hormones, or by clinical administration of digitalis, can be so profound that no further motivation is needed for continued investigation of the control of Na/K pump function in cardiac cells. There is every reason to expect that the next decade will yield information as useful and exciting as that of the past decade.

ACKNOWLEDGMENTS

I am indebted to Dr. Paul F. Cranefield for much helpful discussion. Preparation of this review was supported by USPHS grant HL-14899 and by an Established Fellowship of the New York Heart Association.

Literature Cited

1. Adam-Vizi, V., Seregi, A. 1982. *Biochem. Pharmacol.* 31:2231–36
2. Akasu, T., Ohta, Y., Koketsu, K. 1977. *Jpn. Heart J.* 18:860–66
3. Akasu, T., Ohta, Y., Koketsu, K. 1978. *Experientia* 34:488–90
4. Attwell, D., Cohen, I., Eisner, D., Ohba, M., Ojeda, C. 1979. *Pflügers Arch.* 379:137–42
5. Blaustein, M. P., Lieberman, M., eds. 1984. *Electrogenic Transport: Fundamental Principles and Physiological Implications.* New York: Raven. 416 pp.
6. Borasio, P. G., Vassalle, M. 1974. In *Myocardial Biology*, ed. N. S. Dhalla, 4:41–57. Baltimore: Univ. Park Press
7. Boyden, P. A., Cranefield, P. F., Gadsby, D. C. 1983. *J. Physiol. London* 339:185–206
8. Boyett, M. R., Fedida, D. 1982. *J. Physiol. London* 324:24–25P
9. Boyett, M. R., Jewell, B. R. 1980. *Prog. Biophys. Mol. Biol.* 36:1–52
10. Brink, F. Jr. 1983. *Am. J. Physiol.* 244: C198–204
11. Brinley, F. J., Mullins, L. J. 1968. *J. Gen. Physiol.* 52:181–211
12. Chapman, J. B., Kootsey, J. M., Johnson, E. A. 1979. *J. Theor. Biol.* 80: 405–24
13. Cheng, L. C., Rogus, E. M., Zierler, K. 1977. *Biochim. Biophys. Acta* 464:338–46
14. Cohen, C. J., Fozzard, H. A., Sheu, S.-S. 1982. *Circ. Res.* 50:651–62
15. Cohen, I. S. 1983. *Experientia* 39:1280–82
16. Cohen, I., Daut, J., Noble, D. 1976. *J. Physiol. London* 260:75–103
17. Cohen, I., Falk, R., Kline, R. 1981. *Biophys. J.* 33:281–88
18. Colatsky, T. J., Gadsby, D. C. 1980. *J. Physiol. London* 306:20P
19. Connelly, C. M. 1959. *Rev. Mod. Phys.* 31:475–84
20. Cranefield, P. F. 1977. *Circ. Res.* 41: 415–23
21. Cranefield, P. F., Gadsby, D. C. 1981. *J. Physiol. London* 318:34–35P
22. Daut, J., Rudel, R. 1982. *J. Physiol. London* 330:243–64
23. Deitmer, J. W., Ellis, D. 1978. *J. Physiol. London* 284:241–59
24. Deitmer, J. W., Ellis, D. 1980. *J. Physiol. London* 300:269–82
25. De Weer, P. 1970. *J. Gen. Physiol.* 56:583–620
26. De Weer, P. 1975. In *Neurophysiology, MTP Int. Rev. Sci. Physiol. Ser. 1*, ed. C. C. Hunt, 3:231–78. London: Butterworths
27. De Weer, P. 1984. See Ref. 5, pp. 1–15
28. Dudel, J., Trautwein, W. 1956. *Experientia* 12:396–98
29. Eisenberg, B. R., Cohen, I. S. 1983. *Proc. R. Soc. London Ser. B* 217:191–213
30. Eisner, D. A., Lederer, W. J. 1980. *J. Physiol. London* 303:441–74
31. Eisner, D. A., Lederer, W. J., Sheu, S.-S. 1983. *J. Physiol. London* 340:239–57
32. Eisner, D. A., Lederer, W. J., Vaughan-Jones, R. D. 1981. *J. Physiol. London* 317:163–87
33. Eisner, D. A., Lederer, W. J., Vaughan-Jones, R. D. 1981. *J. Physiol. London* 317:189–205
34. Eisner, D. A., Lederer, W. J., Vaughan-Jones, R. D. 1983. *J. Physiol. London* 335:723–43
35. Eisner, D. A., Lederer, W. J., Vaughan-Jones, R. D. 1984. See Ref. 5, pp. 193–213
36. Ellis, D. 1977. *J. Physiol.* 273:211–40
37. Ellis, D., Deitmer, J. W. 1978. *Pflügers Arch.* 377:209–15
38. Gadsby, D. C. 1980. *Proc. Natl. Acad. Sci. USA* 77:4035–39
39. Gadsby, D. C. 1983. *Nature* 306:691–93
40. Gadsby, D. C., Cranefield, P. F. 1977. *J. Gen. Physiol.* 79:725–46
41. Gadsby, D. C., Cranfield, P. F. 1979. *Proc. Natl. Acad. Sci. USA* 76:1783–87
42. Gadsby, D. C., Cranefield, P. F. 1979. *J. Gen. Physiol.* 73:819–37
43. Gadsby, D. C., Cranefield, P. F. 1982. In *Normal and Abnormal Conduction in the Heart*, ed. A. Paes de Carvalho, B. F. Hoffman, M. Lieberman, pp. 225–47. Mount Kisco, NY: Futura
44. Ghysel-Burton, J., Godfraind, T. 1976. *J. Physiol. London* 266:75–76P
45. Glitsch, H. G. 1979. *Am. J. Physiol.* 236:H189–99
46. Glitsch, H. G. 1982. *Ann. Rev. Physiol.* 44:389–400

47. Glitsch, H. G., Kampmann, W., Pusch, H. 1981. *Pflügers Arch.* 391:28–34
48. Glitsch, H. G., Pusch, H., Schumacher, T., Verdonck, F. 1982. *Pflügers Arch.* 394:256–63
49. Glitsch, H. G., Pusch, H., Venetz, K. 1976. *Pflügers Arch.* 365:29–36
50. Glynn, I. M., Karlish, S. J. D. 1975. *Ann. Rev. Physiol.* 37:13–55
51. Godfraind, T., Ghysel-Burton, J. 1977. *Nature* 265:165–66
52. Goldman, D. E. 1943. *J. Gen. Physiol.* 27:37–60
53. Hamill, O. P., Marty, A., Neher, E., Sakmann, B., Sigworth, F. J. 1981. *Pflügers Arch.* 391:85–100
54. Hamlyn, J. M., Cohen, N., Blaustein, M. P. 1983. *Circulation (Pt. II)* 68:III–63 (Abstr.)
55. Hart, G., Noble, D., Shimoni, Y. 1983. *J. Physiol. London* 334:103–31
56. Hauswirth, O., Noble, D., Tsien, R. W. 1968. *Science* 162:916–17
57. Hobbs, A. S., Dunham, P. B. 1978. *J. Gen. Physiol.* 72:381–402
58. Hodgkin, A. L., Katz, B. 1949. *J. Physiol. London* 108:37–77
59. Hodgkin, A. L., Keynes, R. D. 1956. *J. Physiol. London* 131:592–616
60. Hoffman, B. F., Singer, D. 1967. *Ann. NY Acad. Sci.* 139:914–39
61. Hoffman, J. F., Kaplan, J. H., Callahan, T. J. 1979. *Fed. Proc.* 38:2440–41
62. Hougen, T. J., Spicer, N., Smith, T. W. 1981. *J. Clin. Invest.* 68:1207–14
63. Isenberg, G. 1977. *Pflügers Arch.* 371:71–76
64. Isenberg, G., Trautwein, W. 1974. *Pflügers Arch.* 350:41–54
65. Kline, R. P., Cohen, I., Falk, R., Kupersmith, J. 1980. *Nature* 286:68–71
66. Kracke, G. R., De Weer, P. 1982. *Biophys. J.* 37:220a (Abstr.)
67. Kunze, D. L. 1977. *Circ. Res.* 41:122–27
68. Kurachi, Y., Noma, A., Irisawa, H. 1981. *Pflügers Arch.* 391:261–66
69. Lederer, W. J., Nelson, M. T. 1981. *J. Physiol. London* 319:62–63P
70. Lee, C. O., Dagostino, M. 1982. *Biophys. J.* 40:185–98
71. Lee, C. O., Kang, D. H., Sokol, J. H., Lee, K. S. 1980. *Biophys. J.* 29:315–30
72. Lee, C. O., Vassalle, M. 1983. *Am. J. Physiol.* 244:C110–14
73. Lee, K. S., Weeks, T. A., Kao, R. L., Akaike, N., Brown, A. M. 1979. *Nature* 278:268–71
74. Minota, S., Koketsu, K. 1979. *Experientia* 35:772–73
75. Mullins, L. J. 1979. *Am. J. Physiol.* 236:C103–10
76. Mullins, L. J., Noda, K. 1963. *J. Gen. Physiol.* 47:117–32
77. Noble, D. 1965. *J. Cell Comp. Physiol.* 66(Suppl. 2):127–36
78. Noble, D. 1980. *Cardiovasc. Res.* 14:495–514
79. Noma, A., Irisawa, H. 1974. *Pflügers Arch.* 351:177–82
80. Noma, A., Irisawa, H. 1975. *Pflügers Arch.* 358:289–301
81. Palmer, R. F., Lasseter, K. C., Melvin, S. L. 1966. *Arch. Biochem. Biophys.* 113:629–33
82. Peters, T., Raben, R.-H., Wassermann, O. 1974. *Eur. J. Pharmacol.* 26:166–74
83. Phyllis, J. W., Wu, P. H. 1981. *Prog. Neurobiol. Oxford* 17:141–84
84. Rakowski, R. F., De Weer, P. 1982. *Biol. Bull.* 163:402 (Abstr.)
85. Rang, H. P., Ritchie, J. M. 1968. *J. Physiol. London* 196:183–221
86. Sachs, J. R. 1970. *J. Gen. Physiol.* 56:322–41
87. Sheu, S.-S., Hamlyn, J. M., Lederer, W. J. 1983. *Circulation (Pt. II)* 68:III–63 (Abstr.)
88. Thomas, R. C. 1969. *J. Physiol. London* 201:495–514
89. Thomas, R. C. 1972. *Physiol. Rev.* 52:563–94
90. Thomas, R. C. 1972. *J. Physiol. London* 220:55–71
91. Trautwein, W., Schmidt, R. F. 1960. *Pflügers Arch.* 271:715–26
92. Vassalle, M. 1965. *Am. J. Physiol.* 208:770–75
93. Vassalle, M. 1970. *Circ. Res.* 27:361–77
94. Vassalle, M., Barnabei, O. 1971. *Pflügers Arch.* 322:287–303
95. Vereecke, J., Isenberg, G., Carmeliet, E. 1980. *Pflügers Arch.* 384:207–17
96. Waddell, A. W. 1961. *J. Physiol. London* 155:209–20
97. Wasserstrom, J. A., Schwartz, D. J., Fozzard, H. A. 1982. *Am. J. Physiol.* 243:H670–75
98. Weidmann, S. 1956. *Elektrophysiologie der Herzmuskelfaser*, pp. 74–75. Bern: Huber
99. Wit, A. L., Cranefield, P. F. 1977. *Circ. Res.* 41:435–45
100. Wit, A. L., Cranefield, P. F., Gadsby, D. C. 1980. In *The Slow Inward Current*, ed. D. P. Zipes, pp. 437–54. The Hague: Nijhoff
101. Wit, A. L., Cranefield, P. F., Gadsby, D. C. 1981. *Circ. Res.* 49:1029–42

Ann. Rev. Biophys. Bioeng. 1984. 13: 399–423

SEQUENCE-DETERMINED DNA SEPARATIONS

L. S. Lerman, S. G. Fischer,[1] I. Hurley, K. Silverstein, and N. Lumelsky

Center for Biological Macromolecules, Department of Biological Sciences, State University of New York at Albany, 1400 Washington Avenue, Albany, New York 12222

Introduction

The electrophoretic mobility of DNA in polyacrylamide gels is sensitive to the secondary structure of the molecule with respect to its helicity, partial melting, or complete melting and dissociation of the strands. Partially melted molecules consisting of both double helix and disordered, single-stranded sections move much more slowly than complete double helices or fully melted single strands. As temperature rises, DNA melting proceeds under equilibrium conditions as a series of relatively abrupt transitions of portions of the molecule from helix to random chain. The number of base pairs cooperating to define each section is determined (assuming a specified environment) by the base sequence. When the melting equilibrium controls electrophoretic mobility, it provides a means for sequence-specific separation of DNA molecules and a searching approach to the study of melting.

The use of a gradient of denaturing solvent in a uniform polyacrylamide gel maintained at the temperature of incipient DNA melting provides a convenient experimental system; the solvent gradient is equivalent to a shallow, linear temperature ramp. The simplicity and discrimination of the technique, together with its capability for resolving complex mixtures of molecules, generates a wealth of experimental data. Detailed comparisons among large sets of closely related molecules can be made in a single experiment. The introduction of configurational considerations drawn from polymer theory on the effect of partial melting on mobility establishes a close link between the experimental data and the statistical mechanical

[1] Present address: Actagen Inc., 4 Westchester Plaza, Elmsford, New York 10523.

0084–6589/84/0615–0399$02.00

theory of sequence-specific helix-random chain transitions of Poland (25) and Fixman & Friere (13) (PFF). The results with known sequences demonstrate that the theory provides a high level of predictability and that reliable interpretations of the data in terms of sequence can be made.

Since the physical separations achieved in denaturing gradient electrophoresis are determined by the sequence of the molecule rather than by its length, and the sensitivity of separation can be adjusted over a broad span, various otherwise difficult or inaccessible studies become relatively straightforward. We have detected single base substitutions responsible for genetic anomalies among fragments of phage and whole human genomic DNA. The full range of hybridization techniques are applicable to identify and isolate specific sequences. High sensitivity gradients have been used to isolate new sequences in which base substitutions were introduced at random by chemical mutagenesis. In combination with restriction enzyme fragmentation, the denaturing gradient procedure provides two-dimensional patterns in which essentially all parts of the *Escherichia coli* genome are resolved and recognizable without the need for hybridization. The distinctive melting characteristics of repeated sequences in eukaryotic DNA are recognizable against a diffuse background of unique sequence material, even where the repeated sequence does not provide fragments of uniform length.

Our observations on melting properties include precise localization of the boundary between melting domains, new estimates of the contribution of individual base pairs to helix stability, demonstration of nearest neighbor contributions, measurement of destabilization by mismatched base pairs, the effect of base substitutions on strand dissociation equilibria, and the difference in melting equilibria for a sequence bounded on one or both sides by stable helix.

The Experimental System: Perpendicular Gradients

Denaturing gradients are used in two configurations. Both require gels of uniform polyacrylamide density, maintained at a constant temperature by immersion in a bath near the melting temperature of DNA. In one configuration, which we term a perpendicular gradient, a uniform electrical field is applied perpendicular to the direction of the gradient, and each molecule moves along a straight contour of uniform denaturant concentration. The sample may be applied as a continuous line along the cathode edge of the gel, either from solution or as a long band cut from a preliminary gel. Some sample molecules move through the gel at each level of denaturant concentration; their migration velocity is nominally constant, determined by the equilibrium structure at that concentration and temperature. The electric field is applied until the fastest-moving molecules approach the anode edge of the gel. In the example shown in Figure 1 (*left*), the sample consisted of an unfractionated digest of pBR322 by the restriction endonuclease, *Alu*I, applied as a line of solution at the top of the gel. Migration is from top to bottom; the gradient increases linearly, left to right. At the left edge of the figure, where the denaturant concentration is very low, the fragments are separated according to their length as in conventional gel electrophoresis. Each fragment generates an S-shaped curve with a steep inflection at a characteristic position along the gradient. The apparent mobility is very much smaller at the right, high denaturant edge. The fragments shown here range from 400 to 900 bp; with longer fragments, the ratio of low denaturant to high denaturant mobilities is very much larger, up to 50-fold (9). The details of the sigmoid curves differ

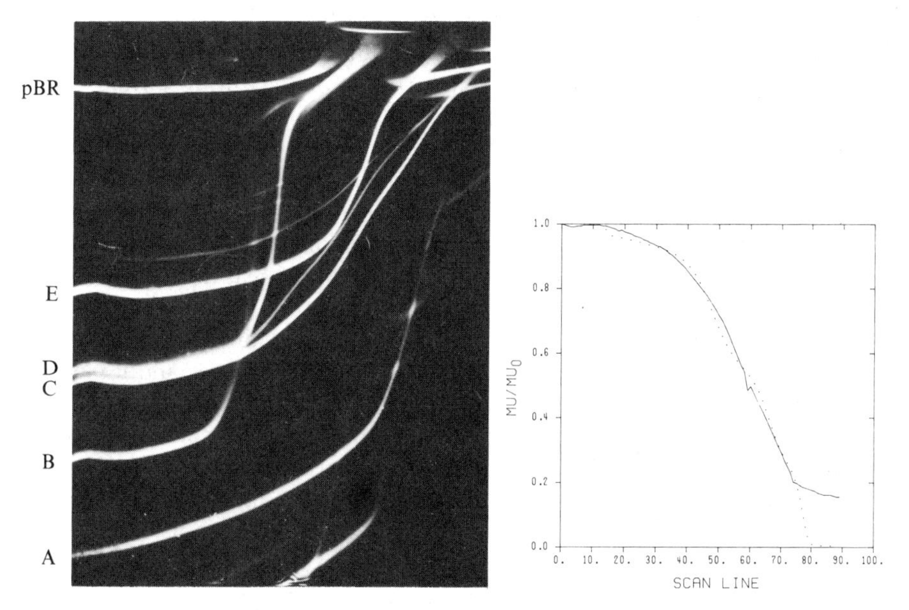

Figure 1 The experimental and calculated mobilities of *Alu*I fragments of pBR322 in a perpendicular gradient gel. The gel contained 5.6 M urea and 32% (v/v) formamide at the right, none at the left, and 65 mg/ml polyacrylamide throughout. *Left:* the ethidium-stained gel, showing the larger fragments. *Right:* the contour of fragment C in the gel (indicated) inferred from a two-dimensional, digital, microdensitometric scan of the photograph is shown as a solid line. The dashed line represents the theoretical mobility of a branched molecule relative to that of an intact double helix, MU/MU_0, calculated using Equation 2. The length of the branched portion as a function of temperature was inferred from the PFF algorithms and the base sequence of fragment C. The length of the flexible unit in melted strands, the temperature offset, and the slope of the temperature-denaturant concentration relation were adjusted to an optimum fit. The abscissa, SCAN LINE, represents denaturant concentration or effective temperature in arbitrary units.

reproducibly from fragment to fragment. Some have a flat baseline before undergoing the sharp change; some change much less sharply, and in some the transition is multiphasic. We will show that the changes in mobility across the gradient closely correspond to the extent of loss of helical structure attributable to the denaturing equilibrium.

Patterns resembling our results with perpendicular gradient gels have been presented by Thatcher & Hodson (29), using a temperature gradient with uniform solvent composition in the gel. Gels containing gradients of urea were introduced by Birnstiel and collaborators (5, 15) for electrophoresis of mRNA at 26–35° to determine the urea concentrations favorable for two-dimensional separations. Large changes in urea concentration effect relatively small changes in mobility of mRNA.

The Experimental System: Parallel Gradients

Migration in the same direction as the gradient, such that the molecule moves continually into a higher concentration of denaturing solvent, has interesting consequences. The initial velocity is determined by the length of the molecule, as in conventional electrophoresis, but when the molecule reaches a depth in the gradient corresponding to the abrupt decline in mobility seen in perpendicular gradients, there is little further travel. In general, further advance effects further reduction in mobility. The bands sharpen with the same ratio of final to original breadth as the ratio of final to original velocity, focusing the band.

This configuration, the parallel gradient, is useful for one-dimensional comparisons of similar molecules in a set of adjacent lanes and for preparing two-dimensional patterns after preliminary length separation in a separate gel. The gradient patterns become almost time-independent after the molecules that are retarded deepest in the gradient have reached that level, and the focusing provides high resolution. The final position, which represents an integral of the mobility as a function of temperature and solvent, is determined principally by the sequence of part of the molecule; it changes relatively little with variation in length of the remaining sequence.

Methodology

The standard system consists of a polyacrylamide gel containing between 40 and 140 mg/ml of acrylamide, $1\frac{1}{2}$–$2\frac{1}{2}$ mm thick, cast between $\frac{1}{4}$-inch glass plates. The solvent gradient is prepared by mixing two outgassed solutions containing the desired boundary solvent concentrations to provide linearly varying fractional composition. Both conventional, gravity-balanced mixing chambers and digitally controlled gradient pumps are in use. The plates enclosing the gel are completely submerged and in contact with the anode electrolyte on both sides. The electrolyte bath is well stirred and controlled at 60°C.

We have avoided the use of any metal parts that might provide current continuity through the thermistor probe, the heater, the stirrer, or other components. A graphite cathode in a small chamber with rapidly circulating electrolyte has proved satisfactory. For further details, see Fischer & Lerman (10).

Because of the sensitivity of the system, precise temperature control is important. Ohmic heating raises the average temperature of the gel about 1.4° above the bath under typical conditions (150 V applied, 0.044 M total cation, 1.5 mm gel). This implies significant nonuniformity across the thickness of the slab and suggests that improvement in resolution may be possible.

We will refer to the environmental variables promoting melting as an equivalent temperature, T', such that $T' = T_0 + gc$, where T_0 is the temperature of the gel determined by the bath and internal heating, c is the concentration of denaturing solvent, and g is the linear coefficient for the first derivative of melting temperature with respect to solvent concentration. Klump & Burkart (18) showed that T_m declines linearly with urea concentration, with $g = 2.5/\text{M}$, and that g is independent of base composition. Since the destabilizing efficacy of closely related substances appears to be similar, based on their concentration as volume fractions, our 100% stock denaturant, 7 M urea with 40% v/v formamide, will be regarded as equivalent approximately to 15.6 M urea. A gradient rising from zero to 80% stock denaturant would represent a T' span of 61.4° to 92.6°, if $g = 2.5/\text{M}$.

Melting of the Double Helix

We have followed the statistical-mechanical algorithms for sequence-specific helix-disorder transitions in DNA developed by Poland (25) and Fixman & Friere (13) in which we assume that (*a*) the status of every base pair is described by the probability of its distribution between one of two states, either helical or melted, where melting implies both unpairing and unstacking; (*b*) the strength of the cooperativity is independent of the sequence; and (*c*) the probability of melting of a sequence bounded by helix at both ends is suppressed according to the inverse 1.5–2 power dependence of the entropy of the loop on its length. The PFF scheme of calculation depends on recursion relations between adjacent base pairs or base pair doublets as it proceeds along the sequence; partition functions and statistical weights are implicit in the probability relations. It offers relatively easy computability for any sequence, and computing time increases linearly with length. There are no a priori assumptions as to where the disordering will begin, nor are any assumptions made with respect to the limits of regions that melt at different temperatures; these are consequences of the calculation.

We have followed Gotoh & Tagashira (14) in supposing that the intrinsic stability of each base pair is influenced by its nearest neighbors but not by interactions over a longer distance. Lacking independent information on the entropy and enthalpy of disordering each type of nearest-neighbor doublet, we have assumed that the entropy change is the same for all base pairs. We assume that the presence of the gel matrix has no significant effect on the melting equilibrium. Values for the intrinsic stabilities of the ten types of nearest-neighbor base pair doublets are taken from Gotoh & Tagashira (GT) (14) for an aqueous environment containing about 0.02 M of sodium ion. This differs slightly from the gel medium which contains, in addition, 0.02 M of Tris ion as well as urea and formamide. The GT stabilities, like others, are underdetermined by the data from which they are inferred, but a similar set of stability values has been derived by R. D. Blake and collaborators (personal communication) using different data. Our value for the cooperativity parameter, σ, represents a consensus of the literature, close to that derived by Amerikyan et al (1) from the effect of a single restriction cleavage on the melting temperature of a well-resolved subtransition in the melting of ColEI. We have chosen a relatively large value, 1.8, for the power dependence of the loop entropy contribution to allow for the stiffness of the melted chain and have not included the arbitrary correction parameter, d, in the loop length.

The result of the calculation is the probability at each temperature of helicity or disorder for each base. Repetition at temperature increments describes the entire progression of the molecule from helix to disorder as the temperature is raised but neglects the probability that the strands completely dissociate. Results of a typical calculation are illustrated in Figure 2 for a portion of lambda DNA lying between 38219 and 38754, inclusive of the first and last paired bases, treated as an isolated fragment of 536 bp. The top panel represents the variation in composition with respect to the local fraction of G+C as a function of position along the sequence. The plot was derived by one cycle of cubic smoothing using a sliding frame of 25 bp, with G·C or C·G as 1.0 and A·T or T·A as 0. It is clear that there are large ups and downs in local fractional G+C content throughout the molecule. The center panel shows the calculated probability of melting at two arbitrarily chosen temperatures. It can be seen that all of the bases from about 30 to 130 are expected to have a 20% probability of helicity at 67.3°C despite the variation of composition within this region—a reflection of cooperativity. The first few bases are completely disordered, and the remainder of the molecule is helical. At 72.5°C, helicity is abolished up to base 145, and the bases from 480 to 535 are also no longer helical, but the central portion remains fully helical. Cooperativity and the improbability of forming melted loops link the base pairs into regions of closely similar melting probability, termed domains. Contour lines showing the temperature at which each base pair reaches a particular melting probability provide a useful summary of the calculations. For most purposes, T_m, the temperature for 50–50 equilibrium between helix and disorder is sufficient, but calculations at other levels convey an impression of the breadth of the transition. Contours for five ratios of the helix-disorder equilibrium, 25 : 1, 5 : 1, 1 : 1, 1 : 5, and 1 : 25, are shown in the bottom panel of Figure 2. The temperature intervals between the lowest and highest contour are relatively narrow, 1.6°, within the lowest melting domain, and broader, perhaps 5°,

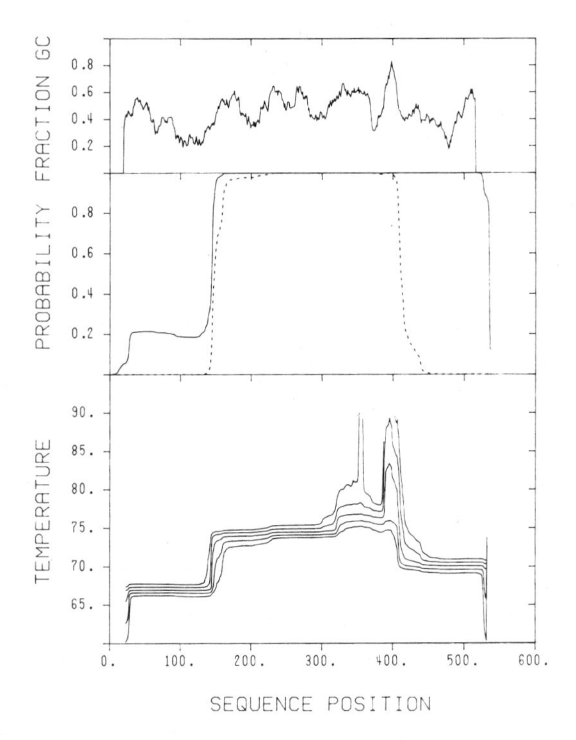

Figure 2 The melting of a 536-bp fragment of lambda DNA. *Top panel:* the variation in local composition with respect to fractional G + C content along the sequence. The value plotted represents the center of a cubic polynomial fitted to 25 bp at each position along the sequence from position 13 to 523. The polynomial provides a smooth, continuous representation of a binary series in which G or C is given the value one and A or T the value 0. *Center panel:* the solid line represents the calculated probability that each base is helical at 67.3°C; the *dashed line* represents the calculated probability that each base is helical at 72.5°C. *Bottom panel:* calculated temperature contours for ratios of 25 : 1, 5 : 1, 1 : 1, 1 : 5, and 1 : 25 helix : random chain. The contours progress from most to least helical from the lower to the upper curves. Strand dissociation equilibria are not included.

among the bases at the boundary between the first domain and the adjacent, higher melting domain. The diagram constitutes a melting map of the sequence.

Calculation of the relation between melting and electrophoretic mobility is based on $\bar{p}$, the total number of bases melted irrespective of site, taken as the sum over all bases of the probability of nonhelicity. If θ is the fractional helicity and N the number of base pairs in the molecule, $\bar{p} = N(1-\theta)$. The derivative, $d\bar{p}/dT$, is expected to parallel conventional hyperchromicity profiles.

These calculations assume that the separation of melted strands does not take place. Specification of the state of dissociation requires additional information, since it depends also on DNA concentration. Results such as those shown in Figure 2 would be appropriate for a system of indefinitely high DNA concentration or one in which the melted strands remain connected. The DNA concentration in electrophoretic gels is neither uniform throughout the band nor necessarily reproducible from experiment to experiment. However, the dissociation constant calculated following Benight & Wartell (3) is a steep function of temperature, and the relevant concentrations define a narrow temperature range on the melting map. The calculated dissociation temperature for the lambda fragment at typical gel concentrations is near 72.6°.

Melting and Mobility

The molecular mechanism of electrophoretic transport of DNA in gels is not fully understood. The estimated pore size of typical polyacrylamide gels is only a few multiples of the diameter of the double helix, much smaller than the root-mean-square radius of the equivalent wormlike chain or its persistence length. In free solution in the absence of a gel, the mobility is independent of the length of DNA molecules (24). Even at low field strengths, well below usual practice, both mobility and its variation with molecular length depend on the voltage gradient. The field strength in gels is several orders of magnitude smaller than the fields needed to effect significant orientation of DNA molecules in solution. Lumpkin & Zimm (22) have shown that the net force on a randomly oriented random chain molecule embedded in a stationary matrix leads to an appropriate relation between mobility and length at zero field. The same relation had been derived previously (21) by means of a more intricate argument concerning the dependence of migration on spontaneous configurational contortions of the molecule. The relation between mobility and the end-to-end length in both formulations agrees with systematic measurements on the low field mobility of DNA in agarose gels over a substantial range of molecular length, gel density, and salt concentration by Hervet & Bean (16). The implied persistence length necessary for good agreement with the data differs from conventional values and changes sharply with gel density. The field strengths in general use are much higher than these and yield length mobility relations that have not yet been rationalized.

We have proposed (21) that partial melting affects the rate of electrophoretic transport by changing the double helix, which can be regarded as two-ended, into a molecule that has a larger number of ends. Where melting has begun from one end of the double helix, the molecule corresponds to the configuration of a star polymer with three arms; if melting has proceeded from both ends, it is a four-armed star in which two pairs of arms are connected by the remaining double helix. If melting has begun centrally, the particle is equivalent to a four-armed star where each of the extra arms consists of a melted loop. In nearly all of the numerous random configurations constituting the equilibrium set, each arm of the partly melted molecule will weave through several pores, but it is unlikely that two arms that have previously been paired will follow the same path. However, unless they follow the same path, there can be no substantial migration; a coincidence in random disposition of the arms is required.

The diffusion of branched polymers in a dense, relatively immobile matrix has been considered analytically by de Gennes (7) and by means of a Monte Carlo simulation of diffusion by Evans (8). Their results agree in that the diffusion of a branched polymer with long, equal arms would be much slower than that of a linear polymer twice as long as any arm, and that the rate of diffusion declines rapidly as the length of the branches increases. However, the Monte Carlo simulation indicates an inverse third-power dependence on branch length. The analytical results suggest a negative exponential dependence on branch length. Measurements on the diffusion of three-arm branched polybutadiene in a matrix of melted polyethylene agree with the exponential dependence (17).

We have suggested (21) that the kinetics of branched polymer diffusion can explain the electrophoretic retardation accompanying partial melting. We find the exponential relation a convenient approximation, but the form of the relation will be unimportant for short arms. The relative mobility of a long helix with a much shorter melted segment would be given by the relation:

$$\mu_0 = \mu_0 e^{-\alpha N} \qquad 1.$$

where μ_0 is the mobility of the unbranched molecule and N is the number of flexible units of each melted arm and α is a constant not far from unity. Representing the length of the flexible unit (in bases) as L_r and the length of the melted segment as the number of nonhelical base pairs, taken as the probability of nonhelicity summed over the entire molecule, $\bar{p}$:

$$\mu(T) = \mu_0 \exp[-\bar{p}(T)/L_r] \qquad 2.$$

where α has been absorbed into L_r. This implies that an intermediate probability of melting is equivalent to a shorter chain of fully melted bases. The progression from negligible to essentially complete melting of a domain takes place over a narrow temperature interval, and the intermediate states make relatively little contribution to the result. If melting proceeds from both ends, retardation would be represented as a product of two exponentials deriving from the right and left ends—equivalent to a single exponential in which $\bar{p}$ is the total melted length including both ends.

When early melting takes place in the interior of a molecule, a similar description appears to be adequate, allowing a different value for the length of the flexible unit. As a star polymer, the molecule has two, rather than one, additional arms, and these consist of melted strands doubled-back to the branch point. The mechanics of partly melted molecules is clearly more complex than that of a multi-armed star polymer in which all arms have the same length, stiffness, frictional properties, and charge density; the present formulation can be regarded as an expedient simplification.

The Mobility of Partially Melted Molecules

The sigmoid pattern produced from a line source of DNA in a perpendicular gel, such as that illustrated in Figure 1, can be expected to represent mobility as a function of temperature within the range of T' and can be compared with Equation 2. However, the approximations and assumptions implicit in matching these curves to a calculated melting progression should be recognized. The contour of the band of DNA in the gel after a fixed period of time is determined not only by the mobility at the nominal concentration of the gradient but perhaps also by a delay in entering the gel, by the diffusion of denaturant out of the top of the gel, by whatever effect the denaturant may have on gel porosity, by electric field inhomogeneity, by viscosity of the denaturant, and by its effect on the charge density of the polynucleotides. It is not clear that high field strengths, which increase the mobility of all lengths by an almost uniform increment, affect the mobility of branched molecules in the same way. The data presently available were taken where the field strength perturbations are significant. Experiments carried out at lower field strengths show that, with at least some molecules, migration under standard conditions is too fast to allow equilibrium with respect to strand dissociation. For this reason, it is appropriate to compare only the portion of the patterns in which strand dissociation is improbable with the first order calculation.

With these reservations, the sigmoid pattern in a perpendicular gradient can be identified with the theoretical melting progression by conventional nonlinear curve fitting procedures, leaving L_r indeterminate. We introduce $\bar{p}$ as a function of T'', such that

$$T'' = T_0 + \Delta T + ghy \qquad 3.$$

where h is the rate of change of denaturant concentration with distance and y is the position along the gradient. ΔT is introduced to allow for a uniform difference between the GT parameters and the values appropriate for our solvent. One example is shown in the right part of Figure 1, corresponding to the fragment of pBR322 running from bases 30 to 684, the third from the bottom in the gel. The flexible unit for the calculated curve is 60 bases. It can be expected to resemble the Kuhn statistical length, twice the persistence length of a wormlike chain (multiplied by alpha).

The persistence length of melted DNA strands can be estimated roughly from the comparison of the intrinsic viscosities of T7 DNA at neutrality and in alkali (26). At 0.04 M sodium, it is about 0.52 of the persistence length of the double helix, or about 262 Å, still substantially larger than the estimated pore size of our polyacrylamide gels. If the monomer length of a fully extended polynucleotide chain is near 7 Å, the persistence length and the Kuhn statistical length would be about 37 and 74 bases.

Another test of the relation between partial single-strand disorder and electrophoretic mobility has been presented by Lyamichev et al (23), who treated partially melted fragments of ColEI DNA with glyoxal at 50°C to prevent reformation of the helix. Where the glyoxal-reacted segment was about 70 bp, the reduction in electrophoretic mobility (as judged from the reproductions of the gels) was approximately 40%; where the estimated reactive length was over 400 bp, the residual mobility is too small to be estimated. The mobility was small, regardless of whether the residual segment of that length was at the end or toward the center of the original molecule. The 70-base result implies a flexible unit of length of about 100 bp, perhaps slightly longer than our estimates, but it is also possible that the glyoxal-treated chain may be partially stacked at ordinary temperatures, since the reaction affects guanine almost exclusively.

Strand Dissociation

Consider the special case for a molecule of 500 bp at a temperature where the melting map implies that 210 bases are melted. With $L = 60$ bases, the expected mobility would be about 4% of μ_0, the mobility of the double helical molecule. However, we find that the mobilities of fully separated strands of this length cluster near 1/4 μ_0. Any strand liberated by dissociation from a severely retarded molecule will migrate ahead more rapidly, thereby lowering the concentration in the region of severe retardation and promoting further dissociation. The severely retarded band may or may not be seen, depending on the rates of dissociation and migration. The effect is demonstrated in perpendicular gradient gels of field strength 10 v/cm or higher and in Figure 1. A number of the DNA molecules we have examined approach very low mobility at an intermediate value of T'. There is then a discontinuous change in slope, and the mobility increases to a new value that remains constant to the highest effective temperature in the gel. The band is typically broadened in the region of discontinuity. At lower field strengths, up to 4 v/cm, there is no discontinuity nor the severe intermediate retardation; the pattern follows a smooth, continuous transition from moderate retardation to the same, slow plateau at high denaturant. If local equilibrium is preserved, the curve will follow a path intermediate between that of fully melted and fully dissociated molecules (see also 6). We think that the lowest mobility before the discontinuity under conditions of rapid transport represents retardation of molecules with long branches that are unstable toward dissociation.

The Retardation Level in Parallel Gradients

To compare related fragments with small sequence differences in two-dimensional gels, migration into an ascending gradient of denaturant is more useful. Each molecule slows and nearly stops at a level in the gradient that remains only slightly affected by longer application of the field. The position as a function of time of a molecule in a parallel gradient will be related to the pattern of retardation by the relation:

$$dy/dt = \mu(T'') \qquad 4.$$

where y is the distance moved from the top of the gel. Incorporating Equations 2 and 3, this relation becomes suitable for numerical integration with any table of the melting function, $\bar{p}$:

$$t(y) = \mu_0^{-1} \int_0^y \exp(p(y)/L_r)\, dy. \qquad 5.$$

Usually it is more convenient to remain in T rather than y units. Then,

$$t(T) = \mu_0^{-1} \int_{T_0}^{T} \exp(p(T)/L_r)\, dT. \qquad 6.$$

Fragments with negligible initial melting leave the top of the gradient at a uniform velocity, μ_0, but decelerate to a lower velocity at some depth where the melted length becomes significant. An example (20) is shown in Figure 3; the top panel presents the melting maps of four different fragments of

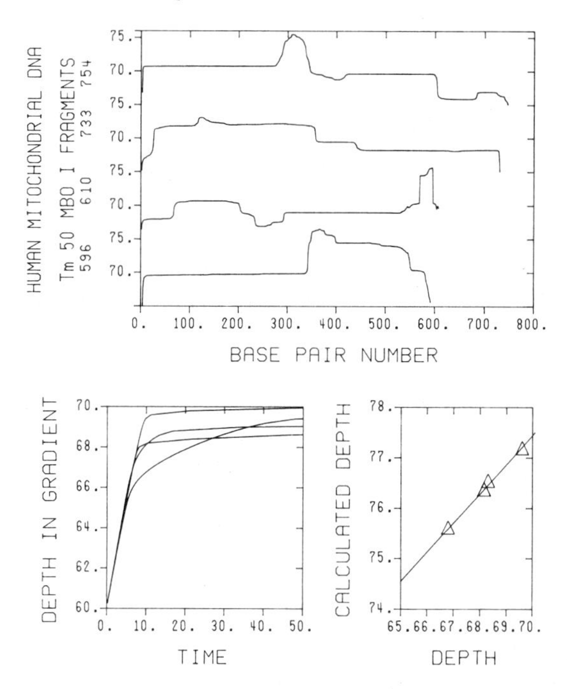

Figure 3 The retardation of fragments of human mitochondrial DNA in a parallel denaturing gradient. *Top:* melting maps of four fragments. Only temperature contours for 1 : 1 helix : random chain equilibria are shown. *Bottom left:* expected travel of each of the four fragments through the gradient, calculated for $L_r = 60$ bases, as a function of time according to Equation 6. *Bottom right:* comparison of the experimental and theoretical position of the bands at 16 hr.

human mitochondrial DNA. Their patterns of melting clearly differ. The middle panel shows the distance travelled, $y(t)$, into the gradient as a function of time for each fragment, calculated from the human mitochondrial sequence (2). The gradient depth at which each fragment changes velocity is different, and they approach different final velocities. The bottom panel compares the experimental depth of fragments of HeLa mitochondrial DNA at the end of the run with the result calculated by Equation 6. Since changes in local stability alter domain boundaries, the equation is also useful for anticipation of the effect of small sequence alterations. An example is given below.

Detection of Single Base Substitutions

We have explored the effect of single base substitutions in a molecule of 536 bp with a fragment of lambda DNA carrying the *y* promotor (12). The results with nearly homologous fragments from a set of 16 mutant strains provide a challenge to the validity of the statistical-mechanical melting theory in both definition of domain geography and quantitative changes in melting temperatures predicted for sequence changes. The compositional map and the melting map of the fragment are shown in the top and bottom panels of Figure 2. The base substitutions examined divide the molecule into two regions, independent of the nature of the base change. The substitutions at all of the 7 sites tested 5′ from base 145 in the fragment alter the equilibrium temperature of the lowest melting domain and result in a displaced final position of the corresponding fragment in the parallel gradient. None of the substitutions 3′ from base 145 affect the melting temperature of the first domain, nor is the retardation of their fragments in the gel different from wild type. Nevertheless, they effect a clear perturbation of the melting map in a high melting domain. The highest substitution site that results in gel displacement is at base 144; the lowest that fails to show a displacement is at 146. The steepest ascent in calculated melting temperatures (that is to say, the calculated domain boundary) falls precisely between these two positions (12).

The differences between retardation levels of the mutant fragments and the parental phage are closely correlated with the change in gradient position predicted from sequence-specific melting theory and Equation 6, the travel function, as shown in Figure 4. Good agreement between theory and experiment supports the analysis of helix stability in terms of nearest-neighbor doublet pairs and the particular values assigned by Gotoh & Tagashira and by Blake (R. D. Blake, personal communication). The experimental and theoretical values demonstrate the effect of a transversion (*cy*2001 as compared with *cy*3071), an interchange of G and C between strands, and a similar effect from a close double substitution (*cin*1 *cnc*1), in which the net base composition remains the same as that of the wild type. If helix stabilities were determined on the basis of local base composition rather than doublet sequence, no effect from these changes would be expected. The double mutant *ctr*5-*c*II3086 perturbs the domain strongly enough to change its length as well as its T_m. Without correction for the full effect through Equation 6 (as in Figure 7; 12), the double mutant does not appear to conform to the line defined by other mutants.

The rule that substitutions can be detected only within the lowest melting domain or domains comprising a small multiple of L_r is illustrated by the properties at sites 424 and 433, which lie at the high-numbered end of the molecule in a domain of 140 bp, melting near 70.2°, the next domain above 30-144. In the full 536-bp molecule, these substitutions have no effect, but in a shorter molecule in which the lower domain is removed by *Alu*I cleavage at 178, both substitutions result in a displaced gradient position (12). The domain containing these sites is lowest melting in the shorter molecule.

We have detected substitutions in the human β-globin sequence in the same way (N. Lumelsky, S. G. Fischer, and L. S. Lerman, unpublished). We have compared the gradient positions of homologous 272-bp fragments carrying the last 11 bases of the first exon and the entire first intron, isolated from persons with β-thalassemia. The gradient displacements with respect to the normal β-globin sequence again agree with calculated changes. The same displacements have been demonstrated by transfer and hybridization after separation in a denaturing gradient in the corresponding fragments of whole genomic DNA.

One substitution outside of the lowest melting domain in the globin fragment has been discerned by its effect on strand dissociation. The smaller velocity of dissociated strands generates displacements in formally the same way as above, though with lower sensitivity. The increment in T', 0.2°, agrees with the calculated effect of the substitution (N. Lumelsky, S. G. Fischer, and L. S. Lerman, unpublished).

The effect of two different single base substitutions, both G to A, on the hyperchromicity profile of a *coli lac* promotor fragment has been reported by Shaeffer et al (27), who found that melting was shifted about 0.2°. A careful theoretical analysis of the changes in melting properties expected for single base substitutions has been presented by Benight & Wartell (3), who

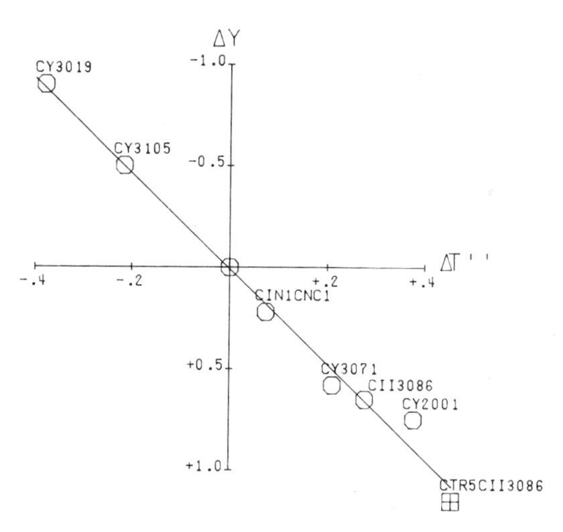

Figure 4 Comparison of experimental and theoretical retardation levels for lambda fragments carrying base substitutions. The point at 0, 0 represents at least seven mutants. The experimental gel displacements of the bands of mutant fragments from the position of the wild type fragment, ΔY, were determined in polyacrylamide gels with a urea/formamide denaturing gradient. For details, see (12). The displacements are in cm, and the expected differences in gel position in °C.

have calculated the changes for 17 different substitutions; they find that shifts in T_m up to about 0.6° can be expected.

Families of Fragments With Closely Distributed End Sites

If a long molecule is subjected to random or quasi-random double-strand scission, the product includes many sets of fragments with most sequences in common but differing in length by small increments.

The distributions in a one-dimensional denaturing gradient given by random shear of each of the *Eco*RI fragments of lambda DNA show intricate patterns of peaks, different for each of the six restriction fragments (11). The sum of the six distributions matches the pattern of sheared, whole lambda DNA. The gradient depth at which each intact *Eco*RI fragment is retarded corresponds closely to the depth of its most easily retarded subfragment. It was inferred that the retardation level of any molecule was determined by the melting of its least stable domain.

Interesting properties are displayed in two-dimensional gels, where length separation is carried out at right angles to the denaturing gradient. The members of a family will be retarded as a long band extended in the direction of length separation at a nearly uniform gradient depth. Roughly 50 such streaks can be discerned after random shear of lambda DNA (S. G. Fischer and L. S. Lerman, unpublished). The sequence intervals between the nearest lower melting domains on the melting map determine the length of the longest fragment on each streak. Thus, the pattern defines a mapping procedure. The two-dimensional pattern of fragments resulting from a single cleavage of pBR322 DNA linearized with *Eco*RI using minimal hydrodynamic shear (19) is shown in Figure 5 (*upper left*). The gel also includes a set of eight fragments of ϕX174 RFDNA from *Hae*III, *Hpa*I, and *Hpa*II cleavage, seen as faint spots at a nearly uniform depth a little below the upper portion of the shear pattern. The relation of the pattern to plasmid geography was calibrated by means of a set of restriction fragments, incorporated a few at a time into repeated two-dimensional separations of the sheared preparation. The restriction fragments are recognized as compact spots against the more diffuse shear bands. The region included in each restriction fragment is shown against the compositional map of pBR322 in Figure 8 (*right*), and the localization of each in the shear pattern is shown in outline in Figure 8 (*lower left*). The isolated spot in the upper right is intact linear pBR322.

Since minimal shear is expected to make only one double-strand break, the pattern can be expected to have only two bands crossing any meridian of fixed length, representing parts from the low-numbered end and from the high-numbered end. Only near the left center is there an appearance of three bands across a meridian of constant length; the additional band is faint,

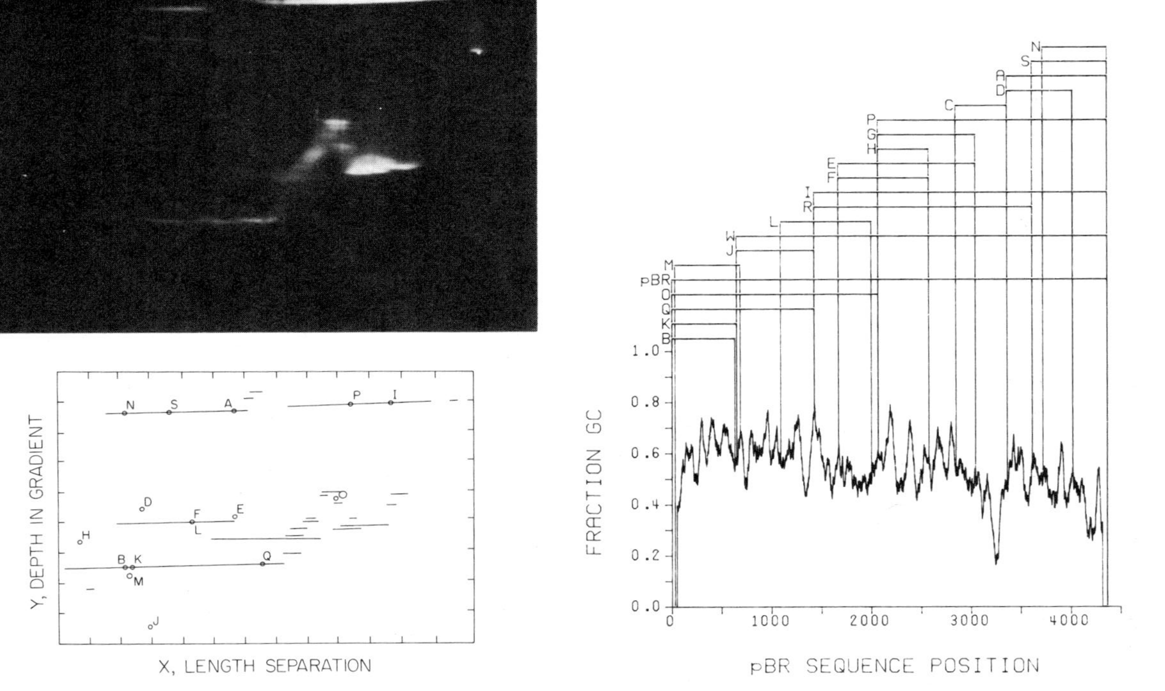

Figure 5 Two-dimensional separation pattern of pBR322 linearized with *Eco*RI and sheared to make one break per molecule. Fragments were separated from right to left according to length by conventional electrophoresis in agarose. The top to bottom position was determined by retardation in a polyacrylamide gel containing a urea/formamide gradient at 61.4°C. The denaturant concentration was zero at the top and 5.6 M urea, 32% (v/v) formamide at the bottom. *Lower left*: a sketch of the gel pattern shown above, together with the experimental positions of various restriction fragments of pBR322 found in gels prepared with a mixture of sheared and restricted fragments. The isolated spot in the upper right corner represents intact, linear pBR322. *Right*: identification of the restriction fragments mapped against the shear pattern in the gel at the left. Each fragment is represented as a horizontal bar with end points projected onto a compositional map of pBR322, numbered according to Sutcliffe (27). The compositional map was derived from the base sequence as described in Figure 2.

corresponding to secondary shear of rare long fragments. Otherwise, all fragments retain one terminus of the original linear pBR322 molecule. The pattern of restriction fragments indicates that the top bands in the low melting portion of the gradient contain all fragments extending to the high-numbered end, while the lower portion represents the complementary fragments with low-numbered ends. The intense streak at the upper right, representing all fragments longer than about 1200 bp carrying the original end at 4362, is retarded at about the same level as the weaker band at the left, which represents all fragments shorter than about 1008 bp carrying the 4362 end. The latter set excludes the AT-dense trough near 3250. We infer that the retardation of long right-end fragments and intact pBR322 is due to melting of the AT-dense region near 4250.

Destabilization Near Ends

It is well known that short double helices of simple polynucleotides melt at lower temperatures than long molecules of the same composition. Similarly, a natural DNA sequence will undergo a thermal subtransition at a lower temperature if it lies near an end of the molecule, rather than separated from the end by a hundred base pairs or more of a higher melting sequence (1, 9). Where the lowest melting domain is surrounded by higher melting domains, the continuous decline to shorter fragments in a randomly cleaved preparation will generate members in which that low melting domain is terminal, rather than central. The shift in melting temperature of the controlling domain to a lower effective temperature for some short fragments is conspicuous as a discontinuous jump in the band. Two spots from fragments of minimally sheared linear pBR322 of about 1200 bp are evident higher in the gel (lower T') than the pair of streaks constituting the main progression of right-end fragments in Figure 5. For a right-end fragment of 1200 bp, the AT-dense section surrounding base 3250 is proximal to the sheared end. While melting near 3250 in longer fragments requires formation on an internal loop, melting in fragments ending near bp 3200 can proceed by unravelling from the end. PFF calculations suggest that the melting temperature is lowered at least 0.7°.

Nicks, Backbone Interruptions in One Strand

We have examined the effects of single-strand interruptions produced by limited deoxyribonuclease action, partly to determine the extent to which nicks may interfere with simple, sequence-determined retardation levels. We find that DNAse scissions near an average of one per 5 kb produce a new broad peak at about 1.5° higher in the gel (lower T') than the original band (S. G. Fischer, K. Silverstein, and L. S. Lerman, unpublished). Assuming that the nicks are reasonably random over the fragments (short-range differences in vulnerability can be disregarded), only nicks occurring within a particular small fraction of the length, roughly one-sixth, change the retardation behavior. The gel distribution of the 4878-bp *Eco*RI fragment of lambda DNA with one random nick per molecule is significantly more strongly shifted and broadened than might be expected if all cooperativity were abolished at each nick. Alternative calculations that preserve cooperativity across the nick fail to approach the experimental results. The changes are large enough to obscure single-strand substitutions.

The Site of Shear Cleavage

The identification of the positions in a double helix where hydrodynamic shear is likely to cause a double-strand break remains poorly defined unless the fragments are scored according to polarity. The appearance of a Gaussian distribution of fragments centered at 1/2 the original length from the application of minimal shear to an initially monodisperse DNA sample has been noted previously (4). However, the failure to distinguish between right- and left-ended fragments in the distribution makes it impossible to specify a unique site-probability distribution. Symmetrical length distributions centered at 1/2 can be derived from unsymmetrical distributions of cleavage probability. We find that the distribution of right-end fragments of DNA along the major upper streak in the gel in Figure 5, representing right-end fragments only, permits inference of the probability of shear cleavage along the pBR sequence (L. S. Lerman, K. Silverstein, S. G. Fischer, unpublished). The part of the distribution near the mode corresponds to a Gaussian centered at 2174, within 7 base pairs of the arithmetic center of pBR322. Cleavage falls to half maximum probability at 0.08 of the molecular length away from the center. Details of the gel in Figure 5, particularly the left-end fragments in the lower section, suggest that there is sequence specificity in shear cleavage, obscured by the heavy loading of the upper streak.

Extending the Detection of Small Changes: Adding a Clamp Sequence

Nearly all of the 38 possible substitutions involving the middle base of a triplet make significant stability changes. The principal limitation in discerning random substitution comes from the restriction that retardation is likely to be determined by only a small part of the total sequence, perhaps 10–30% of the molecule; only changes in that part will be discerned.

In principle, any sequence segment can constitute the lowest melting portion of an artificially constructed, composite molecule in which the sequence of the added portion is more stable. If the difference in stability is large, other details of the attached segment are immaterial, and the same segment can be attached to all fragments of the genome to be examined. If the test fragments are not too long, a small multiple of L_r, each composite molecule will be retarded only by melting of the test DNA. We refer to the attached stable segment as a clamp.

Composites can be constructed by conventional recombinant techniques. A general purpose plasmid can be prepared with a G+C-dense segment adjacent to an insertion site. However, the properties of the composite may be sensitive to the direction of insertion. The bottom panel of Figure 6 shows the calculated melting map of a composite containing the mouse β-globin promotor segment in the conventional 5′–3′ gene orientation. The top panel is similar, but the orientation of the promotor is reversed. The gene orientation in the lower map shows two small domains, but reversal in the top results in a single domain of an intermediate T_m.

The calculated patterns in perpendicular gradient gels of the molecules mapped in Figure 6 are shown in Figure 7 (neglecting the effect of strand dissociation). For the polarity of attachment showing two small domains,

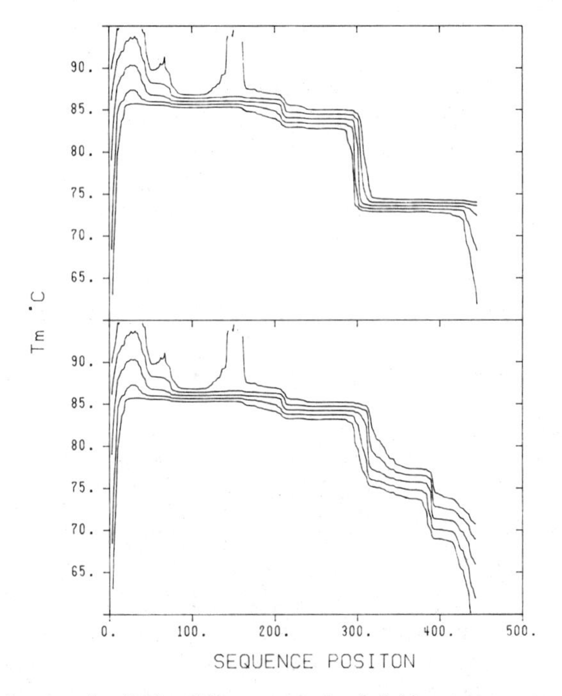

Figure 6 Melting maps for 134 bp of the mouse major β-globin promotor attached to a 316-bp GC-dense clamp. Five ratios of probability of helicity of disordered states are shown. *Lower panel:* the promotor in the gene orientation. *Upper panel:* promotor section reversed.

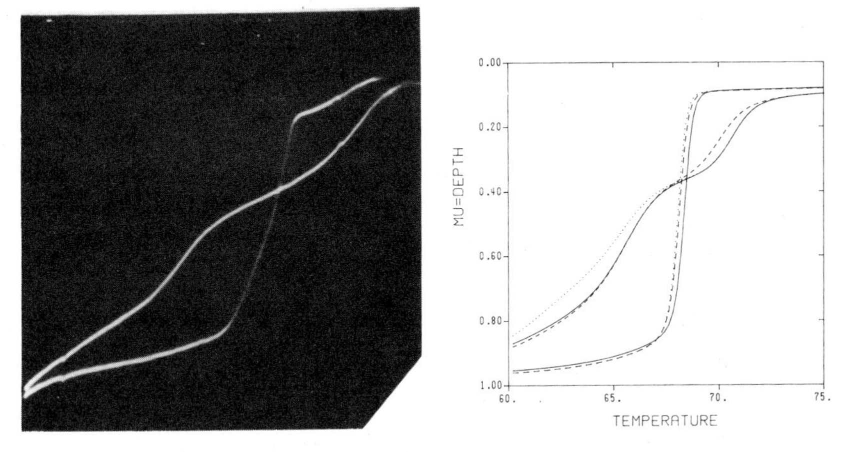

Figure 7 The electrophoretic mobility, MU, of the clamped promotor in both orientations vs the effective temperature, T'. *Left:* a perpendicular gradient gel of the 450-bp molecule consisting of the mouse promotor attached in both orientations to a GC-dense clamp sequence. The curve with the double inflection corresponds to the recombinant carrying the promotor in the gene orientation (lower melting map, Figure 6). The recombinant with the reversed orientation of the promotor (upper melting map, Figure 6) provided the curve with a single inflection. *Right:* theoretical mobility calculated from the base sequence, according to Equation 2. The solid line represents normal globin promotor. Two inflections, gene orientation of attachment; one inflection, reversed gene orientation. The dashed line represents substitution of GAA for GGA at base 47 of the promotor section. The dotted line represents substitution of GTT for GCT at base 114 of the promotor section.

the expected pattern has two inflections; where there is a single domain with an intermediate melting temperature, there is only a single inflection, and the inflection occurs at an intermediate gradient level. The actual patterns given by the clamped β-globin promotor (shown in Figure 7) display the expected features (S. G. Fischer et al, unpublished). The high T' for a strand dissociation constant of 10^{-8} indicates the efficacy of the clamp in maintaining a branched configuration through complete melting of the globin section. An artificial link maintaining covalent continuity between both strands of the helix at one end may be an effective alternative to the clamp sequence.

The efficacy of a clamp in rendering substitutions detectable is shown by the calculated mobility-T' function in Figure 7 for two substitutions, one in the first and another in the second of the short domains of the promotor. The patterns in perpendicular gradients (R. Myers et al, unpublished) correspond to the calculated properties in that each inflection is shifted independently when the promotor is in the gene orientation, but the steeper single inflection is shifted in its entirety by either substitution in the reverse orientation. Mutant fragments are clearly resolved from the normal sequence in parallel gradient gels.

Isolation of Rare Variants

Separations according to base substitution provide a means of isolating molecules that deviate from the standard sequence. DNA can be recovered from the gel in the regions adjacent to but not including the main band. The sharp focusing accompanying retardation facilitates isolation of rare variants. Different classes of mutants can be segregated.

The clamped mouse major β-globin promotor fragments discussed above were subjected to hydroxylamine mutagenesis in vitro, restored to a wild type vector, introduced to a new host, and incubated briefly. After denaturing gradient separation of the mutagenized plasmid DNA fragments from transformed cells, segments of the gel above the main band were pooled, electroeluted, and recloned. A high yield of substitutions in the promotor, characterized by both denaturing gradient properties and detailed sequencing, was found (R. Myers et al, unpublished).

Mispaired Bases

The destabilization of a domain resulting from mismatch of a single base pair is larger than the difference resulting from any substitution that preserves pairing. The two new partners derived by reassortment of two matched duplexes are retarded at different levels. Figure 8 shows the retardation levels of the normal clamped mouse globin promotor and a mutant carrying a CCG to CTG substitution near the junction with the clamp sequence. Reassociation after melting the mixed fragments generates two new bands higher in the gel (R. Myers, T. P. Maniatis and L. S. Lerman, unpublished). We tentatively suppose that the least stable band represents an A · C mismatch, and the next is G · T. Each band contains about the same amount of DNA. The destabilization corresponds to substantially lower GT sums for adjacent doublets, as shown in Table 1. The pyrimidine-pyrimidine and purine-purine mismatches from reassociation of a sequence with its counterpart differing by a transversion result in stronger destabiliz-

Table 1 Stability of triplets containing a mismatched pair

Space 4	Wild type	Mutant	Mismatch	Mismatch
Triplet	CCG	CTG	CGG	CAG
Complement	GGC	GAC	CTG	GCG
Mean GT stability, °C	79.26	55.57	8.6	−8.5

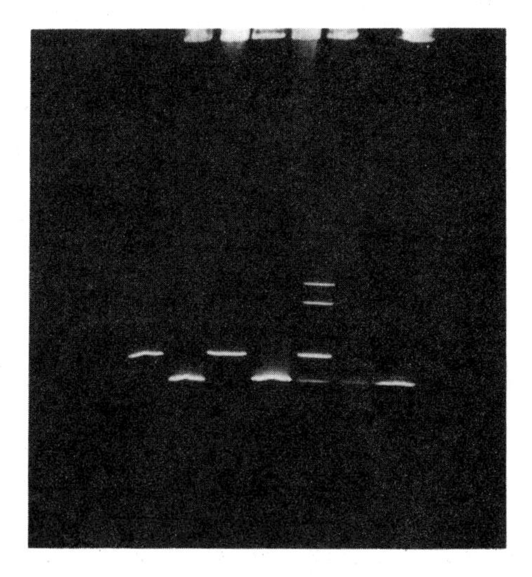

Figure 8 The effect of mispairing. *Left to right:* retardation positions in a parallel gradient: lane 1, the clamped promotor; lane 2, the same with a G to A substitution in the promotor; lanes 3 and 4, the normal and mutant fragments heated and annealed separately; lane 5, both fragments heated and annealed together, and two additional samples.

ation and provide a sensitive means for detection of transversions (N. Lumelsky and L. S. Lerman, unpublished).

Two-Dimensional Restriction Patterns: Detection of Large Insertions and Deletions

Since denaturing gradients separations are only slightly sensitive to molecular length, the combination of gradient separation with orthogonal length separation can be expected to distribute fragments broadly over two-dimensional space. A two-dimensional gel containing *Eco*RI fragments of *E. coli* chromosomal DNA together with ^{32}P-labelled fragments of the same strain carrying an integrated lambda prophage permitted recognition of four well-resolved extra spots attributable to restriction fragments of the prophage (9). The patterns, though complex, are reproducible.

Complicating Factors

Idiosyncrasies occasionally arise and introduce complexity in interpretation. Satellite bands sometimes accompany bands of certain restriction fragments, slightly removed from the main band. While these bands are observed with some restriction cleavages and not others, we have not yet determined whether they are characteristic of particular enzymes. Both the main band and the satellite bands of lambda DNA fragments are shifted by the same increment as a result of base substitutions (12). Some satellites were absent if the phage sequence was cultivated as a recombinant plasmid. The satellites are usually much weaker than the main band, but comparable intensities have been observed. An hypothesis not excluded by the data supposes that the band multiplicity derives from incomplete post-replication modification of DNA at specific sites that may be affected by some substitutions.

Summary

The variation in electrophoretic mobility of DNA under conditions of marginal helix stability provides a useful means for investigation of the relation between the helix-random chain transition and base sequence in natural DNA and a powerful procedure for separation of DNA molecules according to sequence. The use of statistical mechanical theory for analysis of the transition equilibria together with new, simplified theoretical considerations on the effect of strand unravelling on mobility have shown that the gel behavior is predictable for known sequences. A number of the distinctive consequences of the theory and their correspondence with the properties of real molecules have been demonstrated. These include the extremely close cooperative linkage of large blocks of bases into domains, the existence of sharp boundaries between domains, the major role of nearest-neighbor interaction in determining stability, the dependence of domain structures on neighboring and more remote sequences, and the depression of domain melting temperature if the sequence lies at the end of a molecule.

New and unusual applications derive from the possibility of separating DNA molecules by properties of their sequence. Exceedingly complex mixtures, such as the sum of all fragments produced by the action of a six-base specific restriction endonuclease on a complete bacterial genome, can be resolved completely. Additional inserted sequences are easily discerned. The difference of a single base pair in a molecule permits detection and isolation of mutant sequences. The need for full sequential analysis of long molecules for characterization of mutants can be reduced by localizing a change within a small fragment.

Acknowledgments

This work was supported by a grant from the National Institutes of Health, GM2403007. We wish to thank Nelly Brown, Martina Kirstein, Rick Myers, and Ned Seeman for their help.

Literature Cited

1. Amerikyan, B. R., Vologodskii, I. L., Lyubchenko, Y. L. 1981. *Nucleic Acids Res.* 9:5469–82
2. Anderson, S., Bankier, A. T., Barrell, B. G., de Bruijn, M. H. L., Coulson, A. R., Drouin, J., et al. 1981. *Nature* 290:457–65
3. Benight, A. S., Wartell, R. M. 1984. *Biopolymers.* In press
4. Bowman, R. D., Davidson, N. 1972. *Biopolymers* 11:2601–24
5. Burkhardt, J., Birnstiel, M. L. 1978. *J. Mol. Biol.* 118:61–79
6. Cann, J. R. 1968. *Immunochemistry* 5:107–34
7. de Gennes, P. G. 1979. *Scaling Concepts in Polymer Physics.* Ithaca, NY: Cornell Univ. Press
8. Evans, K. E. 1981. *J. Chem. Soc. Faraday Trans. 2* 77:2385–99
9. Fischer, S. G., Lerman, L. S. 1979. *Cell* 16:191–200
10. Fischer, S. G., Lerman, L. S. 1979. *Methods Enzymol.* 68:183–91
11. Fischer, S. G., Lerman, L. S. 1980. *Proc. Natl. Acad. Sci. USA* 77:4420–24
12. Fischer, S. G., Lerman, L. S. 1983. *Proc. Natl. Acad. Sci. USA* 80:1579–83
13. Fixman, M., Friere, J. J. 1977. *Biopolymers* 16:2693–2704
14. Gotoh, O., Tagashira, Y. 1981. *Biopolymers* 20:1033–42
15. Gross, K., Probst, E., Schaffner, W., Birnstiel, M. L. 1976. *Cell* 8:455–69
16. Hervet, H., Bean, C. 1984. *Biopolymers.* In press
17. Klein, J., Fletcher, D., Fetters, L. J. 1983. *Nature* 304:526–27
18. Klump, J. H., Burkart, W. 1977. *Biochim. Biophys. Acta* 475:601–4
19. Lerman, L. S., Fischer, S. G., Bregman, D. B., Silverstein, K. J. 1981. In *Biomolecular Stereodynamics*, ed. R. Sarma, pp. 459–70. Albany, NY: Academic
20. Lerman, L. S., Fischer, S. G., Lumelsky, N. 1983. In *Recombinant DNA and Medical Genetics 13*, ed. A. Messer, I. Porter. New York: Academic. In press
21. Lerman, L. S., Frisch, H. L. 1982. *Biopolymers* 21:995–97
22. Lumpkin, O. J., Zimm, B. H. 1982. *Biopolymers* 21:2315–16
23. Lyamichev, V. I., Panyutin, I. G., Lyubchenko, Y. L. 1982. *Nucleic Acids Res.* 10:4813–28
24. Olivera, B. M., Baine, P., Davidson, N. 1964. *Biopolymers* 2:245–57
25. Poland, D. 1974. *Biopolymers* 13:1859–71; Poland, D. 1978. *Cooperative Equilibria in Physical Biochemistry.* Oxford, England: Oxford Univ. Press
26. Rosenberg, A. H., Studier, F. W. 1969. *Biopolymers* 7:765–74
27. Schaeffer, F., Kolb, A., Buc, H. 1982. *EMBO J.* 1:99–105
28. Sutcliffe, J. G. 1977. *Cold Spring Harbor Symp.* 46:77–90
29. Thatcher, D. R., Hodson, B. 1981. *Biochem. J.* 197:105–9

Ann. Rev. Biophys. Bioeng. 1984. 13: 425–51

BIOPHYSICAL APPLICATIONS OF QUASI-ELASTIC AND INELASTIC NEUTRON SCATTERING

H. D. Middendorf

Department of Biophysics, University of London King's College, 26–29 Drury Lane, London WC2B 5RL, United Kingdom

INTRODUCTION

Thermal and cold neutrons with de Broglie wavelengths in the 1 to 20 Å region are versatile probes of the structure and dynamics of organic molecules. Neutron diffraction, i.e. the study of interference patterns produced by coherently scattered neutrons, has already been used extensively in structural investigations of a wide range of biomolecular samples since the first application to vitamin B_{12} at Harwell in 1967 (59). Because of the large rest mass of the neutron, the particle velocities encountered here are only of the order of 10^3 m/s so that it is easy to energy-analyze the differential intensity scattered into each solid angle element, in addition to determining its dependence on scattering angle. Inelastic scattering experiments of this kind are a valuable source of spatiotemporal information on dynamical processes with scale lengths in the 0.1 to 100 Å region and characteristic times between 10^{-13} and 10^{-6} s. The spin of the neutron provides a third variable that can be exploited in a number of ways (33).

From a biophysical or biochemical point of view, essentially all applications of neutron diffraction to structural analyses and to slowly time-dependent processes are straightforward extensions of X-ray scattering, the principal new features being the emphasis on light elements and the large scattering contrast between hydrogen and deuterium. These applications have been the subject of several reviews (24, 58, 85, 86), and two Brookhaven Symposia (67, 68) give a good cross section of advances in this field. Compared with diffraction work, biomolecular applications of

0084–6589/84/0615–0425$02.00

inelastic scattering (53) have developed much more slowly because of the fact that the basic limitations of neutron techniques, i.e. low flux and relative inaccessibility of instruments, are quite severe in experiments requiring an energy analysis of the scattered radiation.

Much of the inelastic bio-neutron work done during the 1970s has been exploratory, and it is only recently that first results justifying some of the early expectations have been obtained. Despite considerable advances in instrumentation, implemented primarily at the Institut Laue-Langevin (ILL) in Grenoble, the task of collecting neutron spectra from small samples still requires a good deal of patience, since the few high-performance instruments suitable for biomolecular spectroscopy are shared between several scientific disciplines.

For these reasons, the volume of published work is rather small. The papers and reports that have appeared, however, deserve to be scrutinized carefully not only because of the new ground they break individually and the specific questions they address but also because, as a class, they represent first attempts to derive spatiotemporal information on biomolecular processes by means of radiation scattering techniques that will be used more widely from around 1985–1986 onwards. Following a brief introduction to the essentials of neutron spectroscopy, this article aims to combine a review of experimental and theoretical work on biophysical applications of inelastic neutron scattering[1] with an outline of current developments and an assessment of future trends.

THEORETICAL BACKGROUND

The neutron is an uncharged particle of mass $m = 1.675 \times 10^{-24}$ g and spin $\frac{1}{2}$. The particle-like and the wave-like properties of a monoenergetic beam of neutrons travelling with velocity $\mathbf{v}_0$ are linked by de Broglie's relation, $\lambda_0 = 2\pi\hbar/mv_0$, where λ_0 is the wavelength ($\hbar$ = Planck's constant/2π). Momentum and energy are given by $\mathbf{P}_0 = m\mathbf{v}_0 = \hbar\mathbf{k}_0$ and $E_0 = mv_0^2/2 = \hbar^2k_0^2/2m$, respectively, the neutron wavenumber being $k_0 = 2\pi/\lambda_0$. Some representative numerical values are listed in Table 1. In the absence of magnetic interactions, which is normally the case for biological materials, neutrons are scattered only from the atomic nuclei in a sample. Because of the very short range of nuclear forces in relation to λ_0, and the low incident energies E_0, each neutron-nucleus interaction generates a scattered wavelet

[1] More precisely, the scattering processes considered here belong to the class for which radiation field and target nuclei are weakly coupled (i.e. negligible absorption and small multiple scattering); applications that depend on energy absorption through production of secondary particles, such as radiotherapeutic methods using cold neutrons (43a), are beyond the scope of this article.

Table 1 Some neutron wavelengths and velocities with their frequency and energy equivalents[a]

λ (Å)	v (m/s)	ω (THz)	E (meV)	E (cm^{-1})	E (kcal/mol)	E/k_BT
1.0	3956	19.78	81.80	659.8	1.885	3.164
1.78	2224	6.25	25.85	208.5	0.596	1.000
2.5	1582	3.16	13.09	105.6	0.302	0.506
5.0	791	0.79	3.27	26.4	0.075	0.127
10.0	396	0.20	0.82	6.6	0.019	0.032
20.0	198	0.05	0.20	1.6	0.005	0.008

[a] Here and throughout this article, units of cm^{-1} are used simply as a convenient energy measure defined by 1 $cm^{-1} \mathrel{\hat{=}} 0.124$ meV $\mathrel{\hat{=}} 0.02998$ THz, without reference to optical spectroscopy. Thermal energy k_BT for $T = 300$ K (k_B = Boltzmann's constant).

that is spherical and characterized by a single empirical parameter, the scattering length b. Thus the scattering cross section of an isolated stationary nucleus is $\sigma = 4\pi b^2$, a quantity equal to the ratio of the total isotropically scattered intensity to the incident flux.

For a sample comprising N point-like nuclei located at $\mathbf{r}_i$ ($i = 1, 2, \ldots, N$), it is appropriate to write the nuclear scattering density $\rho_s(r)$ as a sum over δ-functions each multiplied by its scattering length b_i. The scattering amplitude for this assembly of static nuclei is then given by

$$A(\mathbf{Q}) = \int_{\text{volume}} \rho_s(\mathbf{r}) \exp(-\mathrm{i}\mathbf{Q}\cdot\mathbf{r})\, d\mathbf{r} = \sum_i b_i \exp(-\mathrm{i}\mathbf{Q}\cdot\mathbf{r}_i) \qquad 1.$$

where $\hbar\mathbf{Q}$ is the momentum transfer[2] and $\hbar\mathbf{Q} = \hbar(\mathbf{k}-\mathbf{k}_0)$ expresses the conservation of momentum (k, λ, v, and E denote the wavenumber, wavelength, velocity, and energy of scattered neutrons, respectively). In elastic scattering there is no exchange of energy so that $k = k_0$ (or $v = v_0$) and $Q = (4\pi/\lambda_0) \sin\theta$, 2θ being the scattering angle. Taking AA^*, the result for the differential cross section with respect to solid angle Ω is

$$\frac{d\sigma}{d\Omega} = \sum_{i,j} b_i b_j \exp(-\mathrm{i}\mathbf{Q}\cdot\mathbf{r}_i) \exp(\mathrm{i}\mathbf{Q}\cdot\mathbf{r}_j) \qquad 2a.$$

$$= Nb_{\text{inc}}^2 + Nb_{\text{coh}}^2 \left\{ \frac{1}{N} \sum_{i,j} \exp(-\mathrm{i}\mathbf{Q}\cdot\mathbf{r}_i) \exp(\mathrm{i}\mathbf{Q}\cdot\mathbf{r}_j) \right\}. \qquad 2b.$$

In the last equation, the averaging over $b_i b_j$ has been performed assuming a single atomic species with coherent and incoherent scattering lengths

[2] Apart from a factor of 2π, $\mathbf{Q}$ is identical with the "reciprocal space vector" used in X-ray scattering.

$b_{coh} = \langle b_i \rangle$ and $b_{inc}^2 = \langle b_i \rangle^2 - \langle b_i^2 \rangle$. Neutron scattering from a natural sample always consists of two parts: (*i*) coherent scattering $\sim \langle b_i \rangle^2$, giving an interference pattern that is due to the superposition of N^2 wavelets with phases according to the separation of pairs of nuclei in the structure (including N terms describing self-scattering); (*ii*) incoherent scattering arising from randomly distributed spin and isotope disorder in the sample, this component being proportional to N and the mean-square deviation from $\langle b_i \rangle$. In units of 10^{-12} cm, values of b_{coh} for the commonly encountered elements H, D, C, O, N are -0.374, 0.667, 0.665, 0.58, 0.94, respectively. The incoherent cross sections $\sigma_{inc} = 4\pi b_{inc}^2$ for H and D are 79.7 and 2.0×10^{-24} cm^2; those for C, O, N are small or zero. Further values and relevant details are given in (26).

In Equation 2b, the expression in curly brackets is identified with the static structure factor $S(\mathbf{Q}) \equiv S_{coh}(\mathbf{Q})$ used in diffraction work. This contains the structural information of interest; the corresponding incoherent structure factor $S_{inc}(\mathbf{Q}) \equiv 1$ represents a unit background and does not appear explicitly in the first term of Equation 2b. For any real sample, because of thermal vibrations and other excitations involving nuclear motions, a differential cross section calculated from Equation 2 can only relate to a "snapshot" of a particular spatial coordination of the N nuclei considered. The pattern actually observed in a diffraction experiment reflects the time-averaged configuration, and an approximate way of accounting for the "blur" introduced by nuclear motions is to apply a Debye-Waller factor to both $S_{coh}(\mathbf{Q})$ and $S_{inc}(\mathbf{Q})$. It is possible, now, by means of the techniques discussed below, to spectrally analyze this "blur" and to gain information on dynamical processes simultaneously with probing their spatial properties by varying $\mathbf{Q}$. The two fundamental variables in radiation scattering events are momentum transfer $\hbar\mathbf{Q}$ and energy transfer $\hbar\omega$, their Fourier transform conjugates being radius vector $\mathbf{r}$ and time t. The conservation of energy is readily expressed by

$$\hbar\omega = E - E_0 = \tfrac{1}{2}m(v^2 - v_0^2) = (\hbar^2/2m)(k^2 - k_0^2) \qquad 3.$$

so that the velocity v (or wavenumber $k = 2\pi/\lambda = mv/\hbar$) of neutrons scattered into $d\Omega$ is analyzed as the second independent variable. An inelastic scattering experiment therefore yields a double differential cross section which is given by (17, 48)

$$\frac{d^2\sigma}{d\Omega dE} = \frac{N}{\hbar}\frac{k}{k_0}[b_{inc}^2 S_{inc}(\mathbf{Q}, \omega) + b_{coh}^2 S_{coh}(\mathbf{Q}, \omega)], \qquad 4.$$

where the dynamic structure factors $S_{inc}(\mathbf{Q}, \omega)$ and $S_{coh}(\mathbf{Q}, \omega)$, or "scattering laws," represent energy-resolved generalizations of $S_{inc}(\mathbf{Q})$ and $S_{coh}(\mathbf{Q})$ in the sense that upon integration over all ω these static structure factors are recovered. Although in a diffraction experiment the incoherent scattering is of no interest since it appears as a diffuse, essentially Q-independent background, it will be seen that its spectral structure, i.e. $S_{\text{inc}}(\mathbf{Q}, \omega)$, is an important source of information on atomic and molecular motions.

The quantitative connection between $S_{\text{coh}}(\mathbf{Q}, \omega)$ or $S_{\text{inc}}(\mathbf{Q}, \omega)$ and the molecular dynamics is established by time correlation functions which for $t \to 0$ reduce to the static structure factors. Short-cutting the rigorous calculation, we may simply insert time-dependent position vectors into Equation 2a such that the product of scattering amplitudes consists of one set taken at time $t = 0$ and the other at a later time t. Averaging these products over an ensemble of trajectories $\mathbf{r}_i(t)$ corresponding to thermal equilibrium then leads to the "intermediate" scattering function

$$F_{\text{coh}}(\mathbf{Q}, t) = \frac{1}{N} \left\langle \sum_{i,j} \exp\left[-\mathrm{i}\mathbf{Q} \cdot \mathbf{r}_i(0)\right] \exp\left[\mathrm{i}\mathbf{Q} \cdot \mathbf{r}_j(t)\right] \right\rangle_{\text{th}} \qquad 5.$$

together with a similar function $F_{\text{inc}}(\mathbf{Q}, t)$ for which only the diagonal terms $i = j$ are summed. Equation 5 relates to coherent scattering as it covers all pair correlations including the "self" terms $i = j$; for $F_{\text{inc}}(\mathbf{Q}, t)$ the single-particle correlations are sufficient because here the scattering is due only to self-interference along $\mathbf{r}_i(t)$. These "intermediate scattering functions" play a central role in inelastic scattering, because on Fourier-transforming them with respect to t the corresponding dynamic structure factors are obtained, whereas Fourier transformation with respect to $\mathbf{Q}$ yields two functions depending only on $\mathbf{r}$ and t. Schematically,

$$S_{\text{coh}}(\mathbf{Q}, \omega) \leftrightarrow F_{\text{coh}}(\mathbf{Q}, t) \leftrightarrow G(\mathbf{r}, t), \qquad 6.$$

$$S_{\text{inc}}(\mathbf{Q}, \omega) \leftrightarrow F_{\text{inc}}(\mathbf{Q}, t) \leftrightarrow G_{\text{s}}(\mathbf{r}, t). \qquad 7.$$

The G functions appearing here are van Hove's space-time correlation functions (48); they connect the experimentally accessible quantities (mainly S, but also F) with a substantial body of theoretical work on the time-dependent statistical mechanics of systems of interacting particles and provide a firm basis for interpretation. Of special interest for our purposes are the fundamental integral relations that define macroscopic transport coefficients (diffusivity, viscosity) in terms of correlation functions describing the underlying microscopic dynamics (17, 31).

INTERPRETATIONAL ASPECTS

Scattering from Molecules

There are two conceptually simple limiting situations for which $d^2\sigma/d\Omega dE$ can be calculated rigorously: the case of perfect disorder, e.g. the $Q \to 0$

limit of pure Brownian motion ($\hbar\omega \lesssim 0.01\ \text{cm}^{-1}$), and that of perfect order in the form of a crystalline solid traversed by $3N$ harmonic lattice waves (phonons) with known dispersion properties ($\hbar\omega \gtrsim 10\ \text{cm}^{-1}$). Proceeding from highly idealized systems such as these, the theory of inelastic scattering has gradually moved into the real world of finite Q, partial order or disorder, molecular interactions, mode coupling, anharmonicity, etc. At high frequencies, therefore, a description in terms of discrete localized modes and phonon-like excitations damped in various ways suggests itself. Towards lower frequencies, the dissipative and diffusive features of a macromolecular system become more pronounced, giving rise to larger amplitudes and involving higher levels of structural organization. Here the van Hove formalism will generally be more appropriate.

Many aspects of the correlation function approach to inelastic neutron scattering from polyatomic systems have been developed in the literature, and those relating to molecular fluids, molecular crystals, and polymers are of particular interest in the present context. Equation 4 is strictly valid only for monatomic scatterers but will be a good approximation in cases where the scattering from a single species of nuclei predominates greatly, the prime example for this being hydrogen in organic molecules. Although $d^2\sigma/d\Omega dE$ can always be written as the sum of an incoherent and a coherent part, it is not in general possible to factorize each of these into a cross section $\sim b^2$ and a scattering law $S(\mathbf{Q}, \omega)$. For a molecular sample, the incoherent part of Equation 4 must be replaced by appropriate sums over atomic species or groups and their particular scattering laws, but the formulation of this is still relatively straightforward. The coherent part, however, because it describes interference effects that reflect the relationship between structure and dynamics much more intimately, will give very complicated expressions which, as a rule, require simplifying assumptions.

Theoretical work on neutron scattering from molecules has developed mainly along five lines: (*a*) approaches based on empirically justified or intuitive assumptions about the functional form of F or G, such as the Gaussian r-dependence of G_s or a convolution ansatz for G (1, 17, 80); (*b*) phenomenological theories incorporating a bimodal description of damped molecular rotations coupled to microscopic density fluctuations, with allowance for internal degrees of freedom (43, 65); (*c*) extensions of the van Hove formalism to systems composed of nonspherical molecules, by means of generalized space-time correlation functions depending explicitly on the center-of-mass coordinates and the Eulerian angles (20, 74); (*d*) investigations of the coupling between rotational and translational degrees of freedom in molecular crystals and rough sphere fluids, with the aim in particular of quantifying deviations from the commonly used convolution assumption (6, 52); (*e*) calculations of correlation functions for coherent and incoherent scattering from chain molecules under conditions of interest in polymer studies (2, 3, 35, 47).

It will be obvious already from this short list that questions relating to the description of the various rotational degrees of freedom (rotational diffusion, reorientational jumps, librations, torsional modes) and their interaction with other modes of motion play an important role in inelastic scattering theory. This is a direct consequence of the fact that neutrons "follow" all nuclear motions in space and time, i.e. give Q-dependent spectra that—although weighted with respect to cross section and amplitude—are not subject to selection rules of the kind operative in infrared and Raman spectroscopy. For a complex biomolecular system composed of many groups, domains, subunits, or larger building blocks (viruses, ribosomes), it would appear useful to pattern further extensions of van Hove's method on the hierarchy of structural organization and to treat each type of "unit building block" as an approximately rigid entity with its own scattering cross section and form factor (18, 19).

Quasi-elastic Scattering

A spectral range of about seven decades centered on $\hbar\omega = 0.1 - 1\ \text{cm}^{-1}$ is accessible to the neutron scattering techniques in current use (Figure 1). This range is of considerable interest in biophysical applications as it bridges the transition region between "discrete" excitations and diffusive processes, thus covering a number of important dynamical phenomena with characteristic times in the ps to ns region. There are two natural reference frequencies in any such scattering experiment: one corresponding to the energy E_0 of the incident beam, the other to the thermal energy $k_B T$ of the sample.

For $\lambda_0 = 2.5$ to 10 Å, the incident wavelengths normally used in the experiments discussed below, the scattering from non-quantized diffusive motions or molecular transitions of low energy satisfies $|\hbar\omega| \ll E_0$ and is therefore called quasi-elastic. Scattering of this kind leads to an essentially symmetric broadening of the elastic peak at $E = E_0$ (or $k = k_0$) as the result of many small positive and negative Doppler-like shifts in the neutron-nucleus interaction (Figure 2). Detailed information on low-energy processes may be obtained from high-resolution studies of the half-width and shape of this peak as a function of $\mathbf{Q}$. Whereas analogous optical techniques are restricted to $2\pi/Q > 1000$ Å because of the long wavelengths employed, quasi-elastic neutron scattering is capable of probing motions with scale lengths in the range 500 Å $\gtrsim 2\pi/Q \gtrsim$ 0.5 Å. At low Q, these are long enough to yield values for the transport coefficients close to those obtained in the continuum limit. As $Q \to 1$ Å^{-1}, the "graininess" of matter becomes increasingly apparent in the spectra, and the transition from medium to short range dynamics is observed. Finally, for $Q > 1$ Å^{-1}, nuclear motions are sampled over distances of the order of 1 Å and strong rotational contributions are seen.

Because of the difficulty of formulating $G(\mathbf{r}, t)$ for a molecular scatterer, the interpretation of neutron spectra in the quasi-elastic region has been developed mainly for incoherent scattering. The motion of a proton or a relatively rigid protonated group occupying a volume element ΔV_p may be decomposed as follows: a vibrational motion relative to its position in the time-averaged structure, a rotational motion relative to the macromolecular center of mass (CM), and a translational motion of the CM relative to a laboratory frame of reference. For reasons of mathematical tractability, it is usually assumed that the various types of motion are dynamically uncoupled. This means that $S_{inc}(\mathbf{Q}, \omega)$ can be expressed as a convolution with respect to ω of the individual scattering laws:

$$S_{inc}(\mathbf{Q}, \omega) = S^{T}_{inc}(\mathbf{Q}, \omega) \otimes S^{R}_{inc}(\mathbf{Q}, \omega) \otimes S^{V}_{inc}(\mathbf{Q}, \omega). \quad 8.$$

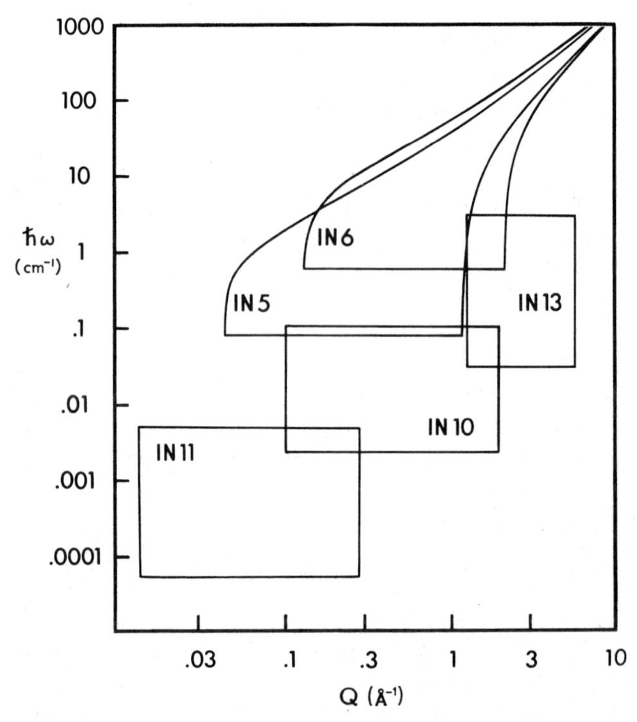

Figure 1 Regions in Q,ω-space covered by current quasi-elastic and inelastic neutron scattering experiments on biomolecular samples at the Institut Laue-Langevin, using spectrometers IN5 and IN6 (time-of-flight), IN10 and IN13 (backscattering), and IN11 (spin-echo).

Equation 8 is obviously a good approximation as long as the characteristic times associated with the three types of motion are well separated, but it will be less satisfactory when the time scales overlap. The high-frequency vibrational motions usually have amplitudes of the order of 0.1 Å, and their effect on the quasi-elastic scattering is reasonably well accounted for by a Debye-Waller factor so that $S^V_{\text{inc}} \approx \exp(-\langle u_p^2 \rangle Q^2)$ (here $\langle u_p^2 \rangle$ is the mean-square proton displacement along **Q**). At energies comparable with $k_B T$ and below, hindered rotational modes of various kinds become more important, with displacements that can be up to one order of magnitude larger. The coupling of these modes to translational motions characterized by diffusive steps d of the order of 1 Å may be appreciable, and at high Q-values it may be difficult in principle to distinguish between translational and rotational contributions. In practice, *faute de mieux*, Equation 8 is used as the basis for almost all model calculations. The formulation of these calculations proceeds from some approximate picture of the relevant low-frequency modes in a sample of known structure, and an overall scattering law is constructed from a number of components. Any such scattering function calculated from a classical model will be symmetrical in ω and therefore violate the quantum-mechanical condition of "detailed balance" (48), i.e. the fact that the probabilities for energy-loss and energy-gain processes are connected by a factor $\exp(\hbar\omega/k_B T)$. At the temperatures considered here, it suffices to apply a correction factor $\exp(\hbar\omega/2k_B T)$ to the

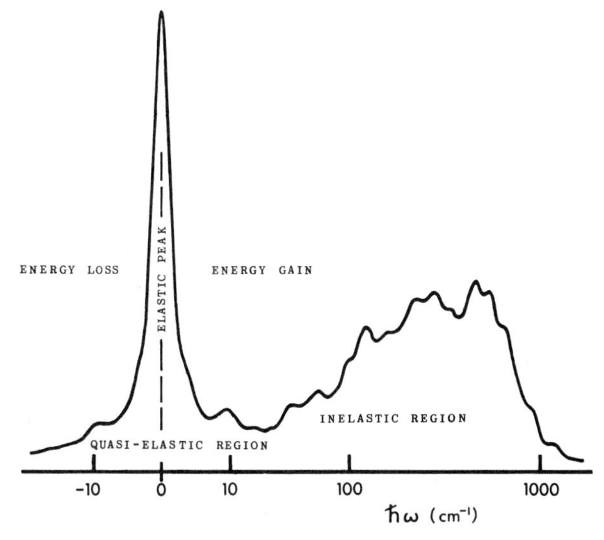

Figure 2 Typical appearance of cold neutron spectrum measured on a high-resolution time-of-flight spectrometer at 2θ = const. (intensities in inelastic region exaggerated relative to elastic peak).

classical scattering law. The resulting expression is finally convoluted with the instrumental resolution function and tested against experimental data.

Numerous scattering laws for particular translational and rotational motions have been derived (9, 14, 15, 44, 70–72). In the continuum limit where $2\pi/Q \gg d$, $G_s(\mathbf{r}, t)$ is obtained as the solution of a Fick's law diffusion equation such that $G_s \to \delta(\mathbf{r})$ as $t \to 0$. The simplest case is that of isotropic, infinite-domain diffusion for which $G_s \sim t^{-3/2} \exp(-r^2/4D_T t)$, $F^T_{inc} = \exp(-D_T t Q^2)$, and $S^T_{inc} = D_T Q^2/\pi(\omega^2 + D_T^2 Q^4)$; here D_T is the translational diffusion coefficient and the spectra are Lorentzians of width $\Delta E_{inc}(Q) = 2\hbar D_T Q^2$. At Q values in the 0.2 to 0.5 $Å^{-1}$ region, a continuum description begins to break down and microscopic features of the diffusion process have to be incorporated into $G_s(\mathbf{r}, t)$. This may be achieved by means of the Langevin equation or similar, more elaborate dynamical equations (17, 22, 27); these typically contain a stochastic forcing term and molecular friction parameters closely related to the viscosity. As in the case of Fickian diffusion, the spatial dependence of G_s in models of this kind often turns out to be Gaussian so that $F^T_{inc} = \exp[-\frac{1}{2}w(t)Q^2]$, where $w(t)$ is a "width function" proportional to the time evolution of the mean-square displacement, $\langle r^2(t) \rangle$. Another important class of models for translational diffusion describes jump processes (17, 71, 72). The physical picture here is that of nuclei, atoms, or molecules diffusing on a regular or random lattice of discrete sites by thermally or nonthermally activated jumps. These "sites" are relatively shallow potential wells in which the scattering particles are trapped or weakly bound for a time τ_0, the average residence time.

A common property of the diffusion processes discussed so far is that $G_s(\mathbf{r}, t) \to 0$ as $t \to 0$, i.e. as time progresses the probability density of the particle considered spreads out over an increasing volume of space. In the case of rotational motions, or of translational Brownian motion restricted to a small volume (14), the self-correlation function approaches a finite time-asymptotic distribution $G^R_s(\mathbf{r}, \infty)$ within a bounded domain ΔV_R. The total rotational self-correlation function may therefore be written as the sum of $G^R_s(\mathbf{r}, \infty)$ and a time-dependent part; after Fourier transformation with respect to $\mathbf{r}$ and t we have

$$S^R_{inc}(\mathbf{Q}, \omega) = \mathscr{A}_0(\mathbf{Q})\delta(\omega) + S^R_{inc*}(\mathbf{Q}, \omega), \qquad 9.$$

and the spectrum is seen to consist of a sharp line $\sim \delta(\omega)$ superimposed on a broader quasi-elastic feature described by the second term. The elastic contribution, i.e. $\mathscr{A}_0(\mathbf{Q})$, is called the elastic incoherent structure factor, or EISF. It represents the square of the spatial Fourier transform of the stationary probability distribution (or probability of occupancy) set up by the ensemble of particle trajectories within ΔV_R as $t \to \infty$. Purely incoherent scattering can thus provide information on the geometry of rotational motions.

Little theoretical work has been done on coherent quasi-elastic scattering from biomolecular systems. Lovesey & Schofield (46) have derived analytical expressions for the scattering from density fluctuations in a spherical virus modelled as a compressible viscous fluid contained in an elastic protein shell. Spectra calculated for $Q \lesssim 0.1$ Å^{-1} and representative parameters show significant structure in the 0.5 to 5 cm^{-1} region, but the intensities are likely to be small. Schofield (69) has developed a general method for writing such scattering laws as the sum of two terms: "contrast" scattering, depending on the difference of the scattering amplitude density of particle and solvent, and "fluctuation" scattering, arising from the dynamics in these regions considered separately. Each contribution can in turn be decomposed into a bulk scattering and a boundary scattering term.

Inelastic Scattering

At energy transfers comparable with or larger than E_0, cold neutrons have everything to gain but not much to lose on colliding with nuclei in thermal motion at $T \approx 300$ K. Therefore, the inelastic spectrum proper[3] is dominated by anti-Stokes scattering, i.e. appears on the energy-gain side of the quasi-elastic peak where $v > v_0$, or $k > k_0$ (Figure 2). Vibrational and rotational modes of motion with distinct energy levels $\hbar\omega_j$ give rise to the scattering in this region which extends up to $\hbar\omega \approx 5k_BT \approx 1000$ cm^{-1}. Because of the large number of modes, the absence of selection rules, and various intrinsic as well as instrumental broadening effects, the spectra actually obtained always show a continuous intensity distribution carrying a number of more or less differentiated features, and any peaks observed will rarely be as sharp as in optical spectra.

There are three basic scattering processes here that provide points of departure for quantitative analyses: inelastic scattering from single nuclei in a harmonic potential, from rotators such as spherical-top or symmetrical-top molecules, and from harmonic lattice waves (phonons) in a crystal (3, 39, 48). The principal factors determining the differential intensity due to a quantized mode with frequency ω_j are the following: (*a*) the coherent or incoherent cross sections, (*b*) the Debye-Waller factor, (*c*) a "polarization" term $(\mathbf{Q}\cdot\mathbf{C}_{pj})^2$ where $\mathbf{C}_{pj}$ is the normalized amplitude vector of nucleus p vibrating in mode j, (*d*) a Bose-Einstein factor

[3] We are now using "inelastic scattering" in its narrow sense; in general, this term embraces all conditions in which some of the scattered radiation differs in energy from E_0, by however small an amount.

$\bar{n}_j \equiv [\exp(\hbar\omega_j/k_B T) - 1]^{-1}$ for thermally populated energy levels, and (*e*) a product of two δ-functions ensuring conservation of energy and momentum.

In the present context, inelastic scattering from phonons is of interest primarily for those fibrous structures and extended arrays of ordered molecules for which normal mode calculations can be performed (62). Based on known or assumed atomic and molecular interaction potentials, such calculations yield a set of dispersion curves $\omega_j(\mathbf{q})$ for the $3LN$ harmonic vibrations in a sample consisting of L unit cells each comprising N atoms, j being the branch index and $\mathbf{q}$ the phonon wave vector. Once $\omega_j(\mathbf{q})$ is known, it is possible to write down explicit expressions for the neutron scattering laws at various levels of approximation and averaging; details of these depend on geometry, coherent/incoherent scattering properties, frequency range studied, experimental conditions, etc. The information content of $d^2\sigma/d\Omega dE$ is highest in the case of coherent scattering where, by virtue of its dependence on both $\delta(\omega \mp \omega_j)$ and $\delta(\mathbf{Q} \mp \mathbf{q} + 2\pi\mathbf{l})$ according to (*e*), it becomes possible to trace out experimentally the dispersion of phonons in $\mathbf{Q}, \omega$-space (upper signs for neutron energy gain, lower for loss; $\mathbf{l}$ is a reciprocal space vector). By comparing the results with calculated dispersion properties, the basic atomic and molecular interaction parameters (force constants) may be tested and improved.

The more commonly encountered situation of dominant incoherent scattering from protons is less informative, but it still gives some insight into the spatial properties of vibration spectra through the $\mathbf{Q}$-dependent polarization term (*c*). The single-phonon approximation to the incoherent scattering law is of the form

$$S^{\mathrm{V}}_{\mathrm{inc}}(\mathbf{Q}, \omega) \sim \sum_{\mathrm{p}} e^{-2W_{\mathrm{p}}} \sum_{j,\mathbf{q}} (\bar{n}_j + \tfrac{1}{2} \mp \tfrac{1}{2})(\mathbf{Q} \cdot \mathbf{C}_{pj})^2 (2m\omega_j)^{-1} \delta(\omega \mp \omega_j) \qquad 10.$$

where $p = 1, 2, \ldots, N_{\mathrm{p}}$ labels the protons in a unit cell or molecule ($\mathbf{C}_{\mathrm{p}j}$, $\bar{n}_j$, and ω_j depend on $\mathbf{q}$, and the double sum extends over all modes). This equation describes the dominant scattering process in which single quanta $\hbar\omega_j$ of phonon energy are gained or lost by the neutrons. In general there will also be scattering events in which quanta of energy equal to integral multiples of $\hbar\omega_j$ are exchanged; these multi-phonon processes are by no means negligible at energies $\lesssim k_B T$ when several or many quantum levels are populated and the resulting nuclear amplitudes are appreciably larger than those associated with high-frequency modes.

It is seen from Equation 10 that the scattering due to $\omega_j(\mathbf{q})$ is essentially proportional to the square of the displacement of p along $\mathbf{Q}$, to Q^2, and to the population density of modes. At high frequencies, we have

$\bar{n}_j \rightarrow \exp(-\hbar\omega_j/k_B T)$ for energy gain and $(\bar{n}_j+1) \rightarrow 1$ for loss; in cold neutron experiments the spectra therefore become unobservable beyond $\hbar\omega \approx 5k_B T$. An important aspect of Equation 10 is that the molecular dynamics contained in $\omega_j(\mathbf{q})$ is largely determined by the heavy atoms (mainly C, O, N), whereas the incoherent scattering comes almost exclusively from hydrogen atoms. Insofar as these can be regarded as rigidly attached to the former, and this is certainly a reasonable approximation at low frequencies, they serve as natural probes of the dynamics.

The Debye-Waller factor multiplying each mode in the $j, \mathbf{q}$-sum in turn depends on all modes in such a way that $2W_p \cong \Sigma(\mathbf{Q} \cdot \mathbf{C}_{pj})^2(2\bar{n}_j+1)/2m\omega_j$ (summed over $j, \mathbf{q}$). Although $2W_p$ is insensitive to details of the phonon spectrum, its interpretation in terms of a mean-square displacement (i.e. $2W_p \approx \langle u_p^2 \rangle Q^2$) requires careful analysis of the averaging implied by $\langle u_p^2 \rangle$. In a system exhibiting a multiplicity of modes over a range of characteristic times from 10^{-11} to 10^{-14} s, the hierarchical time dependence of the self-correlation function $F_{\text{inc}}(\mathbf{Q}, t)$ associated with the "thermal cloud" of proton p has to be considered in relation to the time scale that is being probed (28). This question is closely related to the problem of separating the mean-square displacement due to "conformational" modes from that due to high-frequency vibrations in X-ray temperature factor analyses (13); neutron spectra would allow independent determinations of $\langle u_p^2 \rangle_c$ and $\langle u_p^2 \rangle_v$ to be made and could become a valuable source of complementary information here.

While a detailed evaluation of the dynamical properties of a biomolecular sample in terms of discrete modes may neither be possible nor desirable in many cases, a knowledge of the distribution of its vibrational and librational degrees of freedom over frequency down to $\hbar\omega \approx 1\ \text{cm}^{-1}$ should be extremely useful. This is because, first, much of the current simulation work is on irregularly folded biopolymers, so that frequency distributions provide a convenient basis for comparison. Then there is also the possibility of interpreting spectral information in terms of macroscopic observables, especially in difference experiments where the repartition of degrees of freedom in the important region below $1000\ \text{cm}^{-1}$ can be correlated with thermodynamic and kinetic measurements (75).

Incoherent scattering laws such as Equation 10 obviously contain information that is closely related to a density-of-states function defined by $Z(\omega) = (3LN)^{-1}\Sigma(\omega - \omega_j)$ (summed over all modes), but a rigorously valid expression for $S_{\text{inc}}^{\text{V}}(\mathbf{Q}, \omega)$ in terms of $Z(\omega)$ can be given only for a cubic lattice with $N = 1$. It is possible, however, to extract a hydrogen-amplitude weighted function of this kind from the scattering law, subject to orientational averaging over $(\mathbf{Q} \cdot \mathbf{C}_{pj})^2$ in order to separate Q^2 from the $j, \mathbf{q}$-

sum (e.g. for fiber or powder samples), and subject to assumptions about the dynamical equivalence of the N_p protons in the unit cell or molecule. This averaged frequency distribution will be denoted by $\bar{Z}_p(\omega)$.

Interpretation of Complex Spectra

Because of the closeness of $d^2\sigma/d\Omega dE$ to correlation functions of fundamental significance, and the possibilities inherent in extending familiar methods of isotopic contrast variation to dynamical studies, a good set of neutron data covering a large $\mathbf{Q}, \omega$-domain for a range of sample parameters contains a wealth of information. It is not at all straightforward, however, to interpret data of this kind in terms of frequencies and displacements that can be assigned to identifiable entities in a biomolecular sample of complex structure. Two factors determine the extent to which this is possible.

1. The degree of order in the system investigated. A well-ordered sample makes it possible to fully exploit the spatial information contained in its diffraction pattern and the vectorial nature of $\mathbf{Q}$. In the ideal case of a perfectly crystalline system, it would be feasible in principle to spectrally resolve the diffracted intensity for each Bragg reflection or for certain groups of reflections, and possibly to determine phonon dispersion relations for particular modes. This set of coherent scattering data will be concentrated at low and intermediate values of Q; it must be supplemented by spectral measurements of the incoherent "background" over a wide $\mathbf{Q}$ range.

2. The kind and number of deuterated analogs that can be studied. By selectively substituting deuterium for hydrogen, the change in sign of b_{coh} and that in magnitude of σ_{inc} may be exploited systematically. The sets of $\mathbf{Q}$-dependent difference spectra generated for appropriate sample combinations are then utilized to deduce the dynamical properties of particular groups, segments, or larger parts of a structure. The spatial selectivity thus achieved is of considerable interest in the case of partially or completely disordered samples (solutions, gels, or amorphous powders) for which $d^2\sigma/d\Omega dE$ depends only on the magnitude of $\mathbf{Q}$.

Although there appears to be no fundamental obstacle to *tour de force* projects aiming at a spectrally and spatially resolved description of the molecular dynamics of a crystalline or quasi-crystalline system on the basis of comprehensive $S_{coh}(\mathbf{Q}, \omega)$ and $S_{inc}(\mathbf{Q}, \omega)$ measurements, the practical limitations imposed by neutron flux levels, instrument time, and biochemical effort are very considerable. The price that has to be paid, in terms of $\mathbf{Q}$-resolution and/or selective deuteration, for relating spectral information directly to elements of the time-averaged structure is much too high at present. In current experimental practice, the data are usually obtained in the form of a set of spectra taken along a relatively small number (<20) of ΔQ bands centered on Q_n where $n = 1, 2, 3, \ldots$, and these correspond to scale lengths $2\pi/Q_n$ over which the nuclear motions are sampled.[4] It must be emphasized that this "spectral sorting" with regard to a nonspecific distance variable, because of the wide Q-range covered, represents a valuable source of spatiotemporal information even when only a limited series of measurements can be made. In the important case of dominant incoherent scattering from the protons in the sample, the relation of the incoherent dynamic structure factor $S_{\text{inc}}(Q_n, \omega)$ to the time-averaged structure is somewhat analogous to that of the Patterson function $\mathscr{P}(\mathbf{r})$ used in the analysis of elastic scattering. The latter gives the distribution of distances between all pairs of scattering centers irrespective of the actual location of these vectors in the 3D structure; similarly, $S_{\text{inc}}(Q_n, \omega)$ represents a set of spectra each carrying a distance "label" not specifically related to the structure because $S_{\text{inc}}(Q_n, \omega)$, like $\mathscr{P}(\mathbf{r})$, lacks phase information. If the sample is partially or fully ordered, two or more sets of $S_{\text{inc}}(Q_n, \omega)$ corresponding to different orientations of Q_n with respect to the time-averaged structure can be measured, and appropriate "projections" of **Q**-dependent dynamical properties onto the principal axes may be determined.

Since a self-consistent interpretation of inelastic neutron scattering from biomolecules that relies on exhaustive $d^2\sigma/d\Omega dE$ measurements and variation of sample parameters is not realistic, the approaches taken in current work generally involve a combination of the following: (*a*) semi-quantitative ideas about the dynamics based on the known time-averaged structure and on the properties of systems similar to that being studied; (*b*) analytical model calculations or *ab initio* numerical simulations of the scattering laws; (*c*) gross H/D contrast variation, e.g. by deuterating one of the constituents of a heterogeneous system, by nonspecific biosynthetic incorporation of deuterium, or by reconstituting deuterated with hydrogenous subunits; (*d*) suppression or "freezing" of certain degrees of freedom by means of temperature and viscosity changes, specific ligands or inhibitors, covalent attachment of groups immobilizing particular regions, etc, together with efforts to reduce the level of bulk solvent scattering.

EXPERIMENTAL WORK

At this stage in the application of neutron spectroscopy to biophysical problems, it is important to be aware of the instrumental characteristics and

[4] We assume here, for simplicity, that in the case of time-of-flight experiments a constant-Q interpolation routine has been applied to the data outside the quasi-elastic region where Q_n depends strongly on ω.

limitations when discussing particular experimental results. It is not possible in a short review to adequately cover this aspect; the reader is instead referred to a number of articles and books (23, 33, 35, 72, 84). In the following, aside from mentioning the specific technique used, we shall only quote a rough value for the energy resolution at $\hbar\omega = 0$ (FWHM of elastic peak), denoted by ΔE_0. It should be kept in mind that in the case of time-of-flight (t.o.f.) spectrometers the actual resolution decreases substantially with $\hbar\omega$ and can reach values between 20 and 30% of $\hbar\omega$ at high energy transfers. In hydration experiments, the relative water uptake will be quoted as h = (grams of H_2O or D_2O)/(grams of dry sample). Unless stated otherwise, temperatures are ambient or close to ambient.

Apart from neutron studies of light and heavy water (10, 32, 37, 61), which have a relatively long history and will be mentioned here only in passing, the first inelastic neutron scattering investigations that can claim to be of direct biophysical interest were those on polypeptides during the 1960s. Most of these have been reviewed by Boutin & Yip whose monograph on molecular spectroscopy with neutrons (8) covers much of the early work on polymers, organic compounds, salts, and aqueous solutions.

Polypeptides

In experiments performed between 1965 and 1969, first-generation t.o.f. spectrometers of low resolution were used and the Q-dependence was not analyzed, i.e. spectra were obtained for a single scattering angle (typically $2\theta = 90°$, $Q > 2\ \text{Å}^{-1}$).

Boutin & Whittemore (7) obtained cold neutron spectra from polycrystalline samples of polyglutamic acid in the 30 to 600 cm^{-1} region, comparing $\bar{Z}_p(\omega)$ for the α-helical form with the random coil conformation assumed by the sodium salt. Consistent with normal-mode calculations, low-frequency motions of the secondary structure are expected to be most sensitive to changes in the helical conformation, and this is borne out by the density-of-states changes observed. While the peak assignments made in this low-resolution work were not always justified, it was possible nevertheless to analyze integral properties of the difference spectra, and to derive approximate values for the thermodynamic quantities characterizing the helix → random-coil transition which agreed reasonably well with other data. For the sodium salt of $(\text{Glu})_n$ at $h \approx 0.03$, Whittemore (83) later presented the first t.o.f. hydration difference spectrum of a biopolymer, which showed a slightly "blue-shifted" librational water band centered on 550 cm^{-1}.

Hydrogen-weighted frequency distributions between 5 and 800 cm^{-1} were further derived for polyalanine (29) and for polyglycine in both its β

form, $(Gly)_nI$, and the precipitated form, $(Gly)_nII$, a collagen-like three-fold helix (29, 30). $\bar{Z}_p(\omega)$ for $(Gly)_nI$ was interpreted in conjunction with detailed normal-mode calculations. Interest here focuses on a very strong band at 210 cm^{-1} that is due to torsional motion around the peptide C-N bond (amide VII), on the differences in H-bonding between the two forms, and on their energy differences (conformational and lattice) which can be evaluated from low-frequency spectra.

Because of their large hydrogen amplitudes, torsional vibrations of methyl or phenyl side groups in ordinary polymers have already been investigated extensively by neutron scattering (35, 36, 47). In polypeptides and proteins, the aliphatic side chains possessing terminal methyl groups should exhibit characteristic low-frequency vibrations in the 150 to 400 cm^{-1} region because of torsional oscillations (62). Using a high-resolution spectrometer at ILL, Drexel & Peticolas (16) measured energy-gain spectra for poly(L-alanine) in both its α-helical and β-sheet form. The bands observed were assigned with reference to existing dispersion calculations, and it was shown conclusively that methyl torsion, not α-helical backbone vibration, is responsible for the strong band at 230 cm^{-1}. The data yielded a value of 14 kJ/mol for the potential barrier, assuming that the CH_3 group is a rotor attached to an infinite mass chain.

Hydration Studies

For both fundamental and practical reasons, the majority of biophysical studies using inelastic neutron scattering has been on hydration phenomena in systems ranging from live cells to biochemically pure preparations. The aim here is to characterize in detail the mobility of water closely associated with biomolecules, by means of $d^2\sigma/d\Omega dE$ measurements over a Q, ω-domain matched to the distance and time scales of the microscopic interactions of interest. It is hoped in particular that new information derived from data of this kind will lead to a better understanding of the nature and function of biological water. On the practical side, the H_2O/D_2O composition may be changed with ease in these experiments, sometimes independently of the level of hydration, and it is possible to explore a very wide range of dynamical conditions and H/D contrast (77).

Dahlborg & Rupprecht (12) investigated the hydration dynamics of wet-spun calf thymus NaDNA fibers oriented along $\hat{\mathbf{a}}$ at $h = 0.15$ and 0.45, using a crystal spectrometer for small $\hbar\omega$ ($\Delta E_0 = 0.6-0.9$ cm^{-1}) and a t.o.f. instrument up to ≈ 900 cm^{-1}. The wet-minus-dry structure factor $S_{coh}(Q)$ for D_2O-hydrated NaDNA shows a pronounced and relatively narrow peak at $Q = 1.85$ Å^{-1} when $\mathbf{Q}\cdot\hat{\mathbf{a}} = 1$; this suggests a certain degree of "binding" and also some structuring of D_2O molecules along $\hat{\mathbf{a}}$ because the corresponding peak is weak when $\mathbf{Q} \cdot \hat{\mathbf{a}} = 0$. Anisotropic Debye-Waller factors determined for $\mathbf{Q} \| \hat{\mathbf{a}}$ and $\mathbf{Q} \perp \hat{\mathbf{a}}$ differ by a factor of ≈ 2.5 and demonstrate that $\langle u_p^2 \rangle$ is larger along the DNA helix than across it. Small $\Delta E_{inc}(Q)$ increasing from 0.04 to 0.14 cm^{-1} were measured between 1.1 and 2.0 $Å^{-1}$; these did not depend on $\mathbf{Q}$ relative to $\hat{\mathbf{a}}$ within the error limits. The correlation time for hindered translational and rotational motions was estimated to be 5×10^{-11} s, about one order of magnitude longer than in bulk water. The inelastic spectrum was measured for $\mathbf{Q} \perp \hat{\mathbf{a}}$, and $\bar{Z}_p(\omega)$ for H_2O-hydrated NaDNA shows broad bands centered on 100, 300, and 700 cm^{-1}. The last of these appeared to be "ice-like" in the sense that its centroid is close to that of the librational band in ice; the resolution here was so poor, however, that this conclusion must be very tentative. The low-frequency band for dry NaDNA revealed a distinct peak shifting from 30 to 60 cm^{-1} as a function of orientation.

The dynamics of water in another fibrous system, that of collagen, has been studied by Miller et al (40, 57, 82) at Grenoble, and basic aspects of the coupling of acoustic excitations in a fibrous structure to density fluctuations in a surrounding fluid have been discussed by White (81). Quasi-elastic spectra measured on a t.o.f. spectrometer ($\Delta E_0 = 0.2$ cm^{-1}) for an array of H_2O-hydrated collagen fibers oriented with $\mathbf{Q} \perp \hat{\mathbf{a}}$ showed essentially no broadening below $h \approx 0.25$; the broadening observed above this value gave $D_T \approx 5 \times 10^{-6}$ cm^2/s. For $\mathbf{Q} \| \hat{\mathbf{a}}$, on the other hand, D_T appeared to be the same as that for bulk water. Additional measurements made at very high resolution (0.003 cm^{-1}) were aimed at determining the effect of hydration on the rigidity of the collagen backbone.

Hecht & White (34) used two Harwell t.o.f. spectrometers ($\Delta E_0 = 2-3$ cm^{-1}) to study the mobility of H_2O in a 7% solution of oriented tobacco mosaic virus (TMV) particles. The inelastic spectrum was found to be largely unchanged compared with the buffer solution. The diffusion coefficient $D_T \approx 2 \times 10^{-5}$ cm^2 s^{-1} derived from $\Delta E_{inc}(Q)$ was essentially the same for TMV solution and buffer, consistent with the partial immobilization of no more than 10% of the water fraction closely associated with TMV particles. For $Q \gtrsim 1$ $Å^{-1}$, however, a progressive decrease of ΔE_{inc} relative to buffer was observed. In this region ΔE_{inc} reflects motions with characteristic times $\lesssim 10^{-11}$ s, and the deviations from $\Delta E_{inc} \sim Q^2$ were attributed in part to longer τ_R resulting from the interaction of the macroscopic electric fields between charged TMV particles and the water dipoles.

The controversial question of an "ice-like" component in the water of hydration, and its possible contribution to the librational region between 400 and 900 cm^{-1}, was followed up by Martel (49) who compared spectra from an NaDNA gel ($h = 0.8$) and from rat muscle with those from water and ice (all H_2O). Within the limits of the beryllium-filter technique used, and for an unoriented NaDNA sample, no evidence was found for an ice-like component; the spectrum for ice in fact differed significantly from all others.

More recently, the inelastic scattering properties of cell-associated water above 100 cm^{-1} were studied in greater detail by Martel (50), who measured energy-loss spectra (resolution 35 cm^{-1} at 800 cm^{-1}) for three species of plant leaves ($h \approx 9$). The instrument used here was a beryllium-filter spectrometer at Chalk River; in this high-Q technique $(\bar{n}_j + 1) \approx 1$ and $\hbar\omega \approx -(\hbar^2/2m)Q^2$, so that for counting times $\sim k_0$ the observed cross section is approximately proportional to $\bar{Z}_p(\omega)$. The data were analyzed with reference to spectra from water, ice, and cytosine monohydrate taken under identical conditions. All plant leaf samples gave spectra which for $200 < \hbar\omega < 1100$ cm^{-1} closely resembled those from water at transmissions between 40 and 80%, each consisting of a single broad band centered on $\hbar\omega \approx 580$ cm^{-1} and fitted by a Gaussian (FWHM $= 370-440$ cm^{-1}) on a sloping background. Both the statistical analysis of parameters derived from Gaussian models and direct convolution fits containing a fractional ice contribution lead to the conclusion that overall spectral changes due to a water component with strongly hindered rotational modes are small, and that the "irrotationally bound" water fraction in the plant leaves studied cannot amount to more than a few percent.

Trantham et al (78, 79) are investigating the diffusive properties of H_2O in cysts of the brine shrimp *Artemia* ($h = 0.1-1.2$), and of H_2O and D_2O in 20% agarose gels. Quasi-elastic scattering within ± 6.7 cm^{-1} was measured for $Q = 0.5-1.9$ $Å^{-1}$ on a triple-axis spectrometer ($\Delta E_0 = 0.8$ cm^{-1}) at Oak Ridge. The spectra were interpreted on the basis of a composite scattering law given by $[fS^{b}_{inc}(Q,\omega) + (1-f)S^{m}_{inc}(Q,\omega)]$ where f is the fraction of water molecules that appear to be "bound" with respect to the time scale resolved, and $(1-f)$ is the "mobile" fraction. The motions of the former are approximated by a Debye-Waller factor, whereas the latter, i.e. $S^{m}_{inc}(Q,\omega)$, is expressed as the convolution of a translational jump diffusion model with the first two terms of the expansion of Equation 9 for continuous rotational diffusion. Apart from the "bound" proton parameters f and $\langle u^2_b \rangle$, which include the covalently bound non-water protons, it was possible to evaluate $\langle u^2_m \rangle$, τ_0, and the translational and rotational diffusion coefficients, D_T and D_R. In *Artemia*, $\Delta E_{inc}(Q)$ differs substantially from the broadening observed for pure H_2O, and its Q-dependence can only be understood by a much larger ratio of rotational to translational contributions to the broadening. Relative to H_2O, the values obtained for D_T,

$1/\tau_0$, and D_R at $h = 1.2$ are reduced by factors of 3.4, 3.9, and 13, respectively. The results show quantitatively how the motional freedom of much of the intracellular water is reduced by association with macromolecular constituents. Comparison with parallel NMR and dielectric relaxation measurements leads to the conclusion that this is not due to obstructions, compartments, or exchange with minor phases, but is a consequence of microscopic interactions over distances between 1 and 10 Å.

In the hydration experiments discussed so far, the dynamics of a protonated component undergoing large-amplitude displacements (i.e. H_2O) was observed against a "background" of relatively immobile protons (i.e. those belonging to the macromolecular constituents, including strongly associated water protons). With the partial exception of (12) and (79), no attempt was made to carry out parallel studies of the informative, but interpretationally more difficult, case of hydration with D_2O. Fully complementary H/D contrast variation experiments can only be done with samples that are also available in biosynthetically deuterated form.

High resolution quasi-elastic and inelastic scattering studies of hydrogenous and covalently deuterated proteins extracted from blue-green algae are being pursued by Randall et al (55, 56, 63, 64). The experimental strategy (66) adopted in this work is to perform difference experiments on amorphous powder or film samples, which are being hydrated on-beam in steps from the gently dried state ($h \approx 0.01$) up to levels close to the water content necessary for functional interactions ($h = 0.5 - 1$). By measuring $d^2\sigma/d\Omega dE$ at several points along the sorption isotherm and carrying out separate runs for different H/D contrast combinations, it is possible to examine the structural and dynamical changes accompanying the gradual activation of the various degrees of freedom of a protein and the sequence of hydration events taking place at or near its surface. Studies of this kind are necessarily concerned with surface interactions and the detection of small differences, and are therefore rather demanding of instrument time. Most of the experiments performed to date have been on C-phycocyanin (PC), an oligomeric chromoprotein involved in Photosystem II; a discussion of the current state of this work may be found in (56). A key observation has been that wet-minus-dry difference broadenings $\Delta E_{inc}(Q)$ for H_2O sorbed to deuterated PC ($h < 0.3$) exhibit a hydration-dependent oscillatory structure. The intensity changes seen here are consistent with an interpretation in terms of jump diffusion of water molecules between primary hydration sites. The corresponding t.o.f. spectra reflect the way in which the faster translational and librational degrees of freedom of the protein-water system are repartitioned over $1 < \hbar\omega < 800\ \text{cm}^{-1}$ as a function of hydration, and careful measurements of the dependence on H/D contrast should make it possible eventually to quantify the coupling between interior and surface modes.

Molecular Dynamics of Proteins

Fluctuations involving elements of the secondary, tertiary, and quaternary structure are thought to determine important aspects of the functional properties of proteins (13, 39a, 41, 62). The coupling or competition between cooperative and dissipative modes of motion with frequencies below a few 100 cm^{-1} is of special interest in processes such as ligand binding, subunit interaction and complex formation. Disregarding the much more demanding approach of systematic selective deuteration mentioned earlier, information on the molecular dynamics of processes of this kind may be extracted from incoherent scattering experiments in two ways: (*a*) by analyzing integral properties of the difference spectra obtained from solution scattering data for two states A and A′ of a sample, e.g. by deriving $\bar{Z}_p(\omega)$ and relating overall changes in the "soft modes" distribution to the thermodynamic quantities characterizing the $A \rightarrow A'$ transition (75); (*b*) for crystalline or quasi-crystalline systems, by obtaining extremely well-resolved spectra that can be analyzed in detail with respect to their dependence on **Q**, orientation, and sample parameters, with the aim of testing dynamical models. In both cases, but especially so for (*b*), a full interpretation of the data obviously calls for extensive model calculations or numerical simulations of the scattering laws.

Despite these and other compelling reasons for developing the application of inelastic scattering techniques to globular proteins in parallel with the flourishing theoretical work in this area (11, 41, 45, 62), the experiments that have been performed to date represent only a very modest beginning. Exceptionally good counting statistics, and consequently high incident flux levels, are required to resolve inelastic scattering from intramolecular modes against the dominant background of solvent scattering.

Following early solution scattering experiments by three groups at Grenoble attempting to detect low-frequency fluctuations (>1 cm^{-1}) in lysozyme, hexokinase, and trypsin/trypsinogen, t.o.f. spectra from polycrystalline lysozyme and from lysozyme in concentrated solution (180 mg/ml, D_2O buffer) were first reported by Bartunik et al (5). To improve statistics, the data for each sample were summed over a wide 2θ-range ($0.05 < Q_0 < 0.5$ Å^{-1}). Outside the quasi-elastic region, the summed t.o.f. spectra did not differ significantly and the contribution from crystal lattice modes seemed to be small. The spectra consist of a continuous, rather smooth intensity distribution carrying several weak bands between 25 and 375 cm^{-1}; a few of these were tentatively assigned on the basis of neutron and Raman spectra from polypeptides. At very small $\hbar\omega$ the inelastic intensity appeared to depend on inhibitor binding (*N*-acetylglucosamine) and crystalline environment.

Much improved solution scattering experiments have become possible recently with the availability of a new time-focusing t.o.f. spectrometer at ILL. This is being used by Jacrot et al (21, 38) to study the molecular dynamics of hexokinase and by Middendorf et al (56) to study that of lysozyme. T.o.f. spectra up to around 800 cm^{-1} were obtained for 45 mg/ml solutions (D_2O buffer) of yeast hexokinase and its complex with glucose. The buffer-subtracted spectra show two conspicuous bands centered on 50 and 300 cm^{-1}, the former increasing in intensity in the interval $15° < 2\theta < 110°$, whereas the latter does not change significantly with angle. On glucose binding, the overall intensity is reduced by 5 to 10% such that the difference spectrum is everywhere positive, equivalent to a reduction in the hydrogen-weighted vibrational and librational modes. The spectral structure of this difference profile provides the first quantitative dynamical information on the well-known "stiffening" of an enzyme upon ligation. The low-frequency modes are more markedly affected, and a band at 30 cm^{-1} could be associated with the closing of the cleft around the glucose substrate.

Hexokinase is one of several enzymes whose tertiary structure is known to change appreciably on substrate or inhibitor binding (39a). The intermediate scattering function for an idealized "hinge-bending" model of domain motions in enzymes was calculated by Morgan & Peticolas (60), and a detailed simulation of damped interdomain vibrations in lysozyme was performed by McCammon et al (51). The set of lysozyme t.o.f. spectra (56) obtained from short runs at high concentration (200 mg/ml) again suggest a "tightening" of the structure upon saturation with oligosaccharide inhibitors. At low scattering angles, where the dynamics is sampled over distances in the 10 to 15 Å range, the free lysozyme molecule exhibits substantially enhanced long-range fluctuations relative to the motionally more constrained complexes. Consistent with model calculations (76), the difference patterns show a minimum near 200 cm^{-1} under all conditions; beyond this a broad distribution of more localized high-frequency degrees of freedom is also affected by inhibitor binding.

The t.o.f. spectrometers used in these enzyme studies simultaneously provide data on the quasi-elastic broadening and allow changes in the translational and rotational Brownian motion properties of protein and/or substrate molecules as a whole to be detected, although with a resolution limit of about 0.1 cm^{-1} or $D_T \gtrsim 2 \times 10^{-6}$ cm^2/sec. Much lower energy transfers can be observed by backscattering and spin-echo techniques (33, 73), and these are being employed at ILL in current work on four aspects of the molecular dynamics of proteins as follows: (*a*) low-frequency collective motions of the protein shell of bromegrass mosaic virus as a function of swelling, down to frequencies of about 1 MHz (4); (*b*) contribution of low-frequency modes (1 – 10 MHz) to the diffusion-dominated quasi-elastic scattering at very low Q from concentrated solutions (100 – 280 mg/ml) of oxy and deoxy-hemoglobin (4); (*c*) low-Q coherent scattering due to restricted Brownian motion of the subunits of fully deuterated C-phycocyanin (d-PC) (56); (*d*) diffusive rotational motions (0.1 – 1 cm^{-1}) of hydrophilic side chains at or near the surface of d-PC (56).

Photosensitive Systems

Proton transfer has been shown to be an essential intermediate in a number of basic energy transduction processes; it takes place in complex structures acting cooperatively to allow directional, slowly relaxing degrees of freedom to be sustained within a "sea" of vibrational modes of higher frequency. Several models for proton "pumps" based on H-bond networks between α-helical structures have been proposed, with reference in particular to bacteriorhodopsin, but experimental tests are lacking. Calculations show that the fundamental acoustic and breathing modes of an array of α-helices fall in the 1 to 50 cm^{-1} region (62). Because of the sensitivity to $\mathbf{Q} \cdot \mathbf{u}_p$ and the fact that neutron techniques have already been applied to numerous proton-hopping and tunnelling processes in solid state physics, it is tempting to look for changes in $\Delta E_{inc}(Q)$ and the near inelastic scattering in suitably designed difference experiments. Following an early experiment on hydrated purple membrane stacks at Harwell (54), which seemed to show a low-Q peak near 5 cm^{-1}, the low-frequency scattering from purple membranes is now being reexamined at ILL as a function of illumination and orientation.

The only published neutron study of light-induced changes in the molecular dynamics of a photosensitive system appears to be that by Furrer et al (25). Using a triple-axis spectrometer at Zurich, these authors measured energy-gain and energy-loss spectra ($-160 < \hbar\omega < 160$ cm^{-1}, $Q = 4.2$ $Å^{-1}$) from photosynthetic membranes of *Rhodopseudomonas viridis* at temperatures between 28 and 300 K (20 mg/ml, D_2O buffer). On broad-band irradiation (1 – 2 MLux), significant intensity changes were observed such that the elastic scattering decreased and a substantial peak developed on the energy-gain side at 51 cm^{-1}. The corresponding peak on the other side picked up some intensity but did not change qualitatively. These results were interpreted in terms of the excitation of a nonequilibrium population of vibrational modes via radiationless transitions. For the 28 K sample, an effective temperature of 88 K characterizing this enhanced population of states was obtained by applying Equation 10 to the energy-gain and the energy-loss region. Spectra measured for chlorophyll-protein complexes and other chlorophyll-containing preparations showed similar light-induced features.

CONCLUSIONS AND OUTLOOK

Inelastic neutron scattering is a powerful technique whose potential for biophysical and biochemical applications rests on (*a*) its closeness in conceptual and practical terms to well-established methods of structure determination by diffraction, as a "hybrid" technique capable of probing structural and dynamical detail at the molecular level; (*b*) its spectral range, which centers on processes with time scales in the ns to ps region while providing sufficient overlap with complementary information derived from NMR on the low-frequency side and from optical techniques at high frequencies; (*c*) the interpretational advantage of obtaining nuclear scattering data related by Fourier transformation to fundamental correlation functions, thus facilitating comparison with dynamical simulations; (*d*) the emphasis on proton motions and the possibilities inherent in H/D contrast variation.

Despite these strong assets, neutron spectroscopy has yet to establish itself as one of the three or four major experimental techniques capable of contributing spatially and/or spectrally resolved data to the growing field of macromolecular dynamics. Considered in isolation from other factors, the Q,ω-range and resolving power of instruments available for developing biomolecular applications of inelastic scattering must be regarded as excellent. The flux levels at which these instruments operate, however, are still very low compared with the sources used for photon or electron spectroscopy. This limitation is compounded by the fact that globally there are only a few centralized research establishments equipped with spectrometers of advanced design. Quasi-elastic studies requiring very high resolution, for example, can only be done at the Institut Laue-Langevin in Grenoble.

Neutron scattering is a multidisciplinary science with a very wide range of applications, and the need for higher flux levels and better access is felt in many areas. Substantial investments in new or expanded neutron scattering facilities are being made at all major research centers to alleviate this situation. Large increases in thermal neutron fluxes are predicted for the spallation sources currently under construction or at the design stage (23, 84). Higher rates of data collection will bring run times down into the 10 to 100 min region and will make it possible, in particular, to study changes with temperature that have not been explored at all. These sources, by virtue of their pulsed nature, are also likely to stimulate novel experimental approaches to probing $S(\mathbf{Q}, \omega)$ by intense polychromatic neutron bursts with repetition rates that are comparable with the characteristic relaxation times of several processes of biophysical interest. Although it would be unrealistic to expect a dramatic rise in inelastic bio-neutron work during the next five years, we can certainly look forward to some exciting new results.

ACKNOWLEDGMENTS

I wish to thank many friends and colleagues at several institutions, principally the Institut Laue-Langevin, the Rutherford Appleton Laboratory, AERE Harwell, and King's College, for helpful information and discussions. This work is supported by the UK Science and Engineering Research Council.

Literature Cited

1. Agrawal, A. K., Yip, S. 1969. *Nucl. Sci. Eng.* 37:368–79
2. Akcasu, A. Z., Benmouna, M., Han, C. C. 1980. *Polymer* 21:866–90
3. Allen, G., Higgins, J. S. 1973. *Rep. Prog. Phys.* 36:1073–1133
4. Alpert, Y. 1980. In *Neutron Spin Echo. Lecture Notes in Physics*, ed. F. Mezei, 128:87–93. Berlin: Springer-Verlag
5. Bartunik, H. D., Jollès, P., Berthou, J., Dianoux, A. J. 1982. *Biopolymers* 21:43–50
6. Berne, B. J., Montgomery, J. A. 1976. *Mol. Phys.* 32:363–78
7. Boutin, H., Whittemore, W. L. 1966. *J. Chem. Phys.* 44:3127–28
8. Boutin, H., Yip, S. 1968. *Molecular Spectroscopy with Neutrons*. Cambridge, Mass: MIT Press. 226 pp.
9. Brot, C., Lassier-Govers, B. 1976. *Ber. Bunsenges. Phys. Chem.* 80:31–41
10. Burgman, J. O., Sciesinski, J., Sköld, K. 1968. *Phys. Rev.* 170:808–15
11. Careri, G., Fasella, P., Gratton, E. 1979. *Ann. Rev. Biophys. Bioeng.* 8:69–97
12. Dahlborg, U., Rupprecht, A. 1971. *Biopolymers* 10:849–63
13. Debrunner, P. G., Frauenfelder, H. 1982. *Ann. Rev. Phys. Chem.* 33:283–99
14. Dianoux, A. J., Pineri, M., Volino, F. 1982. *Mol. Phys.* 46:129–37
15. Dianoux, A. J., Volino, F. 1977. *Mol. Phys.* 34:1263–75
16. Drexel, W., Peticolas, W. L. 1975. *Biopolymers* 14:715–21
17. Egelstaff, P. A. 1967. *An Introduction to the Liquid State*. London: Academic. 236 pp.
18. Egelstaff, P. A. 1974. In *Spectroscopy in Biology and Chemistry*, ed. S.-H. Chen, S. Yip, pp. 269–96. New York: Academic
19. Egelstaff, P. A. 1976. See Ref. 67, 27:126–39
20. Egelstaff, P. A., Gray, C. G., Gubbins, K. E., Mo, K. C. 1975. *J. Stat. Phys.* 13:315–30
21. Engelman, D. M., Dianoux, A. J., Cusack, S., Jacrot, B. 1983. See Ref. 68. In press
22. Ermak, D. L., McCammon, J. A. 1978. *J. Chem. Phys.* 69:1352–60
23. Fender, B. E. F., Hobbis, L. C. W., Manning, G. 1980. *Philos. Trans. R. Soc. London Ser. B* 290:657–72
24. Fuess, H. 1979. In *Modern Physics in Chemistry*, ed. E. Fluck, V. I. Goldanskii, 2:1–193. London: Academic
25. Furrer, A., Stöckli, A., Hälg, W., Kühlbrandt, W., Mühlethaler, K., Wehrli, E. 1983. *Helv. Phys. Acta* 56:655–62
26. Garber, D. I., Kinsey, R. R., eds. 1976. *Neutron Cross Sections*, Vol. 2, BNL 325. Upton, NY: Brookhaven Natl. Lab. 3rd ed.
27. Gerling, R. W. 1981. *Z. Phys. B-Condensed Matter* 45:39–48
28. Griffin, A., Jobic, H. 1981. *J. Chem. Phys.* 75:5940–43
29. Gupta, V. D., Boutin, H., Trevino, S. 1967. *Nature* 214:1325–26
30. Gupta, V. D., Trevino, S., Boutin, H. 1968. *J. Chem. Phys.* 48:3008–15
31. Hansen, J. P., McDonald, I. R. 1976. *Theory of Simple Liquids*. London: Academic. 395 pp.
32. Harling, O. K. 1969. *J. Chem. Phys.* 50:5279–96

33. Hayter, J. B. 1981. In *Scattering Techniques Applied to Supramolecular and Nonequilibrium Systems*, ed. S.-H. Chen, B. Chu, R. Nossal, pp. 49–74. New York: Plenum
34. Hecht, A. M., White, J. W. 1976. *J. Chem. Soc. Faraday Trans.* 2 72:439–45
35. Higgins, J. S. 1982. In *Developments in Polymer Characterization*, ed. J. V. Dawkins, 4:131–76. Barking, England: Appl. Sci.
36. Howard, J., Waddington, T. C. 1980. In *Advances in Infrared and Raman Spectroscopy*, ed. R. J. H. Clark, R. E. Hester, 7:86–222. London: Heyden
37. Irish, J. D., Graham, W. G., Egelstaff, P. A. 1978. *Can. J. Phys.* 56:373–80
38. Jacrot, B., Cusack, S., Dianoux, A. J., Engelman, D. M. 1982. *Nature* 300:84–86
39. Janik, J. A., Kowalska, A. 1965. In *Thermal Neutron Scattering*, ed. P. A. Egelstaff, pp. 414–52. London: Academic
39a. Janin, J., Wodak, S. J. 1983. *Prog. Biophys. Mol. Biol.* 42:21–78
40. Jenkin, G. T., Miller, A., White, J. W., White, S. W. 1976, 1977. *ILL Ann. Rep. Annex*, pp. 320, 373 (Abstr.). Grenoble: Inst. Laue-Langevin
41. Karplus, M., McCammon, J. A. 1981. *CRC Crit. Rev. Biochem.* 9:293–349
42. Kossiakoff, A. A. 1983. *Ann. Rev. Biophys. Bioeng.* 12:159–82. See also earlier reviews cited herein
43. Larsson, K. E. 1973. *J. Chem. Phys.* 59:4612–20
43a. Larsson, B., Carlson, J., Börner, H., Forsberg, J., Fourcy, A., Thellier, M. 1982. In *Progress in Radio-Oncology II*, ed. K. H. Kärcher et al, p. 151. New York: Raven
44. Leadbetter, A. J., Lechner, R. E. 1979. In *The Plastically Crystalline State*, ed. J. N. Sherwood, pp. 285–320. New York: Wiley
45. Levitt, M. 1982. *Ann. Rev. Biophys. Bioeng.* 11:251–71
46. Lovesey, S. W., Schofield, P. 1976. *J. Phys. C* 9:2843–56
47. Maconnachie, A., Richards, R. W. 1978. *Polymer* 19:739–62
48. Marshall, W., Lovesey, S. W. 1971. *Theory of Thermal Neutron Scattering*. Oxford: Clarendon. 599 pp.
49. Martel, P. 1980. *J. Biol. Phys.* 8:1–10
50. Martel, P. 1982. *Biochim. Biophys. Acta* 714:65–73
51. McCammon, J. A., Gelin, B. R., Karplus, M., Wolynes, P. G. 1976. *Nature* 262:325–26
52. Michel, K. H., Naudts, J. 1978. *J. Chem. Phys.* 68:216–28
53. Middendorf, H. D. 1983. See Ref. 68. In press
54. Middendorf, H. D., Blaurock, A. E. 1974. *UK Neutron Beam Res. Comm., Ann. Rep.*, p. 35 (Abstr.)
55. Middendorf, H. D., Randall, J. T. 1980. *Philos. Trans. R. Soc. London Ser. B* 290:639–55
56. Middendorf, H. D., Randall, J. T., Crespi, H. L. 1983. See Ref. 68. In press
57. Miller, A., Taylor, A. D., White, J. W. 1983. In preparation
58. Moore, P. B., Engelman, D. M. 1979. *Methods Enzymol.* 59:629–69
59. Moore, F. M., Willis, B. T. M., Hodgkin, D. C. 1967. *Nature* 214:130–32
60. Morgan, R. S., Peticolas, W. L. 1975. *Int. J. Pept. Protein Res.* 7:361–65
61. Page, D. I. 1972. In *Water: A Comprehensive Treatise*, ed. F. Franks, 1:333–62. New York: Plenum
62. Peticolas, W. L. 1979. *Methods Enzymol.* 61:425–58
63. Randall, J. T., Middendorf, H. D. 1982. In *Biophysics of Water*, ed. F. Franks, S. F. Mathias, pp. 15–20. Chichester: Wiley
64. Randall, J. T., Middendorf, H. D., Crespi, H. L., Taylor, A. D. 1978. *Nature* 276:636–38
65. Rościszewski, K. 1974. *Physica* 75:268–81
66. Rupley, J. A., Gratton, E., Careri, G. 1983. *Trends Biochem. Sci.* 8:18–22
67. Schoenborn, B. P., ed. 1976. *Neutron Scattering for the Analysis of Biological Structures. Brookhaven Symp. Biol. 27.* Upton, NY: Brookhaven Natl. Lab.
68. Schoenborn, B. P., ed. 1983. *Neutrons in Biology—Neutron Scattering Analysis for Biological Structures. Brookhaven Symp. Biol. 32.* New York: Plenum. In press
69. Schofield, P. 1978. In *Inelastic Scattering of Neutrons*, pp. 637–47. Vienna: Int. At. Energ. Agency
70. Sjölander, A. 1965. In *Thermal Neutron Scattering*, ed. P. A. Egelstaff, pp. 291–346. London: Academic
71. Springer, T. 1972. In *Springer Tracts in Modern Physics*, ed. G. Höhler, 64:1–100. Berlin: Springer-Verlag
72. Springer, T. 1977. In *Dynamics of Solids and Liquids by Neutron Scattering*, ed. S. W. Lovesey, T. Springer, pp. 255–300. Berlin: Springer-Verlag
73. Springer, T. 1980. *Philos. Trans. R. Soc. London Ser. B* 290:673–81
74. Steele, W. A., Pecora, R. 1965. *J. Chem. Phys.* 42:1863–79
75. Sturtevant, J. M. 1977. *Proc. Natl. Acad. Sci. USA* 74:2236–40
76. Suezaki, Y., Gō, N. 1975. *Int. J. Pept. Protein Res.* 7:333–34

77. Touret-Poinsignon, C., Timmins, P., ed. 1981. *Workshop on Water at Interfaces, ILL Report 81T055S*. Grenoble: Inst. Laue-Langevin
78. Trantham, E. C., Rorschach, H. E., Clegg, J. S., Hazlewood, C. F., Nicklow, R. M. 1982. In *Neutron Scattering—1981*, ed. J. Faber, Jr., pp. 264–66. New York: Am. Inst. Phys.
79. Trantham, E. C., Rorschach, H. E., Clegg, J. S., Hazlewood, C. F., Nicklow, R. M., Wakabayashi, N. 1983. *Biophys. J.* Submitted for publication
80. Vineyard, G. H. 1958. *Phys. Rev.* 110:999–1010
81. White, J. W. 1976. See Ref. 67, pp. VI3–26
82. White, S. W. 1977. PhD thesis. Oxford Univ., England
83. Whittemore, W. L. 1968. In *Inelastic Scattering of Neutrons*, 2:175–84. Vienna: Int. At. Energ. Agency
84. Windsor, C. G. 1981. *Pulsed Neutron Scattering*. London: Taylor & Francis. 423 pp.
85. Worcester, D. L. 1982. See Ref. 78, pp. 368–77
86. Zaccaï, G., Jacrot, B. 1983. *Ann. Rev. Biophys. Bioeng.* 12:139–57. See earlier reviews cited herein

Ann. Rev. Biophys. Bioeng. 1984. 13: 453–92

EVOLUTION AND THE TERTIARY STRUCTURE OF PROTEINS

Mona Bajaj and Tom Blundell

Laboratory of Molecular Biology, Department of Crystallography,
Birkbeck College, London University, Malet Street,
London WC1E 7HX, England

SUMMARY AND PERSPECTIVES

The fact that biological information is stored in the form of DNA has led some molecular biologists and biophysicists to the view that evolutionary history will be most evident in DNA sequences. This view is attractive, for there is a redundancy in the genetic code, and the number of base changes is a sensitive indicator of evolutionary distances between closely related gene sequences. However, selective pressures in evolution will be mainly on the functions of the proteins as manifested in the whole organism. These are dependent on certain critical amino acids, such as those that constitute catalytic groups or substrate binding sites, and on the maintenance of a stable three-dimensional structure. The invariant functionally important amino acids may comprise a very small percentage of the total. Furthermore, because many differing amino acid sequences are able to attain the same three-dimensional arrangement of the polypeptide main-chain, we may conclude that there may be little restraint on amino acid sequences over long periods of time. Variations of the DNA sequences will be even more extensive as a result of codon redundancy, so that there may be little statistically significant resemblance to an ancestral gene sequence. Thus it is in tertiary structures that the most distant relationships between proteins may be evident.

For these reasons this review is unashamedly concerned with tertiary structures in the evolution of proteins. We begin with a discussion of techniques for comparing the relatedness of protein structures, with particular

0084–6589/84/0615–0453$02.00

emphasis on general topology and detailed three dimensional structures. These structures may be compared systematically using methods based on either inter-atomic distances or rotation of a reference frame. For closely related systems the latter may be a root mean square deviation of two optimally oriented sets of coordinates, but for less clearly similar structures it may involve a search with fragments through Eulerian angle space. These techniques then lead naturally to a comparison of amino acid residues at topologically equivalent positions.

The most conserved part of a protein in evolution appears to be the identity and the arrangement of the amino acid residues in the active site, including those binding metal cofactors or coenzymes. Conservation is achieved by maintaining the hydrophobic character of the core and by retaining a topological equivalence of structural elements; but the helices and sheets may translate and rotate relative to each other while maintaining the class of interactions. These characteristics allow a slow change in the core during the evolutionary process. Conservation of secondary structural potentials is much less stringent as indicated by the low success of predictive techniques. Residues in loops or turns undergo substantial insertions, deletions, or conformational changes that not only accommodate relative movements of helices and sheets but also substantially vary in size and surface while leaving essential residues in the same arrangement.

Each protein may have several functions that are optimized in evolution in addition to the main function, which for an enzyme would be a properly arranged set of catalytic groups and a binding site complementary to the transition state. These functions may include the evolution of regulatory functions such as inactive precursors, cooperativity, and positive and negative allosteric controls. There is some evidence that the greater size of the mammalian serine proteinases, compared to the microbial enzymes, is related to evolution of an inactive precursor. Changes in residues between globins may relate to differing allosteric effectors, ATP or DPG, and to variations in the required sensitivity to protons, for example the root effect in teleost fish. Oligomer formation and new specificities may also be new functions that are selected for both in protein differentiation and protein speciation.

Convergent evolution clearly plays a role in the development of active site residues. Several groups of enzymes such as the zinc enzymes (thermolysin, carboxypeptidase, carbonic anhydrase, and alcohol dehydrogenase) have similar arrangements of active site catalytic groups, presumably reflecting a complementarity to similar pentavalent transition states of the zinc ions. The topological similarities of domains such as those binding nucleotides (the dehydrogenases, some kinases), those binding hemes (the globins and cyctochromes *c*′), and those binding polysaccharides (lysozymes from chicken, goose, and T_4 phage) may also be examples of convergent evolution. Strict rules on the assembly of secondary structure, the binding of charged groups, etc, limit the number of structures, which might have a particular function, and make the existence of similar structures more probable. However, it is still not possible to exclude divergent evolution, which can clearly lead to structures with little amino acid homology.

The view that many proteins have evolved by the assembly of domains previously having other functions, a concept largely based on the imaginative work of Michael Rossmann and his colleagues, is now generally accepted. There are more examples of proteins that have internally duplicated structures than can be explained on the basis of convergence. There are many proteins such as the catalases or subunits of spherical viruses where an extra domain has been attached during speciation. Finally, the gene structure of some eukaryotic protein families such as the globins, immunoglobulins, histocompatibility antigens, and the lens β, γ-crystallins shows a relationship between exon structure in the DNA and structural or functional motifs in the proteins. However, not all exon-intron junctions reflect the assembly of structural or functional domains. There is good evidence that some introns may have been inserted recently and may have a selective advantage in allowing, through a sliding exon-intron function, greater exploration of variations in the sequence at surface regions of the proteins.

A knowledge of the tertiary structure of equivalent proteins in different species shows that the accepted mutations do not occur randomly. There is good evidence for evolution of new functions that appear to have been selected for and that, once developed, are often maintained in later evolutionary stages. Although there are also strong arguments for the occurrence of selectively neutral changes, many amino acid mutations must be adaptive. Characteristic rates of accepted mutations for certain classes of proteins—globins, cytochromes, insulins, for example—are held to be dependent on a constant rate of neutral mutation in amino acids for which there are no functional restraints. If selectively advantageous mutations occur more often than has been assumed, the rates may be less constant than has been assumed and their use as biological clocks unreliable, especially over short periods of evolutionary time.

COMPARISON OF PROTEIN STRUCTURES

Amino acid and nucleotide sequences have provided much useful information on the divergent evolution of proteins. The first systematic approach to the use of sequence data was developed by Dayhoff & Eck (36); since then, advances in computer technology have allowed the development of an extensive methodology of sequence alignment and phylogenetic tree construction (48, 58). Computer analysis of tertiary structures began with the systematic comparison of tertiary structures of oxy- and deoxyhemoglobin by least squares fitting of their electron densities (99). Subsequently, comparison of protein tertiary structures was initiated as the results of X-ray analysis became available (41, 70, 107). These methods may identify structural equivalence, which defines those structural elements that are coincident in three dimensions. For evolutionary comparisons, a sequential series of structurally equivalent residues, or run, is required where the chain segments are similarly directed. This is known as topological equivalence. The methods of approach may be categorized as those comparing interatomic distances, which calculate a root mean square deviation $D(\bar{S}, S)$ [defined by Levitt (92)], and those comparing positions (96, 109), which calculate a root mean square deviation $R(\bar{S}, S)$ where the coordinates of each set are defined with respect to their molecular centers of gravity. Both $D(\bar{S}, S)$ and $R(\bar{S}, S)$ methods tend to be oversensitive to the overall agreement of two structures and insensitive to secondary structure agreement, but they are qualitatively similar in behavior (29, 122).

Rossman & Argos (114) have developed a modified $R(\bar{S}, S)$ algorithm for comparison of structures that are less obviously similar. The method involves rotations of the C_α atoms in Eulerian space. Insertions and deletions are made in the two proteins until the number of atoms placed in equivalent positions to a specified degree of precision is maximized. Although the procedure has proved to be very valuable, it is difficult to get an objective measure of the relationship of the two proteins as a whole because the nonequivalent positions are ignored. In order to overcome this difficulty, Remington & Matthews (109) have used the statistical methods of Fitch (47) for amino acid sequences in developing a segment approach. The similarity between two backbone segments of length L residues is determined for each possible segment in the length of the chain. The root mean square error is then analyzed by statistical techniques to assess the probability that a given degree of structural homology could arise by chance. Sippl (122) has recommended a related procedure using interatomic distances.

In the case of identical structures, a comparison of the main-chain atomic positions of the two crystallographically independent molecules of porcine pancreatic kallikrein (22), a specific trypsin-like serine proteinase (2-Å refinement, $R = 0.22$), shows an overall root mean square deviation of all main chain atoms of 0.37 or 0.26 Å if only the internal segments are considered, since the relatively large deviations (up to 2.5 Å) occur in the main chain of surface loop regions. Deviations of this extent might then be expected to result from changing crystallographic environments. A comparison of the two kallikrein molecules with the closely related trypsin structure has shown that the root mean square deviations of the internal segments are 0.68 and 0.69 Å; similar values have been reported for trypsin and chymotrypsin (71) and for γ-crystallin and elastase (28, 118). For the microbial enzymes the relationship to the mammalian enzymes is less close. Thus 63% of the α-carbon atoms (mainly internal segments) of *Streptomyces griseus* proteinase B are topologically equivalent to chymotrypsin with a root mean square deviation of 2.07 Å (38). As expected, the *S. griseus* proteinases A and B are more closely related, with 85% topological equivalence and with a root mean square deviation of 1.46 Å (38).

Once the topologically equivalent residues have been identified, the sequences can be realigned using this information. Such operations have shown that sequence comparisons alone lead to many incorrect equivalences when the homology is low. (In this article we define homology as the degree of amino acid similarity at equivalent positions in protein sequences that are divergently evolved from a common ancestor.) An example is the misalignment of the sequence of α-lytic protease with elastase by McLachlan & Shotton (98) based on both sequences but only the three dimensional structure of elastase. Later, Delbaere et al (38) showed that a misplacement of an insertion had led to the conclusion that a serine was not found at 214 (chymotrypsinogen numbering), a residue that is invariant in serine proteases. Of the 92 residues seen to be aligned between azurin and plastocynan by Dayhoff (35) on the basis of sequence, only 47 were aligned in the same way by structural comparison (25).

Even when the overall sequence homology is high, sequence comparisons can often produce incorrect alignments. Thus for chymotrypsin and trypsin, the alignments given by Dayhoff (35) are as follows:

170						175					180
Chymotrypsin											
Cys	Lys	—	Tyr	Trp	Gly	Thr	Lys	Ile	Lys	Asp	Ala	Met
Trypsin											
Cys	Lys	Ala	Tyr	Pro	Gly	—	Gln	Ile	Thr	Ser	Asn	Met

The chance homology between a tyrosine and a glycine causes a misalignment in a region where the conformations of the two proteins are actually identical (59). In insulins from the hystricomorph rodents, a similar misalignment is usually made on the basis of sequence alone in the receptor-binding region.

	B23	B24	B25	B26	B27
Porcine insulin	Gly	Phe	Phe	Tyr	Thr
Casiragua/Coypu insulin	Gly	Phe	—	Tyr	Arg

In fact, as this is an extended chain, the residues should be aligned so that casiragua and coypu insulin have B25 Tyr and B26 Arg (16; M. Bajaj, unpublished results).

The sequence variation or minimum base change at each topologically equivalent position can also be systematically evaluated. This should give a more reliable indication of sequence homology than when sequences are considered alone. Not surprisingly, close structural similarity is usually reflected in close sequence homology for topologically equivalent structures.

EVOLUTION OF PROTEIN CORES

We first consider variation of the buried, usually hydrophobic residues that form the core of a globular protein. As we are interested in the process of evolution, we will concentrate on closely related members of families of proteins for which both sequence homology and good topological equivalence have been demonstrated. These families include the serine proteinases (59, 63), the aspartyl proteinases (50), the globins (103), the insulins (15, 21, 34), the rubredoxins (51), and the cytochromes *c* (2). The precision of structure analyses before 1975 precluded useful detailed comparison, and it was generally assumed, following an analysis of Lim & Ptitsyn (93), that protein cores evolved with a general retention of the volume and that this retention was achieved by complementary changes in the largely hydrophobic residues. Furthermore, it was assumed that cores would be relatively well conserved compared to external residues.

In fact, some protein families do evolve with a surprising retention of the hydrophobic core. The hydrophobic cores of all known insulins (from the primitive vertebrate, the hagfish, to the most advanced mammals) comprising $\sim 28\%$ of the amino acids of the A- and B-chains are found to be identical (33). High resolution, refined structures of porcine and hagfish insulin have root mean square differences of 0.11 to 0.25 Å between their structurally equivalent C_α atoms when optimally aligned (34). This is a remarkable conservation, given the inherent flexibility in the insulin molecule (26). The hydrophobic core is also conserved in the insulin-like growth factors (13, 14, 17).

A topological comparison of rubredoxins from *Desulphovibrio gigas*, *D. vulgaris*, *Clostridium pasteuriarium* (51) shows that they are similar, with a root mean square deviation of 0.4 Å between the mean three main chains. The hydrophobic core includes six aromatic residues, the side chains of which keep the same atomic arrangement within 0.4 Å. The cores of the chloroplast-type ferredoxins 2Fe-2S are also well conserved in evolution. Fukuyama et al (52) have discussed the sequence variation in 26 proteins in terms of the structure of the ferredoxin from *Spirulina platensis*, shown in

Figure 1. In a polypeptide chain of 99 amino acids there are 22 invariant positions, which are mainly clustered around the 2Fe-2S group and in the core. Thus there has been invariance in evolution for a period of 15 billion years since the time of divergence of blue green algae and higher plants.

In general, however, the hydrophobic cores of proteins are less conserved than these examples. High resolution X-ray analysis of γII crystallin (18, 140) shows two domains, each of which may be considered as a sandwich of two similar, four-stranded antiparallel β-pleated sheets with strands *badc'*. The overall least-squares fitting of one domain to another gives a root mean square deviation of 1.42 Å for all C_α atoms. Residues at one end of each sheet are more variable, and their C_α atoms contribute to the core, but the atoms further down the side chain are increasingly oriented to the solvent

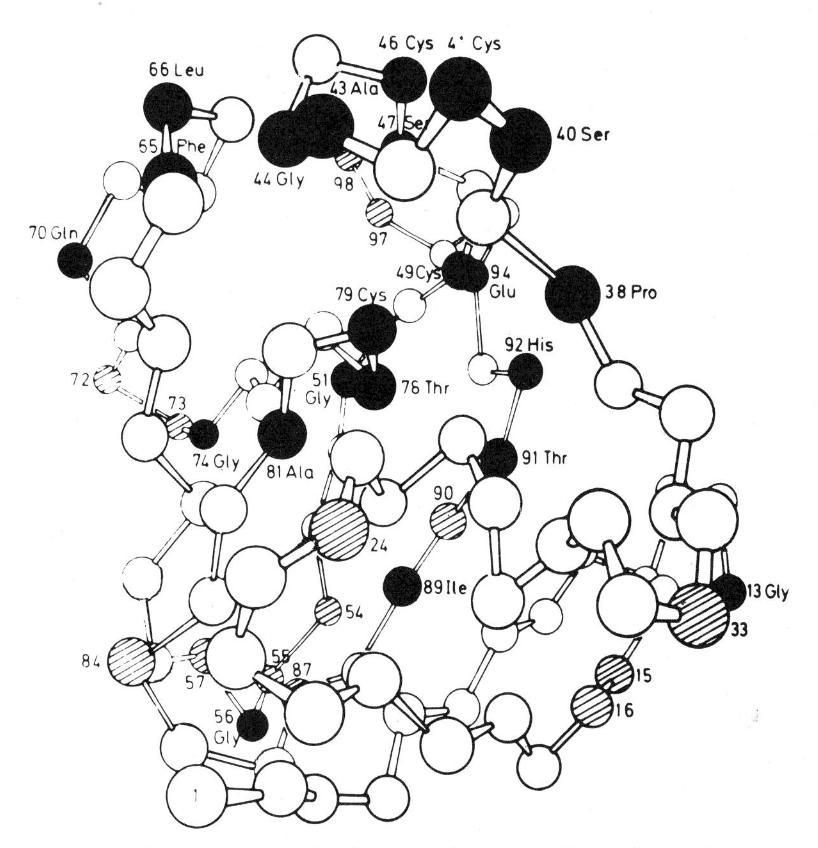

Figure 1 The distribution of invariant (●) and variant (⊘) positions in ferredoxin sequences in the three-dimensional structure of *S. plantensis* ferredoxin. The iron sulphur cluster is not shown. [Reproduced from (52).]

(140). The two sheets form a wedge, with the bottom filled with conserved isoleucine and tryptophan residues. Few of the other residues are completely conserved. There is some evidence of conservation of volume by complementary changes, e.g. from Cys to Phe (total volume = 321 Å) in the NH_2-terminal domains to Leu and Ile (total volume = 337 Å) in the COOH-terminal domains of γ-crystallins. In proteins containing only β-sheet structure, the distance between the sheets tends to be conserved although their relative positions and orientations may change. For example, consider the two copper proteins, popular leaf plastocyanin and *Paeruginosa* azurin, each of which contains two β-sheets packed face to face with the same general topology (Figure 2). One pair of β-sheets is closely homologous, whereas the other pair has phenylalanines and valines/isoleucines interchanged in a complementary way so that the same volume is occupied (25):

Plastocyanin		Azurin	
Ile 27	Val 72	Phe 29	Phe 97
⋮	⋮	⋮	⋮
Phe 29	Phe 70	Val 31	Val 95

In order to accommodate the changes in the same packing mode, sheet I is

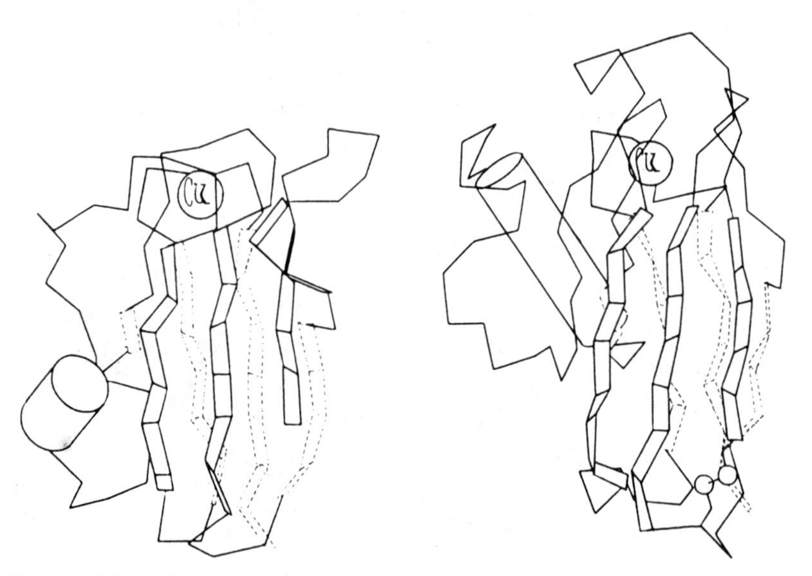

Figure 2 Schematic drawings of a plastocyanin (*left*) and azurin (*right*). Each vertex represents the position of an α-carbon. β-sheets are drawn as ribbons and α-helices as cylinders [Reproduced from (25).]

translated by 3.75 Å and rotated by 15° in azurin relative to plastocyanin when the two β-sheets II are aligned.

In the immunoglobulins, the domains of which are also comprised of a sandwich of antiparallel β-sheets, the two sheets are covalently linked by a disulfide, which forms part of the core. This constrains both the distances between the sheets and to some extent their relative lateral translations. Apart from the two half cystines, only a tryptophan, which is packed against them, is conserved in all immunoglobulin folds. These residues, together with a residue that is Val, Ile, or Leu, and atoms from five further residues form a compact central feature that is very similar in all domains (90). These changes in the surrounding core in general are accommodated not by complementary changes but rather by both shifts of up to 2 Å, by rotations of as much as 20°, and by insertions into the core from residues derived from external loops. More residues are common to all constant domains and the membrane proximal domains of both class I and class II histocompatibility antigens than to both constant and variable domains of immunoglobulins (P. Travers, submitted for publication). These residues include a valine, a leucine, and a tyrosine, which pack against the central invariants Cys Cys Trp groups, but they also include three residues—a histidine (His L199) and two prolines—which form a close cluster in a region topologically close to the highly variable region of V_L and V_H domains.

The most systematic analysis carried out has been on the globins. Lim & Ptitysn (93) concluded that 81% of the core residues have nonpolar side chains in 52 sequences. The 31 core residues have a total volume of 3180 Å with rms deviation of 15 Å in these globins even though there is an homology of only 19% between certain distantly related globins. Gō & Miyazawa (57) and Lesk & Chothia (89) show that local volumes, i.e. between individual helix pairs, vary considerably but cancel out when residues are all taken together. In fact, although the side chains of buried residues remain nonpolar, they vary in size, with a mean change in volume at any position of 56 Å^3. This results in a local change in volume of the residues at certain helix interfaces of up to 57%, which is accommodated by changes of relative position and orientation of helices by as much as 7 Å and 30°. The variation depends on the nature of the contacts between the helices (see Figure 3). In the A/E packings there is some conservation of the size of the two central positions. Further G/H packing variation of the residues can cause a difference of 2.5 Å in the inter axial distance.

In summary, it can be seen that cores of similar proteins are largely conserved as hydrophobic, but there is considerable latitude for variation of the volumes and shapes of individual amino acid side chains. These variations are accommodated mainly by changes in relative orientations and distances between topologically equivalent secondary structures, although the class of interactions between secondary structures tends to be conserved. In retrospect, these generalizations seem sensible. If we assume that the gene must be expressed in terms of a functionally useful protein in all stages of evolution, then all steps in the evolutionary process must lead to a thermodynamically stable protein (89). As the free energy of a folded protein is only 8 to 15 kcal per mole lower than that of the unfolded state (105), then mutations will only be acceptable if they cost less. Large changes in the core would not usually lead to a functional protein; smaller changes leading to readjustments of the sheets or helices may be less energetically unfavorable. Of course where there are families of genes such as for globins, immunoglobins, or histocompatibility antigens, the nonexpression of a gene may not be particularly disadvantageous, and the organism might survive until a complementary mutation occurred. Pseudogenes may also

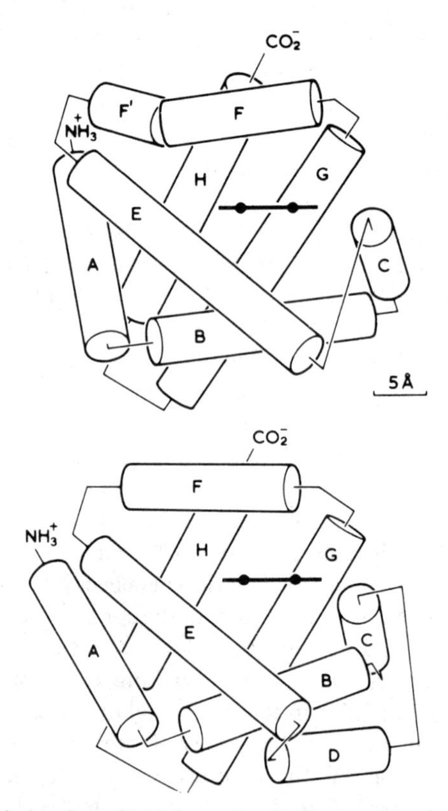

Figure 3 Schematic drawings of (*top*) the α-subunit of human deoxyhemoglobin and (*bottom*) *Chironomus* erythrocruorin. The helices represent cylinders. The heme group is viewed from the same direction in each globin molecule. [Reproduced from (89).]

provide a mechanism for silent mutations leading to complementary changes.

ACTIVE SITE AND COFACTOR BINDING RESIDUES

Although the structural elements may change their relative positions and orientations during divergent evolution of a family of proteins, the arrangement of the side chains that define the active site or bind a cofactor is remarkably conserved in both identity and three-dimensional arrangement.

For example, in the aspartyl proteinases the sequences and three dimensional structures around the two active site aspartates are highly conserved from fungi to mammals (L. Pearl and T. L. Blundell, submitted for publication). In the serine proteinases the catalytic quartet—His 57, Asp 202, Ser 195, and Ser 214 (chymotrypsinogen numbering)—is invariant and the arrangement in space is conserved. In a detailed comparison between the α-lytic protease from myobacter 495 (molecular weight: 19,900) and the much larger porcine elastase (molecular weight: 25,900), James et al (75) have shown that, of the 55% residues of α-lytic protease, which are topologically equivalent with a root mean square deviation of 2.08 Å, most are in the vicinity of the active site (38).

Even though there are significant differences in the relative positions and orientations of the β-sheets of azurin and plastocyanin, the geometry of the ligands of the important copper atom is the same within the accuracy of the present coordinates (25). Histidines (87 in plastocyanin and 117 in azurin) occupy positions on loops of different lengths, but different side chain conformations bring the imidazole nitrogens to the same positions relative to the copper atoms. Again in rubredoxins from *D. gigas*, *D. vulgaris*, and *C. pasteurianium*, the iron sulphur complexes have very similar environments with no significant modification of the geometry, which is nearly a tetrahedron with four Fe-S bonds close to 2.28 Å.

In the globins the fifteen residues that make contact with the heme belong to five regions of the protein (Figure 3). As the heme has little flexibility, most differences must be accommodated in the side chains of the protein. In fact these side chains rarely shift more than 2 Å relative to the heme, and the positions are relatively much more conserved than those between helices. Mutations occur less frequently at the faces of the hemes than on the edges where there are fewer contacts per residue and the ends of the side chains are in contact with the solvent. The hemes have not been entirely passive during evolution. In certain globins the vinyl groups are recruited to fill small cavities resulting from small movements of the C-helices. In erythrocruorin a more radical change occurs: the heme is flipped 180° about an axis in the plane of the heme.

Escherichia coli cytochrome b_{562} and the *Rhodospirillum molischiancum* cytochrome c' are also α-helical proteins that bind hemes. Each comprises a left-twisted bundle of four sequentially connected α-helices in which the heme prosthetic groups are situated in the singly connected end of the divergent helices (Figure 4). Superposition of the heme groups results in a corresponding superposition of most of the α-helices with 14 identities among the 77 topologically equivalent positions. Of these the iron-ligating histidines are equivalent; both show a solvent accessibility, which is unusual among heme ligands. Although cytochrome c' lacks the sixth axial methionine ligand, a methionine close to the iron is conserved in high spin cytochrome. In addition, three aromatics occupy similar spatial arrangements around the heme.

The cytochrome c_2 family includes components from both photosynthetic and respiratory electron transport systems. The structures defined at high resolution include several respiratory cytochromes c (126, 129), the photosynthetic c_2 from the purple nonsulphur bacterium *Rhodospirillum rubrum* (117), c_{550} from the facultative nitrate-respiring bacterium *Paracoccus denitrificans* (132), c_{551} from the O_2-respiring *P. aeruginosa* (39), and c_{555} from the green sulphur bacterium *Chlorobium thiosulphatophilium*. Although these range in size from 82 to 134 amino acids, the internal residues in the vicinity of the heme and some of the surface residues are remarkably similar. Figure 5 illustrates the similar structures of these cytochromes in a proposed family tree.

The conservation of the identities and arrangements of active site residues, and those which bind important cofactors or other molecules, is consistent with a strong evolutionary restraint preventing mutations that would lower the catalytic efficiency, potential oxygen binding, etc, of the protein. Nevertheless, the fact that this can be achieved in so many different, but closely related, ways in different proteins of the same family is further evidence of the versatility of protein structure.

SECONDARY STRUCTURE AND EVOLUTION

In order to retain a stable tertiary structure, the protein must retain not only a close packed hydrophobic core but also a potential to achieve the correct main chain conformation. The low success rate (less than 56%) (80) of even the best secondary structural predictive schemes such as those of Chou & Fasman (27) and Garnier et al (54) indicates that the local, predominantly steric factors which lead to preference of particular conformations may often take second place to tertiary interactions (88).

Some of the reasons for the ability of secondary structures to accommodate unusual residues is becoming apparent from detailed analyses of

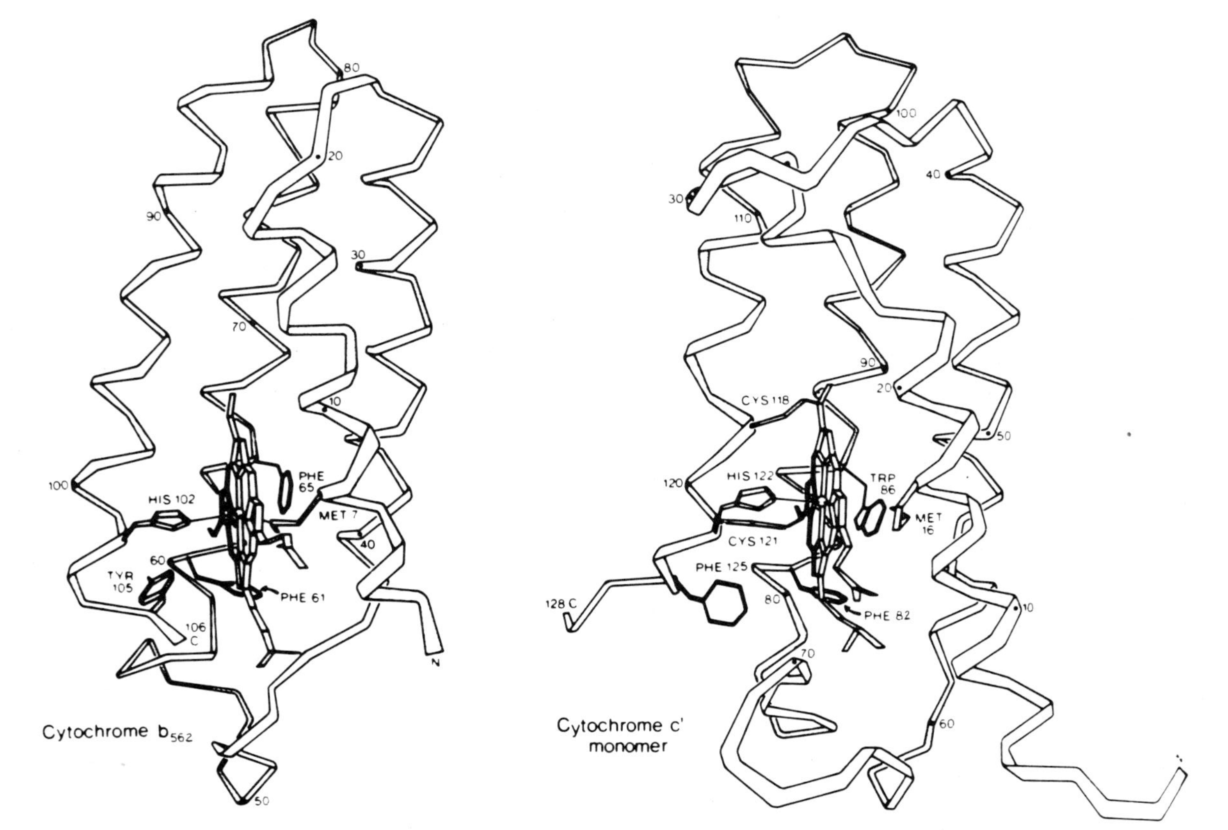

Figure 4 Schematic drawings of *E. coli* cytochrome b_{562} and *R. molischianum* cytochrome *c′* monomer. [Reproduced from (137a).]

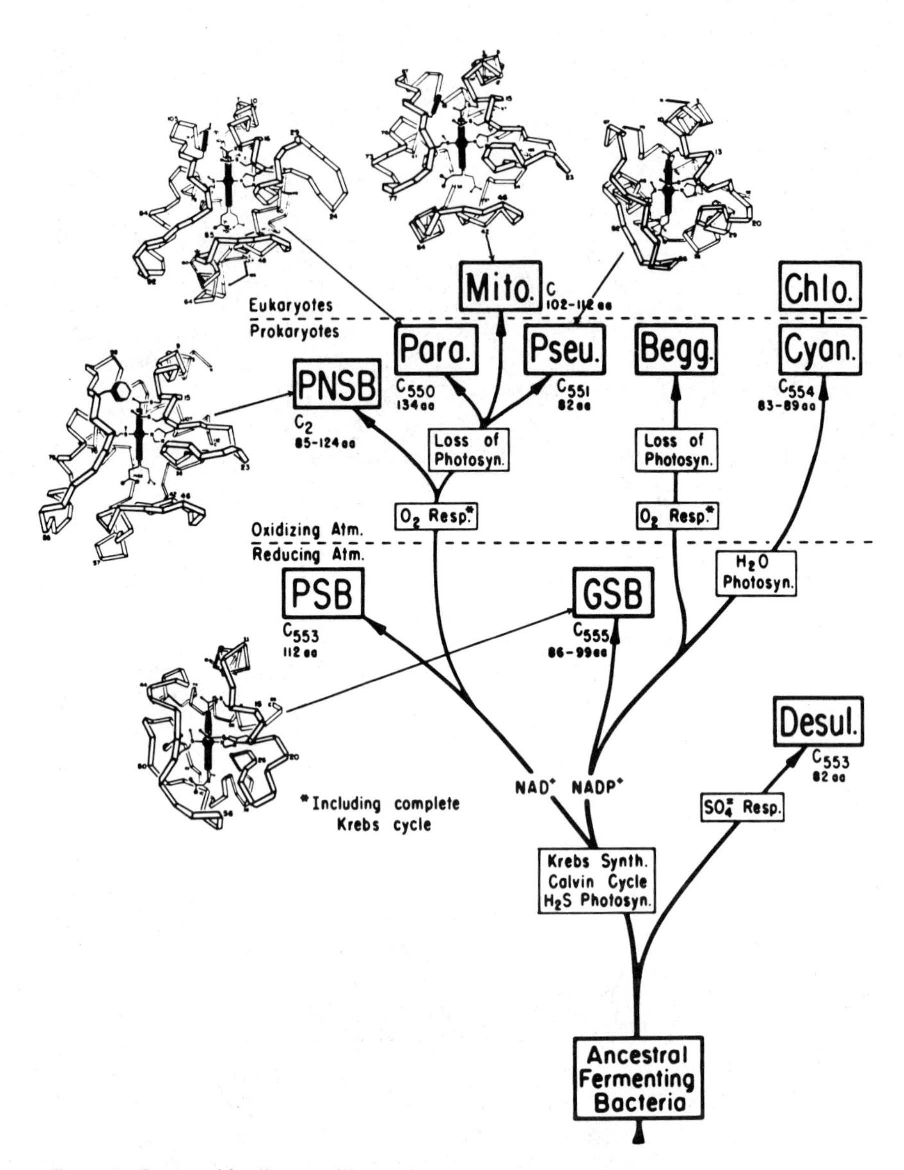

Figure 5 Proposed family tree with cytochrome *c* structures superposed. GSB, green sulphur bacteria; PSB, purple sulphur bacteria; PNSB, purple nonsulphur bacteria; Mito, mitochondria; Para., *Paracoccus*; Pseu., *Pseudomonas*; Begg., *Beggiatoa*; Cyan., *Cyanobacteria*; Chlo., chloroplasts; Desul.; *Desulphovibrio*. [Reproduced from (2).]

these secondary structures. For example, Blundell and co-workers (submitted for publication) have shown that many helices defined by high resolution X-ray analysis are irregular, with sharp discontinuities, and that most regular helices are curved so that even prolines can be accommodated. Of the 47 prolines found in nine globins studied by high resolution X-ray analyses, Lesk & Chothia (89) found that only three occur within helices: the A-helix of lamprey hemoglobin and the G-helices of glycera and legume hemoglobins. Distortions at both the NH_2- and COOH-termini of helices are common. Evolutionary differences include the winding up of α-helices into 3_{10} helices and unwinding to π-helices or rather irregular conformations.

INSERTIONS AND DELETIONS

Insertions may introduce a single amino acid, extend helices or β-strands by several amino acids, introduce new helices on β-strands, or even introduce a new domain. We now consider how the insertions and deletions are accommodated so that the active site residues are conserved in relative positions and orientations.

Insertions of a single residue in β-strands tend to result in β-bulges (111) whereby two residues occupy the position previously occupied by one and the hydrogen bonding of sheet is preserved. Examples in evolution include the one residue insertion in the C_H1 domain of the Fab′ NEW relative to the sequence of McPC603C_H1: the NEW structure has a bulge in the middle of an antiparallel pair of β-strands, whereas McPC603 has regular hydrogen bonding all the way along (110). In the variable domains of immunoglobulins a β-bulge in the G-strands is compensated for by the existence of a glutamine in the A-strand, which hydrogen bonds to the G-strand; however, in the constant domains the G- and F-strands have normal antiparallel β-sheet hydrogen bonding. A similar β-bulge occurs in motifs 2 and 4 of the γ-crystallins in which the sequence is typically Leu-Arg-Arg-Val where the leucine and valine are buried. This compares with a sequence of the type Leu-Asn-Val in motifs 1 and 3, where the leucine and valine are again buried and contribute to a normal antiparallel β-sheet structure.

β-bulges often occur at bends in antiparallel sheets, where they fold orthogonally onto themselves. For example, in the aspartyl proteinases, which contain four repeated $\beta\beta\beta\beta\alpha$-units ($a_N b_N c_N d_N h_N$ and $a'_N b'_N c'_N d'_N h'_N$ in the NH_2-terminal domain and $a_C b_C c_C d_C h_C$ and $a'_C b'_C c'_C d'_C h'_C$ in the COOH-terminal domain), residues 97 and 98 of strand C'_N show this phenomenon (T. L. Blundell, B. L. Sibanda, and L. H. Pearl, unpublished results). Such positions appear to be the sites of further mutations. Thus in a topologically equivalent position in the second domain (strand c'_c residues 208–210) there is a bulge corresponding to three extra residues in mammalian aspartyl proteinases such as renin, pepsin, and chymosin. This is reduced to two extra residues in endothiapepsin and is absent in pencillopepsin, both enzymes being of fungal origin. Interestingly, this insertion is stabilized by a disulfide bridge in the mammalian enzymes. An even larger insertion, which is also stabilized by a disulfide bond, occurs at the bend in strand c'_c of endothiapepsin compared to topologically equivalent positions; this is also a region where insertions occur in pencillo-pepsin, endothiapepsin, and renin relative to procine pepsin.

Insertions and deletions most often occur in regions between $\beta\beta$-, $\beta\alpha$- or $\alpha\alpha$-structures. Figure 6 shows that the great majority of the insertions and

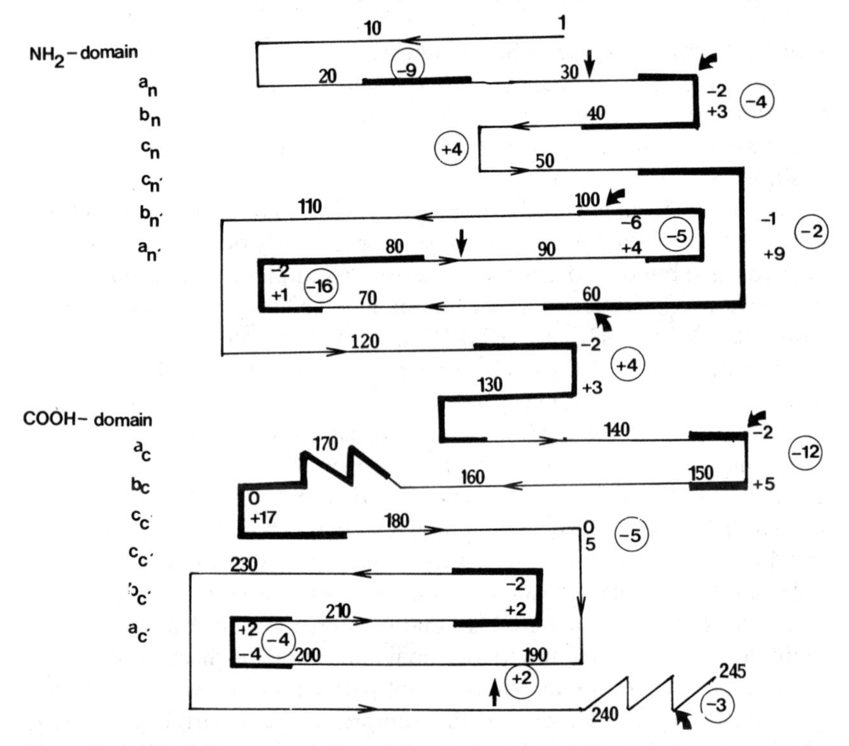

Figure 6 A schematic representation of the arrangement of the secondary structure of chymotrypsin showing highly variable regions in the mammalian serine proteinases in heavy lines. The maximum insertions and deletions for the mammalian serine proteinases (trypsin, chymotrypsin, elastase, kallikrien, Factor $1X_a$, Factor X_a, plasmin, group specific proteinase, and thrombin B-chain) are shown as positive and negative numbers. The greatest change in length of chain in *S. grisens* proteinases A and B is circled. Heavy arrows indicate the position of intron-exon junctions found in the genes of chymotrypsin, elastase, or trypsin.

deletions in serine proteinases occur at such positions between strands. Rather large insertions have occurred in mammalian enzymes such as chymotrypsin relative to the *Streptomyces griseus* proteinases A and B, and almost all the β-turn regions show variations between the mammalian enzymes (Figure 6). A detailed analysis of the structures of chymotrypsin, trypsin, and elastase (59) indicates that highly variable sequences of the same length in closely related enzymes usually have a similar conformation, as for example chymotrypsin and trypsin residues 170–175 (chymotrypsinogen numbering).

In other cases there may be a common structural motif regardless of length. Thus the variable region 36–38 in serine proteinases forms a turn between two strands 26–35 and 39–48, and the 3, 1, and 6 residues of chymotrypsin, trypsin, and elastase respectively are accommodated by variation of the length of the β-strands while maintaining a β-bend.

An interesting compensation to an insertion occurs in the loop at position 203–206 (chymotrypsinogen numbering). In trypsin, kallikrein, and group specific protease, there are no residues equivalent to 203–206, but a glycine is found at 207 and plays a role in the β-bend then formed. When there are 4 or more residues in the region 203–206, residue 207 is usually a tryptophan or a tyrosine, which acts as a spacer between this loop and the residues at positions 27, 29, and 137. A similar situation occurs in βBp- when compared to γII crystallin. Tyr 62 in γII is replaced by Trp in bovine βBp and rodent βB1 while the nearby Trp 64 is conserved (139; D. Mahadevan, unpublished results). With no other change, it would be difficult to accommodate the bulky new side chain without its protruding into the solvent or greatly disturbing the position of the conserved Trp 64. However, the extension of the following loop by insertion of two or more residues in βBp and βB_1 allows the side chain to be protected from the solvent.

There are also many well-documented examples of insertions and deletions in the turns between $\alpha\beta$-structures. A comparison of the four topologically equivalent d(β-strand) h(helix) regions of endothiapepsin shows a typical variation from β-strands running through a tight turn to helices ($d_c h_c$ and $d'_c h'_c$) to arrangements with large loops between the β-strands and helices ($d_N h_N$ and $d_{N'} h_{N'}$). Insertions or deletions among the aspartyl proteinases are found at three of the four dh $\beta\alpha$ turns.

Many examples of insertions and deletions in turn regions between α-helices are found in the globins, which show little conservation in either length, sequence, or conformation except between closely related globins. This arises from the different lengths and relative orientations of the helices. In many cases, evolution introduces new strands of secondary structure. Thus in the mammalian globins, helix F′, which makes an angle of 135° with the F-helix, is introduced in the E-F region as shown in Figure 3. In several globin structures such as the α-subunit of human hemoglobin, the D-helix of myoglobin is completely lost. In azurin an extensive loop of 39 residues contains a helix in positions 54 to 65, the equivalent connecting loop in plastocyanin contains only 27 residues, and the shorter helix is quite differently spatially arranged (Figure 2). In the immunoglobulins, a hairpin loop from the V-domains is deleted so that variable domains comprise a sheet of five strands packed onto four strands while the constant domains have a three-stranded sheet packed onto the four-stranded one. Rearrangements of the chains on the periphery occur between the small cytochromes, such as c_{551} from *Pseudomonas aeruginosa*, compared to the chains of mammalian proteins (Figure 5). Thus the deletion in c_{551} is accommodated by the formation of an α-helix and the rearrangement of a loop.

THE EVOLUTION OF SPECIFICITY

Proteins may acquire a number of new characteristics during speciation and differentiation. One of the most important is the refinement or change of specificity in a family of enzymes that bind proteins or hormones. Such homologous protein families may contain topologically similar structures with identical arrangements of catalytic or active residues but different subsites, which define different specificities.

The high resolution X-ray analyses of serine proteinases from both microbial and mammalian sources have provided a wealth of information about the development of new specificities in evolution. In chymotrypsin the primary specificity site S1 (notation of Schechter & Berger) is mediated by a well-defined pocket involving residues 189–192, 214–220, and 226 (8, 63). This pocket accommodates a tyrosine side chain in chymotrypsin so that its phenolic hydroxyl can hydrogen bond to Oγ of serine 189. In elastase the specificity is for smaller side chains such as alanine, valine, or serine; in this enzyme the specificity pocket is blocked by the presence of Val 216 (otherwise invariant in mammalian enzymes as Gly) and Thr 226 (which is also invariant in mammalian enzymes with the exception of kallikrien as Gly) (120). The microbial α-lytic protease has a similar specificity to elastase, but it is achieved in a quite different way, since the residue topologically equivalent to 216 is glycine (38). In fact, insertion of a dipeptide at 192 in α-lytic protease leads to the occlusion of the specificity pocket by Met 192, whereas the side chain of Gln 192 in elastase is directed away from the pocket. There is also an insertion of a pentapeptide at 217, so that the side chain of Val 217 Å is also directed into the binding pocket. Thus the pocket is filled and the specificity is for smaller side chains.

Trypsin, kallikrien, factor IX_a, factor X_a, plasmin B-chain, thrombin B-chain, urokinase, and nerve growth factor γ-subunit have primary specificities for a basic residue, arginine, or lysine, and this is explained by the existence of an aspartate at 189 with which it can form an ion pair (22, 63, 124). In kallikrien and γ-nerve growth factor, the greater specificity is due to insertions and amino acid changes that make a circular wall around the active site depression and restrict accessibility of the binding site and substrate binding sites (22). In contrast, the microbial enzymes *Streptomyces griseus* proteinases A and B have a shallow surface depression rather than a pocket (38), which leads to a specificity of the Phe > Tyr > Leu compared to Tyr > Trp > Phe for chymotrypsin.

Unlike the serine proteinases, the aspartyl proteinases have an extended, deep cleft that can accommodate seven amino acid residues of the substrate S_4 to S'_3 (69, 125; L. Pearl and T. L. Blundell, manuscript in preparation). The binding sites P_1 and P'_1 of most aspartyl proteinases are relatively similar, which accounts for their similar specificities for large hydrophobic residues at S_1 and S'_1. However, there is a radical difference in mouse renin at P'_3 with a serine at position 189 instead of the phenylalaline or tyrosine of other aspartyl proteinases. This may account for the specificity of renin for rather larger groups at S_3, typically tyrosine or histidine in angiotensinogens. Further specificity in renin and in chymosin may arise from groups along the periphery of the active site (20; S. L. Sibanda and T. L. Blundell, unpublished results).

Structural comparisons of mammalian and yeast alcohol dehydrogenases give further insights into the evolution of different specificities. Differences in the specificities for substrates between the different horse isoenzymes and the rat enzyme appear to derive from the existence of Phe 110 in the horse E-chain in contrast to Leu 110 in the horse S-chain and the rat enzyme, which results in an activity towards some steroid substrates. Further substitutions in the yeast enzyme lead to a more narrow substrate pocket (45).

Homologous series of hormones and growth factors can also have similar tertiary structures and different although sometimes overlapping receptor binding capacities (17). For example, the receptor binding region of insulin (15) is conserved in insulins with high receptor affinities such as bovine and human insulins. The insulin-like growth factors have similar structures to insulin, and the region equivalent to the receptor-binding region of insulin is partly conserved but is partly occluded by the existence of the C- and D-peptide addition as shown in Figure 7 (13).

EVOLUTION OF OLIGOMERS

In this section we consider the evolution of globular monomeric proteins to oligomers, subunits of which are often more hydrophobic and frequently

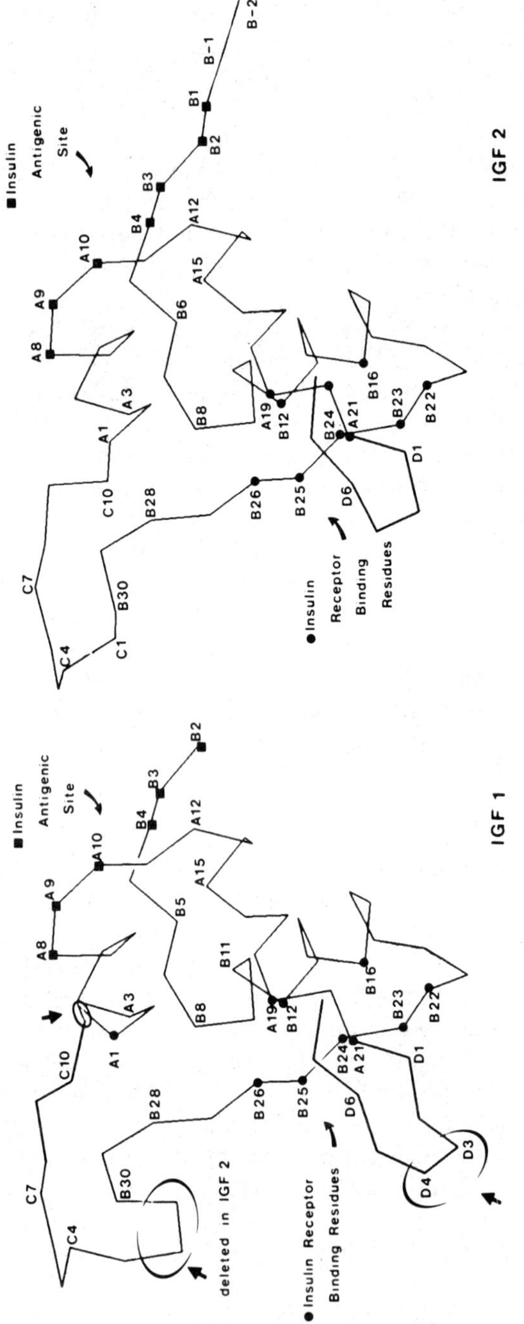

Figure 7 Schematic drawings of insulin-like growth factors. [Reproduced from (13).]

have extensions of the chain termini that can act as arms to wrap around adjacent subunits.

Hagfish insulin, the most phylogenetically original insulin to be sequenced and studied by high resolution X-ray analysis, forms dimers with a perfect dyad (33). This dimeric form is retained in most, but not all vertebrate insulins. In the cartilaginous fishes such as the dogfish shark (6), an accepted mutation to a B10 histidine allows weak association to hexamers via zinc coordination. The 2zinc hexamers are well characterized in the teleost fishes and most higher vertebrates. The porcine insulin 2Zn hexamer has approximate 32 symmetry with the two zinc ions on the threefold axis; each ion is bound to three B10 His, one from each dimer. The contacts between the three dimers are also mainly through hydrophobic side chains. Many hystricomorph rodent insulins form only monomers. In guinea pig (141), casiragua (67), coypu (M. Bajaj, manuscript in preparation), and cuis (M. Bajaj, unpublished results) insulins, lack of hexamer formation results from a loss of the B10 His and from a change of many surface residues involved in intrahexamer interactions to hydrophilic, often charged and larger residues. The formation of dimers in casiragua, coypu, and cuis insulins is prevented by mutation of B26 Tyr to an arginine or serine, whereas in porcupine and guinea pig the conformation is probably changed (68). Interestingly, the porcupine insulin has retained the B10 His and most of the hydrophobic faces used in oligomer formation by other insulins. The formation of oligomers in vertebrate insulins appears to be unnecessary for receptor binding, but zinc hexamers are found in storage granules in the B-cells of the endocrine pancreas. The formation of oligomers appears to endow an extra thermal stability and decreases susceptibility to enzymatic degradation—an advantageous characteristic in hormone storage.

Oligomeric proteins tend to be more hydrophobic than homologous monomers. This is true of the globins, although association can be controlled by conformation; thus lamprey hemoglobin dissociates to monomers in the oxy form (64), a primitive version of the allosteric control that has evolved later in the vertebrates as a two state equilibrium of $\alpha_2\beta_2$-tetramers. Although malate dehydrogenase forms dimers with a perfect dyad and hydrophobic inter subunit interactions, the homologous lactate dehydrogenase forms tetramers of 222 symmetry in which two malate dehydrogenase-like dimers are related by twofold symmetry. They are held together by hydrophobic interactions and extended NH_2-termini, which act as arms holding the tetramers together (65). More recently, NH_2-terminal chain extensions have been found to be important in the formation of $T = 3$ spherical viruses (1, 62). Extended NH_2-termini are found in a number of other oligomeric proteins including alfalfa mosaic virus protein

(135), yeast alcohol dehydrogenase (78), and bovine superoxide dismutase (123). They are found at both the NH_2- and COOH-termini of βBp-crystallin, which forms dimers and higher oligomers (139), while the homologous monomeric γ-crystallins have no extensions at the termini.

Similar, possibly divergently evolved protein subunits may associate in different ways. For example, a "back-to-back" structure is found in the association of the V_L and V_H domains of immunoglobulins (46, 121). The C_L and C_{H_1} domains form the reverse structure with a "front-to-front" arrangement, which is almost certainly found between the proximal domains of the histocompatibility antigens. Pairs of C_{H_2} domains are widely separated and the interactions are mediated by carbohydrate. The fact that these may all have derived from a monomeric structure is strengthened by the discovery of Thy 1, a cell surface antigen that has a sequence consistent with the immunoglobulin fold (28).

EVOLUTION OF REGULATORY MECHANISMS

Apart from the development of new specificities and changes in the quarternary structure, other new functions may be acquired during protein differentiation and even speciation. These include the development of regulatory mechanisms such as precursor activation, cooperativity, and positive or negative allosteric controls.

The development of precursor activation has been well documented for the serine proteinases. The requirement for an inactive zymogen is absent in bacterial enzymes such as the *S. griseus* proteinases A and B but is developed in mammalian enzymes as a self-protective mechanism. For example, chymotrypsinogen A is activated by a single proteolytic cleavage to produce an active enzyme that has an ion pair between Ile 16-NH_3^+ and Asp 194-CO_2^-. In the bacterial serine proteinases an ion pair is formed between Asp 194 and the buried Arg 138. James (74) suggests that evolution to the precursor system necessitated a hydrophobic region to bury the amino group, which is accessible in the bacterial enzymes, and a relatively long stretch of polypeptide chain to allow the NH_2-terminal Ile 16 to approach the carboxylate side chain of Asp 194, changes that account for much of the extra chain in the mammalian enzymes.

Zymogen activation in the aspartyl proteinases is not as well understood. In pepsinogen and prochymosin it involves the addition of about 40 amino acids to the N-terminus, compared to the basic structure found in fungal enzymes such as endothiapepsin or penicillopepsin. The activation process involves acidification, followed by a rearrangement of the precursor and an autolysis in the middle of the added residues. The exact position varies in different pepsinogens and prochymosin. Presumably, the precursor structures are similar for all enzymes, but lowering of the pH breaks ion pairs and leaves the added residues conformationally flexible and able to gain access to the catalytic residues of the now active enzyme. A second cleavage—a biomolecular reaction—releases the remaining precursor residues so that pepsin and endothiapepsin are very similar in size. Another aspartyl proteinase, renin, is active at neutral pH and has evolved a different zymogen activation, which is similar to that of hormone precursors and which involves cleavage at sequences of two basic residues. One cleavage occurs at a position equivalent to the N-terminus of the fungal enzymes while the second occurs at a β-loop in the region of 277, which is extended by four residues in prorenin compared to pepsin. It is possible that after cleavage of the two basic residues from this loop, there is a rearrangement of its position at the edge of the active site cleft.

The close homology and shared biological activities of insulin and the insulin-like growth factors are evidence of their divergent evolution from a common ancestor, probably shortly before the appearance of the first vertebrates (17). The existence of shorter C-peptides in the growth factors, the fact that they are active as single chain molecules, and the higher probability that growth-promoting activity evolved before the complex metabolic control activities of pancreatic insulin make it likely that the growth factors are phylogenetically the most original molecules. As there is no requirement for an inactive precursor of a hormone, whereas a single chain precursor is required for proper folding, we assume that cleavage of the connecting peptide was advantageous as it increased binding to new receptors. The most likely explanation of the very great length of the C-peptide, ~ 30 amino acids in insulin compared to 8 or 12 in the growth factors, lies in the evolution of a storage mechanism for the hormone, insulin. This is not needed in a growth factor that is constantly released and held as a complex with a specific binding protein in the circulation. In the case of proinsulin, the connecting peptide appears to endow it with great solubility so that it may travel along the endoplasmic reticulum until cleavage of the connecting peptide in the golgi results in a much less soluble protein that is able to form compact storage granules by precipitation and crystallization.

Most protein regulatory mechanisms are reversible and involve noncovalent interactions with allosteric inhibitors or effectors. In this respect, hemoglobin has been studied in greatest detail and is well understood. Hemoglobin exists in two states, the deoxy or tense (T) structure, which has a low affinity for oxygen but a high one for protons, chloride, inorganic phosphate, and CO_2, and the oxy or relaxed (R) structure, where the relative affinities are reversed. Perutz and his co-workers (101) have produced a series of detailed papers to show how these structures are exploited in different ways in different animals to optimize adaptation to the particular environment. For example Perutz & Brunori (102) have produced a molecular account of the root effect in certain fish and amphibian hemoglobins. In teleost fish, this effect involves the release of lactic acid into the swim bladder; the consequent release of oxygen arises mainly from the existence of a serine at F_9, a position occupied by a cysteine in mammalian hemoglobins. The cysteine may occupy either of the two alternative positions found in the oxy and deoxy states respectively, whereas the serine prefers to hydrogen bond in a more hydrophilic environment that favors the deoxy state and consequently the binding of hydrogen ions. Perutz & Brunori also discuss the molecular origin of the larger Bohr effect in teleost fish and the use of ATP, GTP, or inositol pentaphosphate as an allosteric effector in preference to D-2,3-diphosphoglycerate (DPG). These effects all appear to be adaptive and have certainly been selected for in evolution. A picture is thus emerging of evolution "tinkering with nature" (73) in a way that needs a detailed biological and structural study for a full appreciation of its molecular origins.

CONVERGENCE OF ACTIVE SITES

In 1971 Kraut et al (84) noted that, although chymotrypsin and subtilisin have quite different tertiary structures, the residues involved in catalysis (serine, histidine, and aspartic acid) have the same arrangement. Superposition of these groups gave a root mean square difference of only 1.0 Å. Since that time, further comparisons have been made showing that convergence of active site geometries is a relatively common phenomenon. For enzymes catalyzing similar reactions, chemically important groups should show the same functionality and spatial arrangements. Thus Argos et al (3) point out that the planar stereochemistry of the carbonyl group and the trigonal-tetrahedral transition are essential features of the hydrolysis of esters and that a limited number of amino acid groups and cofactors can act as nucleophiles or electrophiles under physiological conditions.

In the active centers of zinc-containing enzymes carbonic anhydrase, liver alcohol dehydrogenase, thermolysin, and carboxypeptidase (3, 82), each zinc has a similar coordination by three protein groups. A water and substrate carbonyl may occupy fourth and fifth positions while a proton abstracter group has a constant relationship to the zinc, an invariant histidine ligand, water, and substrate. The remarkable similarity in geometry may be related to a common, transient pentagonal zinc coordination during catalysis. An extension of the comparison of subtilisin and chymotrypsin (both serine proteinases) to the sulphydryl proteinase, papain, has shown that, although all have a charge relay system (Asp-His-Ser or Asn-His-Cys), the tetrahedral intermediates produced during acylation are of opposite hands (40, 53). However, the intermediates of papain and glyceraldehyde-3-phosphate dehydrogenase can be regarded as of the same hand.

Lactate dehydrogenase and glyceraldehyde-3-phosphate dehydrogenase have an orientation of the nicotinamide ring about the glycosidic bond, and thus an A or B side specificity of the nicotinamide, which is related to the substrate stereochemistry. For lactate dehydrogenase and glyceraldehyde-3-phosphate dehydrogenase, the catalytic environments are of different hands. Garavito et al (53) suggest that dehydrogenases, which have divergently evolved from a common precursor, must maintain their nicotinamide specificity if the protein fold of the catalytic domain is to be conserved.

CONVERGENT OR DIVERGENT EVOLUTION OF TERTIARY STRUCTURES

Evidence for the divergent evolution of proteins may be found in their sequences, their three dimensional structures, or their functions. Most families of proteins—the globins, immunoglobulins, serine proteinases, aspartyl proteinases, cytochromes c_2, and insulin-like hormones—have similarities in all these respects, and divergent evolution is probable. Other groups of proteins such as hen egg white lysozyme and lactalbumin (24) or the serine proteinases and haptoglobin heavy chain (85) have closely related sequences, so that similar tertiary structures must be assumed; the probability of divergent evolution is high.

There are, however, many groups of proteins that are similar in tertiary structure and often in function, but where definitive statements concerning their homology are not possible. For example, dehydrogenases, kinases, and flavodoxins have a similar structural domain whose function is to bind nucleotides. The so-called mononucleotide binding domain or Rossmann fold comprises a $\beta\alpha\beta\alpha\beta$-structure in which a nucleotide binds at the –COOH termini of the parallel β-strands. In the dehydrogenases, two such motifs are related by a pseudodyad, and each binds a mononucleotide moiety of NAD (113). Lactate dehydrogenase, liver alcohol dehydrogenase, and glyceraldehyde-3-phosphate dehydrogenase have 94 topologically equivalent positions with a root mean square deviation of between 2.7 and 3.5 Å, but only three glycines and one aspartic acid are conserved in all structures (100). The topologically equivalent domains of kinases (12, 119) and flavodoxins (136) bind nucleotides at similar positions. A less extensive but similar $\beta\alpha\beta$-structure also occurs in the redox protein, thioredoxin (66), which resembles the structure of glutathione peroxidase (86). These extensive homologies have been discussed in terms of divergent evolutionary processes, and the apparent lack of sequence homology has been explained by the time involved since divergence (100, 116). Less than random minimum base changes per codon of between 1.0 and 1.16 for topologically equivalent positions of the dehydrogenases have been held to support these conclusions (116).

Arguments for the alternative proposition that the proteins are convergently evolved are based upon the following observations: (*a*) any parallel β-sheet will be twisted in the same way and will have α-helices connecting the β-strands and $\beta\alpha\beta$-units of the same hand; (*b*) a preference for hydrophobic β-branched amino acids (valines and isoleucines) in parallel β-sheets increases the probability of less than random minimum base changes per codon at topologically equivalent positions; (*c*) charged groups such as the phosphates of the nucleotides tend to bind at NH_2-termini of helices where the charge-dipole interaction is favorable; (*d*) there are no alternatives to one of the invariant glycines (Gly 199 in liver alcohol dehydrogenase) and the invariant aspartates that are required for proper binding of ribose. These observations imply that convergent evolution of similar structures may be less improbable than was first imagined; there may be few structures that can function to bind a nucleotide.

Several other examples of structurally similar proteins with related functions have been identified. These include (*a*) cytochrome b_5 and the hemoglobin β-chain α-proteins, which bind heme (115); (*b*) insulin and relaxin—hormones which bind membrane receptors (17); (*c*) chicken, goose, and phage lysozymes for which topologically equivalent orientations of backbones align their active site clefts so that catalytic groups and saccharide units coincide as shown in Figure 8 (60, 94, 109, 115). (*d*) DNA binding proteins such as the *cro* repressor, catabolite gene activator protein, λ repressor and *lac* repressor have $\sim$25 amino acids arranged as an $\alpha\alpha$-bihelical motif, which binds the major groove of DNA (95).

These examples are all candidates for divergent evolution; they have similar tertiary structures and functions and some limited amino acid identities at topologically equivalent positions. However, in each case convergent evolution cannot be excluded.

Where there is topological equivalence but no convincing sequence or functional similarity, convergent evolution appears more probable. This applies to the immunoglobin fold and Cu, Zn superoxide dismutase (112), triose phosphate isomerase and the first domain of pyruvate kinase (91), snake venom neurotoxins and the wheat germ agglutinin domain (42), and concanavalin A and tomato bushy stunt virus protein (4). In the last examples there is a superficial similarity of function in that they all bind to membranes. However, neurotoxins bind to the acetylcholine receptor, whereas agglutinins bind cell surface *N*-acetyl-D-glucosamine or its oligomers. Both concanavalin A and viruses bind carbohydrate receptors and calcium, but the sites are not topologically equivalent in the two proteins. The structural basis of the similarity of neurotoxins and agglutinins lies in their many disulphide bridges, which help in stabilizing these small extracellular polypeptides in the absence of extensive hydrophobic cores. The viral proteins and concanavalin A share a common structural motif, the jelly roll, which is likely to be preferred both from its ease of folding as well as its thermodynamic stability.

In a sense, discussion of convergent versus divergent evolution is not a useful scientific exercise as it is difficult to propose a hypothesis that is open to falsification, as Karl Popper would wish. Keim & Heinrikson (81) have attempted to examine the expected degree of sequence similarity that might arise in proteins that have converged to similar states. They conclude that secondary structural preferences should not in general give rise to statistically significant sequence similarity at topologically equivalent positions but that some positional amino acid preferences may be found, particularly in supersecondary structures. However, the question of convergent or divergent evolution is still left open when the sequence similarities are not statistically significant.

At present, we understand little about the positional preferences in

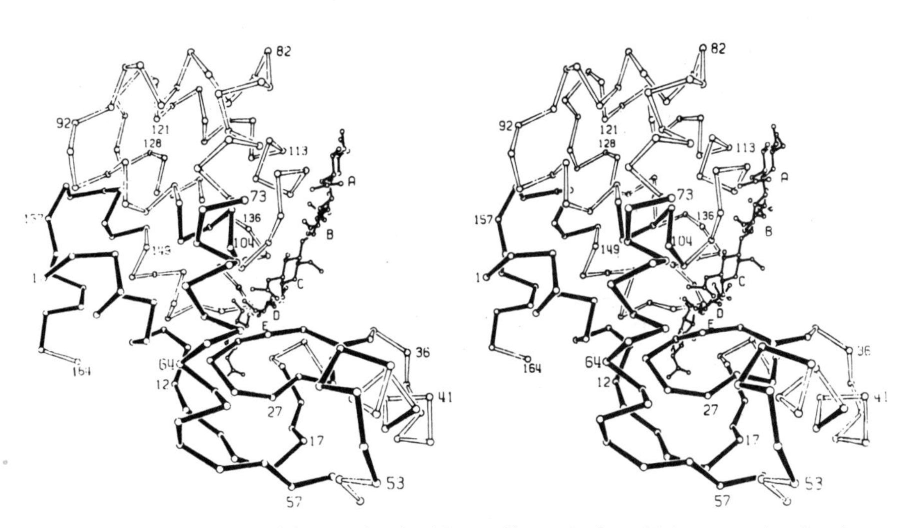

Figure 8 A stereo vew of the α-carbon backbone of bacteriophage T_4 lysozyme showing the location of the bound oligosaccharide. The solid lines connect the 78 α-carbon atoms of phage lysozyme, which have equivalent α-carbons in chicken lysozyme. [Reproduced from (94).]

supersecondary structures. Certain amino acid residues are more likely to be identical in a series of similar proteins. Thus in the microbial and mammalian serine proteinases, with the exception of Ser 195, Ser 214, His 57, Asp 102, and Asp 194, which are conserved at the active site, five glycines and two half-cystines are the invariant residues. In the insulin family, insulin, insulin-like growth factors, and relaxins-glycines are conserved as are the six half-cystines. In β/γ-crystallins only a glycine and a buried serine are conserved in all topologically equivalent motifs (18). Similar observations can be made in other distantly related proteins and almost certainly reflect the unique roles of these amino acids. The absence of a side chain allows glycine unique access to certain regions of conformational space and enables it to participate in interactions with other groups, especially coenzymes, where other amino acids would be structurally unacceptable. Cystine disulfide bridges are unique in providing a covalent interchain link.

INTERNALLY DUPLICATED SEQUENCES AND STRUCTURES

Gene duplication followed by gene fusion has led to a number of proteins with good internal homology. Evolution has exploited this mechanism frequently in the development of fibrous proteins such as collagen, tropomyosin, fibrinogen, and keratin, which have extended helical structures. It has also occurred in the evolution of globular proteins such as bacterial ferredoxin and neurophysin (duplications), serum albumin (triplication), calmodulin and immunoglobulin IgG heavy chains (quadriplication), and so on. X-ray analysis has shown that structural repeats are even more common than identified by sequence alone (97). Repeated helical motifs are found in hemerythrin, paravalbumin, vitamin D-dependent calcium-binding protein (Figure 9), and cytochrome b_5; repeated $\alpha\beta$-motifs

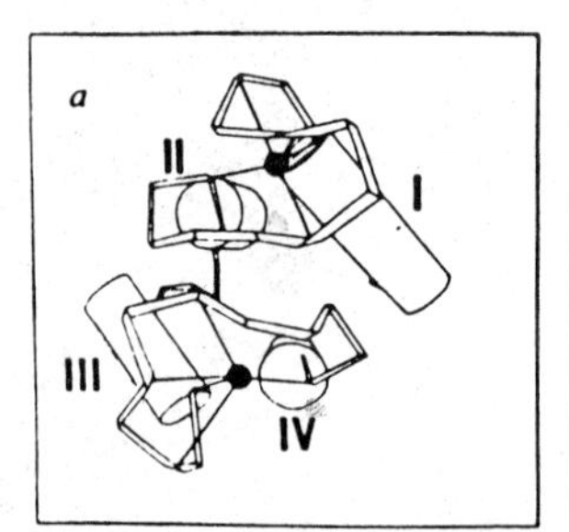

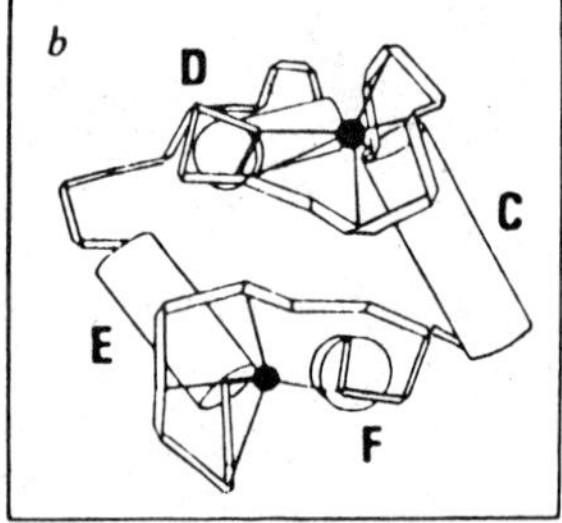

Figure 9 The α-carbon backbone and calcium ions viewed down the approximate dyad of (*a*) intestinal calcium binding protein and (*b*) parvalbumin, residues 38–108. [Reproduced from (128).]

are found in dinucleotide binding domains of dehydrogenases, rhodanese, L-arabinose-binding protein, hexokinase, and glutathione reductase; and repeated β-structures are found in serine proteinases, aspartyl proteinases, β/γ-crystallins, immunoglobulins, and soybean trypsin inhibitor.

In some duplicated structures, the repeated motifs are arranged as a string of more or less separate domains. This is true of the immunoglobulins and wheat germ agglutinin. However, in most cases the motifs are arranged symmetrically in a way that is reminiscent of protein quaternary structure. Thus, most are dyad related as, for example, the calcium-binding motifs of parvalbumin and vitamin D dependent calcium-binding protein (104, 128) (shown in Figure 9) and the domains of rhodanase and L-arabinose binding protein (106). As shown in Figure 10, the aspartyl proteinases and β/γ-crystallins each comprise four repeated motifs organized as two pseudodyad-related domains, each with an internal pseudodyad (18, 19). For the crystallins, there are three coplanar dyads; the two intra-domain dyads make an angle of about 20° with the inter-domain dyad (18). It has

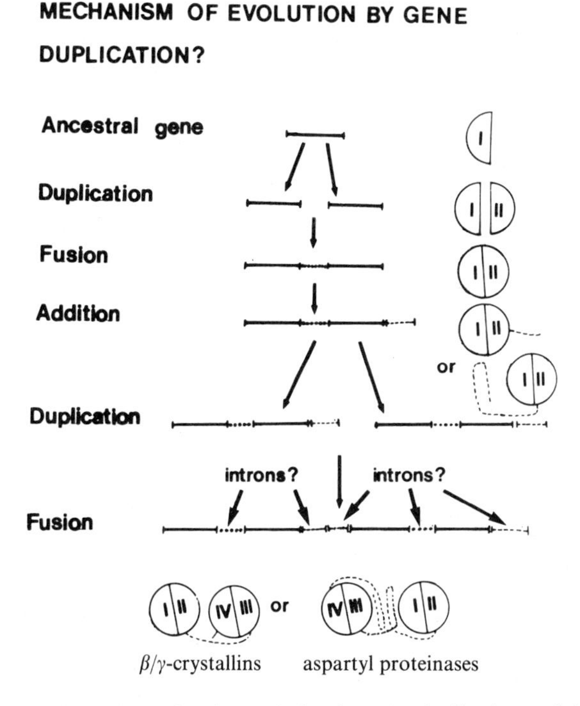

Figure 10 A possible scheme for the evolution by gene duplication and fusion of β/γ-crystallins and aspartyl proteinases. Introns are found where indicated in the β-crystallins, but not in the aspartyl proteinases.

been suggested that the four calcium-binding motifs of calmodulin or troponin *c* may be organized with 222 symmetry (128). Soybean trypsin inhibitor appears to contain a pseudotriad (96).

The oligomer-like organization of the structural motifs or domains in these internal repeated proteins suggests that they have evolved by gene duplication and fusion from proteins that could form oligomers. Of course, the evidence for divergent evolution in these structures rests to some extent on the same criteria of sequence and structural similarity as was discussed in the previous section. However, we can assume that gene duplication would be more likely to give rise to clusters of related gene sequences juxtaposed in the genome than dispersed through it (23). In this case, we would expect and indeed find a high frequency of internally repeated structures among the large number of multidomain proteins. The structure of the γ-crystallins offers further evidence for gene duplication and divergence. Motifs 1 and 2 or 3 and 4 are less similar in sequence and topology than are 1 and 3 or 2 and 4. This would be expected on the basis of the two-stage evolutionary process illustrated in Figure 10, where an initial duplication of a structural motif leads to a domain (1 and 2 or 3 and 4) and is followed by a further duplication to give the two domain structure.

Duplication and fusion may have had several functional advantages. The closely interacting dyad-related domains may have conferred structural stability and tuned up the folding process. In several duplicated structures such as the ferredoxins or the structural motifs (half domains) of aspartyl proteinases or crystallins, the isolated motifs would be unlikely to maintain a stable tertiary structure. In other cases such as the immunoglobulins, more complex structures are advantageously built from covalently bonded domains. In many cases, the duplication of two domains has led to a bilobal protein with a cleft suitable for binding a substrate, such as is found in the aspartyl proteinases (130).

ASSEMBLY OF FUNCTIONAL AND STRUCTURAL DOMAINS

Apart from point mutations, insertions, and deletions, it is becoming evident that nature can assemble a series of structural and functional modules to provide complex multidomain proteins with modified or new functions. The assembly of similar domains considered in the previous section is a common but special example of this phenomenon. Thus in the immunoglobulins, extra domains can be recruited to the heavy chains, which allow them to bind to the membrane. In class II histocompatibility antigens, immunoglobulin-like domains (α_2 and β_2) are linked on one side to a transmembrane domain and on the other to further nonhomologous domains (α_1 and β_1), which presumably mediate cell recognition. In certain β-crystallins, NH_2-terminal extensions are recruited to bind membranes, whereas in others the extensions favor oligomers (72).

The identification of domains in the above examples depends on their having a structural identity and a special function (76). In many cases comparison of two related structures shows that one has an extra region, which is comprised of a continuous polypeptide folded into a globular domain. Figure 11 shows the structure of *Penicillium vitale* catalase (134). Comparison with the three dimensional structure of beef liver catalase (108) shows that there is a close similarity in the domain binding the heme group, but a COOH-terminal $\alpha\beta$-domain is completely absent. A similar evolutionary process involving addition of a domain appears to have occurred in the virus proteins. Southern bean mosaic virus comprises a single, jelly roll-type domain organized as a T = 3 quasi-equivalent icosahedral structure (1). The tomato bushy stunt virus has an identical organization of similar domains, to each of which is attached a further protruding domain (61, 62). In pyruvate kinase, an eight-stranded $\alpha\beta$-barrel, similar to that found in triose phosphate isomerase, is linked to two other domains. The substrates of the two enzymes are bound in similar positions with respect to the barrels (91).

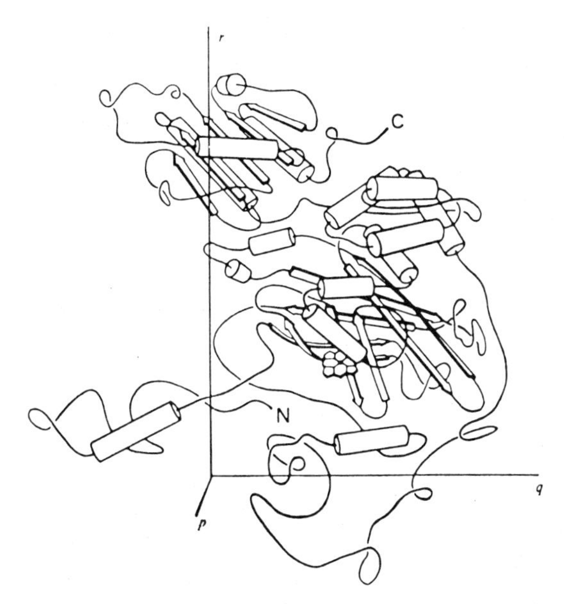

Figure 11 The structure of *P. vitale* catalase, showing the heme-binding domain and a smaller COOH-terminal domain, which is not present in the beef liver catalase. [Reproduced from (134).]

In many complex proteins the globular domains are associated with distinct functions and are candidates for modules, which may have been assembled in evolution. Thus Rossmann and his colleagues (113) showed that several dehydrogenases comprise a nucleotide-binding domain linked to a further domain, which provides many of the residues binding the substrate. In phosphorylase (127, 137) there is a glycogen-binding domain, an NH_2-terminal nucleotide-binding domain, and a COOH-terminal domain that contributes to the catalytic site.

EXON-INTRON JUNCTIONS AND TERTIARY STRUCTURE

Recent developments in genetic manipulation and DNA sequencing have revealed much about the organization of the genome, which frequently reflects the modular organization of proteins and provides interesting insights into their evolution.

Much can be learned by considering the immunoglobulin gene sequences (133). In the mouse germ cell, the coding potential for the λ_1 light chain lies in three coding segments or exons. These are the V- and smaller J-exons, which code for the variable region, and the C-exon, which codes for the constant region. The exons are interspaced with long intervening sequences, or introns, which do not code for proteins. For the heavy chains the third complementarity-determining or highly variable regions are coded by separate D-exons, and there are, in addition, exons corresponding to each of the constant domains for every immunoglobulin class. Diversity is generated by four somatic mechanisms. The first is combinatorial: the joining together of various segments or exons. The second is a junctional site diversifier arising because the V_L–J_L, V_H–D, and D–J_H joining ends are imprecise. The third is a junctional insertion occuring only in the V_H–D and D–J_H junctions where one to several nucleotides are inserted, apparently in a template-independent fashion. Fourth, there is somatic mutation, which is not based on recombination and which alters bases throughout the sequences coding for the variable region. All these mechanisms may be exploited in the evolution of other proteins in addition to reciprocal recombinations between homologous genes, gene conversion, and re-incorporation of processed mRNA into the genome to give pseudogenes.

Given that noncoding intervening sequences or introns do not occur in most present day bacteria, it is not surprising that relationships between exon structures and modules of protein structure are found in proteins, for which there is good evidence of evolution in eukaryotes. Thus, histocompatibility antigens are also coded for by exons corresponding to the intramembrane regions, the immunoglobulin-like membrane proximal domains, and the unrelated membrane distal domains. In the β-crystallins, four exons correspond exactly to the repeated structural motifs as shown in Figure 12 (72), but in the γ-crystallins, the two domains, each comprising two structural motifs, are coded for by single exons (J. Schoenmaker, unpublished results). In collagen (37) the triple helical structure is coded by about 50 exons. Each appears to derive from a prototype 54 base pair exon coding for six Gly-X-Y triplets, although some exons are roughly twice as long, whereas others are 9 base pairs longer or shorter. Thus immunoglobulins, histocompatibility antigens, β/γ-crystallins, and collagen may have evolved by linking together in a modular fashion exons corresponding to the assembly of domains. The existence of introns would make it easier to link these together in evolution, as the exact joining points are less critical.

Much effort has been expended in finding a correlation between exons and functional domains in other globular proteins. It is suggested that if the exons correspond to structural domains, they are more likely to assemble to give viable proteins (10, 11, 55). Thus for mouse α-globin genes, the three

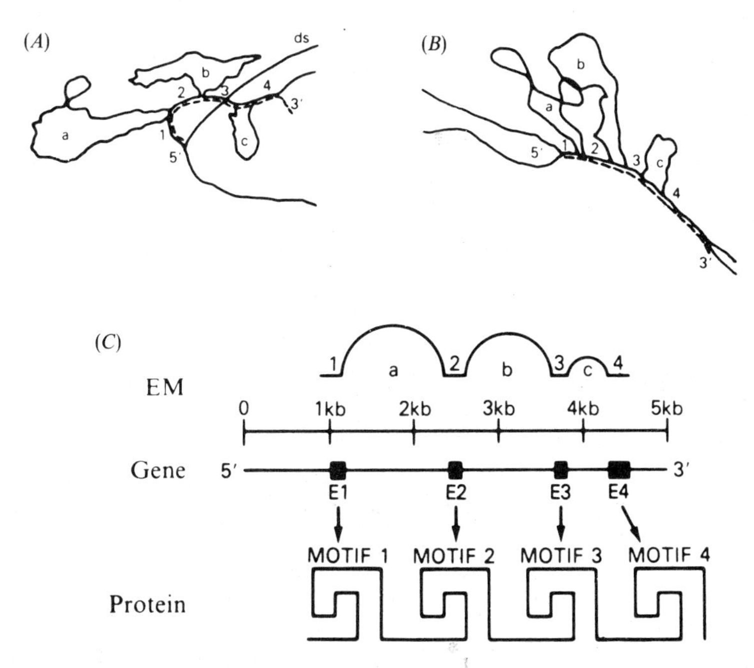

Figure 12 Electron microscopic analysis of the hybridization of the 23Kβ-crystallin gene with RNA (*A*) and cDNA (*B*). A schematic diagram (*C*) illustrates the electron micrograph (*top*), the deduced structure of the crystallin gene (*middle*), and its relationship to the proposed structure of the polypeptide, which is divided into four structural motifs (*bottom*). a, b, and c are introns, and 1, 2, 3, and 4 are exons. [Reproduced from (72).]

coding sequences correspond to the sequences 1–31, 32–99, and 100–141, and a similar arrangement occurs in the β-globin and myoglobin genes. Eaton (43) has suggested that the exons may correspond to functional units; the heme and $\alpha_1\beta_2$-contacts are mainly concentrated in exon 2, the residues concerned with the Bohr effect and diphosphoglycerate binding are in exons 1 and 3, and the $\alpha_1\beta_2$-contacts cluster in exon 3.

Although the exons in mammalian globins do not correspond to obvious domains, Gō (56) has used a diagonal plot to indicate four compact units, two of which correspond to the smaller exons while the larger exons corresponds to two such compact units. Thus she predicts a further intron that is not found in the mammalian globins between residues 68 and 69. Interestingly, an intron at such a position was later identified in the soybean leghemoglobin (77). A similar analysis has been carried out for lysozymes (5, 60). Figure 13 shows the relationships between exons and structural/functional domains of chicken, goose, and T_4 lysozymes. Exons 2 and 3 correspond to the regions shared between the lysozymes. The much expanded region at the C-terminus of phage lysozyme in the approximate region of exon 4 of chicken lysozyme may have been added to bind cross-linked peptide of *E. coli* cell walls, which is not necessary for chicken lysozyme.

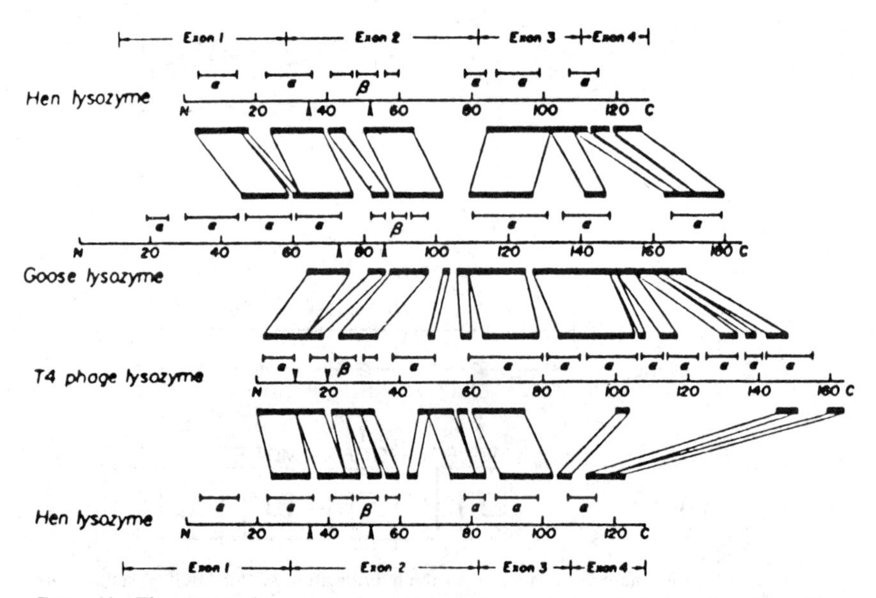

Figure 13 The structural correspondance between goose, chicken, and phage lysozymes. The connected solid bars indicate topologically equivalent parts of the polypeptides. The locations of the α-helices and β-strands are shown, as are the exons of chicken lysozyme. The arrowheads show locations of residues presumed to be involved in catalysis. [Reproduced from (60).]

The association of exons to structural or functional domains, and their assembly in a modular fashion, corresponds to the combinatorial role of diversification. Other exon-intron junctions occur at points coding for surface residues and may play a role in creating diversity through mutations or insertions. Thus, Figure 6 shows the positions of the intron-exon junctions in three serine proteinases superposed on the secondary structure diagram, which also indicates the highly variable regions in mammalian compared to microbial enzymes (31, 32). It is possible that, in order to develop an inactive percursor, the large insertions occurring in the mammalian enzymes were achieved by sliding the intron-exon junctions. A similar relationship of the intron-exon junctions to surface regions also occurs in dihydrofolate reductase (31) and the aspartyl proteinase renin (W. Brammar and D. Burt, manuscript in preparation). In the latter case, the surface turn of prorenin, which has an insertion when compared to pepsin and which is cleaved in activation, corresponds to an exon-intron junction. However, the exon structure has no relationship to the internally repeated three dimensional structure. This may indicate that aspartyl proteinases evolved before eukaryotes or that the introns have been lost in evolution.

The fact that related genes can have unrelated introns is further illustrated by human α_1-antitrypsin and chicken olvalbumin, which have no introns in common but nevertheless have 24% amino acid sequence homology (30, 87). The most probable interpretation is that introns are inserted in evolution rather than from an ancestoral gene lost by chance. For some proteins there is good evidence that gene duplication can duplicate introns within the region coding for the repeated domain. Thus the gene for α-fetoprotein, which like serum albumin has a threefold repeat structure (44), contains the same pattern of introns within each repeat in the DNA.

NEUTRALIST VERUS NEODARWINIST EVOLUTION

We finally consider what proportion of the accepted mutations in the evolution of proteins are adaptive or selectively advantageous for the organism and what proportion are selectively neutral and accumulate during evolutionary divergence as a result of genetic drift (79, 83). That neutral or near neutral changes occur in evolution cannot be doubted, since silent base changes lead to divergence of closely related, duplicated genes, e.g. the β-globin genes (79). However, the extreme neutralist view that all amino acid changes occurring in orthologous genes during protein speciation are neutral conflicts with our present detailed understanding of protein structure and function. The evolution of new functions and oligomer formation, changing specificity, and development of precursors and other regulatory functions during speciation must have been selectively advantageous.

The extreme neutralist view holds that after protein differentiation, constraints due to tertiary structure and function throw out certain mutations that are deleterious but allow in other areas those mutations that are neutral. The percentage of amino acids that can be mutated depends on the constraints on the protein, which are constant from species to species. The percentage is likely to decrease in the series histones, cytochromes *c*, globins, and fibrinopeptides. The extreme neutralist view makes the following predictions: (*a*) that the rate of evolution is constant for any protein and is independent of the species generation time; (*b*) that the sites of mutations are randomly distributed in evolution among the amino acids, which are not constrained by the function; and (*c*) that acceleration in the otherwise even rate of accepted mutations must be explained by release of constraints.

On the first point there has been much discussion relating to the manner of calculation of rates and estimation of branch times in the phylogenetic trees (58, 138). Fitch & Langely (49) find that individual proteins are rather poor clocks. Goodman (58) shows that dates for the rodent ancestor are very recent (in the range of 0 to 5 MyBP) when calculated from cytochrome *c* and α-lens crystallin sequences and are ancient (in the range of 63 to 71 MyBP) when calculated from α- and β-hemoglobin sequences. There are many examples of accelerated evolution such as the α-hemoglobins of the old monkey tribe or the insulins of the hystricomorph rodents.

The absence of the predicted random distribution of sites for accepted mutations during different stages of evolution must be the strongest argument against the extreme neutralist view (7, 21). We have shown that different areas of the insulin molecule are accepting mutations at different stages in evolution. In the early vertebrates such as the cyclostomes, elasmobranchs, and the teleosts, the regions involved in hexamer formation are evolving the fastest. This cannot be due to release of constraints because a new function is evolving. However, the surface region most quickly evolving in mammals—A8, A9, and A10—is relatively slow in accepting mutations within the teleosts. In the hystricomorph rodents the rate of accepted mutations is much more than ten times that in other mammals. The accepted mutations mainly relate to loss of oligomerization, but the changes are not random; they are to a different set of amino acids (hydrophobic to hydrophilic; small to large side-chains). In a similar way the transition of the monomeric hemoglobin to a tetramer with a sigmoidal oxygen equilibrium curve in the early vertebrates led to an elevated rate of mutations in prospective $\alpha_1\beta_2$ contact sites and in the salt bridge forming sites associated with the Bohr effect, which have been only conservatively changed from the bird-mammal ancestor to the present. There have also been specific changes in the diphosphoglycerate (DPG)-binding residues in species such as the llama, which needs increased oxygen binding at high altitudes and thus reduced DPG interactions. In birds and teleosts, specific conserved residues allow adaption to binding of differing allosteric effectors.

Thus, mutations have not been accepted at the same sites at different stages of evolution in either the hemoglobins or the insulins. Similar conclusions appear to hold for the cytochromes *c* and the calmodulins (58). Furthermore, these differences are related to well-defined changes of function in evolution and appear to be adaptive. There is an associated variation in the rate of accepted mutations; adaptive changes seem to be accompanied by an acceleration of the rate. These observations argue for an important role for selectively advantageous mutations in the evolution of proteins.

The extent of neutral changes remains unclear, but their occurrence appears more limited as we learn more of the details of structure and function in evolution. It is probable that both neutral and adaptive mutations are accepted in evolution. Indeed, some neutral changes may even become adaptive at later stages in evolution, thus smoothing out the rates of change. Changes that might otherwise be neutral or even disadvantageous can certainly occur when there is a cluster of genes or if a gene is silent (a pseudogene) for a period of time.

Whatever the relative roles of neutral and adaptive changes, even the globins make reasonable biological clocks when comparisons are made between distantly related species and there is a "levelling in the number of amino acid replacements despite differences in rate during some periods of time" (131). However, biological clocks based on single proteins for short periods of time must be inherently very unreliable.

ACKNOWLEDGMENTS

We are grateful to David Penny, Lynn Sibanda, and Lawrence Pearl for making available unpublished manuscripts and to Jim Pitts and Willie Taylor for helpful comments.

Literature Cited

1. Abad-Zapatero, C., Abdel-Meguid, S. S., Johnson, J. E., Leslie, A. G. W., Rayment, I., Rossmann, M. G., et al. 1980. *Nature* 286:33–39
2. Almassy, R. J., Dickerson, R. E. 1978. *Proc. Natl. Acad. Sci. USA* 75:2674–78
3. Argos, P., Garavito, R. M., Eventoff, W., Rossmann, M. G. 1978. *J. Mol. Biol.* 126:141–58
4. Argos, P., Tsukihara, T., Rossmann, M. G. 1980. *J. Mol. Evol.* 15:169–79
5. Artymiuk, P. J., Blake, C. C. F., Sippel, A. E. 1981. *Nature* 290:287–88
6. Bajaj, M., Blundell, T. L., Pitts, J. E.,

Wood, S. P., Tatnell, M., Gowan, L. K., et al. 1983. *Eur. J. Biochem.* 135:535–42
7. Beintema, J. J., Gasstra, W., Lenstra, J. A., Welling, G. W., Fitch, W. M. 1977. *J. Mol. Evol.* 10:49–71
8. Birktoft, B., Blow, D. M. 1972. *J. Mol. Biol.* 68:187–240
9. Birktoft, J. J., Blow, D. M., Henderson, R., Steitz, T. A. 1970. *Philos. Trans. R. Soc. London Ser. B* 257:67–73
10. Blake, C. C. F. 1978. *Nature* 273:267
11. Blake, C. C. F. 1983. *Trends Biochem. Sci.* 8:11–13
12. Blake, C. C. F., Evans, P. R. 1974. *J. Mol. Biol.* 84:585–601
13. Blundell, T. L., Bedarkar, S., Humbel, R. E. 1983. *Fed. Proc.* 42:2592–97
14. Blundell, T. L., Bedarkar, S., Rinderknect, E., Humbel, R. E. 1978. *Proc. Natl. Acad. Sci. USA* 75:180–84
15. Blundell, T. L., Dodson, G. G., Hodgkin, D. C., Mercola, D. 1972. *Adv. Protein Chem.* 26:279–402
16. Blundell, T. L., Horuk, R. 1981. *Hoppe-Seyler's Z. Physiol. Chem.* 362:727–37
17. Blundell, T. L., Humbel, R. E. 1980. *Nature* 287:781–87
18. Blundell, T. L., Lindley, P., Miller, L., Moss, D., Slingsby, C., Tickle, I. J., et al. 1981. *Nature* 289:771–77
19. Blundell, T. L., Sewell, B. T., McLachlan, A. D. 1979. *Biochim. Biophys. Acta* 580:24–31
20. Blundell, T. L., Sibanda, L. B., Pearl, L. 1983. *Nature* 304:273–75
21. Blundell, T. L., Wood, S. P. 1975. *Nature* 257:197–203
22. Bode, W., Zhongguo, C., Bartels, K., Kutzbach, C., Schmidt-Kastner, G., Bartunik, H. 1983. *J. Mol. Biol.* 164:237–82
23. Bodmer, W. F. 1981. *Am. J. Hum. Genet.* 33:664–82
24. Browne, W. J., North, A. C. T., Phillips, D. C., Brew, K., Vanamann T. C., Hill, R. L. 1969. *J. Mol. Biol.* 42:65–86
25. Chothia, C., Lesk, A. M. 1982. *J. Mol. Biol.* 160:309–23
26. Chothia, C., Lesk, A. M., Dodson, G. G., Hodgkin, D. C. 1983. *Nature* 302:500–5
27. Chou, P. Y., Fasman, G. D. 1974. *Biochemistry* 13:222–45
28. Cohen, F. E., Novotny, J., Sternberg, M. J. E., Campbell, D. G., Williams, A. F. 1981. *Biochem. J.* 195:31–40
29. Cohen, F. E., Sternberg, M. J. E. 1980. *J. Mol. Biol.* 138:321–33
30. Cornish-Bowden, A. 1982. *Nature* 297: 625–28
31. Craik, C. S., Rutter, W. J., Fletterick, R. 1983. *Science* 220:1125–29
32. Craik, C. S., Sprang, S., Pletterick, R., Rutter, W. J. 1982. *Nature* 299:180–82
33. Cutfield, J. F., Cutfield, S. M., Dodson, E. J., Dodson, G. G., Emdin, S. F., Reynolds, C. D. 1979. *J. Mol. Biol.* 132:85–100
34. Cutfield, J. F., Cutfield, S. M., Dodson, E. J., Dodson, G. G., Reynolds, C. D., Vallely, D. 1981. In *Structural Studies on Molecules of Biological Interest*, ed. G. Dodson, J. P. Glusker, D. Sayre, pp. 527–46. Oxford: Clarendon
35. Dayhoff, M. O. 1978. *Atlas of Protein Sequence and Structure*, Vol. 5 (Suppl. 3). Washington DC: Natl. Biomed. Res. Found.
36. Dayhoff, M. O., Eck, R. V. 1966. *Atlas of Protein Sequence and Structure*, Vol. 2. Washington DC: Natl. Biomed. Res. Found.
37. deCrombrugge, B., Pastan, I. 1982. *Trends Biochem. Sci.* 7:11–13
38. Delbaere, L. T. J., Brayer, G. D., James, M. N. G. 1979. *Can. J. Biochem.* 57:135–43
39. Dickerson, R. E., Timkovich, R., Almassy, R. J. 1976. *J. Mol. Biol.* 100:473–91
40. Drenth, J., Enzung, C. M., Kalk, K. H., Vessies, J. C. A. 1976. *Nature* 264:373–77
41. Drenth, J., Hol, W. G. J., Jansonius, J. N., Koekkoek, R. 1972. *Eur. J. Biochem.* 26:177–81
42. Drenth, J., Low, B. W., Richardson, J. S., Wright, C. S. 1980. *J. Biol. Chem.* 255:2652–55
43. Eaton, W. A. 1980. *Nature* 284:183–85
44. Eiferman, F. A., Young, P. R., Scott, R. W., Tilghman, S. M. 1981. *Nature* 294:715–18
45. Eklund, M., Branden, C.-I., Jornvall, H. 1976. *J. Mol. Biol.* 102:61–73
46. Epp, O., Colman, P., Fehlhammer, H., Bode, W., Schiffer, M., Huber, R., Palm, W. 1974. *Eur. J. Biochem.* 45:513–24
47. Fitch, W. M. 1970. *Syst. Zool.* 19:99–107
48. Fitch, W. M. 1977. *Am. Nat.* 111:223–57
49. Fitch, W. M., Langley, C. H. 1976. In *Molecular Anthropology*, ed. M. Goodman, R. E. Tashian, p. 197. New York: Plenum
50. Foltmann, B., Pederson, V. B. 1977. *Adv. Exp. Med. Biol.* 95:3–22
51. Frey, M., Pepe, G., Adman, E. T., Sieker, L. C. 1983. *8th Eur. Crystallogr. Meet., Liege, Aug. 8–12, Abstr.*, 1.09P:22
52. Fukuyama, K., Hase, T., Matsumoto, S., Tsukihara, T., Katsube, Y., Tanaka, N., et al. 1980. *Nature* 286:522–23
53. Garavito, R. M., Rossman, M. G.,

Argos, P., Eventoff, W. 1977. *Biochemistry* 16:5065–71
54. Garnier, J., Osguthorpe, D. J., Robson, B. 1978. *J. Mol. Biol.* 120:97–120
55. Gilbert, W. 1978. *Nature* 271:501
56. Gō, M. 1981. *Nature* 291:90–91
57. Gō, M., Miyazawa, S. 1978. *Int. J. Pept. Protein Res.* 12:237–41
58. Goodman, M. 1981. *Prog. Biophys. Mol. Biol.* 38:105–64
59. Greer, J. 1981. *J. Mol. Biol.* 153:1027–42
60. Grutter, H. G., Weaver, L. H., Matthews, B. W. 1983. *Nature* 303: 828–30
61. Harrison, S. C. 1980. *Nature* 286:558–59
62. Harrison, S. C., Olson, A. J., Schutt, C. E., Winkler, F. K., Bricogne, G. 1978. *Nature* 276:368–73
63. Hartley, B. 1974. In *Evolution in the Microbial World*, ed. M. J. Carlisle, J. J. Skehel, pp. 151–82. Cambridge, England: Cambridge Univ. Press
64. Hendrickson, W. A., Love, W. E. 1971. *Nature* 232:197–203
65. Holbrook, J., Liljas, A., Steindel, S. J., Rossmann, M. G. 1975. In *The Enzymes*, ed. P. D. Boyer, 11:191–292. New York: Academic. 3rd ed.
66. Holmgren, A. A., Soderberg, B. O., Eklund, H., Branden, C. J. 1975. *Proc. Natl. Acad. Sci. USA* 72:2305–9
67. Horuk, R. 1979. *2nd Int. Insulin Symp., Aachen, FRG*, pp. 67–68 (Abstr.)
68. Horuk, R., Blundell, T. L., Lazarus, N. R., Neville, R. W. J., Stone, D., Wollmer, A. 1980. *Nature* 286:822–24
69. Hsu, I.-N., Delbaere, L. T. J., James, M. N. G., Hofmann, T. 1977. *Nature* 266:140–45
70. Huber, R., Epp, O., Steigemann, W., Formanek, H. 1971. *Eur. J. Biochem.* 19:42–50
71. Huber, R., Kukla, D., Bode, W., Schwager, P., Bartels, K., Deisenhofer, J., Steigemann, W. 1974. *J. Mol. Biol.* 89:73–101
72. Inana, G., Piatigorsky, J., Norman, B., Slingsby, C., Blundell, T. L. 1983. *Nature* 302:310–15
73. Jacob, F. 1982. In *The Possible and the Actual*. Seattle: Univ. Wash. Press
74. James, M. N. G. 1980. *Can. J. Biochem.* 58:251–71
75. James, M. N. G., Delbaere, L. T. J., Brayer, G. D. 1978. *Can. J. Biochem.* 56:396–402
76. Janin, J., Wodak, S. 1983. *Prog. Biophys. Mol. Biol.* 42:21–78
77. Jensen, E. O., Paludan, K., Hyldig-Nielsen, J. J., Jorgensen, P., Marcker, K. A. 1981. *Nature* 291:677–79
78. Jörnvall, H. 1977. *Eur. J. Biochem.* 72:425–42
79. Jukes, T. H. 1980. *Science* 210:973–78
80. Kabsch, W., Sander, C. 1983. *FEBS Lett.* 155:179–82
81. Keim, P., Heinrikson, R. L. 1981. *J. Mol. Biol.* 151:179–97
82. Kester, W. R., Matthews, B. W. 1977. *J. Biol. Chem.* 252:7704–10
83. Kimura, M. 1979. *Sci. Am.* 241:94–104
84. Kraut, J., Robertus, J. D., Birktoft, J. J., Alden, R. A., Wilcox, P. E., Powers, J. C. 1971. *Cold Spring Harbor Symp. Quant. Biol.* 36:117–23
85. Kurosky, A., Barnett, D. R., Lee, T. H., Touchstone, B., Hay, R. E., Arnott, M. S., et al. 1980. *Proc. Natl. Acad. Sci. USA* 77:3388–92
86. Ladenstein, R., Otto, E., Bartels, K., Jones, A., Huber, R. 1979. *J. Mol. Biol.* 134:199–218
87. Leicht, M., Long, G. L., Chandra, T., Kurachi, K., Kidd, V. J., Mace, M. Jr., et al. 1982. *Nature* 297:655–64
88. Lenstra, J. A., Hofsteenage, J., Beintema, J. J. 1977. *J. Mol. Biol.* 109:185–93
89. Lesk, A. M., Chothia, C. 1980. *J. Mol. Biol.* 136:225–70
90. Lesk, A. M., Chothia, C. 1982. *J. Mol. Biol.* 160:325–42
91. Levine, M., Muirhead, H., Stammers, D. K., Stuart, D. I. 1978. *Nature* 271:626–30
92. Levitt, M. 1976. *J. Mol. Biol.* 104:95–116
93. Lim, V. I., Ptitsyn, O. B. 1970. *Mol. Biol. USSR* 4:327–82
94. Matthews, B. W., Grutter, M. G., Anderson, W. F., Remington, S. J. 1981. *Nature* 290:334–35
95. Matthews, B. W., Ohlendorf, D. H., Anderson, W. F., Fisher, R. G., Takeda, Y. 1983. *Trends Biochem. Sci.* 8:25–29
96. McLachlan, A. D. 1979. *J. Mol. Biol.* 133:557–63
97. McLachlan, A. D. 1980. In *Protein Folding*, ed. R. Jaenicke. Amsterdam: Elsevier/North Holland Biomed. Press
98. McLachlan, A. D., Shotton, D. M. 1971. *Nature* 229:202–5
99. Muirhead, H., Cox, J., Mazzarella, L., Perutz, M. F. 1967. *J. Mol. Biol.* 28:117–56
100. Ohlisson, I., Nordstrom, B., Branden, C. I. 1974. *J. Mol. Biol.* 89:339–54
101. Perutz, M. F., Bauer, C., Gros, G., Leclercq, F., Vandecasserie, C., Schnek, A. G., et al. 1981. *Nature* 291:67–69
102. Perutz, M. F., Brunori, M. 1982. *Nature* 299:421–26
103. Perutz, M. F., Kendrew, J. C., Watson, H. C. 1965. *J. Mol. Biol.* 13:669–78

104. Ploegman, J. H., Drent, G., Kalk, K. H., Mol, W. G. J., Heinrickson, R. L., Keim, P., et al. 1978. *Nature* 273:124–29
105. Privalov, P. L., Khechinashvili, N. N. 1974. *J. Mol. Biol.* 86:665–84
106. Quiocho, F. A., Gilliland, G. L., Philips, G. N. 1977. *J. Biol. Chem.* 252:5142–49
107. Rao, S. T., Rossmann, M. G. 1973. *J. Mol. Biol.* 76:241–56
108. Reid, J. R. III, Murthy, M. R. N., Sicignano, A., Tanaka, N., Musick, W. D. L., Rossmann, M. G. 1981. *Proc. Natl. Acad. Sci. USA* 78:4767–71
109. Remington, S. J., Matthews, B. W. 1978. *Proc. Natl. Acad. Sci. USA* 75:2180–84
110. Richardson, J. S. 1981. *Adv. Protein Chem.* 34:167–339
111. Richardson, J. S., Getzoff, E. D., Richardson, D. C. 1978. *Proc. Natl. Acad. Sci. USA* 75:2574–78
112. Richardson, J. S., Richardson, D. C., Thomas, K. A., Silverton, E. W., Davies, D. R. 1976. *J. Mol. Biol.* 102:221–35
113. Rossmann, M. G., Argos, P. 1975. *J. Biol. Chem.* 250:7525–32
114. Rossmann, M. G., Argos, P. 1976. *J. Mol. Biol.* 105:75–95
115. Rossmann, M. G., Argos, P. 1977. *J. Mol. Biol.* 109:99–129
116. Rossmann, M. G., Moras, D., Olsen, K. W. 1974. *Nature* 250:194–99
117. Salemme, F. R., Miller, M. D., Jordan, S. R. 1977. *Proc. Natl. Acad. Sci. USA* 74:2820–24
118. Sawyer, L., Shotton, D. M., Campbell, J. W., Wendell, P. L., Muirhead, H., Watson, H. C., et al. 1978. *J. Mol. Biol.* 118:137–208
119. Schulz, G. E., Elzinga, M., Marx, F., Schrimer, R. M. 1974. *Nature* 250:120–23
120. Shotton, D. M., Watson, H. C. 1970. *Nature* 225:811–16
121. Silverton, E. W., Navia, M. A., Davies, D. R. 1977. *Proc. Natl. Acad. Sci. USA* 74:5140–44
122. Sippl, M. J. 1982. *J. Mol. Biol.* 156:359–88
123. Steinman, H. M., Naik, V. R., Abernathy, J. L., Hill, R. L. 1974. *J. Biol. Chem.* 249:7326–38
124. Strasssburger, W., Wollmer, A., Pitts, J. E., Glover, I. D., Tickle, I. J., Blundell, T. L., et al. 1983. *FEBS Lett.* 157:219–23
125. Subramanian, E., Swan, I. D. A., Lieu, M., Davies, D. R., Jenkins, J. A., Tickle, I. J., Blundell, T. L. 1977. *Proc. Natl. Acad. Sci. USA* 74:556–59
126. Swanson, R., Trus, B. L., Mandel, N., Mandel, G., Kallai, O. B., Dickerson, R. E. 1977. *J. Biol. Chem.* 252:759–75
127. Sygusch, J., Madsen, N. B., Kasvinsky, P. J., Fletterick, R. J. 1977. *Proc. Natl. Acad. Sci. USA* 74:4757–61
128. Szebenyi, D. M. E., Obendorf, S. K., Moffat, K. 1981. *Nature* 294:327–32
129. Takano, T., Dickerson, R. E., Schichman, S. A., Meyer, T. E. 1979. *J. Mol. Biol.* 133:185–88
130. Tang, J., James, M. N. G., Hsu, I. N., Jenkins, J. A., Blundell, T. L. 1978. *Nature* 271:618–21
131. Thomson, E. O. P. In *Evolution of Protein Structure and Function.* New York: Academic
132. Timkovich, R., Dickerson, R. E. 1976. *J. Biol. Chem.* 251:4033–46
133. Tonegawa, S. 1983. *Nature* 302:575–81
134. Vainstein, B. K., Melik-Adamyan, W. R., Barynin, V. V., Vagin, A. A., Grebenko, A. I. 1981. *Nature* 293:411–14
135. Van Beynum, G. M. A., De Graat, J. M., Castel, A., Kraal, B., Bosch, L. 1977. *Eur. J. Biochem.* 72:63–78
136. Watenpaugh, K. D., Sieker, L. C., Jensen, L. H. 1973. *Proc. Natl. Acad. Sci. USA* 70:3857–60
137. Weber, I. T., Johnson, L. N., Wilson, K. S., Yeates, D. G. R., Wild, D. L., Jenkins, J. A. 1978. *Nature* 274:433–37
137a. Weber, P. C., Salemme, F. R., Mathews, F. S., Bethge, P. H. 1981. *J. Biol. Chem.* 256:7702–4
138. Wilson, A. C., Carlson, S. S., White, T. J. 1977. *Ann. Rev. Biochem.* 46:573–639
139. Wistow, G., Slingsby, C., Blundell, T., Driessen, H., de Jong, W., Bloemendal, H. 1981. *FEBS Lett.* 133:9–16
140. Wistow, G., Turnell, B., Summers, L., Slingsby, C., Moss, D., Miller, L., et al. 1983. *J. Mol. Biol.* 170:175–202
141. Wood, S. P., Blundell, T. L., Wollmer, A., Lazarus, N. R., Neville, R. W. J. 1975. *Eur. J. Biochem.* 55:532–42

Ann. Rev. Biophys. Bioeng. 1984. 13: 493–521

DETAILED ANALYSIS OF PROTEIN STRUCTURE AND FUNCTION BY NMR SPECTROSCOPY: Survey of Resonance Assignments

John L. Markley and Eldon L. Ulrich

Department of Chemistry, Purdue University, West Lafayette, Indiana 47907

PERSPECTIVES AND SUMMARY

The first reported nuclear magnetic resonance (NMR) study of a protein was the publication in 1957 of a 40 MHz ^{1}H NMR spectrum of bovine pancreatic ribonuclease A (497), which revealed four distinct regions of overlapping peaks. These features were interpreted later the same year (267) in terms of the spectra of the constituent amino acids of the protein molecule. It was soon recognized that NMR spectroscopy offered the potential for providing information about the structural environment, chemical properties, and motional characteristics of defined atoms and functional groups in protein molecules. The logical prerequisites for obtaining this information would be the resolution of spectral signals from individual groups and their assignment to particular sites in a defined amino acid residue or prosthetic group of the protein. That this could be accomplished, at least in a limited sense, was demonstrated by the detection of hyperfine-shifted ^{1}H resonances from individual groups in oxidized cytochrome *c* (309), ring-current shifted ^{1}H resonances in spectra of lysozyme and cytochrome *c* (391), and individual histidine C_ε–H groups in ribonuclease A (69) along with other proteins (395). The pH titration shifts of the histidine peaks provided pK'_a values for known individual side chains in the molecule (review 377). Detection of site-specific chemical shifts demonstrated that NMR could be used to follow protein conformational

0084–6589/84/0615–0493$02.00

changes including denaturation-renaturation processes (527); and the sharpening of resonances on denaturation (178) indicated that NMR would be useful for quantitating mobilities of protein constituents.

The early NMR experiments considered only chemical shifts, peak areas, and peak widths. Measurable parameters have been expanded to include relaxation times, nuclear Overhauser enhancements (NOEs), NOE connectivities, coupling constants, and coupling connectivities. Nuclei in proteins that are accessible to NMR analysis have been extended from ^{1}H to ^{31}P (227) (phosphoproteins), ^{13}C (9, 340), ^{15}N (211), ^{2}H (441, 599) and ^{3}H (596) (both with enrichment), plus several metal nuclei in metalloproteins. The development of techniques for the assignment of individual resonances to specific protein groups has progressed in parallel with the ability to detect and resolve them. Milestones along this path were the development of superconducting solenoids for high-field NMR (178), which now go as high as 600 MHz for ^{1}H NMR (63), selective isotopic substitution strategies allowing the resolution of individual resonances used first for ^{2}H-labeling (385) and later for ^{13}C-, ^{15}N-, or ^{3}H-enrichment, application of pulse-Fourier transform NMR in one (9) and two dimensions (421), and spectral-simplification and resolution-enhancement methods discussed below. Incorporation of ^{19}F-labeled amino acids into a protein (538) and insertion of $^{113}Cd^{3+}$ into a protein metal binding site (28) have led to numerous studies involving substitutions of these atoms that are not normally found in biomolecules. New approaches for the insertion of non-perturbing NMR probes appear regularly. For example, the rare ^{57}Fe isotope has been substituted recently into the heme of myoglobin, and a ^{13}C-^{57}Fe coupling constant has been observed on binding ^{13}CO (338).

The current literature reveals a number of trends. (*a*) Two-dimensional NMR methods for extending assignments are now yielding nearly complete assignments in ^{1}H NMR spectra of small proteins (molecular weight under 7,000). (*b*) NMR studies of families of closely related proteins are allowing assignments to be made in larger proteins and provide the potential for comprehensive investigations of the structural and dynamical consequences of sequence variations. (*c*) Multinuclear NMR approaches are generating information pertinent to mechanisms of enzyme catalysis and inhibition. (*d*) NMR results have given valuable insights into the origin of cooperative (positive and negative) effects in proteins. (*e*) Recent detection of resonances from "exotic" metal nuclei such as ^{25}Mg (179), ^{43}Ca (15, 180), and ^{6}Li or ^{7}Li (468) bound to proteins (along with ^{113}Cd mentioned above) is opening up new possibilities in the study of the role of metal ions in proteins. (*f*) NMR is being used to investigate very large protein-containing assemblies such as protein aggregates, multienzyme complexes, ribosomes, and virus particles. Highly mobile groups in these complexes give rise to sharp high-resolution NMR signals; and with suitable enrichment, solid state NMR techniques can be applied to the study of their more-rigid regions. (*g*) Considerable success has been achieved in NMR investigations of interactions between proteins and other classes of biomolecules: for example, nucleic acids, lipids, and carbohydrates.

The repertoire of techniques and number of applications has expanded to the point where a review of this size can touch on only a few highlights. Our literature search identified over 200 kinds of proteins, including 75 enzymes, whose structures have been investigated by NMR spectroscopy (over 450 different proteins have been studied if species variants are counted). We include here a catalog of recent reviews, a brief survey of current methodology of protein NMR spectroscopy in solution, and a discussion of proteins for which extensive NMR assignments have been made. The major classes of proteins that have been investigated in detail are presented in Table 1. We have excluded from this review studies of peptides (although the distinction we make between proteins and peptides is somewhat arbitrary), nonfunctional fragments or constituents of proteins, and protein model compounds. The field has reached the stage where it would be valuable to institute a data bank for the deposition of protein NMR parameters for individual, assigned resonances obtained under defined conditions: included should be chemical shifts, coupling constants, NOE and relaxation values along with the dependence of these parameters on variables such as pH, temperature, and, for redox proteins, oxidation state.

OTHER REVIEWS

In focusing this review on structural studies of the proteins themselves in solution, we only cite recent reviews on the related topics of the use of NMR to determine the kinetics of chemically exchanging systems involving proteins (465) or to follow enzyme-catalyzed isotopic exchange reactions (123). Solid-state NMR investigations of proteins have been reviewed recently (192, 439, 473).

The general topic of protein NMR spectroscopy has been covered in several older reviews (92, 114, 349, 354, 462, 463, 463a, 486, 602, 607). Recent monographs that treat NMR in molecular biology (270) or NMR studies of the conformations of biomolecules (204) contain sections on proteins. Wüthrich's book on peptide and protein NMR (601), although dated, still presents a useful compendium of chemical shifts, pH titration shifts, and coupling constants of ^{1}H, ^{13}C, and ^{15}N nuclei in peptides. Two series of volumes (44, 121) regularly have chapters on NMR applications to proteins. Recent edited volumes (64, 154, 440, 518) also contain relevant chapters. NMR spectroscopy of specific kinds of nuclei in proteins is

Table 1 Survey of proteins studied by NMR spectroscopy[a]

Proteins[b]	Nuclei studied	References[c]
A. Proteinase inhibitors and related proteins		
1. Bovine pancreatic trypsin inhibitor (PTI) (Kunitz) family		
PTI	^{1}H, ^{13}C	567, 608, *605*, 571, 572, 568, 420, 475, 611, 471, 152, 613, 474, 617, 415, 79, 476, 454, 213, 78, 459, 79, 522
colostrum inhibitor	^{1}H	574
toxin E	^{1}H	35
toxin K	^{1}H	288
isoinhibitor K	^{1}H	575
2. Pancreatic secretory trypsin inhibitor (PSTI) (Kazal) family		
PSTI [3]	^{1}H	139, 138, 137
seminal plasma inhibitor	^{1}H	375, 532, 531, 530
ovomucoid [7]	^{1}H, ^{13}C	387, 132, 584, 431
3. *Streptomyces* subtilisin inhibitor (SSI) family		
SSI	^{1}H, ^{13}C, ^{15}N	277, 5, 3, 4, 187, 186
4. Soybean trypsin inhibitor (STI) (Kunitz) family		
STI	^{1}H, ^{13}C	39, 241, 382
B. Toxins		*88, 89, 252*
1. Bee venom toxins		
apamin	^{1}H, ^{13}C	582, *89*, 90, 433, 87
melittin	^{1}H	77, 81, 80, 342, 341, 75, 529
2. *Laticauda semifasciata* toxins		
erabutoxins [3]	^{1}H	402, 251, 246, 254, 542, 253, 250
laticauda semifasciata III	^{1}H	248, 247
3. *Cobra* toxins		
Naja toxins [11]	^{1}H	237, 402, 218, 170, *89*, 33, 524, 169, 168, 190, 34, 32
4. *Bungarus* toxins		
α-bungarotoxin	^{1}H	402, 169
5. Lytic factor		
Haemachatus	^{1}H	524
C. Enzymes		
1. Oxidoreductases		
[1.1.1.1] alcohol dehydrogenase [2]	^{113}Cd	53
[1.1.3.4] glucose oxidase	^{31}P	260
[1.5.1.3] tetrahydrofolate dehydrogenase (dihydrofolate reductase) [6]	^{1}H, ^{13}C, ^{19}F	*478*, 364, 207, 175, 206, 177, 176, 388, 457, 458, 301, 48, 119

Table 1—*continued*

Proteins[b]	Nuclei studied	References[c]
[1.11.1.5] cytochrome *c* peroxidase	^1H	496, 495, 494, 328, 493, 325
[1.11.1.7] peroxidase [4]	^1H	329, 327, 328, 332, 325, 330, 331, 326
[1.15.1.1] superoxide dismutase [5]	^1H	41, 219, 528, 102, 359, 101
2. Transferases		
[2.4.1.1] phosphorylase, glycogen [3]	^{31}P	232
[2.7.3.2] creatine kinase	^1H	*298*, 487
[2.7.3.9] PEP-phosphotransferase [2]	^1H	500, 484
[2.7.4.3] adenylate kinase [3]	^1H	280, 393
[2.7.5.1] phosphoglucomutase	^1H, ^6Li, ^7Li, ^{31}P, ^{113}Cd	470, 469
3. Hydrolases		
[3.1.1.4] phospholipase a_2 [5]	^1H, ^{13}C, ^{43}Ca	163, 16, 164, 264, 263
[3.1.3.1] alkaline phosphatase	^{13}C, ^{31}P, ^{113}Cd	*127*, 200, 445, 444, 442, 54, 116, 240, 28, 50
[3.1.27.3] ribonuclease T_1 [2]	^1H, ^{13}C	249, 22, 21, 302, 189
[3.1.27.5] ribonuclease (A, B, S, S′) [7]	^1H, ^{13}C, ^{19}F	316, 317, 318, 17, 46, 578, 122, 492, 255, 352, 259, 579, 428, 51, 515, 162, 18, 67, 516, 76, 514, 70, 449, 450, 451, 378, 377, 104, 398, 394
[3.1.31.1] micrococcal nuclease (staphylococcal nuclease) [2]	^1H	269, 172, 268
[3.2.1.17] lysozyme [3]	^1H, ^2H, ^{13}C	*147*, 503, 235, 581, 468, 312, 142, 461, 599, 351, 515, 144, 11, 453, 93, 438, 435, 8
[3.4.21.1] chymotrypsin	^1H, ^{13}C, ^{19}F	415, 460, 381, 374, 481, 482, 480
[3.4.21.4] pancreatic trypsin [2]	^1H, ^{13}C	415, 455, 475, 460, 459, 384, 481
[3.4.21.5] thrombin (kringle)	^1H	546, 231, 135
[4.21.12] α-lytic proteinase	^1H, ^{13}C, ^{15}N	585, 37, 460, 38, 481, 242
[3.4.21.14] subtilisin	^1H	276, 481
[3.4.22.2] papain	^1H	357, 272
[3.5.2.6] β-lactamase	^1H	40
4. Lyases		
[4.2.1.1] carbonic anhydrase [6]	^1H, ^{113}Cd	286, 96, 534, 28, 95
[4.2.1.14] D-serine dehydratase	^{31}P	501
5. Isomerases		
[5.3.1.1] triosephosphate isomerase [2]	^1H, ^{13}C, ^{31}P	274, 82
6. Ligases		
[6.2.1.5] succinyl-CoA synthetase	^{31}P	564, 563

Table 1—*continued*

Proteins[b]	Nuclei studied	References[c]
D. Cofactor proteins		
pancreatic colipase	^{1}H	99, 587, 97, 98, 588
α-lactalbumin [5]	^{1}H	43, 541
E. Heme proteins		*339*, *294*
1. Cytochromes		*619*, *404*
a. *c*-type		
cytochrome *c* [19]	^{1}H, ^{2}H, ^{13}C	506, 58, 591, 600, 85, 291, 59, 408, 410, 409, 411, 499, 129, 292, 203, 293, 413, 438, 148, 155, 143, 289, 435, 392, 434, 526, 466
cytochrome c_{551} [4]	^{1}H	507, 291, 293, 129, 406, 295
cytochrome c_{552} [2]	^{1}H	233, 291, 129, 296
cytochrome c_{553} [5]	^{1}H	551
cytochrome c_{556} [2]	^{1}H	405
cytochrome c_{557}	^{1}H	290, 289
cytochrome cd_1	^{1}H	543
cytochrome c' [3]	^{1}H	405, 167, 166
cytochrome c_2	^{1}H	521
b. other cytochromes		
cytochrome P-450	^{3}H	596
cytochrome b_5	^{1}H	324
cytochrome b_{562}	^{1}H	620
2. Myoglobins and hemoglobins		
myoglobin [9]	^{1}H, ^{2}H, ^{13}C, ^{19}F	*210*, 197, 297, 329, 328, 66, 303, 325, 488, 133, 333, 65, 321, 322, 432, 326, 243, 60, 61, 323, 590, 595, 589, 438, 389, 519
leghemoglobin [2]	^{1}H	368, 307, 19, 66, 333, 545, 273
Chironomus hemoglobin [3]	^{1}H	*198*, 320, 337, 335
human hemoglobin [19]	^{1}H, ^{13}C	*229*, 256, 401, 489, 540, 491, 416, 502, 539, 490, 369, 334, 555, 36, 436, 191, 299
F. Copper proteins		*552*
plastocyanin [10]	^{1}H, ^{13}C, ^{113}Cd	306, 171, 303, 128, 184, 386
azurin [2]	^{1}H, ^{13}C, ^{113}Cd	171, 400, 1, 49, 221, 550, 549, 220
stellacyanin	^{1}H, ^{113}Cd	171

Table 1—*continued*

Proteins[b]	Nuclei studied	References[c]
G. Non-heme Iron Proteins		*91*
1. Ferredoxins		
2Fe·2S ferredoxin [13]	^{1}H, ^{13}C	419a, 109, 107, 110, 112, 111, 113
4Fe·4S ferredoxin [9]	^{1}H, ^{13}C	419a, 447, 412, 448, 413
2. Respiratory pigments		
hemerythrin	^{1}H	624
H. Nucleic acid binding proteins		
gene 5 protein [2]	^{1}H, ^{19}F	430, 13, 12, 194, 193, 126, 125
lac repressor	^{1}H, ^{19}F	*285*, 498, 31, 271, 57, 472, 30, 84
cro protein	^{1}H, ^{19}F	29, 315, 304, 257
elongation factor Tu [2]	^{1}H	425, 424
adenosine cyclic 3′,5′-phosphate receptor protein (CRP or CAP)	^{1}H	118
ubiquitin	^{1}H	100
high mobility group chromosomal protein	^{1}H	100
I. Metal binding proteins		
calmodulin [9]	^{1}H, ^{113}Cd	13a, 13b, 245, 244, 311, 14a, 15, 224, 223, 222, 181, 504
parvalbumin [7]	^{1}H, ^{13}C, ^{113}Cd	347a, 15, 347, 346, 427, 426
metallothionein [12]	^{113}Cd	*27*, *446*, 73, 443
transferrin [2]	^{1}H	598
S-100 protein [2]	^{1}H	372, 305
intestinal calcium binding protein	^{1}H	511
J. Hydrophobic and membrane-associated proteins		
crambin	^{1}H	343, 136
thionin	^{1}H	344
myelin basic protein [5]	^{1}H	397a, 397, 396, 239, 361, 360
lipoprotein from *Escherichia coli*	^{13}C, ^{19}F	348
acyl carrier protein	^{1}H	389a, 191a
phosphocarrier protein, HPR [5]	^{1}H, ^{31}P	279, 278, 483, 500, 195
K. Rhodopsins		
bacteriorhodopsin	^{2}H, ^{13}C	212, 623
rhodopsin	^{13}C	517

Table 1—*continued*

Proteins[b]	Nuclei studied	References[c]
L. Hormone binding proteins and protein hormones		
neurophysin [2]	^{1}H	560, 366, 52, 365
lutropin [2]	^{1}H	370, 74
insulin [3]	^{1}H, ^{113}Cd	533, 68, 594
choriogonadotropin	^{1}H	183
serum gonadotropin	^{1}H	74
epidermal growth factor	^{1}H	139a
M. Flavoproteins		*414*
flavodoxin [8]	^{1}H, ^{13}C	156, 403, 554, 553, 261
N. Plasma proteins		
immunoglobulins [16]	^{1}H, ^{31}P	*150*, 512, 199, 201, 258, 20, 513, 24, 25, 577, 202
plasma albumin	^{13}C	115
O. Glycoproteins		
antifreeze glycoproteins [4]	^{13}C	45
ovalbumin	^{31}P	562
P. Muscle proteins		
myosin	^{1}H, ^{31}P	216, 308
tropomyosin [3]	^{1}H, ^{31}P	160, 159, 157, 158, 371
troponin-C [3]	^{1}H, ^{19}F, ^{31}P, ^{25}Mg, ^{43}Ca	541a, 15, 224, 223, 14, 222, 523, 182, 157, 47, 505

[a] Texts or abstracts of over 1700 publications on NMR of proteins were screened in constructing this table. Unfortunately, each paper could not be read thoroughly; and unintentional omissions are bound to have occurred. References were included if they met two criteria: (*i*) A second stage resonance assignment was made (or corrected) or detailed studies of a protein were carried out using previous assigments. (*ii*) The assigned resonance in question arises from a natural part of the protein; prosthetic groups were included, but inhibitors or other small ligands were not.

[b] Enzymes are identified by numbers (in parentheses) and names recommended by the Nomenclature Committee of the International Union of Biochemistry (1978). (Nomenclature Committee of the International Union of Biochemistry, 1979. *Enzyme Nomenclature 1978*. New York: Academic. 606 pp.) Commonly used names are given in parentheses. Numbers in square brackets indicate the number of species variants that have been studied by NMR; references to all of these may not be included.

[c] References are listed chronologically beginning with the most recent; references to review articles are in italics.

covered in a number of reviews: nuclei other than ^{1}H (120), ^{13}C (6, 7, 10, 161, 238, 363, 486), ^{31}P (561), ^{15}N (282), ^{19}F (196, 535, 537), ^{113}Cd (26, 151), ^{17}O and ^{18}O labeling (547, 548), ^{43}Ca in proteins that bind calcium ions (180), the study of various metal nuclei in metalloproteins (345, 556, 558, 559), and ^{3}H (106). General reviews of protein dynamics (390), hydrogen exchange (597), and protein folding (300) contain discussions of contributions of NMR spectroscopy. A plethora of articles have reviewed the important contributions of NMR to detecting and quantitating internal mobility in proteins by NMR relaxation (265, 362) or by a combination of techniques (592, 593) with the primary subjects being PTI, the bovine pancreatic trypsin inhibitor (Kunitz) (566, 573, 606, 612, 614–616), and lysozyme (141, 147).

Recent NMR texts (e.g. 214, 510) discuss modern pulse-Fourier transform techniques or instrumentation (188). Several methods of relevance to protein NMR have been treated in depth in separate monographs or reviews: two-dimensional Fourier transform NMR (42, 185), multiple quantum NMR (165), and the nuclear Overhauser effect (62, 429, 485). NMR methods as applied specifically to proteins have been reviewed in a number of cases: two-dimensional Fourier transform NMR (417, 422, 610) ring current shifts and their calculation (452), NMR in aqueous solutions (464), photochemically induced nuclear polarization (234, 283, 284), uses of chromium and cobalt nucleotide complexes (557), carbamate reactions with proteins (209), and the sequential assignment of individual ^{1}H resonances in proteins (609).

Detailed reviews have appeared on NMR studies of particular classes of proteins: collagen and elastin (544), membrane proteins (437), organization and dynamics of plasma lipoproteins (287)— including low-density lipoproteins (106, 217) and high-density lipoproteins (71), flavins and flavoproteins (414), active centers of iron-sulfur proteins (91), electron carrier proteins from sulfate reducing bacteria (622), blue-copper proteins (552), neurotoxins (88, 89, 252), histones (130), iron-containing proteins (173), and NMR studies of metal nuclei in metalloenzymes (345, 558). NMR studies of heme proteins have been reviewed extensively (225, 228, 294, 336, 339, 603) along with several subtopics: model compounds as aids in interpreting NMR spectra of heme proteins (319), ^{1}H NMR of hemoglobins (198, 226, 229), NMR studies of myoglobins (210), and cytochromes (404, 407, 619, 621). NMR investigations of calcium-binding proteins (310, 355, 565) and in particular parvalbumin (103) have been summarized. Reviews on individual proteins include PTI (608), its internal motions (567), and a comparison of its solution and crystal structures (604, 605), the *lac* repressor (367) and its interaction with *lac* operator DNA (285), gene 5 protein and its interaction with oligonucleotides (125), metallothionein (27, 446, 583), and an antibody binding site (150, 153).

General reviews on NMR of enzymes have treated the geometry of enzyme-bound substrates and analogues (208, 399) and the active sites of enzyme complexes (124) or phosphoryl transfer enzymes (558). The serine proteinases have been the subject of three reviews (379, 383, 525). Reviews that focus on individual enzymes and describe NMR studies include those on alkaline phosphatase (127), dihydrofolate reductase (174, 477–479), thymidylate synthase (86), creatine kinase (298), succinyl-coenzyme A synthetase (564), and lysozyme (146) and its hydrogen exchange properties (147).

CURRENT NMR METHODOLOGY

One-Dimensional NMR

The methods used for one-dimensional NMR spectroscopy of proteins are now fairly well standardized. When ^2H_2O is used as the solvent, pulse-Fourier transform spectroscopy is generally the method of choice. Selective enhancement of sharp lines over broader lines (resolution enhancement) can be achieved by use of a Carr-Purcell sequence (94) or more commonly by apodization (point-by-point multiplication of the free induction decay) with a sine bell (sinusoidal window function) (140) or similar function. Complex spectra may be simplified by selecting subspectra on the basis of their coupling multiplicity, by use of a J-modulated spin echo pulse sequence (94) or by the related APT (attached proton test) pulse sequence for heteronuclear NMR (113), by use of a summed spin-echo spectrum (55), or by multiple quantum coherence (236). Interproton distance information can be obtained from nuclear Overhauser experiments (528). Truncated-driven NOE measurements (152) are usually superior to steady-state NOE measurements because the method provides a way of limiting spin-diffusion effects (2, 281, 536). Factors influencing the time development of NOEs in proteins have been investigated (149). Solvent suppression for ^1H NMR in solutions containing ^1H_2O is commonly achieved by single-frequency irradiation to saturate the water signal, which, however, leads to transfer of saturation to signals from rapidly exchanging –OH and –NH groups. In cases where the rapidly exchangeable protons are the objects of study, their resonances can be detected through use of the pulse sequences developed by Redfield and co-workers (467) or Plateau & Gueron (456). Rapid-scan correlation spectroscopy (134) also may be used conveniently in situations where there are large solvent or buffer peaks; improvements have been suggested (23) for the processing of correlation NMR data sets. The standard relaxation methods used to determine correlation times (see the reviews above) have been augmented recently by a technique involving the measurement of the rotating frame spin-lattice relaxation in the presence of an off-resonance radio-frequency field (262), which extends the measurements to longer correlation times. It has been pointed out that interatomic distances used for dipolar relaxation calculations should include vibrational corrections and that such corrections can influence calculated correlation times by as much as a factor of 2 (145). The importance of spin diffusion (cross relaxation) on NMR relaxation in paramagnetic proteins has been delineated (520).

Newer multipulse strategies for broadband decoupling incorporating cyclic sequences of a train of composite spin inversion pulses (580), such as the "MLEV" (508) or "WALTZ" (509) supercycles, require less power than older methods. This is important for heteronuclear NMR studies of proteins at high field strengths where sample heating by the proton decoupler can present serious problems.

A novel NMR variant of solvent-perturbation spectroscopy is afforded by photochemical dynamic polarization (CIDNP) experiments using photo-excited external dyes to polarize transitions from the aromatic side chains of histidine, tyrosine, and tryptophan (for a review see 284) or from other groups by means of covalently attached dye molecules (353). The technique has been applied to a large number of proteins and has been used to correct and extend assignments as well as to probe the solvent accessibility of protein groups under various conditions.

Two-Dimensional NMR

Many of the two-dimensional NMR techniques originally developed with organic molecules have been applied to small proteins. The first experiments with proteins utilized two-dimensional J-resolved ^{1}H spectroscopy (421) along with cross-sections and projections (418). Heteronuclear (^{1}H, ^{13}C) two-dimensional J-spectroscopy has been demonstrated with a protein (113), but since the ^{1}H–^{13}C coupling constants are similar for nearly all residues, its utility is limited to detection of abnormal coupling constants such as for the histidine ring C–H groups. Heteronuclear (^{1}H, ^{113}Cd) 2D J-spectroscopy has been carried out with ^{113}Cd, Zn metallothionein (26) with the advantage that a projection could be used to eliminate ^{113}Cd–^{113}Cd coupling.

The various kinds of two-dimensional spectroscopy based on chemical shift connectivity can be broken down into two classes: J-connectivity and NOE- or exchange-connectivity experiments. Homonuclear (^{1}H, ^{1}H) J-connectivity methods include two-dimensional spin-echo correlated spectroscopy (SECSY) (423), foldover-corrected correlated spectroscopy (FOCSY) (419), and two-dimensional correlated spectroscopy (COSY)

(571). COSY is the most versatile method where data accumulation is not limited by available digitization. Three-bond couplings normally are observed in COSY, but longer-range couplings can be studied. The four-bond connectivity between the C_δ–H and C_ε–H of histidine rings in proteins has been reported (303). Phase-sensitive COSY, although less sensitive, appears to afford superior resolution of peaks near the diagonal and may be advantageous for measurement of ^1H–^1H spin-spin coupling constants (375). Two-dimensional double-quantum ^1H NMR spectroscopy has been applied to a protein (576). This method does not give rise to peaks on the diagonal and therefore is ideally suited for resolving connectivities between resonances with very similar chemical shifts. It is also useful for analyzing remote connectivities. Another approach to determining remote connectivities is two-dimensional relayed coherence transfer spectroscopy, which has been used to assign spin systems in proteins (303a, 566a).

Of the various heteronuclear J-connectivity experiments, only (^1H, ^{13}C) J-connectivity has been demonstrated thus far with a protein. The earlier applications used ^{13}C-enrichment (108, 306), but the experiment can be performed at natural abundance provided that sufficient protein is available (387). Two-dimensional (^{13}C, ^{13}C) J-connectivity (387) and (^{15}N, ^1H) and (^{15}N, ^{13}C) (205) J-connectivity experiments have been carried out with amino acids or peptides. Experimental evidence showing a large dispersion in (^1H, ^{15}N) (350) and (^{13}C, ^{15}N) (277) connectivities of peptide units in proteins (obtained by one-dimensional NMR experiments) indicates that these heteronuclear two-dimensional NMR approaches should be of great value if they become feasible for proteins.

Two-dimensional NOE spectroscopy has been shown only for proton-proton interactions (NOESY) in proteins (313), where it is of utility for elucidating cross-relaxation pathways or extending NMR assignments, particularly in ^1H_2O solution (314); the solvent resonance can be suppressed by saturation methods (586). A comparison has been made of selective NOE measurements made by one-dimensional and two-dimensional methods (56). Spin-diffusion problems appear to limit present methods for two-dimensional NOE spectroscopy to proteins with molecular weights below 20,000.

RESONANCE ASSIGNMENTS IN PROTEINS

Assignment Strategies

A useful methodological distinction (92, 146) has been made between first stage assignments, i.e. determination of the kind of amino acid giving rise to the resonance, and second stage assignments, i.e. identification of the particular residue in the peptide sequence giving rise to the resonance. At present, an independently determined and accurate amino acid sequence is required before NMR assignments can be contemplated. First stage assignments usually can be made on the basis of characteristic chemical shifts (and occasionally their pH dependence) and coupling pattern (spin system). Second stage assignments generally require additional information if more than one residue of a given type is present in the protein. Numerous strategies can be followed in making assignments. Isotopic substitution offers a reliable, but somewhat tedious method. Specific labeling can be achieved by isotopic exchange (380, 394), chemical modification and reversal (259, 275, 499), peptide synthesis (105), enzymatic semisynthesis (39, 475), or biosynthesis (83, 376). A [^{13}C, ^{15}N] double-labeling procedure for peptide groups where assignments are made on the basis of dipeptide patterns appears very attractive (277). Selective ^{17}O labeling of an enzymic phosphate has been used to assign a ^{31}P resonance in the NMR spectrum of a phosphoprotein (470). With the advent of NMR methods for extending assignments, it is convenient to distinguish primary resonance assignments, that is assignments made directly to particular atoms or groups in the protein (the second stage assignment above), from secondary assignments, assignments made by extensions of these primary assignments by NMR methods (generally NOE or J-connectivity experiments) (618).

At present nearly all of the ^{1}H resonances in the spectrum of a small protein that gives sharp NMR peaks can be resolved and assigned. The paradigm for complete ^{1}H NMR spectral analysis of a protein is the bovine pancreatic trypsin inhibitor (6,500 mol wt), which has been studied for several years and served as the model for the development of homonuclear (^{1}H) two-dimensional NMR techniques for the identification of spin systems of amino acid side chains (420) and peptide sequential extensions of NMR assignments (568).

Homologous or Variant Proteins

Protein variants provide a reliable means of making second stage (primary) assignments in larger proteins that still yield resolvable peaks, as shown first with staphylococcal nuclease (16,800 mol wt) (268). Modern genetic methods offer the means of extending this assignment approach to virtually any residue (271) as well as providing ways of producing large quantities of interesting proteins. From the standpoint of elucidating the function of single residues in proteins, a comparison of proteins with single residue replacements is ideal (229, 387). Several amino acid replacements can be acceptable, or even desirable in certain cases, if the object is to use the spectral differences to assign resonances. More distantly related proteins have been compared in order to determine common features such as the pK_a' value of the active site residues in serine proteinases (383) or the pattern of electron delocalization or ligand stereochemistry in cytochromes (292, 405, 506, 507, 551). Several interesting families of proteins or protein domains are beginning to be studied in detail: proteins homologous to the Kazal secretory proteinase inhibitor (138, 387, 431, 530–532); "kringle" fragments of plasminogen (135, 231, 546); thionins and crambin (344); myoglobins from various species (60, 61); myelin basic proteins (397a); ferredoxins (107, 109–111, 419a); and PTI and related proteinase inhibitors and toxins (35, 288, 571, 575).

INTERPRETATION OF NMR PARAMETERS

The availability of large numbers of assigned peaks in proteins whose structures have been determined by X-ray crystallography should shortly lead to advances in our understanding of the origins of chemical shifts and coupling constants in proteins. Preliminary attempts have been made to use protein NMR results to determine secondary and tertiary structural features of proteins in solution. One approach has been to use distance information generated by two-dimensional NOE experiments as input data (72) for the distance geometry algorithm of Crippen and co-workers (131, 215). Another method has been to rely on short-range distances determined by NOE data along with amide proton exchange rates and ϕ-angle-dependent $^3J_{^1HN\text{–}C_\alpha{}^1H}$ coupling constants (582). Complete determination of the solution structure of a protein by NMR means still is an unrealized goal.

A sound theoretical basis for NMR chemical shifts in proteins is still lacking. In spite of the fact that progress has been made in evaluating ring-current contributions to chemical shifts (452), a general means of predicting diamagnetic anisotropy or electric field shifts of peptide groups is unavailable (117). A severe complication in calculating NMR parameters arises from the mobility of protein groups (266). Picosecond molecular dynamics simulations will not prove helpful here because NMR chemical shifts and coupling constants are averaged over a much longer time scale (230). Environmental shifts play a proportionately smaller role in ^{13}C and ^{15}N chemical shifts than in ^1H shifts. Large ^{13}C chemical shift differences can be interpreted in terms of differences in bond order (39); and ^{13}C chemical shifts can be more reliable than ^1H NMR chemical shifts for assignment purposes (306). It has been postulated that the chemical shift of protein-bound ^{113}Cd can be used to distinguish types of metal ligands, but recent results with blue-copper proteins casts some doubt on this (171). Several pK_a' values of surface groups determined experimentally by NMR methods have been evaluated rather successfully by a modified form of the Tanford-Kirkwood electrostatic theory (373).

Progress has been made in incorporating the results of molecular dynamics simulations of protein motions into an analysis of NMR relaxation rates. The molecular dynamics trajectories carried out to date have been limited by the speed and cost of computations to about 100 ps. Nevertheless, the rapid motions modeled by these trajectories are expected to have a measurable effect on relaxation of protonated (358) and nonprotonated (356) carbons.

The sensitivity of NMR spectral parameters to minor changes in protein sequence or structure or the presence of contaminants presents special demands on the purity of protein samples. Apart from this difficulty, multiple (interconvertible) forms of proteins have been reported on several occasions on the basis of NMR evidence. Three of the better-understood examples are (*a*) a pH-dependent conformational equilibrium in azurin (1, 550), which is dependent on the protonation state of His-35 (49) and appears to explain anomalies in the kinetics of electron transfer between azurin and cytochrome c_{551}; (*b*) the detection of doubling of ^{13}C resonances from two tryptophan residues of *Streptococcus faccium* dihydrofolate reductase (364), which indicates the presence of three interconvertible forms of the enzyme and may explain the complex kinetics of coenzyme and inhibitor binding observed for the homologous enzymes from other species (478); and (*c*) the observation of "heme disorder" in hemoproteins that stems from the existence of two molecular forms, which differ by a 180° rotation of the heme in its binding pocket (323).

SUMMARY

Techniques are now available for making extensive assignments in NMR spectra of proteins of moderate size (molecular weight 20,000 or less). Such assignments provide the first step for experiments designed to extract the full complement of NMR parameters for each group in a protein. The stage is set for exciting research scenarios in protein chemistry involving, for example, the determination of hydrogen exchange kinetics at all exchangeable positions whose half times are on the order of 100 ms (277) or longer than a few minutes (316, 569, 570); the characterization of intermediates in protein folding pathways (318); measurement of the distribution of internal motions within a protein molecule (573); a detailed description of the biophysical consequences of single amino acid replacements in small proteins (387); elucidation of the mechanisms of conformational transitions in proteins; and multiparametric characterization of the parts of an enzyme that participate in catalytic mechanisms. Small proteins for which extensive ^{1}H NMR assignments have been made include lysozyme, several cytochromes, ferredoxins, myelin basic proteins, PTI and related proteinase inhibitors, proteinase inhibitors from seminal plasma and avian eggs, apamin, and several snake venom neurotoxins. (References are given in Table 1.)

ACKNOWLEDGMENTS

We thank numerous colleagues for supplying reprints and preprints of their work. Preparation of this review has been supported by the National Institutes of Health (GM 09077, RR 01077) and the United States Department of Agriculture Competitive Research Grants Office, Cooperative State Research Service, Science and Education (82-CRCR-1-1045).

Literature Cited

1. Adman, E. T., Canters, G. W., Hill, H. A. O., Kitchen, N. A. 1982. *FEBS Lett.* 143:287–92
2. Akasaka, K. 1983. *J. Magn. Reson.* 51:14–25
3. Akasaka, K., Fujii, S., Hatano, H. 1982. *J. Biochem. Tokyo* 92:591–98
4. Akasaka, K., Fujii, S., Kaptein, R. 1981. *J. Biochem. Tokyo* 89:1945–49
5. Akasaka, K., Hatano, H., Tsuji, T., Kainosho, M. 1982. *Biochim. Biophys. Acta* 704:503–8
6. Allerhand, A. 1978. *Acc. Chem. Res.* 11:469–74
7. Allerhand, A. 1979. *Methods Enzymol.* 61:458–549
8. Allerhand, A., Childers, R. F., Oldfield, E. 1973. *Biochemistry* 12:1335–41
9. Allerhand, A., Cochran, D. W., Doddrell, D. 1970. *Proc. Natl. Acad. Sci. USA* 67:1093–96
10. Allerhand, A., Dill, K., Goux, W. J. 1979. See Ref. 440, pp. 31–50
11. Allerhand, A., Norton, R. S., Childers, R. F. 1977. *J. Biol. Chem.* 252:1786–94
12. Alma, N. C. M., Harmsen, B. J. M., Hilbers, C. W., Van der Marel, G., Van Boom, J. H. 1981. *FEBS Lett.* 135:15–20
13. Alma, N. C. M., Harmsen, B. J. M., Van Boom, J. H., Van der Marel, G., Hilbers, C. W. 1982. *Eur. J. Biochem.* 122:319–26
13a. Andersson, A., Drakenberg, T., Thulin, E., Forsén, S. 1983. *Eur. J. Biochem.* 134:459–65
13b. Andersson, A., Forsén, S., Thulin, E., Vogel, H. J. 1983. *Biochemistry* 22:2309–13
14. Andersson, T., Drakenberg, T., Forsén, S., Thulin, E. 1981. *FEBS Lett.* 125:39–43
14a. Andersson, T., Drakenberg, T., Forsén, S., Thulin, E. 1982. *Eur. J. Biochem.* 126:501–5
15. Andersson, T., Drakenberg, T., Forsén, S., Thulin, E., Swärd, M. 1982. *J. Am. Chem. Soc.* 104:576–80
16. Andersson, T., Drakenberg, T., Forsén, S., Wieloch, T., Lindstrom, M. 1981. *FEBS Lett.* 123:115–17
17. Andini, S., D'Alessio, G., Di Donato, A., Paolillo, L., Piccoli, R., Trivellone, E. 1983. *Biochim. Biophys. Acta* 742:530–38
18. Antonov, I. V., Gurevich, A. Z., Dudkin, S. M., Karpeiskii, M. Ya., Sakharovskii, V. G., Yakovlev, G. I. 1978. *Eur. J. Biochem.* 87:45–54
19. Appleby, C. A., Blumberg, W. E., Bradbury, J. H., Fuchsman, W. H., Peisach, J., Wittenberg, B. A., et al. 1982. See Ref. 228, pp. 435–41
20. Arata, Y., Honzawa, M., Shimizu, A. 1980. *Biochemistry* 19:5130–35
21. Arata, Y., Kimura, S., Matsuo, H., Narita, K. 1979. *Biochemistry* 18:18–24
22. Arata, Y., Kimura, S., Matsuo, H., Narita, K. 1980. *Biochem. Biophys. Res. Commun.* 73:133–40
23. Arata, Y., Ozawa, H., Ogino, T., Fujiwara, S. 1978. *Pure Appl. Chem.* 50:1273–80
24. Arata, Y., Shimizu, A. 1979. *Biochemistry* 18:2513–20
25. Arata, Y., Shimizu, A., Matsuo, H. 1978. *J. Am. Chem. Soc.* 100:3230–32
26. Armitage, I. M., Otvos, J. D. 1982. See Ref. 44, 4:79–144
27. Armitage, I. M., Otvos, J. D., Briggs, R. W., Boulanger, Y. 1982. *Fed. Proc.* 41:2974–80
28. Armitage, I. M., Pajer, R. T., Schoot Uiterkamp, A. J. M., Chlebowski, J. F., Coleman, J. E. 1976. *J. Am. Chem. Soc.* 98:5710–12
29. Arndt, K., Boschelli, F., Cook, J., Takeda, Y., Tecza, E., Lu, P. 1983. *J. Biol. Chem.* 258:4177–83

30. Arndt, K. T., Boschelli, F., Lu, P., Miller, J. H. 1981. *Biochemistry* 20:6109–18
31. Arndt, K., Nick, H., Boschelli, F., Lu, P., Sadler, J. 1982. *J. Mol. Biol.* 161:439–57
32. Arseniev, A. S., Balashova, T. A., Utkin, Y. N., Tsetlin, V. I., Bystrov, V. F., Ivanov, V. T., Ovchinnikov, Yu. A. 1976. *Eur. J. Biochem.* 71:595–606
33. Arseniev, A. S., Pashkov, V. S., Pluzhnikov, K. A., Rochat, H., Bystrov, V. F. 1981. *Eur. J. Biochem.* 118:453–62
34. Arseniev, A. S., Surin, A. M., Utkin, Y. N., Tsetlin, V. I., Bystrov, V. F., Ivanov, V. T., Ovchinnikov, Yu. A. 1978. *Bioorg. Khim.* 4:197–207
35. Arseniev, A. S., Wider, G., Joubert, F. J., Wüthrich, K. 1982. *J. Mol. Biol.* 159:323–51
36. Asakura, T., Adachi, K., Wiley, J. S., Fung, L. W.-M., Ho, C., Kilmartin, J. V., Perutz, M. F. 1976. *J. Mol. Biol.* 104:185–95
37. Bachovchin, W. W., Kaiser, R., Richards, J. H., Roberts, J. D. 1981. *Proc. Natl. Acad. Sci. USA* 78:7323–26
38. Bachovchin, W. W., Roberts, J. D. 1978. *J. Am. Chem. Soc.* 100:8041–47
39. Baillargeon, M. W., Laskowski, M. Jr., Neves, D. E., Porubcan, M. A., Santini, R. E., Markley, J. L. 1980. *Biochemistry* 19:5703–10
40. Baldwin, G. S., Galdes, A., Hill, H. A. O., Smith, B. E., Waley, S. G., Abraham, E. P. 1978. *Biochem. J.* 175:441–47
41. Bannister, J. V., Bannister, W., Cass, A. E. G., Hill, H. A. O., Johansen, J. T. 1980. *Dev. Biochem.* 11A:284–89
42. Bax, A. D. 1982. *Two-Dimensional Nuclear Magnetic Resonance in Liquids.* Amsterdam: Reidel. 200 pp.
43. Berliner, L. J., Kaptein, R. 1981. *Biochemistry* 20:799–807
44. Berliner, L. J., Reuben, J., eds. 1978, 1980, 1981, 1982. *Biological Magnetic Resonance*, Vols. 1, 2, 3, 4. New York: Plenum. 345 pp., 351 pp., 268 pp., 340 pp.
45. Berman, E., Allerhand, A., DeVries, A. L. 1980. *J. Biol. Chem.* 255:4407–10
46. Biringer, R. G., Fink, A. L. 1982. *J. Mol. Biol.* 160:87–116
47. Birnbaum, E. R., Sykes, B. D. 1978. *Biochemistry* 17:4965–71
48. Blakley, R. L., Cocco, L., London, R. E., Walker, T. E., Matwiyoff, N. A. 1978. *Biochemistry* 17:2284–93
49. Blaszak, J. A., Ulrich, E. L., Markley, J. L., McMillin, D. R. 1982. *Biochemistry* 21:6253–58
50. Block, J. L., Sheard, B. 1975. *Biochem. Biophys. Res. Commun.* 66:24–30
51. Blum, A. D., Smallcombe, S. H., Baldwin, R. L. 1978. *J. Mol. Biol.* 118:305–16
52. Blumenstein, M., Hruby, V. J., Viswanatha, V. 1979. *Biochemistry* 18:3552–57
53. Bobsein, B. R., Myers, R. J. 1980. *J. Am. Chem. Soc.* 102:2454–55
54. Bock, J. L., Kowalsky, A. 1978. *Biochim. Biophys. Acta* 526:135–46
55. Bolton, P. H. 1981. *J. Magn. Reson.* 45:418–21
56. Bösch, C., Kumar, A., Baumann, R., Ernst, R. R., Wüthrich, K. 1981. *J. Magn. Reson.* 42:159–63
57. Boschelli, F., Jarema, M. A. C., Lu, P. 1981. *J. Biol. Chem.* 256:11595–99
58. Boswell, A. P., Eley, C. G. S., Moore, G. R., Robinson, M. N., Williams, G., Williams, R. J. P., et al. 1982. *Eur. J. Biochem.* 124:289–94
59. Boswell, A. P., Moore, G. R., Williams, R. J. P., Chien, J. C. W., Dickinson, L. C. 1980. *J. Inorg. Biochem.* 13:347–52
60. Botelho, L. H., Friend, S. H., Matthew, J. B., Lehman, L. D., Hanania, G. I. H., Gurd, F. R. N. 1978. *Biochemistry* 17:5197–5205
61. Botelho, L. H., Gurd, F. R. N. 1978. *Biochemistry* 17:5188–96
62. Bothner-By, A. A. 1979. See Ref. 518, pp. 177–219
63. Bothner-By, A. A., Dadok, J. 1979. See Ref. 440, pp. 169–202
64. Bothner-By, A. A., Glickson, J. D., Sykes, B. D., eds. 1982. *Biochemical Structure Determination by NMR*. New York: Dekker. 232 pp.
65. Bradbury, J. H., Carver, J. A. Parker, M. W. 1981. *J. Chem. Soc. Chem. Commun.* 1981:208–9
66. Bradbury, J. H., Carver, J. A., Parker, M. W. 1982. *FEBS Lett.* 146:297–301
67. Bradbury, J. H., Crompton, M. W., Teh, J. S. 1977. *Eur. J. Biochem.* 81:411–22
68. Bradbury, J. H., Ramesh, V., Dodson, G. 1981. *J. Mol. Biol.* 150:609–13
69. Bradbury, J. H., Scheraga, H. A. 1966. *J. Am. Chem. Soc.* 88:4240–46
70. Bradbury, J. H., Teh, J. S. 1975. *J. Chem. Soc. Chem. Commun.* 1975:936–37
71. Brasure, E. B., Henderson, T. O. 1981. In *High-Density Lipoproteins*, ed. C. F. Day, pp. 73–93. New York: Dekker
72. Braun, W., Bösch, C., Brown, L., Go, N., Wüthrich, K. 1981. *Biochim. Biophys. Acta* 667:377–96
73. Briggs, R. W., Armitage, I. M. 1982. *J. Biol. Chem.* 257:1259–62
74. Brown, F. F., Parsons, T. F., Sigman, D. S., Pierce, J. G. 1979. *J. Biol. Chem.* 254:4335–38
75. Brown, L. R. 1979. *Biochim. Biophys. Acta* 557:135–48

76. Brown, L. R., Bradbury, J. H. 1976. *Eur. J. Biochem.* 68:227–35
77. Brown, L. R., Braun, W., Kumar, A., Wüthrich, K. 1982. *Biophys. J.* 37:319–28
78. Brown, L. R., De Marco, A., Richarz, R., Wagner, G., Wüthrich, K. 1978. *Eur. J. Biochem.* 88:87–95
79. Brown, L. R., De Marco, A., Wagner, G., Wüthrich, K. 1976. *Eur. J. Biochem.* 62:103–7
80. Brown, L. R., Lauterwein, J., Wüthrich, K. 1980. *Biochim. Biophys. Acta* 622:231–44
81. Brown, L. R., Wüthrich, K. 1981. *Biochim. Biophys. Acta* 647:95–111
82. Browne, C. A., Campbell, I. D., Kiener, P. A., Phillips, D. C., Waley, S. G., Wilson, I. A. 1976. *J. Mol. Biol.* 100:319–43
83. Browne, D. T., Kenyon, G. L., Packer, E. L., Sternlicht, H., Wilson, D. M. 1973. *J. Am. Chem. Soc.* 95:1316–23
84. Buck, F., Rüterjans, H., Kaptein, R., Beyreuther, K. 1980. *Proc. Natl. Acad. Sci. USA* 77:5145–48
85. Burns, P. D., La Mar, G. N. 1981. *J. Biol. Chem.* 256:4934–39
86. Byrd, R. A., Dawson, W. H., Ellis, P. D., Dunlap, R. B. 1978. *Dev. Biochem.* 4:367–70
87. Bystrov, V. F., Arseniev, A. S., Gavrilov, Yu. D. 1978. *J. Magn. Reson.* 30:151–84
88. Bystrov, V. F., Arseniev, A. S., Kondakov, V. I., Maiorov, N., Okhanov, V., Ovchinnikov, Yu. A. 1983. In *Neurotoxins as Tools in Neurochemistry, Proc. USSR-Berlin (West) Symp., March 22–24*, ed. Yu. Ovchinnikov, F. Hucho. Berlin: de Gruyter
89. Bystrov, V. F., Ivanov, V. T., Okhanov, V. V., Miroshnikov, A. I., Arseniev, A. S., Tsetlin, V. I., Pashkov, V. S. 1981. *Advances in Solution Chemistry*, ed. I. Bertini, L. Lunazzi, A. Dei, pp. 231–51. New York: Plenum
90. Bystrov, V. F., Okhanov, V. V., Miroshnikov, A. I., Ovchinnikov, Yu. A. 1980. *FEBS Lett.* 119:113–17
91. Cammack, R., Dickson, D. P. E., Johnson, C. E. 1977. In *Iron-Sulfur Proteins*, ed. W. Lovenberg, 3:283–330. New York: Academic
92. Campbell, I. D. 1977. See Ref. 154, pp. 33–49
93. Campbell, I. D., Dobson, C. M., Williams, R. J. P. 1975. *Proc. R. Soc. London Ser. A* 345:23–40
94. Campbell, I. D., Dobson, C. M., Williams, R. J. P., Wright, P. E. 1975. *FEBS Lett.* 57:96–99
95. Campbell, I. D., Lindskog, S., White, A. I. 1975. *J. Mol. Biol.* 98:597–614
96. Campbell, I. D., Lindskog, S., White, A. I. 1977. *Biochim. Biophys. Acta* 484:443–52
97. Canioni, P., Cozzone, P. J. 1979. *Biochimie* 61:343–54
98. Canioni, P., Cozzone, P. J. 1979. *FEBS Lett.* 97:353–57
99. Canioni, P., Cozzone, P. J., Sarda, L. 1980. *Biochim. Biophys. Acta* 621:29–42
100. Cary, P. D., King, D. S., Crane-Robinson, C., Bradbury, E. M., Rabbani, A., Goodwin, G. H., Johns, E. W. 1980. *Eur. J. Biochem.* 112:577–88
101. Cass, A. E. G., Hill, H. A. O., Smith, B. E., Bannister, J. V., Bannister, W. H. 1977. *Biochemistry* 16:3061–66
102. Cass, A. E. G., Hill, H. A. O., Smith, B. E., Bannister, J. V., Bannister, W. H. 1977. *Biochem. J.* 165:587–89
103. Cavé, A., Parello, J. 1981. In *Immun. Intercell., Les Houches, Ec. Ete Phys. Theor., 3rd*, ed. R. Balian, M. Chabre, P. F. Devaux, pp. 197–227. Amsterdam: North-Holland
104. Chaiken, I. M., Cohen, J. S., Sokoloski, E. A. 1974. *J. Am. Chem. Soc.* 96:4703–5
105. Chaiken, I. M., Freedman, M. H., Lyerla, J. R. Jr., Cohen, J. S. 1973. *J. Biol. Chem.* 248:884–91
106. Chambers, V. M. A., Evans, E. A., Elvidge, J. A., Jones, J. R. 1978. *Rev. Radiochem. Cent.* 19:1–68
107. Chan, T.-M., Hermodson, M. A., Ulrich, E. L., Markley, J. L. 1983. *Biochemistry* 22:5988–95
108. Chan, T.-M., Markley, J. L. 1982. *J. Am. Chem. Soc.* 104:4010–11
109. Chan, T.-M., Markley, J. L. 1983. *Biochemistry* 22:5982–87
110. Chan, T.-M., Markley, J. L. 1983. *Biochemistry* 22:5996–6002
111. Chan, T.-M., Markley, J. L. 1983. *Biochemistry* 22:6008–10
112. Chan, T.-M., Ulrich, E. L., Markley, J. L. 1983. *Biochemistry* 22:6002–7
113. Chan, T.-M., Westler, W. M., Santini, R. E., Markley, J. L. 1982. *J. Am. Chem. Soc.* 104:4008–10
114. Chapman, G. E. 1977. *Nucl. Magn. Reson.* 6:154–73
115. Chen, T.-C., Knapp, R. D., Rohde, M. F., Brainard, J. R., Gotto, J. M. Jr., Sparrow, J. T., Morrisett, J. D. 1980. *Biochemistry* 19:5140–46
116. Chlebowski, J. M., Armitage, I. M., Coleman, J. E. 1977. *J. Biol. Chem.* 252:7053–61
117. Clayden, N. J., Williams, R. J. P. 1982. *J. Magn. Reson.* 49:383–96
118. Clore, G. M., Gronenborn, A. M. 1982. *Biochemistry* 21:4048–53

119. Cocco, L., Blakley, R. L., Walker, T. E., London, R. E., Matwiyoff, N. A. 1978. *Biochemistry* 17:4285–90
120. Cohen, J. S. 1978. *CRC Crit. Rev. Biochem.* 5:25–43
121. Cohen, J. S., ed. 1980. *Magnetic Resonance in Biology*, Vol. 1. New York: Wiley. 309 pp.
122. Cohen, J. S., Niu, C.-H., Matsuura, S., Shindo, H. 1980. *Dev. Biochem.* 10:3–16
123. Cohn, M. 1982. *Ann. Rev. Biophys. Bioeng.* 11:23–42
124. Cohn, M., Reed, G. H. 1982. *Ann. Rev. Biochem.* 51:365–94
125. Coleman, J. E., Armitage, I. M. 1977. See Ref. 154, pp. 171–200
126. Coleman, J. E., Armitage, I. M. 1978. *Biochemistry* 17:5038–45
127. Coleman, J. E., Armitage, I. M., Chlebowski, J. F., Otvos, J. D., Schoot Uiterkamp, A. J. M. 1979. See Ref. 518, pp. 345–95
128. Cookson, D. J., Hayes, M. T., Wright, P. E. 1980. *Nature* 283:682–83
129. Cookson, D. J., Moore, G. R., Pitt, R. C., Williams, R. J. P., Campbell, I. D., Ambler, R. P., et al. 1978. *Eur. J. Biochem.* 83:261–75
130. Crane-Robinson, C. 1978. See Ref. 44, 1:33–90
131. Crippen, G. M. 1979. *Int. J. Pept. Protein Res.* 13:320–26
132. Croll, D. H. 1982. PhD thesis. Purdue Univ., West Lafayette, Ind. 218 pp.
133. Cutnell, J. D., La Mar, G. N., Kong, S. B. 1981. *J. Am. Chem. Soc.* 103:3567–72
134. Dadok, J., Sprecher, R. F. 1974. *J. Magn. Reson.* 13:243–48
135. De Marco, A., Hochschwender, S. M., Laursen, R. A., Llinás, M. 1982. *J. Biol. Chem.* 257:12716–21
136. De Marco, A., Lecomte, J. T. J., Llinás, M. 1981. *Eur. J. Biochem.* 119:483–90
137. De Marco, A., Menegatti, E., Guarneri, M. 1979. *Eur. J. Biochem.* 102:185–94
138. De Marco, A., Menegatti, E., Guarneri, M. 1982. *Biochemistry* 21:222–29
139. De Marco, A., Menegatti, E., Guarneri, M. 1982. *J. Biol. Chem.* 257:8337–42
139a. De Marco, A., Menegatti, E., Guarneri, M. 1983. *FEBS Lett.* 159:201–6
140. De Marco, A., Wüthrich, K. 1976. *J. Magn. Reson.* 24:201–4
141. Delepierre, M., Dobson, C. M., Hoch, J. C., Olejniczak, E. T., Poulsen, F. M., Ratcliffe, R. G., Redfield, C. 1981. In *Biomol. Stereodyn., Proc. Symp.*, ed. R. H. Sarma, 2:237–53. Guilderland, NY: Adenine Press
142. Delepierre, M., Dobson, C. M., Poulsen, F. M. 1982. *Biochemistry* 21:4756–61
143. Dickenson, L. C., Chien, J. C. W. 1975. *Biochemistry* 14:3534–42
144. Dill, K., Allerhand, A. 1977. *J. Am. Chem. Soc.* 99:4508–11
145. Dill, K., Allerhand, A. 1979. *J. Am. Chem. Soc.* 101:4376–78
146. Dobson, C. M. 1977. See Ref. 154, pp. 77–94
147. Dobson, C. M. 1982. *Jerusalem Symp. Quantum Chem. Biochem.* 15:481–95
148. Dobson, C. M., Moore, G. R., Williams, R. J. P. 1975. *FEBS Lett.* 51:60–65
149. Dobson, C. M., Olejniczak, E. T., Poulsen, F. M., Ratcliffe, R. G. 1982. *J. Magn. Reson.* 48:97–110
150. Dower, S., Dwek, R. A. 1979. See Ref. 518, pp. 271–303
151. Drakenberg, T., Lindman, B., Cavé, A., Parello, J. 1978. *FEBS Lett.* 92:346–50
152. Dubs, A., Wagner, G., Wüthrich, K. 1979. *Biochim. Biophys. Acta* 577:177–94
153. Dwek, R. A. 1977. See Ref. 154, pp. 125–56
154. Dwek, R. A., Campbell, I. D., Richards, R. E., eds. 1977. *NMR in Biology, Proc. Br. Biophys. Soc., March.* London: Academic. 381 pp.
155. Eakin, R. T., Morgan, L. O., Matwiyoff, N. A. 1975. *Biochem. J.* 152:529–35
156. Edmondson, D. E., James, T. L. 1982. *Dev. Biochem.* 21:111–18
157. Edwards, B. F. P., Lee, L., Sykes, B. D. 1978. In *Biomol. Struct. Funct., Symp.*, ed. P. F. Agris, pp. 275–93. New York: Academic
158. Edwards, B. F. P., Sykes, B. D. 1978. *Biochemistry* 17:684–89
159. Edwards, B. F. P., Sykes, B. D. 1980. *Biochemistry* 19:2577–83
160. Edwards, B. F. P., Sykes, B. D. 1981. *Biochemistry* 20:4193–98
161. Egan, W., Shindo, H., Cohen, J. S. 1977. *Ann. Rev. Biophys. Bioeng.* 6:383–417
162. Egan, W., Shindo, H., Cohen, J. S. 1978. *J. Biol. Chem.* 253:16–17
163. Egmond, M. R., Hore, P. J., Kaptein, R. 1983. *Biochim. Biophys. Acta* 744:23–27
164. Egmond, M. R., Slotboom, A. J., De Haas, G. H., Dijkstra, K., Kaptein, R. 1980. *Biochim. Biophys. Acta* 623:461–66
165. Emid, S. 1983. *Bull. Magn. Reson.* 4:99–104
166. Emptage, M. H., Xavier, A. V., Wood, J. M. 1980. *Cienc. Biol. Coimbra* 5:133–35
167. Emptage, M. H., Xavier, A. V., Wood, J. M., Alsaadi, B. M., Moore, G. R., Pitt, R. C., et al. 1981. *Biochemistry* 20:58–64
168. Endo, T., Inagaki, F., Hayashi, K.,

Miyazawa, T. 1979. *Eur. J. Biochem.* 102:417–30
169. Endo, T., Inagaki, F., Hayashi, K., Miyazawa, T. 1981. *Eur. J. Biochem.* 120:117–24
170. Endo, T., Inagaki, F., Hayashi, K., Miyazawa, T. 1982. *Eur. J. Biochem.* 122:541–47
171. Engeseth, H. R., Otvos, J. D. 1983. *Am. Chem. Soc. Abstr. Pap., 186th Meet., Aug. 28–Sep. 21, Inorg. Div., Abstr. No. 53*
172. Epstein, H. F., Schechter, A. N., Cohen, J. S. 1971. *Proc. Natl. Acad. Sci. USA* 68:2042–46
173. Fairhurst, S. A., Sutcliffe, L. H. 1978. *Prog. Biophys. Mol. Biol.* 34:1–79
174. Feeney, J. 1978. *Jerusalem Symp. Quantum Chem. Biochem.* 11:297–310
175. Feeney, J., Birdsall, B., Albrand, J. P., Roberts, G. C. K., Burgen, A. S. V., Charlton, P. A., Young, D. W. 1981. *Biochemistry* 20:1837–42
176. Feeney, J., Roberts, G. C. K., Kaptein, R., Birdsall, B., Gronenborn, A., Burgen, A. S. V. 1980. *Biochemistry* 19:2466–72
177. Feeney, J., Roberts, G. C. K., Thomson, J. W., King, R. W., Griffiths, D. V., Burgen, A. S. V. 1980. *Biochemistry* 19:2316–21
178. Ferguson, R. C., Phillips, W. D. 1967. *Science* 157:257–67
179. Forsén, S., Andersson, A., Drakenberg, T., Teleman, E., Thulin, E., Vogel, H. J. 1983. *4th Int. Symp., Calcium Binding Proteins in Health and Disease, Trieste, Italy, May 16–19*, ed. B. de Bernard. Amsterdam: Elsevier
180. Forsén, S., Andersson, T., Drakenberg, T., Thulin, E., Swärd, M. 1982. *Fed. Proc.* 41:2981–86
181. Forsén, S., Thulin, E., Drakenberg, T., Krebs, J., Seamon, K. 1980. *FEBS Lett.* 117:189–94
182. Forsén, S., Thulin, E., Lilja, H. 1979. *FEBS Lett.* 104:123–26
183. Frankenne, F., Maghuin-Rogister, G., Birdsall, B., Roberts, G. C. K. 1983. *FEBS Lett.* 151:197–200
184. Freeman, H. C., Norris, V. A., Ramshaw, J. A. M., Wright, P. E. 1978. *FEBS Lett.* 86:131–35
185. Freeman, R., Morris, G. A. 1979. *Bull. Magn. Reson.* 1:5–26
186. Fujii, S., Akasaka, K., Hatano, H. 1980. *J. Biochem.* 88:789–96
187. Fujii, S., Akasaka, K., Hatano, H. 1981. *Biochemistry* 20:518–23
188. Fukushima, E., Roeder, S. B. W. 1981. *Experimental Pulse NMR. A Nuts and Bolts Approach.* Reading, Mass: Addison-Wesley. 539 pp.
189. Fülling, R., Rüterjans, H. 1978. *FEBS Lett.* 88:279–82
190. Fung, C. H., Chang, C. C., Gupta, R. K. 1979. *Biochemistry* 18:457–60
191. Fung, L. W.-M., Ho, C. 1975. *Biochemistry* 14:2526–35
191a. Galley, H. U., Spencer, A. K., Armitage, I. M., Prestegard, J. H., Cronan, J. E. Jr. 1978. *Biochemistry* 17:5377–82
192. Ganesh, K. N. 1982. *Curr. Sci.* 51:866–74
193. Garssen, G. J., Kaptein, R., Schoenmakers, J. G. G., Hilbers, C. W. 1978. *Proc. Natl. Acad. Sci. USA* 75:5281–85
194. Garssen, G. J., Tesser, G. I., Schoenmakers, J. G. G., Hilbers, C. W. 1980. *Biochim. Biophys. Acta* 607:361–71
195. Gassner, M., Stehlik, D., Schrecker, O., Hengstenberg, W., Maurer, W., Rüterjans, H. 1977. *Eur. J. Biochem.* 75:287–96
196. Gerig, J. T. 1978. See Ref. 44, 1:139–203
197. Gerig, J. T., Klinkenborg, J. C., Nieman, R. A. 1983. *Biochemistry* 22:2076–87
198. Gersonde, K. 1978. In *Symp. Pap.-IUPAC Int. Symp. Chem. Nat. Prod., 11th*, ed. N. Marekov, I. Ognyanov, A. Orahovats, 4:90–104. Sofia, Bulgaria: Izd. BAN
199. Gettins, P., Boyd, J., Glaudemans, C. P. J., Potter, M., Dwek, R. A. 1981. *Biochemistry* 20:7463–69
200. Gettins, P., Coleman, J. E. 1983. *J. Biol. Chem.* 258:396–407
201. Gettins, P., Dwek, R. A. 1981. *FEBS Lett.* 124:248–52
202. Gettins, P., Potter, M., Rudikoff, S., Dwek, R. A. 1977. *FEBS Lett.* 84:87–91
203. Gordon, S. L., Wüthrich, K. 1978. *J. Am. Chem. Soc.* 100:7094–96
204. Govil, G., Hosur, R. V. 1982. *Conformation of Biological Molecules, NMR Basic Principles and Progress*, Vol. 20. Berlin: Springer-Verlag. 216 pp.
205. Gray, G. A. 1983. *Org. Magn. Reson.* 21:111–18
206. Gronenborn, A., Birdsall, B., Hyde, E. I., Roberts, G. C. K., Feeney, J., Burgen, A. S. V. 1981. *Biochemistry* 20:1717–22
207. Gronenborn, A., Birdsall, B., Hyde, E. I., Roberts, G. C. K., Feeney, J., Burgen, A. S. V. 1981. *Nature* 290:273–74
208. Gupta, R. K., Mildvan, A. S. 1978. *Methods Enzymol.* 54:151–92
209. Gurd, F. R. N., Matthew, J. B., Wittebort, R. J., Morrow, J. S., Friend, S. H. 1980. In *Biophys. Physiol. Carbon Dioxide Symp.*, ed. C. Bauer, G. Gros, H. Bartels, pp. 89–101. Berlin: Springer-Verlag

210. Gurd, F. R. N., Wittebort, R. J., Rothgeb, T. M., Neireiter, G. Jr. 1982. See Ref. 64, pp. 1–29
211. Gust, D., Moon, R. M., Roberts, J. D. 1975. *Proc. Natl. Acad. Sci. USA* 72:4696–700
212. Harbison, G. S., Herzfeld, J., Griffin, R. G. 1983. *Biochemistry* 22:1–5
213. Harina, B. M., Dyckes, D. F., Willcott, M. R. III, Jones, W. C. Jr. 1978. *J. Am. Chem. Soc.* 100:4897–99
214. Harris, R. K. 1983. *Nuclear Magnetic Resonance. A Physiochemical View.* London: Pitman. 250 pp.
215. Havel, T. F., Crippen, G. M., Kuntz, I. D. 1979. *Biopolymers* 18:73–81
216. Henry, G. D., Dalgarno, D. C., Marcus, G., Scott, M., Levine, B. A., Trayer, I. P. 1982. *FEBS Lett.* 144:11–15
217. Herak, J. N., Pifat, G., Brnjas-Kralijevic, J., Jurgens, G., Holasek, A. 1980. *Period. Biol.* 82:351–55
218. Hider, R. C., Drake, A. F., Inagaki, F., Williams, R. J. P., Endo, T., Miyazawa, T. 1982. *J. Mol. Biol.* 158:275–91
219. Hill, H. A. O., Lee, W. K., Bannister, J. V., Bannister, W. H. 1980. *Biochem. J.* 185:245–52
220. Hill, H. A. O., Leer, J. C., Smith, B. E., Storm, C. B., Ambler, R. P. 1976. *Biochem. Biophys. Res. Commun.* 70: 331–38
221. Hill, H. A. O., Smith, B. E. 1979. *J. Inorg. Biochem.* 11:79–93
222. Hincke, M. T., Hagen, S., Sykes, B. D., Kay, C. M. 1980. *Dev. Biochem.* 14: 315–17
223. Hincke, M. T., Sykes, B. D., Kay, C. M. 1981. *Biochemistry* 20:3286–94
224. Hincke, M. T., Sykes, B. D., Kay, C. M. 1981. *Biochemistry* 20:4185–93
225. Ho, C., Fung, L. W.-M., Wiechelman, K. J. 1978. *Methods Enzymol.* 54:192–223
226. Ho, C., Fung, L. W.-M., Wiechelman, K. J., Pifat, G., Johnson, M. F. 1975. In *Erythrocyte Structure and Function*, pp. 43–64. New York: Liss
227. Ho, C., Kurland, R. J. 1966. *J. Biol. Chem.* 241:3002–7
228. Ho, C., Lam, C. H. J., Takahashi, S., Viggiano, G. 1982. In *Hemoglobin and Oxygen Binding, Int. Symp. Interact. Iron Proteins Oxygen Electron Transp.*, ed. C. Ho, pp. 141–49. New York: Elsevier
229. Ho, C., Russu, I. M. 1981. *Methods Enzymol.* 76:275–312
230. Hoch, J. C., Dobson, C. M., Karplus, M. 1982. *Biochemistry* 21:1118–25
231. Hochschwender, S. M., Laursen, R. A., De Marco, A., Llinás, M. 1983. *Arch. Biochem. Biophys.* 223:58–67
232. Hoerl, M., Feldmann, K., Schnackerz, K. D., Helmreich, E. J. M. 1979. *Biochemistry* 18:2457–64
233. Hon-Nami, K., Kihara, H., Kitagawa, T., Miyazawa, T., Oshima, T. 1980. *Eur. J. Biochem.* 110:217–23
234. Hore, P. J., Kaptein, R. 1982. In *ACS Symp. Ser. 191, NMR Spectrosc.: New Methods Appl.*, pp. 285–318
235. Hore, P. J., Kaptein, R. 1983. *Biochemistry* 22:1906–11
236. Hore, P. J., Scheek, R. M., Kaptein, R. 1983. *J. Magn. Reson.* 1983:339–42
237. Hosur, R. V., Wider, G., Wüthrich, K. 1983. *Eur. J. Biochem.* 130:497–508
238. Howarth, O. W., Lilley, D. M. J. 1978. *Prog. Nucl. Magn. Reson. Spectrosc.* 12:1–40
239. Hughes, D. W., Stollery, J. G., Moscarello, M. A., Deber, C. M. 1982. *J. Biol. Chem.* 257:4698–700
240. Hull, W. E., Halford, S. E., Gutfreund, H., Sykes, B. D. 1976. *Biochemistry* 15:1547–61
241. Hunkapiller, M. W., Forgac, M. D., Yu, E. H., Richards, J. H. 1979. *Biochem. Biophys. Res. Commun.* 87:25–31
242. Hunkapiller, M. W., Smallcombe, S. H., Whitaker, D. R., Richards, J. H. 1973. *J. Biol. Chem.* 248:8306–8
243. Ikeda-Saito, M., Inubushi, T., McDonald, G. G., Yonetani, T. 1978. *J. Biol. Chem.* 253:7134–37
244. Ikura, M., Hiraoki, T., Hikichi, K., Mikuni, T., Yazawa, M., Yagi, K. 1983. *Biochemistry* 22:2573–79
245. Ikura, M., Toshifumi, H., Hikichi, K., Mikuni, T., Yazawa, M., Yagi, K. 1983. *Biochemistry* 22:2568–72
246. Inagaki, F., Boyd, J., Campbell, I. D., Clayden, N. J., Hull, W. E., Tamiya, N., Williams, R. J. P. 1982. *Eur. J. Biochem.* 121:609–16
247. Inagaki, F., Clayden, N. J., Tamiya, N., Williams, R. J. P. 1981. *Eur. J. Biochem.* 120:313–22
248. Inagaki, F., Clayden, N. J., Tamiya, N., Williams, R. J. P. 1982. *Eur. J. Biochem.* 123:99–104
249. Inagaki, F., Kawano, Y., Shimada, I., Takahashi, K., Miyazawa, T. 1981. *J. Biochem.* 89:1185–95
250. Inagaki, F., Miyazawa, T., Hori, H., Tamiya, N. 1978. *Eur. J. Biochem.* 89:433–42
251. Inagaki, F., Miyazawa, T., Tamiya, N., Williams, R. J. P. 1982. *Eur. J. Biochem.* 123:275–82
252. Inagaki, F., Miyazawa, T., Williams, R. J. P. 1981. *Biosci. Rep.* 1:743–55
253. Inagaki, F., Tamiya, N., Miyazawa, T. 1980. *Eur. J. Biochem.* 109:129–38
254. Inagaki, F., Tamiya, N., Miyazawa, T., Williams, R. J. P. 1981. *Eur. J. Biochem.* 118:621–25

255. Inagaki, F., Watanabe, K., Miyazawa, T. 1979. *J. Biochem.* 86:591–94
256. Inubushi, T., Ikeda-Saito, M., Yonetani, T. 1983. *Biochemistry* 22:2904–7
257. Iwahashi, H., Akutsu, H., Kobayashi, Y., Kyogoku, Y., Ono, T., Koga, H., Horiuchi, T. 1982. *J. Biochem.* 91:1213–21
258. Jackson, W. R. C., Leatherbarrow, R. J. 1981. *Biochemistry* 20:2339–45
259. Jaeck, G., Benz, F. W. 1979. *Biochem. Biophys. Res. Commun.* 86:885–92
260. James, T. L., Edmondson, D. E. 1981. *Biochemistry* 20:617–21
261. James, T. L., Ludwig, M. L., Cohn, M. 1973. *Proc. Natl. Acad. Sci. USA* 70:3292–95
262. James, T. L., Matson, G. B., Kuntz, I. D. 1978. *J. Am. Chem. Soc.* 100:3590–94
263. Jansen, E. H. J. M., Meyer, H., De Haas, G. H., Kaptein, R. 1978. *J. Biol. Chem.* 253:6346–47
264. Jansen, E. H. J. M., Van Scharrenburg, G. J. M., Slotboom, A. J., De Haas, G. H., Kaptein, R. 1979. *J. Am. Chem. Soc.* 101:7397–99
265. Jardetzky, O. 1979. See Ref. 440, pp. 141–67
266. Jardetzky, O. 1980. *Biochim. Biophys. Acta* 621:227–32
267. Jardetzky, O., Jardetzky, C. D. 1957. *J. Am. Chem. Soc.* 79:5322–23
268. Jardetzky, O., Markley, J. L. 1970. *Il Farmaco, Ed. Sci.* 25:894–97
269. Jardetzky, O., Markley, J. L., Thielmann, J., Arata, Y., Williams, M. W. 1972. *Cold Spring Harbor Symp. Quant. Biol.* 36:257–61
270. Jardetzky, O., Roberts, G. C. K. 1981. *NMR in Molecular Biology*. New York: Academic. 681 pp.
271. Jarema, M. A. C., Lu, P., Miller, J. H. 1981. *Proc. Natl. Acad. Sci. USA* 78:2707–11
272. Johnson, F. A., Lewis, S. D., Shafer, J. A. 1981. *Biochemistry* 20:44–48
273. Johnson, R. N., Bradbury, J. H., Appleby, C. A. 1978. *J. Biol. Chem.* 253:2148–54
274. Jones, R. B., Waley, S. G. 1979. *Biochem. J.* 179:623–30
275. Jones, W. C. Jr., Rothgeb, T. M., Gurd, F. R. N. 1976. *J. Biol. Chem.* 251:7452–60
276. Jordan, F., Polgar, L. 1981. *Biochemistry* 20:6366–70
277. Kainosho, M., Tsuji, T. 1982. *Biochemistry* 21:6273–79
278. Kalbitzer, H. R., Deutscher, J., Hengstenberg, W., Roesch, P. 1981. *Biochemistry* 20:6178–85
279. Kalbitzer, H. R., Hengstenberg, W., Roesch, P., Muss, P., Bernsmann, P., Engelmann, R., et al. 1982. *Biochemistry* 21:2879–85
280. Kalbitzer, H. R., Marquetant, R., Roesch, P., Schirmer, R. H. 1982. *Eur. J. Biochem.* 126:531–36
281. Kalk, A., Berendsen, H. J. C. 1976. *J. Magn. Reson.* 24:343–66
282. Kanamori, K., Roberts, J. D. 1983. *Acc. Chem. Res.* 16:35–41
283. Kaptein, R. 1980. *Curso Reson. Magn. Nucl.: Reson. Magn. Nucl. Pulsos, Alta Resoluc., 1st, Madrid*, pp. 385–407
284. Kaptein, R. 1982. See Ref. 44, 4:145–91
285. Kaptein, R., Scheek, R. M., Zuiderweg, E. R. P., Boelens, R., Klappe, K. J. M., van Boom, J. H., et al. 1983. In *Structure and Dynamics: Nucleic Acids and Proteins*, ed. E. Clementi, R. H. Sarma, pp. 209–25. New York: Adenine Press
286. Kaptein, R., Wyeth, P. 1980. *Cienc. Biol.* 5:125–27
287. Keim, P. 1979. *Biochem. Discuss.* 7:9–50
288. Keller, R. M., Baumann, R., Hunziker-Kwik, E. H., Joubert, F. J., Wüthrich, K. 1983. *J. Mol. Biol.* 163:623–46
289. Keller, R. M., Pettigrew, G. W., Wüthrich, K. 1973. *FEBS Lett.* 36:151–56
290. Keller, R. M., Picot, D., Wüthrich, K. 1979. *Biochim. Biophys. Acta* 580:259–65
291. Keller, R. M., Schejter, A., Wüthrich, K. 1980. *Biochim. Biophys. Acta* 626:15–22
292. Keller, R. M., Wüthrich, K. 1978. *Biochim. Biophys. Acta* 533:195–208
293. Keller, R. M., Wüthrich, K. 1978. *Biochem. Biophys. Res. Commun.* 83:1132–39
294. Keller, R. M., Wüthrich, K. 1981. See Ref. 44, 3:1–52
295. Keller, R. M., Wüthrich, K., Pecht, I. 1976. *FEBS Lett.* 70:180–84
296. Keller, R. M., Wüthrich, K., Schejter, A. 1977. *Biochim. Biophys. Acta* 491:409–15
297. Keniry, M. A., Rothgeb, T. M., Smith, R. L., Gutowsky, H. S., Oldfield, E. 1983. *Biochemistry* 22:1917–26
298. Kenyon, G. L., Reed, G. H. 1982. *Adv. Enzymol.* 54:367–426
299. Kilmartin, J. V., Breen, J. J., Roberts, G. C. K., Ho, C. 1973. *Proc. Natl. Acad. Sci. USA* 70:1246–49
300. Kim, P. S., Baldwin, R. L. 1982. *Ann. Rev. Biochem.* 51:459–89
301. Kimber, B. J., Feeney, J., Roberts, G. C. K., Birdsall, B., Griffiths, D. V., Burgen, A. S. V., Sykes, B. D. 1978. *Nature* 271:184–85
302. Kimura, S., Matsuo, H., Narita, K. 1979. *J. Biochem.* 86:301–10

303. King, G., Wright, P. E. 1982. *Biochem. Biophys. Res. Commun.* 106:559–65
303a. King, G., Wright, P. E. 1983. *J. Magn. Reson.* 54:328–32
304. Kirpichnikov, M. P., Kurochkin, A. V., Skryabin, K. G. 1982. *FEBS Lett.* 150:407–10
305. Klevit, R. E., Girard, P., Esnouf, M. P., Williams, R. J. P. 1981. In *Calcium Phosphate Transp. Biomembrane, Int. Workshop*, ed. F. Bronner, M. Peterlik, pp. 25–29. New York: Academic
306. Kojiro, C. L., Markley, J. L. 1983. *FEBS Lett.* 162:52–56
307. Kong, S. B., Cutnell, J. D., La Mar, G. N. 1983. *J. Biol. Chem.* 258:3843–49
308. Koppitz, B., Feldmann, K., Heilmeyer, L. M. G. Jr. 1980. *FEBS Lett.* 117:199–202
309. Kowalsky, A. 1965. *Biochemistry* 4:2382–88
310. Krebs, J. 1981. *Cell Calcium* 2:295–311
311. Krebs, J., Carafoli, E. 1982. *Eur. J. Biochem.* 124:619–27
312. Krishnamoorthy, G., Prabhananda, B. S. 1982. *Biochim. Biophys. Acta* 709:53–57
313. Kumar, A., Ernst, R. R., Wüthrich, K. 1980. *Biochem. Biophys. Res. Commun.* 95:1–6
314. Kumar, A., Wagner, G., Ernst, R. R., Wüthrich, K. 1980. *Biochem. Biophys. Res. Commun.* 96:1156–63
315. Kurochkin, A. V., Kirpichnikov, M. P. 1982. *FEBS Lett.* 150:411–15
316. Kuwajima, K., Baldwin, R. L. 1983. *J. Mol. Biol.* 169:281–97
317. Kuwajima, K., Baldwin, R. L. 1983. *J. Mol. Biol.* 169:299–323
318. Kuwajima, K., Kim, P. S., Baldwin, R. L. 1983. *Biopolymers* 22:59–67
319. La Mar, G. N. 1979. See Ref. 518, pp. 305–43
320. La Mar, G. N., Anderson, R. R., Budd, D. L., Smith, K. M., Langry, K. C., Gersonde, K., Sick, H. 1981. *Biochemistry* 20:4429–36
321. La Mar, G. N., Budd, D. L., Smith, K. M. 1980. *Biochim. Biophys. Acta* 622:210–18
322. La Mar, G. N., Budd, D. L., Smith, K. M., Langry, K. C. 1980. *J. Am. Chem. Soc.* 102:1822–27
323. La Mar, G. N., Budd, D. L., Viscio, D. B., Smith, K. M., Langry, K. C. 1978. *Proc. Natl. Acad. Sci. USA* 75:5755–59
324. La Mar, G. N., Burns, P. D., Jackson, J. T., Smith, K. M., Langry, K. C., Strittmatter, P. 1981. *J. Biol. Chem.* 256:6075–79
325. La Mar, G. N., Cutnell, J. D., Kong, S. B. 1981. *Biophys. J.* 34:217–25
326. La Mar, G. N., De Ropp, J. S. 1979. *Biochem. Biophys. Res. Commun.* 90:36–41
327. La Mar, G. N., De Ropp, J. S. 1982. *J. Am. Chem. Soc.* 104:5203–6
328. La Mar, G. N., De Ropp, J. S., Chacko, V. P., Satterlee, J. D., Erman, J. E. 1982. *Biochim. Biophys. Acta* 708:317–25
329. La Mar, G. N., De Ropp, J. S., Latos-Grazynski, L., Balch, A. L., Johnson, R. B., Smith, K. M., et al. 1983. *J. Am. Chem. Soc.* 105:782–87
330. La Mar, G. N., De Ropp, J. S., Smith, K. M., Langry, K. C. 1980. *J. Am. Chem. Soc.* 102:4833–35
331. La Mar, G. N., De Ropp, J. S., Smith, K. M., Langry, K. C. 1980. *J. Biol. Chem.* 255:6646–52
332. La Mar, G. N., De Ropp, J. S., Smith, K. M., Langry, K. C. 1981. *J. Biol. Chem.* 256:237–43
333. La Mar, G. N., Kong, S. B., Smith, K. M., Langry, K. C. 1981. *Biochem. Biophys. Res. Commun.* 102:142–48
334. La Mar, G. N., Nagai, K., Jue, T., Budd, D. L., Gersonde, K., Sick, H., et al. 1980. *Biochem. Biophys. Res. Commun.* 96:1172–77
335. La Mar, G. N., Overkamp, M., Sick, H., Gersonde, K. 1978. *Biochemistry* 17:352–61
336. La Mar, G. N., Smith, K. M., De Ropp, J. S., Burns, P. D., Langry, K. C., Gersonde, K., et al. 1982. See Ref. 228, pp. 419–24
337. La Mar, G. N., Smith, K. M., Gersonde, K., Sick, H., Overkamp, M. 1980. *J. Biol. Chem.* 255:66–70
338. La Mar, G. N., Viscio, D. B., Budd, D. L. 1978. *Biochem. Biophys. Res. Commun.* 82:19–23
339. Latos-Grazynski, L., Balch, A. L., La Mar, G. N. 1982. *Adv. Chem. Ser.* 201:661–74
340. Lauterbur, P. C. 1970. *Appl. Spectrosc.* 24:450–52
341. Lauterwein, J., Bösch, C., Brown, L. R., Wüthrich, K. 1979. *Biochim. Biophys. Acta* 556:244–64
342. Lauterwein, J., Brown, L. R., Wüthrich, K. 1980. *Biochim. Biophys. Acta* 622:219–20
343. Lecomte, J. T. J., De Marco, A., Llinás, M. 1982. *Biochim. Biophys. Acta* 703:223–30
344. Lecomte, J. T. J., Jones, B. L., Llinás, M. 1982. *Biochemistry* 21:4843–49
345. Lee, L., Sykes, B. D. 1980. *Adv. Inorg. Biochem., Methods for Determining Metal Ion Environments in Proteins: Structure and Functions of Metalloproteins*, ed. D. W. Darnall, R. G. Wilkins, 2:183–210. New York: Elsevier. 324 pp.

346. Lee, L., Sykes, B. D. 1980. *Biophys. J.* 32:193–210
347. Lee, L., Sykes, B. D. 1980. *J. Magn. Reson.* 41:512–14
347a. Lee, L., Sykes, B. D. 1983. *Biochemistry* 22:4366–73
348. Lee, N., Inouye, M., Lauterbur, P. C. 1977. *Biochem. Biophys. Res. Commun.* 78:1211–18
349. Leigh, J. S. 1976. In *Introduction to the Spectroscopy of Biological Polymers*, ed. D. W. Jones, pp. 189–219. London: Academic
350. LeMaster, D., Richards, F. M. 1983. *Proc. 8th Meet. Int. Soc. Magn. Reson., August 22–26*, p. 104 (Abstr.)
351. Lenkinski, R. E., Dallas, J. L., Glickson, J. D. 1979. *J. Am. Chem. Soc.* 101:3071–77
352. Lenstra, J. A., Bolscher, B. G. J. M., Stob, S., Beintema, J. J., Kaptein, R. 1979. *Eur. J. Biochem.* 98:385–97
353. Lerman, C. L., Cohn, M. 1980. *Biochem. Biophys. Res. Commun.* 97:121–25
354. Levine, B. A., Moore, G. R., Ratcliffe, R. G., Williams, R. J. P. 1979. *Int. Rev. Biochem.* 24:77–141
355. Levine, B. A., Williams, R. J. P., Fullmer, C. S., Wasserman, R. H. 1977. *Calcium-Binding Proteins Calcium Funct., Proc. Int. Symp., 2nd*, ed. R. H. Wasserman, pp. 29–37. New York: Elsevier
356. Levy, R. M., Dobson, C. M., Karplus, M. 1982. *Biophys. J.* 39:107–13
357. Lewis, S. D., Johnson, F. A., Shafer, J. A. 1981. *Biochemistry* 20:48–51
358. Lipari, G., Szabo, A., Levy, R. M. 1982. *Nature* 300:197–98
359. Lippard, S. J., Burger, A. R., Ugurbil, K., Pantoliano, M. W., Valentine, J. S. 1977. *Biochemistry* 16:1136–41
360. Littlemore, L. A. T. 1978. *Aust. J. Chem.* 31:2387–98
361. Littlemore, L. A. T., Ledeen, R. W. 1979. *Aust. J. Chem.* 32:2631–36
362. London, R. E. 1980. See Ref. 121, pp. 1–69
363. London, R. E., Avitabile, J. 1978. *J. Am. Chem. Soc.* 100:7159–65
364. London, R. E., Groff, J. P., Cocco, L., Blakley, R. L. 1982. *Biochemistry* 21:4450–58
365. Lord, S. T., Breslow, E. 1978. *Biochem. Biophys. Res. Commun.* 80:63–70
366. Lord, S. T., Breslow, E. 1979. *Int. J. Pept. Protein Res.* 13:71–77
367. Lu, P., Jarema, M. A., Rackwitz, H. R., Friedman, R. L. 1979. See Ref. 440, pp. 59–66
368. Mabbutt, B. C., Wright, P. E. 1983. *Biochim. Biophys. Acta* 744:281–90
369. Maciel, G. E., Shatlock, M. P., Houtchens, R. A., Caughey, W. S. 1980. *J. Am. Chem. Soc.* 102:6884–85
370. Maghuin-Rogister, G., Degelaen, J., Roberts, G. C. K. 1979. *Eur. J. Biochem.* 96:59–68
371. Mak, A., Smillie, L. B., Bárány, M. 1978. *Proc. Natl. Acad. Sci. USA* 75:3588–92
372. Mani, R. S., Shelling, J. G., Sykes, B. D., Kay, C. M. 1983. *Biochemistry* 22:1734–40
373. March, K. L., Maskalick, D. G., England, R. D., Friend, S. H., Gurd, F. R. N. 1982. *Biochemistry* 21:5241–51
374. Mariano, P. S., Glover, G. I., Petersen, J. R. 1978. *Biochem. J.* 171:115–22
375. Marion, D., Wüthrich, K. 1983. *Biochem. Biophys. Res. Commun.* 113:967–74
376. Markley, J. L. 1972. *Methods Enzymol.* 26:605–27
377. Markley, J. L. 1975. *Acc. Chem. Res.* 8:70–80
378. Markley, J. L. 1975. *Biochemistry* 14:3546–54
379. Markley, J. L. 1979. See Ref. 518, pp. 397–461
380. Markley, J. L., Cheung, S.-M. 1973. *Proc. Int. Conf. on Stable Isotopes in Chem., Biol., Med., Argonne, Ill.*, pp. 103–18
381. Markley, J. L., Ibañez, I. B. 1978. *Biochemistry* 17:4627–40
382. Markley, J. L., Kato, I. 1975. *Biochemistry* 14:3234–37
383. Markley, J. L., Neves, D. E., Westler, W. M., Ibañez, I. B., Porubcan, M. A., Baillargeon, M. W. 1980. *Dev. Biochem.* 10:31–61
384. Markley, J. L., Porubcan, M. A. 1976. *J. Mol. Biol.* 102:487–509
385. Markley, J. L., Putter, I., Jardetzky, O. 1968. *Science* 161:1249–51
386. Markley, J. L., Ulrich, E. L., Krogmann, D. W. 1977. *Biochem. Biophys. Res. Commun.* 78:106–14
387. Markley, J. L., Westler, W. M., Chan, T.-M., Kojiro, C. L., Ulrich, E. L. 1984. *Fed. Proc.* In press
388. Matthews, D. A. 1979. *Biochemistry* 18:1602–10
389. Mayer, A., Ogawa, S., Shulman, R. G., Yamane, T., Calvero, J. A., Gonsalves, A. M. d'A. R., et al. 1974. *J. Mol. Biol.* 86:749–56
389a. Mayo, K. H., Tyrell, P. M., Prestegard, J. H. 1983. *Biochemistry* 22:4485–93
390. McCammon, J. A., Karplus, M. 1983. *Acc. Chem. Res.* 16:187–93
391. McDonald, C. C., Phillips, W. D. 1967. *J. Am. Chem. Soc.* 89:6332–41
392. McDonald, C. C., Phillips, W. D. 1973. *Biochemistry* 12:3170–86
393. McDonald, G. C., Cohn, M., Noda, L. 1975. *J. Biol. Chem.* 250:6947–54
394. Meadows, D. H., Jardetzky, O., Epand,

R. M., Rüterjans, H. H., Scheraga, H. A. 1968. *Proc. Natl. Acad. Sci. USA* 60:766–72

395. Meadows, D. H., Markley, J. L., Cohen, J. S., Jardetzky, O. 1967. *Proc. Natl. Acad. Sci. USA* 58:1307–13
396. Mendz, G. L., Moore, W. J., Carnegie, P. R. 1982. *Aust. J. Chem.* 35:1979–2006
397. Mendz, G. L., Moore, W. J., Martenson, R. E. 1983. *Biochim. Biophys. Acta* 742:215–23

397a. Mendz, G. L., Moore, W. J., Martenson, R. E. 1983. *Biochim. Biophys. Acta* 748:168–75

398. Migchelsen, C., Beintema, J. J. 1973. *J. Mol. Biol.* 79:25–38
399. Mildvan, A. S., Gupta, R. K. 1978. *Methods Enzymol.* 49:322–59
400. Mitra, S., Bersohn, R. 1982. *Proc. Natl. Acad. Sci. USA* 79:6807–11
401. Miura, S., Ho, C. 1982. *Biochemistry* 21:6280–87
402. Miyazawa, T., Endo, T., Inagaki, F., Hayashi, K., Tamiya, N. 1982. *Biopolymers* 22:139–45
403. Moonen, C. T. W., Hore, P. J., Mueller, F., Kaptein, R., Mayhew, S. G. 1982. *FEBS Lett.* 149:141–46
404. Moore, G. R., Huang, Z.-X., Eley, C. G. S., Barker, H. A., Williams, G., Robinson, M. N., Williams, R. J. P. 1982. *Faraday Discuss. Chem. Soc.* 74:311–29
405. Moore, G. R., McClune, G. J., Clayden, N. J., Williams, R. J. P., Alsaadi, B. M., Angstrom, J., et al. 1982. *Eur. J. Biochem.* 123:73–80
406. Moore, G. R., Pitt, R. C., Williams, R. J. P. 1977. *Eur. J. Biochem.* 77:53–60
407. Moore, G. R., Williams, R. J. P. 1977. *FEBS Lett.* 79:229–32
408. Moore, G. R., Williams, R. J. P. 1980. *Eur. J. Biochem.* 103:493–502
409. Moore, G. R., Williams, R. J. P. 1980. *Eur. J. Biochem.* 103:503–12
410. Moore, G. R., Williams, R. J. P. 1980. *Eur. J. Biochem.* 103:533–41
411. Moore, G. R., Williams, R. J. P., Chien, J. C. W., Dickson, L. C. 1980. *J. Inorg. Biochem.* 12:1–15
412. Moura, J. J. G., Xavier, A. V., Bruschi, M., Le Gall, J. 1977. *Biochim. Biophys. Acta* 459:278–89
413. Moura, J. J. G., Xavier, A. V., Cookson, D. J., Moore, G. R., Williams, R. J. P., Bruschi, M., Le Gall, J. 1977. *FEBS Lett.* 81:275–80
414. Mueller, F., Van Schagen, C. G., Van Berkel, W. J. H. 1980. In *Flavins Flavoproteins, Proc. Int. Symp., 6th*, ed. K. Yagi, T. Yamano, pp. 359–71. Tokyo: Jpn. Sci. Soc.
415. Muszkat, K. A., Weinstein, S., Khait, I., Vered, M. 1982. *Biochemistry* 21:3775–79
416. Nagai, K., La Mar, G. N., Jue, T., Bunn, H. F. 1982. *Biochemistry* 21:842–47
417. Nagayama, K. 1981. *Adv. Biophys.* 14:139–204
418. Nagayama, K., Bachmann, P., Wüthrich, K., Ernst, R. R. 1978. *J. Magn. Reson.* 31:133–48
419. Nagayama, K., Kumar, A., Wüthrich, K., Ernst, R. R. 1980. *J. Magn. Reson.* 40:321–34

419a. Nagayama, K., Ozaki, Y., Kyogoku, Y., Hase, T., Matsubara, H. 1983. *J. Biochem. Tokyo* 94:893–902

420. Nagayama, K., Wüthrich, K. 1981. *Eur. J. Biochem.* 114:365–74
421. Nagayama, K., Wüthrich, K., Bachmann, P., Ernst, R. R. 1977. *Biochem. Biophys. Res. Commun.* 78:99–105
422. Nagayama, K., Wüthrich, K., Bachmann, P., Ernst, R. R. 1977. *Naturwissenschaften* 64:581–83
423. Nagayama, K., Wüthrich, K., Ernst, R. R. 1979. *Biochem. Biophys. Res. Commun.* 90:305–11
424. Nakano, A., Miyazawa, T., Nakamura, S., Kaziro, Y. 1979. *Arch. Biochem. Biophys.* 196:233–38
425. Nakano, A., Miyazawa, T., Nakamura, S., Kaziro, Y. 1980. *Biochemistry* 19:2209–15
426. Nelson, D. J., Opella, S. J., Jardetzky, O. 1976. *Biochemistry* 15:5552–60
427. Nelson, D. J., Theoharides, A. D., Nieburgs, A. C., Murray, R. K., Gonzalez-Fernandez, F., Brenner, D. S. 1979. *Int. J. Quantum. Chem.* 16:159–74
428. Niu, C.-H., Matsuura, S., Shindo, H., Cohen, J. S. 1979. *J. Biol. Chem.* 254:3788–96
429. Noggle, J. H., Schirmer, R. E. 1971. *The Nuclear Overhauser Effect*. New York: Academic. 259 pp.
430. O'Conner, T. P., Coleman, J. E. 1983. *Biochemistry* 22:3375–81
431. Ogino, T., Croll, D. H., Kato, I., Markley, J. L. 1982. *Biochemistry* 21:3452–60
432. Ohms, J. P., Hagenmaier, H., Hayes, M. B., Cohen, J. S. 1979. *Biochemistry* 18:1599–1602
433. Okhanov, V. V., Afanasev, V. A., Gurevich, A. Z., Elyakova, E. G., Miroshinikov, A. I., Bystrov, V. F., Ovchinnikov, Yu. A. 1980. *Sov. J. Bioorg. Chem.* (English transl.) 6:840–60
434. Oldfield, E., Allerhand, A. 1973. *Proc. Natl. Acad. Sci. USA* 70:3531–35
435. Oldfield, E., Allerhand, A. 1975. *J. Am. Chem. Soc.* 97:221–24
436. Oldfield, E., Allerhand, A. 1975. *J. Biol. Chem.* 250:6403–7

437. Oldfield, E., Janes, N., Kinsey, R., Kintanar, A., Lee, R. W. K., Rothgeb, T. M., et al. 1981. *Biochem. Soc. Symp.* 46:155–81
438. Oldfield, E., Norton, R. S., Allerhand, A. 1975. *J. Biol. Chem.* 250:6381–402
439. Opella, S. J. 1982. *Ann. Rev. Phys. Chem.* 33:533–62
440. Opella, S. J., Lu, P., eds. 1979. *NMR and Biochemistry: A Symposium Honoring Mildred Cohn.* New York: Dekker. 434 pp.
441. Oster, O., Neireiter, G. W., Clouse, A. O., Gurd, F. R. N. 1975. *J. Biol. Chem.* 250:7990–96
442. Otvos, J. D., Alger, J. R., Coleman, J. F., Armitage, I. M. 1979. *J. Biol. Chem.* 254:1778–80
443. Otvos, J. D., Armitage, I. M. 1979. *Experientia Suppl.* 34:249–57
444. Otvos, J. D., Armitage, I. M. 1980. *Biochemistry* 19:4021–30
445. Otvos, J. D., Armitage, I. M. 1980. *Biochemistry* 19:4031–43
446. Otvos, J. D., Armitage, I. M. 1982. See Ref. 64, pp. 65–96
447. Packer, E. L., Rabinowitz, J. C., Sternlicht, H. 1978. *J. Biol. Chem.* 253:7722–30
448. Packer, E. L., Sweeney, W. V., Rabinowitz, J. C., Sternlicht, H., Shaw, E. N. 1977. *J. Biol. Chem.* 252:2245–53
449. Patel, D. J., Canuel, L. L., Bovey, F. A. 1975. *Biopolymers* 14:987–97
450. Patel, D. J., Canuel, L. L., Woodward, C., Bovey, F. A. 1975. *Biopolymers* 14:959–74
451. Patel, D. J., Woodward, C., Canuel, L. L., Bovey, F. A. 1975. *Biopolymers* 14:975–86
452. Perkins, S. J. 1982. *Biol. Magn. Reson.* See Ref. 44, 4:193–336
453. Perkins, S. J., Johnson, L. N., Phillips, D. C., Dwek, R. A. 1977. *FEBS Lett.* 82:17–22
454. Perkins, S. J., Wüthrich, K. 1978. *Biochim. Biophys. Acta* 536:406–20
455. Perkins, S. J., Wüthrich, K. 1980. *J. Mol. Biol.* 138:43–64
456. Plateau, P., Gueron, M. 1982. *J. Am. Chem. Soc.* 104:7311–12
457. Poe, M., Hoogsteen, K., Matthews, D. A. 1979. *J. Biol. Chem.* 254:8143–52
458. Poe, M., Wu, J. K., Short, C. Jr., Florance, J., Hoogsteen, K. 1979. *Dev. Biochem.* 4:483–88
459. Porubcan, M. A., Neves, D. E., Rausch, S. K., Markley, J. L. 1978. *Biochemistry* 17:4640–47
460. Porubcan, M. A., Westler, W. M., Ibañez, I. B., Markley, J. L. 1979. *Biochemistry* 18:4108–16
461. Poulson, F. M., Hoch, J. C., Dobson, C. M. 1980. *Biochemistry* 19:2597–2607
462. Rattle, H. W. E. 1976. *Amino-Acids Pept. Proteins* 8:199–208
463. Rattle, H. W. E. 1979. *Amino-Acids Pept. Proteins* 10:221–37
463a. Rattle, H. W. E. 1983. *Amino-Acids Pept. Proteins* 14:242–61
464. Redfield, A. G. 1978. *Methods Enzymol.* 49:253–70
465. Redfield, A. G. 1978. *Methods Enzymol.* 49:359–69
466. Redfield, A. G., Gupta, R. K. 1972. *Cold Spring Harbor Symp. Quant. Biol.* 36:405–11
467. Redfield, A. G., Kunz, S. D., Ralph, E. K. 1975. *J. Magn. Reson.* 19:114–17
468. Redfield, C., Poulsen, F. M., Dobson, C. M. 1982. *Eur. J. Biochem.* 128:527–31
469. Rhyu, G. I., Markley, J. L., Ray, W. J. Jr. 1983. See Ref. 350, p. 110 (Abstr.)
470. Rhyu, G. I., Ray, W. J. Jr., Markley, J. L. 1984. *Biochemistry.* 23: In press
471. Ribeiro, A. A., King, R., Restivo, C., Jardetzky, O. 1980. *J. Am. Chem. Soc.* 102:4040–51
472. Ribeiro, A. A., Wemmer, D., Bray, R. P., Wade-Jardetzky, N. G., Jardetzky, O. 1981. *Biochemistry* 20:823–29
473. Rice, D. M., Blume, A., Herzfeld, J., Wittebort, R. J., Huang, T. H., DasGupta, S. K., Griffin, R. G. 1981. See Ref. 141, pp. 255–70
474. Richarz, R., Tschesche, H., Wüthrich, K. 1979. *Eur. J. Biochem.* 102:563–71
475. Richarz, R., Tschesche, H., Wüthrich, K. 1980. *Biochemistry* 19:5711–15
476. Richarz, R., Wüthrich, K. 1978. *Biochemistry* 17:2263–69
477. Roberts, G. C. K. 1977. In *Drug Action Mol. Level, Rep. Symp.*, ed. G. C. K. Roberts, pp. 127–50. Baltimore: Univ. Park Press
478. Roberts, G. C. K. 1983. In *Pteridines and Folic Acid Derivatives, Chemical, Biological and Chemical Aspects*, ed. J. A. Blair. New York: de Gruyter
479. Roberts, G. C. K., Feeney, J., Birdsall, B., Kimber, B., Griffiths, D. V., King, R. W., Burgen, A. S. V. 1977. See Ref. 154, pp. 95–109
480. Robillard, G., Shulman, R. G. 1972. *J. Mol. Biol.* 71:507–11
481. Robillard, G., Shulman, R. G. 1974. *J. Mol. Biol.* 86:519–40
482. Robillard, G., Shulman, R. G. 1974. *J. Mol. Biol.* 86:541–58
483. Roesch, P., Kalbitzer, H. R., Schmidt-Aderjan, U., Hengstenberg, W. 1981. *Biochemistry* 20:1599–1605
484. Roossien, F. F., Dooyewaard, G., Robillard, G. T. 1979. *Biochemistry* 18:5793–97

485. Roques, B. P., Rao, R., Marion, D. 1980. *Biochimie* 62:753–73
486. Rosenthal, S. N., Fendler, J. H. 1976. *Adv. Phys. Org. Chem.* 13:279–424
487. Rosevear, P. R., Desmeules, P., Kenyon, G. L., Mildvan, A. S. 1981. *Biochemistry* 20:6155–64
488. Rothgeb, T. M., Oldfield, E. 1981. *J. Biol. Chem.* 256:1432–46
489. Russu, I. M., Ho, C. 1982. *Biochemistry* 21:5044–51
490. Russu, I. M., Ho, N. T., Ho, C. 1980. *Biochemistry* 19:1043–52
491. Russu, I. M., Ho, N. T., Ho, C. 1982. *Biochemistry* 21:5031–43
492. Santoro, J., Juretschke, H. P., Rüterjans, H. 1979. *Biochim. Biophys. Acta* 578:346–56
493. Satterlee, J. D., Erman, J. E. 1981. *J. Am. Chem. Soc.* 103:199–200
494. Satterlee, J. D., Erman, J. E. 1983. *Biochim. Biophys. Acta* 743:149–54
495. Satterlee, J. D., Erman, J. E., La Mar, G. N., Smith, K. M., Langry, K. C. 1983. *Biochim. Biophys. Acta* 743:246–55
496. Satterlee, J. D., Erman, J. E., La Mar, G. N., Smith, K. M., Langry, K. C. 1983. *J. Am. Chem. Soc.* 105:2099–2104
497. Saunders, M., Wishnia, A., Kirkwood, J. G. 1957. *J. Am. Chem. Soc.* 79:3289–90
498. Scheek, R. M., Zuiderweg, E. R. P., Klappe, K. J. M., Van Boom, J. H., Kaptein, R., Rüterjans, H., Beyreuther, K. 1983. *Biochemistry* 22:228–35
499. Schejter, A., Lanir, A., Vig, I., Cohen, J. S. 1978. *J. Biol. Chem.* 253:3768–70
500. Schmidt-Aderjan, U., Roesch, P., Frank, R., Hengstenberg, W. 1979. *Eur. J. Biochem.* 96:43–48
501. Schnackerz, K. D., Feldmann, K., Hull, W. E. 1979. *Biochemistry* 18:1536–39
502. Scholberg, H. P. F., Fronticelli, C., Bucci, E. 1980. *J. Biol. Chem.* 255:8592–98
503. Schramm, S., Oldfield, E. 1983. *Biochemistry* 22:2908–13
504. Seamon, K. B. 1980. *Biochemistry* 19:207–15
505. Seamon, K. B., Hartshorne, D. J., Bothner-By, A. A. 1977. *Biochemistry* 16:4039–46
506. Senn, H., Eugster, A., Wüthrich, K. 1983. *Biochim. Biophys. Acta* 743:58–68
507. Senn, H., Wüthrich, K. 1983. *Biochim. Biophys. Acta* 743:69–81
508. Shaka, A. J., Frenkiel, T., Freeman, R. 1983. *J. Magn. Reson.* 52:159–63
509. Shaka, A. J., Keeler, J., Frenkiel, T., Freeman, R. 1983. *J. Magn. Reson.* 52:335–38
510. Shaw, D. 1976. *Fourier Transform NMR Spectroscopy.* Amsterdam: Elsevier. 357 pp.
511. Shelling, J. G., Sykes, B. D., O'Neil, J. D. J., Hofmann, T. 1983. *Biochemistry* 22:2649–54
512. Shimizu, A., Honzawa, M., Ito, S., Miyazaki, T., Matsumoto, H., Nakamura, H., et al. 1983. *Mol. Immunol.* 20:141–48
513. Shimizu, A., Honzawa, M., Yamamura, Y., Arata, Y. 1980. *Biochemistry* 19:2784–90
514. Shindo, H., Cohen, J. S. 1976. *J. Biol. Chem.* 251:2648–52
515. Shindo, H., Egan, W., Cohen, J. S. 1978. *J. Biol. Chem.* 253:6751–55
516. Shindo, H., Hayes, M. B., Cohen, J. S. 1976. *J. Biol. Chem.* 251:2644–47
517. Shriver, J., Mateescu, G., Fager, R., Torchia, D., Abrahamson, E. W. 1977. *Nature* 270:271–74
518. Shulman, R. G., ed. 1979. *Biological Applications of Magnetic Resonance.* New York: Academic. 595 pp.
519. Shulman, R. G., Wüthrich, K., Yamane, T., Antonini, E., Brunori, M. 1969. *Proc. Natl. Acad. Sci. USA* 63:623–28
520. Sletten, E., Jackson, J. T., Burns, P. D., La Mar, G. N. 1983. *J. Magn. Reson.* 52:492–96
521. Smith, G. M. 1979. *Biochemistry* 18:1628–34
522. Snyder, G. H., Rowan, R. III, Karplus, S., Sykes, B. D. 1975. *Biochemistry* 14:3765–77
523. Sperling, J. E., Feldman, K., Meyer, H., Jahnke, U., Heilmeyer, L. M. G. Jr. 1979. *Eur. J. Biochem.* 101:581–92
524. Steinmetz, W. E., Moonen, C., Kumar, A., Lazdunski, M., Visser, L., Carlsson, F. H. H., Wüthrich, K. 1981. *Eur. J. Biochem.* 120:467–75
525. Steitz, T. A., Shulman, R. G. 1982. *Ann. Rev. Biophys. Bioeng.* 11:419–44
526. Stellwagen, E., Shulman, R. G. 1973. *J. Mol. Biol.* 75:683–95
527. Sternlicht, H., Wilson, D. 1967. *Biochemistry* 6:2881–92
528. Stoesz, J. D., Malinowski, D. P., Redfield, A. G. 1979. *Biochemistry* 18:4669–75
529. Strom, R., Crigo, C., Viti, V., Guidoni, L., Podo, F. 1978. *FEBS Lett.* 96:45–50
530. Štrop, P., Čechova, D., Wüthrich, K. 1983. *J. Mol. Biol.* 166:669–76
531. Štrop, P., Wider, G., Wüthrich, K. 1983. *J. Mol. Biol.* 166:641–67
532. Štrop, P., Wüthrich, K. 1983. *J. Mol. Biol.* 166:631–40
533. Sudmeier, J. L., Bell, S. J., Storm, M. C., Dunn, M. F. 1981. *Science* 212:560–62
534. Sudmeier, J. L., Perkins, T. G. 1977. *J. Am. Chem. Soc.* 99:7732–33

535. Sykes, B. D., Hull, W. E. 1978. *Methods Enzymol.* 49:270–95
536. Sykes, B. D., Hull, W. E., Snyder, G. H. 1978. *Biophys. J.* 21:137–46
537. Sykes, B. D., Weiner, J. H. 1980. In *Magnetic Resonance in Biology*, ed. J. S. Cohen, pp. 171–96. New York: Wiley
538. Sykes, B. D., Weingarten, H. I., Schlesinger, M. J. 1974. *Proc. Natl. Acad. Sci. USA* 71:469–73
539. Takahashi, S., Lin, A. K. L. C., Ho, C. 1980. *Biochemistry* 19:5196–5202
540. Takahashi, S., Lin, A. K. L. C., Ho, C. 1982. *Biophys. J.* 39:33–40
541. Takesada, H., Nakanishi, M., Tsuboi, M., Ajisaka, K. 1976. *J. Biochem.* 80: 969–74
541a. Teleman, O., Drakenberg, T., Forsén, S., Thulin, E. 1983. *Eur. J. Biochem.* 134:453–57
542. Thiery, C., Nabedryk-Viala, E., Menez, A., Fromageot, P., Thiery, J. M. 1980. *Biochem. Biophys. Res. Commun.* 93:889–97
543. Timkovich, R., Cork, M. S. 1982. *Biochemistry* 21:5119–23
544. Torchia, D. A., Batchelder, L. S., Fleming, W. W., Jelinski, L. W., Sarkar, S. K., Sullivan, C. E. 1983. *Ciba Found. Symp.* 1983(93):98–111
545. Trewhella, J., Wright, P. E. 1980. *Biochim. Biophys. Acta* 625:202–20
546. Trexler, M., Banyai, L., Patthy, L., Pluck, N. D., Williams, R. J. P. 1983. *FEBS Lett.* 154:311–18
547. Tsai, M.-D. 1982. *Methods Enzymol.* 87:235–79
548. Tsai, M.-D., Bruzik, K. 1984. In *Biological Magnetic Resonance*, Vol. 5, ed. L. J. Berliner, J. Reuben. New York: Plenum. In press
549. Ugurbil, K., Bersohn, R. 1977. *Biochemistry* 16:3016–23
550. Ugurbil, K., Norton, R. S., Allerhand, A., Bersohn, R. 1977. *Biochemistry* 16: 886–94
551. Ulrich, E. L., Krogmann, D. W., Markley, J. L. 1982. *J. Biol. Chem.* 257:9356–64
552. Ulrich, E. L., Markley, J. L. 1978. *Coord. Chem. Rev.* 27:109–40
553. Van Schagen, C. G., Mueller, F. 1981. *Eur. J. Biochem.* 120:33–39
554. Van Schagen, C. G., Mueller, F. 1981. *FEBS Lett.* 136:75–79
555. Viggiano, G., Wiechelman, K. J., Chervenick, P. A., Ho, C. 1978. *Biochemistry* 17:795–99
556. Villafranca, J. J. 1982. *Fed. Proc.* 41: 2959–60
557. Villafranca, J. J. 1982. *Methods Enzymol.* 87:180–97
558. Villafranca, J. J., Raushel, F. M. 1980. *Ann. Rev. Biophys. Bioeng.* 9:363–92
559. Villafranca, J. J., Raushel, F. M. 1982. *Fed. Proc.* 41:2961–73
560. Virmani-Sardana, V., Breslow, E. 1983. *Int. J. Pept. Protein. Res.* 21:182–89
561. Vogel, H. J. 1984. In *Phosphorus-31 NMR, Principles and Applications*, ed. D. Gorenstein. New York: Academic
562. Vogel, H. J., Bridger, W. A. 1982. *Biochemistry* 21:5825–31
563. Vogel, H. J., Bridger, W. A. 1982. *J. Biol. Chem.* 257:4834–42
564. Vogel, H. J., Bridger, W. A. 1983. *Biochem. Soc. Trans.* 11:315–23
565. Vogel, H. J., Drakenberg, T., Forsén, S. 1983. In *NMR of Newly Accessible Nuclei*, ed. P. Laszlo. New York: Academic. In press
566. Wagner, G. 1982. *Comments Mol. Cell. Biophys.* 1:261–80
566a. Wagner, G. 1983. *J. Magn. Reson.* 55: 151–56
567. Wagner, G. 1983. *Q. Rev. Biophys.* 16: 1–57
568. Wagner, G., Kumar, A., Wüthrich, K. 1981. *Eur. J. Biochem.* 114:375–84
569. Wagner, G., Wüthrich, K. 1979. *J. Mol. Biol.* 130:31–37
570. Wagner, G., Wüthrich, K. 1979. *J. Mol. Biol.* 134:75–94
571. Wagner, G., Wüthrich, K. 1982. *J. Mol. Biol.* 155:347–66
572. Wagner, G., Wüthrich, K. 1982. *J. Mol. Biol.* 160:343–61
573. Wagner, G., Wüthrich, K. 1983. *Naturwissenschaften* 70:105–14
574. Wagner, G., Wüthrich, K., Tschesche, H. 1978. *Eur. J. Biochem.* 86:67–76
575. Wagner, G., Wüthrich, K., Tschesche, H. 1978. *Eur. J. Biochem.* 89:367–77
576. Wagner, G., Zuiderweg, E. R. P. 1983. *Biochim. Biophys. Acta* 113:854–60
577. Wain-Hobson, S., Dower, S. K., Gettins, P., Givol, D., McLaughlin, A. C., Pecht, I., et al. 1977. *Biochem. J.* 165:227–35
578. Walters, D. E., Allerhand, A. 1980. *J. Biol. Chem.* 255:6200–4
579. Wang, F.-F. C., Hirs, C. H. W. 1979. *J. Biol. Chem.* 254:1090–93
580. Waugh, J. S. 1982. *J. Magn. Reson.* 50:30–49
581. Wedin, R. E., Delepierre, M., Dobson, C. M., Poulsen, F. M. 1982. *Biochemistry* 21:1098–1103
582. Wemmer, D., Kallenbach, N. R. 1983. *Biochemistry* 22:1901–6
583. Weser, U., Rupp, H. 1979. *Top. Environ. Health* 2:267–83
584. Westler, W. M., Bogard, W. C. Jr., Laskowski, M. Jr., Markley, J. L. 1982. *Fed. Proc.* 41:1188 (Abstr.)
585. Westler, W. M., Markley, J. L.,

Bachovchin, W. W. 1982. *FEBS Lett.* 138:233–35
586. Wider, G., Hosur, R. V., Wüthrich, K. 1983. *J. Magn. Reson.* 52:130–35
587. Wieloch, T., Borgstrom, B., Falk, K.-E., Forsén, S. 1979. *Biochemistry* 18:1622–28
588. Wieloch, T., Falk, K.-E. 1978. *FEBS Lett.* 85:271–74
589. Wilbur, D. J., Allerhand, A. 1976. *J. Biol. Chem.* 251:5187–94
590. Wilbur, D. J., Allerhand, A. 1977. *FEBS Lett.* 79:144–46
591. Williams, G., Eley, C. G. S., Moore, G. R., Robinson, M. N., Williams, R. J. P. 1982. *FEBS Lett.* 150:293–99
592. Williams, R. J. P. 1978. *Biochem. Soc. Trans.* 6:1123–26
593. Williams, R. J. P. 1981. *Biochem. Soc. Symp.* 46:57–72
594. Williamson, K. L., Williams, R. J. P. 1979. *Biochemistry* 18:5966–72
595. Wittebort, R. J., Rothgeb, T. M., Szabo, A., Gurd, F. R. N. 1976. *Proc. Natl. Acad. Sci. USA* 76:1059–63
596. Woods, L. F. J., Wiseman, A., Libor, S., Jones, J. R., Elvidge, J. A. 1980. *Biochem. Soc. Trans.* 8:98–99
597. Woodward, C., Simon, I., Tüchsen, E. 1982. *Mol. Cell. Biochem.* 48:135–60
598. Woodworth, R. C., Williams, R. J. P., Alsaadi, B. M. 1977. In *Proteins Iron Metab., Proc. Int. Meet., 3rd*, ed. E. B. Brown, P. Aisen, J. Fielding, pp. 211–18. New York: Grune & Stratton
599. Wooten, J. B., Cohen, J. S. 1979. *Biochemistry* 18:4188–91
600. Wooten, J. B., Cohen, J. S., Vig, I., Schejter, A. 1981. *Biochemistry* 20:5394–5402
601. Wüthrich, K. 1976. *NMR in Biological Research: Peptides and Proteins.* New York: Am. Elsevier. 379 pp.
602. Wüthrich, K. 1977. In *NATO Adv. Study Inst. Ser., Ser. B*, pp. 347–60
603. Wüthrich, K. 1977. In *NATO Adv. Study Inst. Ser., Ser. B*, pp. 361–74
604. Wüthrich, K. 1977. See Ref. 154, pp. 51–62
605. Wüthrich, K. 1980. In *Front. Biorg. Chem. Mol. Biol., Proc. Int. Symp.*, ed. S. N. Anachenko, pp. 161–68. Oxford: Pergamon
606. Wüthrich, K. 1981. *Biochem. Soc. Symp.* 46:17–37
607. Wüthrich, K. 1981. *Macromol. Chem. Phys. Suppl.* 5:234–52
608. Wüthrich, K. 1982. *NATO Adv. Study Inst. Ser., Ser. A 45*, pp. 215–35
609. Wüthrich, K. 1983. *Biopolymers* 22:131–38
610. Wüthrich, K., Nagayama, K., Ernst, R. R. 1979. *Trends Biochem. Sci.* 4:N178–81
611. Wüthrich, K., Roder, H., Wagner, G. 1980. In *Protein Folding, Proc. Conf. Ger. Biochem. Soc., 28th*, ed. R. Jaenicke, pp. 549–64. Amsterdam: Elsevier
612. Wüthrich, K., Wagner, G. 1978. *Trends Biochem. Sci.* 3:227–30
613. Wüthrich, K., Wagner, G. 1979. *J. Mol. Biol.* 130:1–18
614. Wüthrich, K., Wagner, G. 1981. In *Biomol. Struct. Conform. Funct. Evol., Proc. Int. Symp.*, ed. R. Srinivasan, E. Subramanian, N. Yathindra, 2:23–29. Oxford: Pergamon
615. Wüthrich, K., Wagner, G. 1983. *Ciba Found. Symp.* 93:310–28
616. Wüthrich, K., Wagner, G., Richarz, R., Braun, W. 1980. *Biophys. J.* 32:549–60
617. Wüthrich, K., Wagner, G., Richarz, R., Perkins, S. J. 1978. *Biochemistry* 17:2253–63
618. Wüthrich, K., Wider, G., Wagner, G., Braun, W. 1982. *J. Mol. Biol.* 155:311–19
619. Xavier, A. V. 1983. *NATO Adv. Study Inst. Ser., Ser. C* 100:291–311
620. Xavier, A. V., Czerwinski, E. W., Bethge, P. H., Mathews, F. S. 1978. *Nature* 275:245–47
621. Xavier, A. V., Moura, I., Moura, J. J. G., Santos, M. H., Villalain, J. 1982. *NATO Adv. Study Inst. Ser., Ser. C* 89:127–41
622. Xavier, A. V., Moura, J. J. G. 1978. *Biochimie* 60:327–38
623. Yamaguchi, A., Unemoto, T., Ikegami, A. 1981. *Photochem. Photobiol.* 33:511–16
624. York, J. L., Millett, F. S., Minor, L. B. 1980. *Biochemistry* 19:2583–88

SUBJECT INDEX

F

G

Q

R

CUMULATIVE INDEXES

CONTRIBUTING AUTHORS, VOLUMES 9–13

CHAPTER TITLES, VOLUMES 9–13

CONCEPTUAL AND PHYSICAL TOOLS FOR ANALYSIS

PARTICULAR CONSTITUENTS, ASSEMBLIES, AND RELATIONS AMONG THEM

MOLECULAR MECHANISMS UNDERLYING PROCESSES THAT OCCUR IN CELLS OR AMONG CELLS

NEW BOOKS
FROM
ANNUAL REVIEWS INC.

NOW YOU CAN
CHARGE THEM
TO

Annual Reviews Inc.

A NONPROFIT SCIENTIFIC PUBLISHER

4139 EL CAMINO WAY • PALO ALTO, CA 94306-9981 • (415) 493-4400

·rders for Annual Reviews Inc. publications may be placed through your bookstore; subscription agent; par-
·ipating professional societies; or directly from Annual Reviews Inc. by mail or telephone (paid by credit
.rd or purchase order). Prices subject to change without notice.

dividuals: Prepayment required in U.S. funds or charged to American Express, MasterCard, or Visa.
stitutional Buyers: Please include purchase order.
udents: Special rates are available to qualified students. Refer to Annual Reviews ***Prospectus*** or contact
nnual Reviews Inc. office for information.
·ofessional Society Members: Members whose professional societies have a contractural arrangement with
nnual Reviews may order books through their society at a special discount. Check with your society for
formation.

·egular orders: When ordering current or back volumes, please list the volumes you wish by volume number.
anding orders: (New volume in the series will be sent to you automatically each year upon publication. Can-
·llation may be made at any time.) Please indicate volume number to begin standing order.
·epublication orders: Volumes not yet published will be shipped in month and year indicated.
alifornia orders: Add applicable sales tax.
ostage paid (4th class bookrate /surface mail) by Annual Reviews Inc.

ANNUAL REVIEWS SERIES

		Prices Postpaid per volume USA/elsewhere	Regular Order Please send: Vol. number	Standing Order Begin with: Vol. number
nnual Review of ANTHROPOLOGY				
Vols. 1-10	(1972-1981)	$20.00/$21.00		
Vol. 11	(1982)	$22.00/$25.00		
Vols. 12-13	(1983-1984)	$27.00/$30.00		
Vol. 14	(avail. Oct. 1985)	$27.00/$30.00	Vol(s).______	Vol.______
nnual Review of ASTRONOMY AND ASTROPHYSICS				
Vols. 1-19	(1963-1981)	$20.00/$21.00		
Vol. 20	(1982)	$22.00/$25.00		
Vols. 21-22	(1983-1984)	$44.00/$47.00		
Vol. 23	(avail. Sept. 1985)	$44.00/$47.00	Vol(s).______	Vol.______
nnual Review of BIOCHEMISTRY				
Vols. 29-34, 36-50	(1960-1965; 1967-1981)	$21.00/$22.00		
Vol. 51	(1982)	$23.00/$26.00		
Vols. 52-53	(1983-1984)	$29.00/$32.00		
Vol. 54	(avail. July 1985)	$29.00/$32.00	Vol(s).______	Vol.______
nnual Review of BIOPHYSICS				
Vols. 1-10	(1972-1981)	$20.00/$21.00		
Vol. 11	(1982)	$22.00/$25.00		
Vols. 12-13	(1983-1984)	$47.00/$50.00		
Vol. 14	(avail. June 1985)	$47.00/$50.00	Vol(s).______	Vol.______
nnual Review of CELL BIOLOGY				
Vol. 1	(avail. Nov. 1985)	est. $27.00/$30.00	Vol. ______	Vol.______
nnual Review of EARTH AND PLANETARY SCIENCES				
Vols. 1-9	(1973-1981)	$20.00/$21.00		
Vol. 10	(1982)	$22.00/$25.00		
Vols. 11-12	(1983-1984)	$44.00/$47.00		
Vol. 13	(avail. May 1985)	$44.00/$47.00	Vol(s).______	Vol.______
nnual Review of ECOLOGY AND SYSTEMATICS				
Vols. 1-12	(1970-1981)	$20.00/$21.00		
Vol. 13	(1982)	$22.00/$25.00		
Vols. 14-15	(1983-1984)	$27.00/$30.00		
Vol. 16	(avail. Nov. 1985)	$27.00/$30.00	Vol(s).______	Vol.______

SEE ORDERING INFORMATION ON PAGE 4

		Prices Postpaid per volume USA/elsewhere	Regular Order Please send:	Standing Order Begin with:
			Vol. number	Vol. number
Annual Review of ENERGY				
Vols. 1-6	(1976-1981)	$20.00/$21.00		
Vol. 7	(1982)	$22.00/$25.00		
Vols. 8-9	(1983-1984)	$56.00/$59.00		
Vol. 10	(avail. Oct. 1985)	$56.00/$59.00	Vol(s).______	Vol.______
Annual Review of ENTOMOLOGY				
Vols. 8-16, 18-26	(1963-1971; 1973-1981)	$20.00/$21.00		
Vol. 27	(1982)	$22.00/$25.00		
Vols. 28-29	(1983-1984)	$27.00/$30.00		
Vol. 30	(avail. Jan. 1985)	$27.00/$30.00	Vol(s).______	Vol.______
Annual Review of FLUID MECHANICS				
Vols. 1-5, 7-13	(1969-1973; 1975-1981)	$20.00/$21.00		
Vol. 14	(1982)	$22.00/$25.00		
Vols. 15-16	(1983-1984)	$28.00/$31.00		
Vol. 17	(avail. Jan. 1985)	$28.00/$31.00	Vol(s).______	Vol.______
Annual Review of GENETICS				
Vols. 1-15	(1967-1981)	$20.00/$21.00		
Vol. 16	(1982)	$22.00/$25.00		
Vols. 17-18	(1983-1984)	$27.00/$30.00		
Vol. 19	(avail. Dec. 1985)	$27.00/$30.00	Vol(s).______	Vol.______
Annual Review of IMMUNOLOGY				
Vols. 1-2	(1983-1984)	$27.00/$30.00		
Vol. 3	(avail. April 1985)	$27.00/$30.00	Vol(s).______	Vol.______
Annual Review of MATERIALS SCIENCE				
Vols. 1-11	(1971-1981)	$20.00/$21.00		
Vol. 12	(1982)	$22.00/$25.00		
Vols. 13-14	(1983-1984)	$64.00/$67.00		
Vol. 15	(avail. Aug. 1985)	$64.00/$67.00	Vol(s).______	Vol.______
Annual Review of MEDICINE: Selected Topics in the Clinical Sciences				
Vols. 1-3, 5-15	(1950-1952; 1954-1964)	$20.00/$21.00		
Vols. 17-32	(1966-1981)	$20.00/$21.00		
Vol. 33	(1982)	$22.00/$25.00		
Vols. 34-35	(1983-1984)	$27.00/$30.00		
Vol. 36	(avail. April 1985)	$27.00/$30.00	Vol(s).______	Vol.______
Annual Review of MICROBIOLOGY				
Vols. 17-35	(1963-1981)	$20.00/$21.00		
Vol. 36	(1982)	$22.00/$25.00		
Vols. 37-38	(1983-1984)	$27.00/$30.00		
Vol. 39	(avail. Oct. 1985)	$27.00/$30.00	Vol(s).______	Vol.______
Annual Review of NEUROSCIENCE				
Vols. 1-4	(1978-1981)	$20.00/$21.00		
Vol. 5	(1982)	$22.00/$25.00		
Vols. 6-7	(1983-1984)	$27.00/$30.00		
Vol. 8	(avail. March 1985)	$27.00/$30.00	Vol(s).______	Vol.______
Annual Review of NUCLEAR AND PARTICLE SCIENCE				
Vols. 12-31	(1962-1981)	$22.50/$23.50		
Vol. 32	(1982)	$25.00/$28.00		
Vols. 33-34	(1983-1984)	$30.00/$33.00		
Vol. 35	(avail. Dec. 1985)	$30.00/$33.00	Vol(s).______	Vol.______

SEE ORDERING INFORMATION ON PAGE 4